CLEAN ENERGY FOR SUSTAINABLE DEVELOPMENT

CLEAN ENERGY FOR SUSTAINABLE DEVELOPMENT

Comparisons and Contrasts of New Approaches

Edited by

MOHAMMAD G. RASUL
ABUL KALAM AZAD
SUBHASH C. SHARMA
School of Engineering and Technology
Central Queensland University
Queensland, Australia

Amsterdam • Boston • Heidelberg • London
New York • Oxford • Paris • San Diego
San Francisco • Singapore • Sydney • Tokyo
Academic Press is an imprint of Elsevier

Academic Press is an imprint of Elsevier
125 London Wall, London EC2Y 5AS, United Kingdom
525 B Street, Suite 1800, San Diego, CA 92101-4495, United States
50 Hampshire Street, 5th Floor, Cambridge, MA 02139, United States
The Boulevard, Langford Lane, Kidlington, Oxford OX5 1GB, United Kingdom

Notices

Knowledge and best practice in this field are constantly changing. As new research and experience broaden
our understanding, changes in research methods, professional practices, or medical treatment may become
necessary.

Practitioners and researchers must always rely on their own experience and knowledge in evaluating and
using any information, methods, compounds, or experiments described herein. In using such information or
methods they should be mindful of their own safety and the safety of others, including parties for whom they
have a professional responsibility.

To the fullest extent of the law, neither the Publisher nor the authors, contributors, or editors, assume any
liability for any injury and/or damage to persons or property as a matter of products liability, negligence or
otherwise, or from any use or operation of any methods, products, instructions, or ideas contained in the
material herein.

Library of Congress Cataloging-in-Publication Data
A catalog record for this book is available from the Library of Congress

British Library Cataloguing-in-Publication Data
A catalogue record for this book is available from the British Library

ISBN: 978-0-12-805423-9

For information on all Academic Press publications
visit our website at https://www.elsevier.com/

Working together
to grow libraries in
developing countries

www.elsevier.com • www.bookaid.org

Publisher: Joe Hayton
Acquisition Editor: Lisa Reading
Editorial Project Manager: Maria Convey
Production Project Manager: Mohanambal Natarajan
Cover Designer: Greg Harris

Typeset by TNQ Books and Journals

CONTENTS

SECTION 3: WIND ENERGY SYSTEMS

18. Potential of Biodiesel as Fuel for Diesel Engine   557

O.M. Ali, R. Mamat, M.G. Rasul, G. Najafi

LIST OF CONTRIBUTORS

A. Adnan
Government College University Lahore, Lahore, Pakistan

S. Ahmed
Islamic University of Technology, Dhaka, Bangladesh

S.S.U. Ahmed
Islamic University of Technology, Gazipur, Bangladesh

S.I. Al-Resayes
King Saud University, Riyadh, Saudi Arabia

O.M. Ali
Northern Technical University, Kirkuk, Iraq

A. Allali
University of Science and Technology of Oran, Oran, Algeria

M. Arshad
University of Veterinary and Animal Sciences Lahore, Lahore, Pakistan

A.E. Atabani
Erciyes University, Kayseri, Turkey
BioGreen Power Arge, Erciyes Teknopark, Kayseri, Turkey

A.K. Azad
Central Queensland University, Rockhampton, QLD, Australia

H.M. Boulouiha
University of Relizane, Relizane, Algeria

M. Danish
University of Gujrat, Gujrat, Pakistan

M. Denai
University of Hertfordshire, Hatfield, United Kingdom

M.M. El-Sheekh
Tanta University, Tanta, Egypt

P.M. Gresshoff
The University of Queensland, Brisbane, QLD, Australia

P.K. Halder
Jessore University of Science and Technology, Jessore, Bangladesh

M. Hasan
The University of Sydney, Sydney, NSW, Australia

N.M.S. Hassan
Central Queensland University, Rockhampton, QLD, Australia

D. Honnery
Monash University, Clayton, VIC, Australia

M.S. Hossain
UMPEDAC, University of Malaya, Kuala Lumpur, Malaysia

A. Indrasumunar
The University of Queensland, Brisbane, QLD, Australia

M.U.H. Joardder
Queensland University of Technology, Brisbane, QLD, Australia
Rajshahi University of Engineering and Technology, Rajshahi, Bangladesh

M.A. Kalam
University of Malaya, Kuala Lumpur, Malaysia

M.M.K. Khan
Central Queensland University, Rockhampton, QLD, Australia

G. Kumar
National Institute for Environmental Studies (NIES), Tsukuba, Japan

T.A.G. Langrish
The University of Sydney, Sydney, NSW, Australia

W.G. Le Roux
University of Pretoria, Pretoria, South Africa

H.M. Mahmudul
University Malaysia Pahang, Pahang, Malaysia

R. Mamat
Universiti Malaysia Pahang, Pekan, Pahang, Malaysia

H.H. Masjuki
University of Malaya, Kuala Lumpur, Malaysia

M.H. Masud
Queensland University of Technology, Brisbane, QLD, Australia

J.P. Meyer
University of Pretoria, Pretoria, South Africa

M. Mofijur
Central Queensland University, Rockhampton, QLD, Australia

P. Moriarty
Monash University, Caulfield East, VIC, Australia

H. Mukhtar
Government College University Lahore, Lahore, Pakistan

M.W. Mumtaz
University of Gujrat, Gujrat, Pakistan
Government College University Lahore, Lahore, Pakistan

G. Najafi
Tarbiat Modares University, Tehran, Iran

R. Narayanan
Central Queensland University, Bundaberg, QLD, Australia

I.A. Nehdi
King Saud University, Riyadh, Saudi Arabia

A.K. Pandey
UMPEDAC, University of Malaya, Kuala Lumpur, Malaysia

M.A. Quader
University of Malaya, Kuala Lumpur, Malaysia

M.A. Rahim
Queensland University of Technology, Brisbane, QLD, Australia

N.A. Rahim
UMPEDAC, University of Malaya, Kuala Lumpur, Malaysia

M.M. Rahman
Islamic University of Technology, Gazipur, Bangladesh

U. Rashid
Universiti Putra Malaysia (UPM), Serdang, Selangor, Malaysia
King Saud University, Riyadh, Saudi Arabia

M.G. Rasul
Central Queensland University, Rockhampton, QLD, Australia

A.K.M. Sadrul Islam
Islamic University of Technology, Gazipur, Bangladesh

S. Salehin
Islamic University of Technology, Gazipur, Bangladesh

P.T. Scott
The University of Queensland, Brisbane, QLD, Australia

J. Selvaraj
UMPEDAC, University of Malaya, Kuala Lumpur, Malaysia

S.C. Sharma
Central Queensland University, Rockhampton, QLD, Australia

S. Shobana
Aditanar College of Arts and Science, Tirchendur, India

S. Soltani
Universiti Putra Malaysia (UPM), Serdang, Selangor, Malaysia

M.A. Tunio
Mehran University of Engineering Technology, Khairpur Mir's, Sindh, Pakistan

ABOUT THE EDITORS

Mohammad G. Rasul

Mohammad Rasul is an Associate Professor of Mechanical Engineering at the School of Engineering and Technology, Central Queensland University, Australia.

Professor Rasul specializes in clean and sustainable energy technologies and their applications in industry. His research focuses on renewable energy (solar, wind, biomass, and biofuels), building energy (domestic, institutional, and commercial buildings), industrial energy (process and resource industries), and thermochemical conversion of energy (combustion, gasification, and pyrolysis).

His achievements, contributions, and recognition in research have been strongly demonstrated by significant impact and the large number of publications (385 research articles), and citations (h-index 19 and citations 1694 in Scopus) by relevant professionals, both nationally and internationally.

Abul Kalam Azad

Abul Kalam Azad is a Research Scholar at the School of Engineering and Technology, Central Queensland University, Australia.

Azad's research focuses on advanced biofuel production and combustion, energy extraction and processing, advanced computational fluid dynamics, modeling and analysis, renewable energy (bioenergy, wind energy, and so on), natural gas processing, and oil refinery.

Subhash C. Sharma

Dr. Subhash Sharma is a Senior Lecturer and Discipline Leader of Asset and Maintenance Management programs at the School of Engineering and Technology, Central Queensland University, Australia.

Dr. Sharma's main research focuses on conservation of energy and physical assets by applying tribological and asset management principles. His recent works deal with alternative fuels including biofuels, and their applications in industry.

PREFACE

Since the past decades, the energy sector is facing a number of challenges; however, the fundamental challenge is meeting the growing energy demand in sustainable, environment-friendly, efficient, and cleaner ways. *Clean Energy for Sustainable Development: Comparisons and Contrasts of New Approaches* updates the industry and academia with recent developments in this field. The book primarily focuses on developments in the fields of energy technology, clean, low emission, and sustainable energy, energy efficiency, and energy and environmental sustainability to academics, researchers, practicing engineers, technologists, and students. The major themes included in the book are as follows:

- Clean and sustainable energy sources and technologies
- Renewable energy technologies and their applications
- Biomass and biofuels for sustainable environment
- Energy system and efficiency improvement
- Solar thermal applications
- Environmental impacts of sustainable energy systems

The book helps develop understanding the relevant concepts and solutions to the global issues to achieve clean energy and sustainable development in medium- and large-scale industries.

It was a challenging task to arrange and define sections of the book because of the variety of high-quality contributions received from the authors. The book comprises 18 chapters, which we have divided into four sections. Each section has a specific theme that describes about what is contained in that section, thus providing continuity to the book.

The first section introduces clean, renewable, and sustainable energy resources and technologies, and their prospects and policies. Environmental impact assessment of different renewable energy resources and future prospects of carbon-negative technologies are also explained in this section.

The second section presents applications of solar energy in cooling technologies, power plants, and agricultural and forest industries. More specifically, solar energy applications in thermochemical conversion of waste into energy, solar air conditioning, solar kilns for agricultural and forest industries, and solar power generation using Brayton cycle are discussed.

The third section explains recent developments in wind energy systems. Issues on control design, stability, and power qualities in grid-integrated wind energy systems, and analysis of hybrid solar and wind energy systems for power generation are presented and discussed.

The fourth section focuses on potential and applications of biodiesel as an alternative fuel for diesel engine. Biodiesel production from first (edible), second (nonedible), and third (advanced biodiesel) generation feedstocks, such as bush nut (*Macadamia integrifolia*), legume tree (*Pongamia pinnata*), and microalgae, using different additives and transesterification processes (chemical and biochemical) are discussed.

While the titles of these four sections may be, in some cases, a bit unorthodox for the book, we believe that the flow of the material will feel comfortable to both students and practicing engineers in the area of clean and sustainable energy.

All the chapters were peer reviewed and the authors addressed the comments and suggestions of the reviewers and editors before contributions were accepted for publication.

The editors of this book would like to express their sincere thanks to all the authors for their high-quality contributions. The successful completion of this book has been the result of the cooperation of many people. We would like to express our sincere thanks and gratitude to all of them.

We have been supported by Maria Convey, Editorial Project Manager at Elsevier, for completing the publication process. We would like to express our deepest sense of gratitude and thanks to Maria for assisting and guiding us for this publication.

Mohammad G. Rasul
Abul Kalam Azad
Subhash C. Sharma
School of Engineering and Technology
Central Queensland University, Rockhampton
Queensland 4702, Australia

Clean and Sustainable Energy Resources and Technologies

CHAPTER ONE

Sustainable Energy Resources: Prospects and Policy

P. Moriarty[1] and D. Honnery[2]
[1]Monash University, Caulfield East, VIC, Australia
[2]Monash University, Clayton, VIC, Australia

1.1 INTRODUCTION

According to the analysis of Marchetti [1], over the long term, energy sources replace each other in a regular fashion. Thus the millennia-long dominance of biomass ended in the 19th century, and was replaced by coal, which in turn was replaced by oil. But given the problems faced by these fossil fuels, we argue that the world could well see an eventual reversion to renewable energy (RE). But at least for the coming decades, RE must compete with its rival energy sources, fossil fuels and nuclear energy, for share in the global energy market. At present, fossil fuels dominate global energy supply as they have done for over a century, and are only very slowly losing share in global commercial (i.e., excluding fuel wood) energy consumption. The primary energy output in 2014 is shown in Table 1.1 [2].

The global values for energy shares (and energy use per capita) conceal large differences between nations. A number of considerations are important in selecting the energy types used in a given region or country:

- *Local availability of the energy resource.* Using locally available energy resources can improve energy security, save foreign exchange, and provide local employment opportunities. For example, bioethanol production in the United States and Brazil is regarded as a means for raising rural incomes and employment.
- *The costs of each energy type, which includes construction, operation and maintenance, fuel costs, and decommissioning.* As is shown in the following sections, fossil fuels enjoy massive subsidies, although all three energy groups are subsidized to some extent, either through monetary subsidies of various types, or because external costs are not paid. An obvious and important unpaid external cost is the CO_2 emissions from fossil fuel combustion, although the carbon pricing schemes being introduced in some countries partly address this issue.
- *Environmental considerations.* These may range from loss of scenic amenity (as in resident opposition to wind turbines), air pollution from coal-burning power stations, or fears about radiation leakage or reactor accidents in the case of nuclear

Clean Energy for Sustainable Development
ISBN 978-0-12-805423-9, http://dx.doi.org/10.1016/B978-0-12-805423-9.00001-6

Table 1.1 Primary Commercial Energy Output, 2014 (EJ)

Energy Source	Energy Output (EJ)	% of Total Energy
Fossil fuels	467.2	86.3
Nuclear energy	24.0	4.4
All RE sources	50.1	9.3

Note: EJ = exajoule = 10^{18} J.
Data from BP. BP statistical review of world energy 2015. London: BP; 2015.

power. In fact, public opposition has been a major factor in limiting the growth (or even outright moratoria) for nuclear power in OECD countries.

- *Available financial and technical resources.* For example, much of Africa's hydro and geothermal energy potential remains undeveloped because of lack of financial resources. Also many countries do not yet have the technical capacity to start a nuclear power program. A further point is that for some countries the entire national grid may be too small to support even a single nuclear reactor for baseload power.

These considerations are often in conflict with each other. Many claim that the large US corn ethanol program, while undoubtedly beneficial to farmers, is not economic. Also for the United States, shale gas has helped reduce US energy imports and so improved energy security, but some argue that the economics of shale gas are fragile, and that production will drop dramatically in few years [3]. Different countries give different weight to these factors, which helps explain why the energy mix varies so widely from country to country.

In this chapter we first examine in turn the prospects for the two rival fuel groups to RE. We examine the future difficulties these rival fuels are likely to face in a changing world, including the issues of climate change and possible resource depletion. We then use this analysis as a basis for evaluating the prospects for RE, paying particular attention to which RE types are likely to exhibit the greatest growth in the coming decades. In the final section, we look at the policies required to best encourage the needed growth in RE in what may well be an era of continuing financial constraints. We find that removing the vast subsidies to fossil fuels represents the single most effective policy for RE development.

1.2 FOSSIL FUELS

Fossil fuels occupy an entrenched position in the global fuel mix, having dominated energy supply for over a century. In 2014, global consumption was 176.3 EJ, 162.5 EJ, and 130.9 EJ for oil, coal, and gas, respectively [2]. A number of energy researchers doubt that this level of fossil fuel energy use can be maintained for much longer. Schindler [4], for example, regarded oil supply as having been on an (undulating) plateau since 2004, with output decline imminent. For coal, he envisaged that "global coal production will peak around 2025 at about 30% above the current production rate—this being the upper boundary of the possible development." He considered that growth in natural gas can potentially continue for another 5—15 years, also rising to a peak about 30% beyond the current global production rate. Höök and Tang [5] reached similar conclusions, and further argued that global depletion would impose limits on fossil fuel carbon emissions, and hence on their climate change impact. In contrast, McGlade and Ekins [6] believed that "globally, a third of oil reserves, half of gas reserves, and over 80 % of current coal reserves should remain unused from 2010 to 2050 in order to meet the target of 2°C." Clearly, for these authors, global climate change concerns, not fossil fuel depletion, is the decisive factor. If such levels of fossil fuel reserves did remain unused, it would have serious and global financial implications.

Other authorities also assume few supply constraints on fossil fuel use in the coming decades. The US Energy Information Administration (EIA) [7] regularly publishes forecasts for global energy consumption by fuel type. Apart from a reference scenario, the EIA scenarios include high and low economic growth cases, and high and low oil price cases. The high and low economic growth cases are the most and least favorable cases for fossil fuel energy use (and energy growth in general). The EIA forecast that coal, natural gas, and (all) liquid fuels will still account for between 77.3% and 80.3% in 2040.

The latest Intergovernmental Panel on Climate Change (IPCC) reports [8,9] assumed that carbon sequestration can allow fossil fuels to supply carbon-free or green energy. The integrated assessment modelling by van Vuuren et al. [10] examined four representative concentration pathways (RCPs) for the IPCC. The four RCPs were termed 2.6, 4.5, 6, and 8.5, where these numbers refer to the climate forcing (in W/m^2) compared with that for the preindustrial era. Fossil fuels were assumed to supply anywhere between 478 and 1200 EJ by the year 2100, compared with 2014 global consumption of 467 EJ [2,10].

If the low production estimates of the more pessimistic researchers prove accurate, this would mean that RE would eventually be left with only one major competitor, nuclear power. However, most think that global *nonconventional* resources of fossil fuels are large (see, e.g., Refs. [2,9]). Such nonconventional resources—tar sands, oil shales, and various forms of "tight" gas—have much lower energy return on energy invested (EROI) values than conventional fossil fuels, and accordingly have much higher CO_2 and other environmental costs, as well as higher economic costs per EJ of energy delivered to the consumer. Such high costs will, unfortunately, not necessarily prevent them being exploited. Nevertheless, fossil fuels are a finite resource; sooner or later the world will need to shift to alternative energy sources.

As of mid-2016, despite much talk about the need to drastically limit CO_2 emissions, consumption of fossil fuels, and with it their CO_2 emissions, are still growing [2]. Assuming that the annual supply of fossil fuels is adequate to meet the demand over the coming decades, then only concerns about climate change (and, to a lesser extent, regional air pollution) will curb their use. But many see high consumption of fossil fuels continuing even in a carbon-constrained world, because of two possible technological solutions to the CO_2 emissions problem, carbon sequestration and geoengineering.

1.2.1 Carbon Sequestration

Two general approaches are possible for carbon sequestration: biological and mechanical. With biological sequestration, the approach is to enhance soil carbon or carbon storage in biomass, particularly by afforestation and reforestation. Since the carbon stored in biomass is estimated to have fallen by around 45% over the past two millennia, such carbon sequestration would merely help restore the status quo ante [11]. Bio-sequestration is also thought to be fairly cheap compared with other carbon mitigation alternatives; Marshall [12] gave a cost of $20–$100 per tonne of CO_2 captured, with a potential of around three billion tonnes annually (3 Gt/year) (and as much as 6 Gt/year for bioenergy carbon capture and storage (BECCS). Total fossil fuel carbon emissions in 2014 were 9.7 Gt, or 35.5 Gt of CO_2 [2].

Nevertheless, this approach faces two serious problems. The first is the question of how much carbon could potentially be sequestered. Although Marshall suggests that potentially several billion tonnes of carbon could be sequestered annually by the end of this century, other researchers have suggested much lower potentials. Putz and Redford [13] have argued that maximizing carbon storage may conflict with biodiversity conservation. Mature forests are better for biodiversity maintenance, but actively managed forests can store more carbon.

Smith and Torn [14] regard estimates such as those of Marshall for biosequestration as far too optimistic. They argue than even 1.0 Gt/year of carbon sequestration from combined afforestation and BECCS would represent "a major perturbation to land, water, nitrogen, and phosphorous stocks and flows." The reason for their pessimism is that only marginal land would be available, given humanity's already high demands on Earth's net primary production (NPP). Such land would need major inputs of water and fertilizer for the necessary biomass growth.

The second problem concerns the net climate change effects of such tree planting. On the one hand, forest growth in all regions will draw down CO_2 from the atmosphere, with positive climate mitigation benefit. But on the other hand, Keller et al. [15] and Arora and Montenegro [16] have shown that increasing forest area in boreal regions will lower the albedo of such regions, because tree foliage absorbs more insolation than snow-covered ground. Climate forcing (in W/m^2) is thereby raised. Reforestation in tropical areas does not lower albedo, hence reforestation—or rather, preventing further deforestation—should be a priority. Arora and Montenegro have claimed that per unit area, tropical afforestation can give three times the warming reductions compared with boreal or temperate region afforestation.

Mechanical sequestration—carbon capture and storage (CCS)—can also take two forms. First, capturing CO_2 from the exhaust stacks of large fossil fuel power plants or oil refineries, followed by burial in, for example, disused oil and gas fields, or saline aquifers. Second, direct air capture (DAC) of CO_2, again followed by burial. DAC is not limited to national emissions, or even emissions from that year, and can be done anywhere, although areas with good wind speeds will help CO_2 absorption. It can also usually be carried out closer to CO_2 burial sites than is the case for exhaust stack capture. The crucial disadvantage is that it is very energy intensive [17,18], because of the low CO_2 concentration in ambient air. The 2011 report by the American Physical Society [19] summed it up succinctly:

> *In a world that still has centralized sources of carbon emissions, any future deployment that relies on low-carbon energy sources for powering DAC would usually be less cost-effective than simply using the low-carbon energy to displace those centralized carbon sources. Thus, coherent CO_2 mitigation postpones deployment of DAC until large, centralized CO_2 sources have been nearly eliminated on a global scale.*

The burial phase of mechanical sequestration is also not without its problems. Zoback and Gorelick [20] concluded that: "(…) there is a high probability that earthquakes will be triggered by injection of large volumes of CO_2 into the brittle rocks commonly found in continental interiors. Because even small-to moderate-sized earthquakes threaten the seal integrity of CO_2 repositories, in this context, large-scale CCS is a risky, and likely unsuccessful, strategy for significantly reducing greenhouse gas emissions." A further problem with the long-term integrity of CO_2 storage is its potential conflict with "fracking" for natural gas in shale formations. As Elliot and Celia [21] have pointed out, CO_2 sequestration needs a deep permeable formation, overlain with an impermeable one to provide a good caprock. Shale formations are ideal for this purpose, and potentially provide a large storage capacity in the United States. However, they showed that 80% of this storage capacity overlaps with "potential shale-gas production regions," and the fracturing of the shale for gas extraction would conflict with sequestration. Both these papers have proved controversial, but mainly for the extent of the problems they identify, rather than their existence.

1.2.2 Geoengineering

Geoengineering can be defined as the planned modification of the environment on a very large scale, often globally. The idea is not new, but has received recent attention because of the perceived urgent need to avoid dangerous climate change, and has even been cautiously endorsed by the Royal Society in the United Kingdom [22]. Advocates have stressed that conventional mitigation methods, such as energy efficiency improvements and greater use of nonfossil fuel energy sources, have not so far stemmed CO_2 emissions. The most discussed form of geoengineering would mimic the cooling effects of major volcanic eruptions, such as Mount Pinatubo in 1991, and involves placing sulfate aerosols in the lower stratosphere to increase Earth's albedo. Marine cloud brightening is an alternative strategy for increasing albedo. The albedo represents the Earth-averaged percentage of short-wave insolation reflected directly back into space, and is presently around 30%. By increasing Earth's albedo, SRM can counteract the global warming resulting from greenhouse gases (GHGs) absorbing and reemitting long wave radiation back to the Earth's surface. Since the term geoengineering is sometimes taken to include large-scale biosequestration, we use the more specific term solar radiation management (SRM).

SRM would produce a number of benefits. Because the aerosols would be rained out within a year or so, continuous placement of aerosols—perhaps by using civilian airliner flights or military aircraft—would be needed. But this is also an advantage of aerosol placement, since if unanticipated side effects were discovered, the project could be quickly terminated. Further, the temperature reduction benefits would appear within a year, as happened with the Mount Pinatubo natural global cooling. Another important benefit of this approach to SRM is its reported very low (annual) costs [23], especially compared with other climate mitigation methods. SRM also retains the benefits from CO_2 fertilization, which would be lost with CO_2 removal policies discussed earlier.

Nevertheless, there are a number of serious problems already recognized with SRM, and perhaps also presently unknown ones. First, emissions of CO_2 increase the CO_2 content of the oceans as well as that of the atmosphere. The result—which, unlike climate change, is not contested—is the steady acidification of the ocean waters. This acidification could inhibit, or at least slow, calcification in a variety of marine organisms, such as coral and foraminifera [24]. Unlike carbon sequestration, SRM would not address this problem. Second, climate forcing from CO_2 from fossil fuel combustion and land use change would continue, but would be offset by matching aerosol placement. However, if serious problems appeared with SRM, and the climate forcing offset was terminated, temperatures would subsequently rise rapidly because of the sudden increase in net climate forcing. Ecosystems would have difficulty adjusting to such unprecedented rates of temperature change. Also, the low *annual* cost of SRM may be more than offset by the need to continue aerosol placement for the lifetime of excess atmospheric CO_2, which could take until the year 3000 [25].

Third, with SRM, it may be possible to keep global average surface temperatures at their present (or lower) level, *or* globally averaged precipitation, but not both [25]. It is possible for example, that the Asian monsoons could be adversely affected, with grave consequences in an already water-stressed world. Fourth, given the problems as well as the benefits of SRM, there may well be intractable problems in obtaining global political consensus for action. Some countries will gain net benefit, others net losses, from a global SRM program, and so it is unlikely that the net losers will agree to SRM.

A fifth problem is *moral hazard*. Because of the advantages of SRM—its low annual cost, its short lead times for implementation, and the ability to rapidly terminate SRM if serious unforeseen consequences arise—it could prove very attractive if other methods fail to avert global climate change. Conventional mitigation, such as the replacement of fossil fuels by RE, will be a slow and expensive process, if only because remaining fossil fuel reserves would be deemed worthless. Also, successful mitigation requires concerted effort by all large emitting countries. But global SRM could be implemented by one group of countries, or even one major country.

One response to the problems discussed is to limit SRM to certain regions, and at certain times of the year only, with the aim of mitigating specific climate problems. Thus the proposal to concentrate aerosol placement in the Arctic stratosphere, in order to prevent further loss of sea ice and to arrest Greenland icecap melting. But modeling by Tilmes et al. [26] concluded that: "A 4 times stronger local reduction in solar radiation compared to a global experiment is required to preserve summer Arctic sea ice area." With the necessary high aerosol concentrations needed, the effects would spread well beyond the Arctic.

There are also proposals to limit SRM to within national boundaries, such as proposals to paint urban roads and roofs white, cover deserts with reflective material, and even change crop reflectivity, to increase local or regional albedo [22]. But even these more modest measures will either have only minor global cooling potential, or will face serious environmental problems. Given the remaining uncertainties, it would be unwise to rely on this untried technology. As the National Research Council [25] put it: "Intervening in the climate system through albedo modification therefore does not constitute an 'undoing' of the effects of increased CO_2 but rather a potential means of damage reduction that entails novel and partly unknown risks and outcomes." Perhaps for this reason the latest IPCC reports do discuss SRM in some detail, but unlike CCS, do not include SRM in any of their future scenarios.

1.2.3 Discussion

The view of many official organizations, such as the EIA and the IPCC, is that fossil fuels will continue to dominate global fuel supply for many decades. The implicit assumption is that dwindling reserves (or very high extraction costs) will not limit demand. But as Anderson [27] has stressed, CCS is an untested technology at the very large scale needed, so it would be unwise to base future energy policy around it. At present, only about 10 million tonnes of CO_2 are sequestered annually, whereas several *billion* tonnes would be needed for CCS (or air capture) to be a major mitigation technology. Since CO_2 capture from exhaust stacks can only offset around one quarter of all GHG emissions [24], air capture would need to be deployed on a very large scale. Its very high-energy costs would thus lead to an even more rapid depletion of fossil fuels.

1.3 NUCLEAR ENERGY

The first nuclear energy plants went online during the 1950s. After experiencing fairly rapid growth in the two following decades, growth slowed after the Chernobyl accident in 1986. In fact, nuclear energy's share of the global electricity market peaked in 1996 at 17.6%, and by 2014 had fallen to 10.8% [2], although this share can be expected to rise a little as Japan restarts its reactors in the wake of the 2011 Fukushima accident. While all of the 200 plus countries in the world use some fossil fuels, only 31 have nuclear power, mostly countries in the OECD, although China, Russia, and India have important nuclear programs. Hence in most countries, nuclear energy has zero share of the electricity market, but at the other extreme is France, where its share is presently 78.4% [2].

What are the future prospects for nuclear energy? One view is that nuclear power output will be limited by dwindling uranium supplies. Dittmar [28] has argued that even modest growth in nuclear power output will soon be constrained. His forecast, based on historical data from existing and former uranium mines, was that annual global production will soon peak at around 58 kilotons (kt), and that by 2030 will have declined to only about 41 kt. Output will not be enough to sustain even a modest growth rate of nuclear power production of 1% annually, well below the forecast growth rate for global electricity [7].

Other researchers envision either breeder reactors (perhaps using thorium) or even fusion reactors overcoming any possible uranium fuel constraint [29,30]. World reserves of thorium are thought to be around four times those for uranium [31]. Breeder reactors were early on recognized as necessary to extend limited uranium supplies, as they can convert the fertile isotope uranium-238 (U-238) into fissile plutonium-239, compared with conventional reactors that can only use fissile U-235. (The U-235 isotope only forms 0.7% of the naturally occurring uranium, with U-238 accounting for nearly all the remainder.) In conventional "once-through reactors" using fuel enriched to around 2—4% U-235, about 99% of the potential energy content goes unused, as the current plan is to bury the spent fuel rods after treatment.

However, experience with full-scale breeder reactors have shown that they are difficult to operate. France's 1200 MWe Superphénix breeder reactor only operated for a decade at low reliability before being permanently shut down in 1996, and Japan's Monju reactor, after being shut down from 1995 to 2010, may not operate again [32].

Hopes for *fusion energy* are mainly placed in the International Thermonuclear Experimental Reactor (ITER) presently under construction in France, and financed by a multinational consortium. But the date for completion remains uncertain after repeated postponements, and costs have tripled since initial estimates, with further rises likely [33]. And even if successful in its aims, it will still not demonstrate that commercial fusion energy is feasible.

Another reason why growth in growth nuclear output will likely be at a low level is that the present reactor fleet is aging. According to an analysis by Froggatt and Schneider [34]: "the unit-weighted average age of the world operating nuclear reactor fleet continues to increase and by mid-2014 stood at 28.5 years." They further add that over 170 of the global 388-strong reactor fleet have run for 30 years or more, and 39 of these for over 40 years. Thus, a substantial reactor-building program will soon be needed merely to maintain nuclear power's present output. Also, worldwide construction costs and construction times appear to be rising.

Even official projections for nuclear power do not envision large growth rate increases. The EIA [7] forecast that globally, nuclear output will increase by 2.4% and 2.6% annually between 2010 and 2040 in the low and high economic growth cases, respectively. The International Atomic Energy Association (IAEA) [35], an organization charged with promoting nuclear energy, has forecast the share of nuclear energy in global electricity production out to 2050. They envisaged this share rising from 12.3% in 2011 to between 12.8% and 13.9% by 2020, but thereafter *declining* to between 5.0% and 12.2% by 2050. This decline may be in recognition of the aging reactor fleet discussed earlier. A third forecast, based on the integrated assessment modelling by van Vuuren et al. [10] on the four RCPs, showed nuclear power in the year 2100 supplying between 4.1% of global energy in the worst case (RCP 6.0), and 11.3% in the most favorable case (RCP 8.5).

The most serious nuclear accidents so far have occurred in the United States (Three Mile Island, 1979), the former USSR (Chernobyl, 1986), and Japan (Fukushima, 2011), all technologically sophisticated nations. In each case, the accident had major repercussions for nuclear power worldwide. Given that some technically advanced nations are phasing out their nuclear power programs, major global growth in nuclear energy will necessarily mean programs in countries with lower nuclear expertise and regulation. Nuclear power also must soon face the decades-old problem of waste disposal. The conclusion that can be drawn from this brief survey is that nuclear energy cannot be expected to supply more than its present share of global primary energy, and could supply much less, given widespread public opposition.

1.4 RENEWABLE ENERGY

RE sources differ from their main competitor, fossil fuels, in several important ways. The energy from fossil fuels resides in chemical bonds embodied in matter, enabling fossil fuels to be stored above ground, or left underground until needed. On the other hand, RE energy (except for bioenergy and to some extent for geothermal energy) exists only as *flows*—so RE energy not used is lost forever, and with it the chance to reduce carbon emissions. Also, RE sources, except for hydro, are estimated to have a much lower EROI than present fossil fuels [36].

However, RE has important advantages over fossil fuels. As shown earlier, there are two major question marks over the future of fossil fuels. First, there are doubts about how long present, let alone increased, production levels can be maintained at anywhere near present unit costs. Secondly, their combustion produces the long-lived major GHG, CO_2, much of which will remain in the atmosphere for thousands of years [9]. As already shown, technical fixes in the form of carbon sequestration and SRM face several serious problems, probably explaining why neither has been taken up, although discussion on both go back several decades. Finally, the combustion products NO_x, SO_x, and particulates produce serious air health problems, particularly in the densely populated megacities of the world. While RE sources can also produce GHGs and air pollutants, their emissions output per unit of energy are far smaller than for fossil fuels.

At first glance it might appear that rising use of RE on its own will not cut fossil fuel use. Since 1960s, global RE output has grown in step with fossil fuel use [2]. But when the experience of individual countries is examined, a different picture emerges. A number of European countries, including Germany and the United Kingdom, have experienced long-term declines in fossil fuel consumption along with rising RE output. Nevertheless, the various RCP scenarios assumed that, globally, RE output would grow in step with growth in both total energy and fossil fuel energy output, with fossil fuel CO_2 emissions greatly reduced through CCS and especially BECCS [10,37].

1.4.1 Earth Energy Flows and Renewable Energy Potential

The energy potentially available to us in the form of RE comes from three sources: the sun, Earth's interior, and tidal energy (see Fig. 1.1). By far the largest is the low-wavelength radiation from the sun. As already discussed, about 30% is reflected, unchanged, back into space from our planet. The remainder, about 3.9 million EJ/year, or 3900 ZJ/year (ZJ = zettajoule = 10^{21} J), is absorbed by the land, oceans, and clouds. It is this energy that ultimately drives the atmospheric circulation and hydrological systems, and through photosynthesis, plant growth. The energy diverted to atmospheric circulation is subject to a wide range of estimates, but for the entire atmosphere might be 1200 TW, or roughly 38,000 EJ/year [38]. Most of this energy is accounted for by the high-altitude jet stream, and is not available (at least in the foreseeable future) as a human energy source. About one-third of Earth's insolation is diverted to drive the Earth's hydrological system [39]. However, the power of all the world's river runoff is a vastly smaller amount, about 3 TW or 95 EJ/year [40].

The second energy source is a result of the mutual gravitational attraction between Earth and its much smaller but close-by moon, and to a lesser extent, between Earth and the sun. This tidal energy amounts to about 76 EJ/year. Almost all of this energy is dissipated in the oceans, and fortunately for us, mainly along coastlines.

The third energy source is internal to Earth. Geothermal energy is simply the residual heat left over from the violent impacts involved in the formation of our planet around

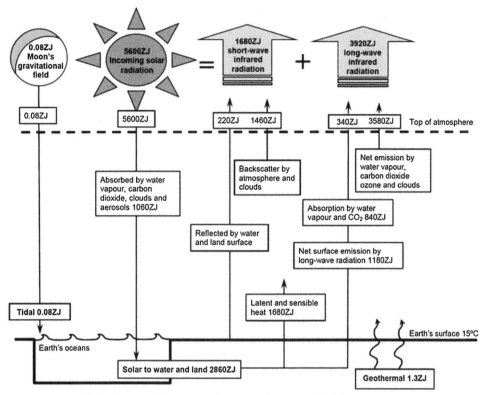

Figure 1.1 Simplified diagram of annual Earth energy flows. *Modified from Moriarty P, Honnery D. Rise and fall of the carbon civilisation. London: Springer; 2011; Trenberth KE. An imperative for climate change planning: tracking Earth's global energy. Curr Opin Environ Sustain 2009;1:19—27.*

4.5 billion years ago, together with heat energy derived from the slow radioactive decay of various isotopes of uranium, thorium, and potassium both in the Earth's core and crust. (The 235–uranium isotope also presently provides the fuel for nuclear fission energy.) Compared with insolation intensity at the top of the atmosphere of 1366 watt/square meter (W/m^2), geothermal output averaged over the Earth's surface is only about 0.08 W/m^2 or about 1300 EJ/year. Fortunately, energy flows are much higher near regions of high tectonic activity, such as plate boundaries.

In the very long term, all three energy flows are only temporary. The sun, a main sequence star, will expand in volume to become a red giant in a billion years or so, engulfing the Earth. Similarly, tidal energy is slowly decreasing, and will eventually fall to zero when the moon and our planet become locked. Geothermal energy will likewise eventually dwindle away; the primeval heat is slowly being lost from Earth, and the radioactive elements are slowly decaying. But compared to fossil fuels, where the difference in remaining lifetimes between the "peak theorists" and the more optimistic experts is measured in mere decades, these flows can be regarded as permanent. Geothermal energy is a partial exception. Geothermal plants are more economical if the accumulated heat in the field is "mined" at a rate faster than replenishment. Fields can then take several decades to recover.

Compared with global commercial primary energy use in 2014 of 541.3 EJ [2], the theoretical availability of the various RE sources, as just discussed, are very high. But a series of constraints limit the amount of each RE source that can actually be tapped. The first major constraint is that not all areas of Earth with suitable RE flows can be developed for energy. The deep oceans, the ice caps and high mountain ranges are obviously unsuitable, but some areas may be off-limits for various environmental reasons, while other areas may simply be too distant from energy markets. Together, these land constraints limit RE theoretical potential to the *geographical potential* [38].

Apart from the use of passive solar energy for space heating and cooling, and wind for drying clothes and crops, Earth's energy flows are not usually used in their crude form. Instead, the natural flows are converted by devices such as photovoltaic (PV) cells or wind turbines into more useful forms of energy, usually electricity. Such conversion is far less than 100% efficient, entailing further energy loss. The energy that can be harvested from the various conversion devices from geographically suitable areas using currently available technology is termed the *technical potential*. Again, not all technical potential is necessarily *economic potential*, which de Vries et al. [42] define as: "The economic potential is the technical potential up to an estimated production cost of the secondary energy form which is competitive with a specified, locally relevant alternative." But as we discuss in detail later in this chapter, large energy subsidies make economics alone a poor guide for selecting energy types.

Many researchers have argued that the technical potential for RE is so vast that it will not possibly constrain any conceivable global energy use level (e.g., Refs. [43,44]). However, the published literature on the technical potential on the main RE types in use today and for the foreseeable future—solar, wind, hydro, biomass, and geothermal—show a range sometimes spanning two orders of magnitude [45–47]. Only for hydroelectricity are estimates fairly tightly constrained at about 30–60 EJ.

Table 1.2 shows the upper and lower limits for technical potential reported for each of the five leading RE types, as published since the year 2000. For geothermal energy, only the electricity potential is shown, but estimates for the global technical potential for lower temperature heat are very large, with estimates as high as 310,000 EJ. One of the reasons for the large range in global RE potential is that few RE technical potential estimates are based on EROI. The EROI is the ratio of the lifetime output energy to input energy for RE device (for construction, operation, and maintenance over the life of the project, and finally decommissioning), both measured in compatible units, for example, primary energy units. The acid test for any new energy project is that the EROI

Table 1.2 Range of Global Renewable Energy Technical Potential Estimates (EJ)

RE Source	Technical Potential Range (EJ)
Biomass	27–1,500
Geothermal[a]	1.1–22
Hydro[a]	19–95
Solar[a]	63.0–15,500
Wind[a]	31.5–3,000

[a]Electric output.
Data from de Castro C, Mediavilla M, Miguel LJ, Frechoso F. Global solar electric potential: a review of their technical and sustainable limits. Renew Sustain Energy Rev 2013;28:824–35; Lu X, McElroy MB, Kiviluoma J. Global potential for wind-generated electricity. Proc Natl Acad Sci 2009;106:10933–8; Moriarty P, Honnery D. What is the global potential for renewable energy? Renew Sustain Energy Rev 2012;16:244–52; de Castro C, Mediavilla M, Miguel LJ, Frechoso F. Global wind power potential: physical and technological limits. Energy Policy 2011;39:6677–82.

must be greater than 1.0. If it is less, the energy project is an energy sink, not a net addition to energy supply. Only in the development stage of a new energy source can an EROI < 1.0 be tolerated.

The main conclusion from the data in Table 1.2 is that the technical potential for the leading RE sources is not known with any accuracy. However, even the minimum values for RE potential are many times the current RE annual production. Further, apart from the biomass values in Table 1.2, the figures are for electricity, and should be multiplied by 2.6 [2] to better indicate potential in primary energy terms.

1.4.2 Present and Future Use of Renewable Energy

Table 1.3 shows the global RE electricity output from various sources for 1990 and 2012 in TWh. For completeness, it also includes the category "ocean energy"; at present, nearly all of this is the output of the tidal power station on the Rance Estuary in France. Although the growth in wind and solar energy has been rapid, hydro still dominates RE electricity production, and will for decades to come. What are the prospects for the various RE sources in the coming decades?

Although biomass has only minor electric power output, it dominates RE, with perhaps 50 EJ worldwide, mainly fuel wood burnt at low efficiency in industrializing countries. Along with oil, it is the only energy source that is used in virtually all countries. But although many see a very large technical potential, its future is uncertain because the human appropriation of the Earth's terrestrial NPP (HANPP) is already very large, with estimates as high as 40% [11]. As human population increases, and with it the demand for food, forage for livestock, fibers (cotton, wool), and timber, HANPP can only rise. Already, given the unprecedented high extinction rates, the natural world is

Table 1.3 Global Renewable Energy Electric Output in 2012 (TWh)

RE Source	Electric Output (TWh)	
	1990	2012
Biomass	88.9	326.2
Geothermal	36.1	70.4
Hydro	2163.3	3663.4
Solar	0.4	104.5
Wind	3.6	534.3
Ocean	0.5	0.5
All RE sources	2292.8	4447.5

Data from BP. BP statistical review of world energy 2015. London: BP; 2015; Moriarty P, Honnery D. Preparing for a low-energy future. Futures 2012;44:883–92; Electricite de France (EdF). Worldwide electricity production from renewable energy sources, Fifteenth Inventory, 2013. http://www.energies-renouvelables.org/observ-er/html/inventaire/pdf/15e-inventaire-Chap01-Eng.pdf.

under stress. Adding another heavy levy on NPP, in the form of bioenergy, will add to this stress, and may even be counterproductive, in the sense of decreasing the absolute levels of biomass available for humans [51].

Table 1.2 shows that optimistic estimates for geothermal *electricity* potential are still small, although many times the existing output from the 45 countries that presently operate geothermal electricity plants [52]. The real potential lies in direct use of geothermal heat, which has a very large potential, and is available in many more countries. Hydro is already heavily exploited, with few suitable sites left in OECD countries; most of the remaining potential is in Asia, Africa, and South America. Already about 125 countries have hydro plants [52]. Although to a lesser extent than biomass, hydro development can also be at the expense of other ecosystem services [53,54]. In any case the total technical potential is probably no larger than about 30 EJ.

Wind turbines presently operate in around 100 countries, and many more have at least some technical wind potential [52]. Wind turbines are available in several standard sizes, and wind farms can also vary in size, from a single turbine to several hundred. They can thus be utilized by countries with small grids, or even by single households. Another advantage of wind energy is that it is compatible with some existing land uses, such as crops or grazing. Their biggest disadvantage, which they share with solar energy, is the intermittent nature of their output. Ways of overcoming this problem are discussed in the next section.

Modern solar energy conversion devices are a relatively recent addition to electricity production (Table 1.3), but output is growing very rapidly. Although solar thermal energy conversion (STEC) is only suitable at utility scale, PV cells can be installed as large arrays by utilities with 100 MW or more output, or by individual households, with output measured in kW. Germany is a leader in rooftop PV installations. Together with a storage battery, they are also very useful for off-grid households, such as the majority of tropical African households.

So far, we have only considered active solar energy, the type delivered by dedicated conversion devices, such as PV cell arrays or solar hot water heaters. But passive solar energy is an important, if relatively neglected, RE source with high technical potential and low costs. It can be used for heating and cooling buildings, and for lighting. Solar heating and cooling have been used for millennia, and are most effective if they are incorporated into the design of buildings and the selection of building materials. However, some passive solar heating and cooling practices can be back fitted to existing buildings. Similarly, solar lighting can be as simple as drawing the curtains in a room in the morning, but light can also be channeled to illuminate interior rooms with no windows, such as in commercial buildings. It is often difficult to separate out passive solar energy from building energy conservation. In any case, to be effective, it requires the active participation of the occupants in opening and closing doors, windows, and sun shades at appropriate times.

What do official projections see as the future for RE in the coming decades? The four RCPs considered by van Vuuren et al. [10] foresaw all RE primary energy growing to between roughly 135 EJ to 335 EJ—but only by the year 2100. Their share of global primary energy would then be between about 16% (RCP6) and 37% (RCP2.6). These low absolute and % values for RE occur because of the major emphasis the RCP scenarios place on CO_2 removal through CCS and BECCS.

1.4.3 Coping With Intermittent Renewable Energy Sources

It is generally agreed that the greatest technical potential for RE lies with solar and wind energy. But these two energy sources, along with wave energy (if ever commercialized), are *intermittent* sources of energy, which could represent a barrier to their large-scale uptake in electricity grids. Electricity grids have always had to deal with variable demand—for instance, demand is much lower late at night than in the mornings or early evenings. They cope with such fluctuations by having standby power units of known output, which can be rapidly brought on line to meet rises in load. But intermittent electricity adds a further uncertainty—this time on the electricity supply side.

There are several ways of overcoming such intermittency as wind and solar inevitably assume progressively larger shares of electricity in a given grid. One approach is simply to build overcapacity, so that even during low periods of wind and/or insolation, there is sufficient power to meet demand. But such an approach is wasteful, as electricity will have to be discarded at times of peak intermittent RE output. Another approach tries to avoid this waste by expanding the grid, both to include more nonintermittent electricity sources, such as hydroelectricity, but also to help even out the fluctuations from wind/solar by drawing on a wider area. Chatzivasileiadis, Ernst and Andersson [54] have discussed the Desertec proposal, which envisaged bringing both solar and wind energy thousands of kilometers from the deserts of North Africa and the Middle East to Europe. Apart from the cost, another disadvantage with this proposal is the question of energy security for European countries. Seboldt [55] has even proposed a truly *global grid* spanning both northern and southern hemispheres and the various time zones. This approach would even out seasonal and diurnal insolation fluctuations, but energy security and cost problems would remain.

Another method for dealing with intermittency is energy storage. Pickard [56] has argued that an RE electric grid will need very large storage capacity. California, a leader in RE electricity, has even mandated that the state must have 1.32 GW of storage capacity by the year 2020 [57]. Storage could take the form of pumped hydro storage, compressed air, batteries, and even conversion to another energy carrier, such as hydrogen. Some storage of intermittent RE cannot be avoided even with a global grid, because not all final energy demand is for electricity—aircraft and freight ships cannot be feasibly run on electricity. Converting intermittent electricity to hydrogen or methanol to fuel such uses, followed by storage and transport to final users, will also entail high energy costs.

Grid expansion and energy storage (or, alternatively, the need for backup fossil fuel energy) are to some extent substitutes. Steinke et al. [58] looked at the trade-offs between the two for a 100% intermittent RE electricity grid in Europe. They found that a European-wide grid expansion could cut the backup energy needed from 40% to 20% of annual electricity consumption. Only for a truly global grid (or satellite solar power transmitted to Earth receiving stations [55]) could the need for energy storage or backup power for the electricity grid be completely avoided. Finally, because of the intermittent nature of power output, transmission lines for RE must be of higher capacity than those for fossil fuel power of the same output [59].

1.5 PROSPECTS AND POLICIES FOR RENEWABLE ENERGY

Previous sections have characterized the situation for rivals to RE in the global energy market, and argued that the futures for both fossil fuels and nuclear energy are far from certain. We then discussed the technical potential of RE, as well as some of its advantages and disadvantages. In this section, we look at policies that, if adopted, could best aid a more rapid uptake of RE globally. We argue that by far the most important action would be to remove the vast subsidies presently given to fossil fuels.

1.5.1 Fossil Fuel Energy Subsidies Must Be Cut

As mentioned earlier, all energy sources receive subsidies, sometimes very heavy ones, depending on the country and the fuel type. But present subsidies for alternative fuels pale beside the global subsidies for all fossil fuels as calculated by the International Monetary Fund (IMF) for 2015 [60]. The IMF considers two types of subsidies. The more easily calculated type consists of consumer subsidies defined as the difference between the international price and the price charged to the consumer; the IMF estimates this subsidy at less than 20% of the total. Most of this subsidy went to consumers in the oil-exporting countries themselves. Darmstadter [61] has pointed out that in Venezuela, motorists were recently paying only 10 cents per gallon (2.6 cents/liter) for petrol.

The second, larger, but less well-defined, subsidy consists of negative externalities that energy use inflicts on society and the environment (such as the health effects of air pollution, or the cost of CO_2 emissions), and accounts for over 80% of the total. The breakdown of these subsidies are shown in Table 1.4. The total subsidy amounted to 6.5% of global gross domestic product, even with the large drop in energy prices in 2015. But large as these values are, they may still be underestimates. As an example of a less obvious subsidy that could be considered, Delucchi and Murphy [62] looked at the reduction in just the US military expenditures that would be possible if there were no oil in the Persian Gulf states, and found that roughly $27–$73 billion (2004 values) per year could be saved in the long run.

The costs of climate change impacts from fossil fuels may even be underestimated in the IMF study. The values calculated there for the negative externalities from fossil fuel CO_2 emissions (the social cost of carbon (SCC) in US$ per tonne carbon) used the values from the US government Interagency Working Group on Social Cost [63]. This Working Group estimated that "a tonne of carbon dioxide emitted now will cause future harms worth US$37 in today's dollars" [64]. However, the latter authors and others (e.g., Refs. [65–67]) have criticized such values as being far too low. Indeed, Ackerman and Stanton [65] have argued that values of SCC of $1000 or even much higher could easily be justified. If this were the case, SCC would dominate the total cost of negative externalities, and total subsidies overall would rise steeply.

Subsidies to nuclear and RE were not considered in the IMF study. Nuclear energy subsidies were fairly small in 2015. However, in the past, nuclear power received very large subsidies, which were vital for its commercial introduction. According to a study by Ward [68] "it is estimated that the US nuclear energy sector received financial support to the tune of $15.3 per kWh in the first 15 years of its development (1947–1961),

Table 1.4 Estimated Subsidies to Fossil Fuels, 2015 (US $)

Energy Type	Subsidy US $ Billions	% of Total Subsidy
Coal	3147	59.4
Oil	1497	28.2
Natural gas	510	9.6
Electricity	148	2.8
All energy	5302	100

Data from Coady D, Parry I, Sears S, Shang B. How large are global energy subsidies? IMF Working Paper WP/15/105; 2015. Available at: https://www.imf.org/external/pubs/ft/wp/2015/wp15105.pdf.

compared to wind energy, which received just $0.46 per kWh in its first 15 years." US support for the nuclear energy was thus around 30 times the rate for wind energy.

One possible defense of consumer energy subsidies is that it promotes equity. But as Edenhofer [69] has stressed: "Energy subsidies are typically captured by rich households in low-income countries and do little to support the poor." Similarly for the indirect subsidies: climate change impacts of fossil fuel use will heavily impact low-income communities [70], and urban air pollution, largely from fossil fuel combustion, is likewise more serious in lower-income areas [71].

The most important point about these huge energy subsidies is that they lead to massive *overuse* of energy in general. If consumer prices were not subsidized, and the full health and environmental (e.g., global climate change) costs were met by users, global energy use would be far smaller. In other words, it does not particularly matter if RE cannot meet present (or future projected) global energy levels. RE will in future be the major energy source (in terms of energy share) partly by absolute growth in RE output, but also by a rapid reduction in fossil fuel use. The most important energy policy—and one that would be of the largest benefit to RE—would be the removal of fossil fuel subsidies.

1.5.2 Policies Needed to Support Renewable Energy

Unruh [72] has introduced the notion of carbon "lock-in," the idea that, as he puts it: "industrial economies have become locked into fossil fuel—based energy and transportation systems through path-dependent processes driven by technological and institutional increasing returns to scale." A 2010 insight on carbon lock-in is provided by Davis et al. [73]. They calculated the cumulative CO_2 emissions emitted if all fossil fuel power stations and vehicles operating in 2010 continued in use until their economic lives ended. Their estimates centered on roughly 500 Gt of CO_2, about 14 times the 35.5 Gt emitted by all fossil fuels in 2014 [2]. At the least, drastic reductions in fossil fuel use would entail the premature retirement of much of the fossil fuel power stations and vehicle fleets.

Fossil fuel energy subsidies will probably not be heavily reduced any time soon, given not only the carbon lock-in just discussed, but also the vested interests and economic power of fossil fuel producers. Another factor is the popularity of direct fuel subsidies to consumers, such as low petrol prices for motorists. We have already discussed the high subsidies for nuclear power in its early years. Promotion of RE will inevitably require some subsidies if it is to become the dominant energy provider. Koseoglu et al. [74] have grouped the vast number of the various specific policies possible into two general approaches:

- market instruments for encouraging greater RE use, the approach favored by Germany and many other countries;
- emphasis on, and support for, R&D for RE, the approach favored in California and some other US states.

However, most countries use a mix of policies to support RE growth, and further, vary in the level of direct government control.

Germany, a leader in RE, has relied heavily on feed-in tariffs, which vary for different forms of RE. They are much higher for off-shore than for on-shore wind electricity, reflecting its higher costs. Support for PV electricity was in turn nearly five times higher in 2009 than for on-shore wind, perhaps because insolation levels in Germany are relatively low. The German experience is considered as effective in supporting rapid RE growth, but not very cost-effective. There is also the danger of "technological lock-in" with technology-specific tariffs. As an alternative to supporting RE market applications, as in Germany, governments can also subsidize R&D for RE. Koseoglu et al. found that supporting R&D was a better use of scarce resources for immature (and rapidly developing) RE technologies. For more mature technologies, such as hydro, however, this approach will be of limited effectiveness.

In China, and to a lesser extent in other lower-income countries, the clean development mechanism (CDM) has been important in encouraging the growth of RE. The CDM allows these countries to offset the CO_2 emissions from industrial countries. It has been criticized for being relatively ineffective, insofar as in many cases the low-carbon projects (such as hydroelectricity projects) would have been undertaken in any case. Others, like Newell [75], have gone much further, and argued that market mechanisms, such as carbon markets, will only have limited success in reducing carbon emissions.

There are evidently advantages and disadvantages with each of the approaches used, and no global recommendation is possible; it can and should vary from country to country, depending, among other factors, on the particular RE source involved and level of technological development in the country. PV cell arrays and solar water heaters have been installed on the roofs of millions of private residences worldwide. Similarly, geothermal heat pumps are rapidly expanding in use in many countries. But private residences cannot install a hydro, STEC, or geothermal plant. Different support is needed for households compared with utility-scale plants.

The support needed for bioenergy is likewise very different in low-income countries compared with high-income countries. Firewood and crop residues are the cheapest fuel in poor countries, which explains their very high level of use, even if this level of use is environmentally unsustainable. Incentives are needed, not to encourage but to *reduce* this form of bioenergy use, through the widespread use of more fuel-efficient cooking stoves.

1.5.3 Which Sources of Renewable Energy Should Be Supported?

There are a very large number of possible sources of RE, apart from the five most commonly used. These include:
- ocean thermal energy conversion (OTEC);
- ocean currents;

- wave energy;
- tidal energy;
- higher altitude wind energy collected from turbines on air balloons or even kites;
- osmotic energy at the fresh water/salt water interface at river estuaries.

The question arises as to whether to support these RE sources in hopes of a break-through, or to avoid spreading limited R&D resources too thinly, and concentrate on improving the mainstream RE types. Sometimes, novel energy sources can be ruled out, at least as major energy sources, by using energy analysis.

For example, OTEC will have a thermal efficiency of only 3—4% [76], given that the temperature difference between tropical surface waters and ocean depths is at most around 20°C, and a 1000-m pipe will be needed to draw up deep colder water as a heat sink. It is possible that small-scale OTEC plants, with distilled water as a coproduct, could be feasible at coastal locations on small islands. However, for large-scale energy production, the OTEC plants would need to move over the ocean to maintain a temperature differential, and the electricity generated converted into an energy carrier, such as ammonia or hydrogen, stored, and periodically shipped to shore. When these conversion, storage, and transport energy costs are considered along with the low Carnot efficiency, it is doubtful that any net energy would result.

However, it is possible to select which RE types should be generally supported. Solar energy is the obvious candidate, for two important reasons. First, it has by far the highest technical potential of any RE source. Second, there are likely to be further technical breakthroughs, not only leading to reduced production costs for existing PV cell types, but also to novel PV materials. As for other new RE technologies, it will be important to choose materials (as well as installation sites) that do not compromise environmental sustainability in general [46,77,78].

For RE to be a major force in energy production, the intermittent RE sources, wind, solar, and perhaps wave energy, will need to supply most RE. But, as we have seen, this will inevitably mean that large amounts of energy storage will be needed. Conversion of intermittent RE electricity to an alternative energy carrier, such as hydrogen or methanol, then storage, followed by reconversion to electricity, will greatly reduce the net energy output from these sources, and hence lower the EROI and raise costs. Other approaches are of course possible if the output is also to be electricity, such as pumped hydro and battery storage. The scope for further conventional pumped storage is low [56].

A large variety of battery types are being investigated, with lithium batteries the most popular for consumer electronics and electric vehicles. Lemmon [57] has argued that "(…) batteries cannot provide rapid (less than a second) high-power responses and supply energy for long periods. Batteries degrade and are expensive to replace." Instead, he argued for new types of fuel cells that "can be modified to store energy and produce liquid fuels, such as methanol, thanks to breakthroughs in materials and designs." But

Soloveichik [79] has argued that *flow batteries* can overcome many of the problems of other batteries, such as limited life, even if they are not suitable for mobile applications. As with different RE sources, it is likely that different applications will need different energy storage systems.

1.6 DISCUSSION

As output of RE rises to make it the dominant energy source, it may prove necessary to move away from the fossil fuel—era idea of energy available in any desired amount, at any time, at any place. *Demand management* of energy is not new: for many decades off-peak energy (usually at night) has been cheaper than peak rates. With the advances in information technology, it is now possible to price electricity according to instantaneous supply. Further, some load-shedding is possible at times of low supply without causing disruption to domestic users. For example, freezers, refrigerators, and hot water systems can be turned off for limited periods without any ill effects.

But we may need to go well beyond this and change policies not directly connected to the energy sector. Future RE potential may well be located at sites remote from existing grids. One possibility is to move some industries to these locations, as is already done with aluminum smelters, sometimes being located near cheap hydropower. Population growth could also be encouraged in these areas. This shift would be especially valuable for low-temperature geothermal heat energy, which has a very large global potential. The problem is that it is neither energetically nor economically feasible to pipe such heat more than a few kilometers. In the United States at least, such geothermal heat (mainly in the western Rocky Mountain states) is poorly matched to population [80].

The climate change problem will need to be decisively tackled in the coming decade or two. We have argued that attempts at "greening" fossil fuels through CO_2 capture are likely to be marginal, or in the case of geoengineering, unlikely to be attempted at all. There will probably be a large mismatch between the rate at which fossil fuel use will have to be reduced, and the maximum rate at which RE can be introduced as replacement energy. At present, except for solar energy, which is still growing exponentially from a small base, all other RE sources are at best growing only linearly [2]. RE sources will likely only come to dominate the future energy mix if absolute global energy levels are greatly reduced.

GLOSSARY

BECCS	Bioenergy carbon capture and storage
CCS	Carbon capture and storage
CDM	Clean development mechanism
CO_2	Carbon dioxide
EdF	Electricite de France
EIA	Energy Information Administration (US)

EJ	Exajoule (10^{18} J)
EROI	Energy return on energy invested
GHG	Greenhouse gas
GW	Gigawatt (10^9 W)
GDP	Gross domestic product
HANPP	Human appropriation of net primary production
IAEA	International Atomic Energy Association
IMF	International Monetary Fund
IPCC	Intergovernmental Panel on Climate Change
ITER	International Thermonuclear Experimental Reactor
MWe	Megawatt (10^6 W) electric
NRC	National Research Council (US)
NO_x	Oxides of nitrogen
NPP	Net primary production
OECD	Organization for Economic Cooperation and Development
OTEC	Ocean thermal energy conversion
PV	Photovoltaic
RE	Renewable energy
RCP	Representative concentration pathway
SCC	Social cost of carbon
SO_x	Oxides of sulphur
SRM	Solar radiation management
STEC	Solar thermal energy conversion
TW	Terawatt (10^{12} W)
U-235	Uranium isotope 235
W/m^2	Watt/square meter
ZJ	Zettajoule (10^{21} J)

REFERENCES

[1] Marchetti C. On energy systems historically and in the next centuries. Global Bioethics 2009;22(1−4):53−65.

[2] BP. BP statistical review of world energy 2015. London: BP; 2015.

[3] Inman M. The fracking fallacy. Nature 2014;56:28−30.

[4] Schindler J. The availability of fossil energy resources' in factor X: policy, strategies and instruments for a sustainable resource use. In: Angrick M, et al., editors. Eco-efficiency in industry and Science, vol. 29. Dordrecht, NL: Springer; 2014. p. 19−38.

[5] Höök M, Tang X. Depletion of fossil fuels and anthropogenic climate change—a review. Energy Policy 2013;52:797−809.

[6] McGlade C, Ekins P. The geographical distribution of fossil fuels unused when limiting global warming to 2°C. Nature 2015;517:187−90.

[7] Energy Information Administration (EIA). International energy outlook 2013: US Department of Energy. 2013. http://www.eia.gov/forecasts/archive/ieo13/pdf/0484(2013).pdf.

[8] Intergovernmental Panel on Climate Change (IPCC). Climate change 2014: mitigation of climate change. Cambridge UK: Cambridge University Press; 2014.

[9] Intergovernmental Panel on Climate Change (IPCC). Climate change 2014: synthesis report. Cambridge UK: Cambridge University Press; 2015.

[10] van Vuuren DP, Edmonds J, Kainuma M, Riahi K, Thomson A, Hibbard K, et al. The representative concentration pathways: an overview. Climatic Change 2011;109:5−31.

[11] Schramski JR, Gattie DK, Brown JH. Human domination of the biosphere: rapid discharge of the earth-space battery foretells the future of humankind. Proceedings of the National Academy of Sciences of the United States of America 2015. http://dx.doi.org/10.1073/pnas.1508353112.

[12] Marshall M. Transforming earth. New Scientist 10–11 October, 10–11, 2013;12.

[13] Putz FE, Redford KH. Dangers of carbon-based conservation. Global Environmental Change 2009;19:400–1.

[14] Smith LJ, Torn MS. Ecological limits to terrestrial biological carbon dioxide removal. Climatic Change 2013;118:89–103.

[15] Keller DP, Feng EY, Oschlies A. Potential climate engineering effectiveness and side effects during a high carbon dioxide-emission scenario. Nature Communications 2014;5(3304). http://dx.doi.org/10.1038/ncomms4304.

[16] Arora VK, Montenegro A. Small temperature benefits provided by realistic afforestation efforts. Nature Geoscience 2011;4:514–8.

[17] House KZ, Baclig AC, Ranjanc M, Van Nierop EA, Wilcox J, Herzog HJ. Economic and energetic analysis of capturing CO_2 from ambient air. Proceedings of the National Academy of Sciences of the United States of America 2011;108(51):20428–33.

[18] Moriarty P, Honnery D. A human needs approach to reducing atmospheric carbon. Energy Policy 2010;38:695–700.

[19] American Physical Society (APS). Direct air capture of CO_2 with chemicals. 2011. http://www.aps.org/policy/reports/assessments/upload/dac2011.pdf.

[20] Zoback MD, Gorelick SM. Earthquake triggering and large-scale geologic storage of carbon dioxide. Proceedings of the National Academy of Sciences of the United States of America 2012;109(26):10164–8.

[21] Elliot TR, Celia MA. Potential restrictions for CO_2 sequestration sites due to shale and tight gas production. Environmental Science and Technology 2012;46:4223–7.

[22] Royal Society. Geoengineering the climate: science, governance and uncertainty. London: Royal Society; 2009.

[23] McClellan J, Keith DW, Apt J. Cost analysis of stratospheric Albedo modification delivery systems. Environmental Research Letters 2012;7(034019):8.

[24] Moriarty P, Honnery D. Rise and fall of the carbon civilisation. London: Springer; 2011.

[25] National Research Council (NRC). Climate intervention: reflecting sunlight to cool earth. Washington, DC: National Academies Press; 2015.

[26] Tilmes S, Jahn A, Kay JE, Holland M, Lamarque J-F. Can regional climate engineering save the summer Arctic sea ice? Geophysical Research Letters 2014;41:880–5. http://dx.doi.org/10.1002/2013GL058731.

[27] Anderson K. Duality in climate science. Nature Geoscience 2015. http://dx.doi.org/10.1038/ngeo2559.

[28] Dittmar M. The end of cheap uranium. Science of the Total Environment 2013;461–462:92–798.

[29] Moriarty P, Honnery D. Intermittent renewable energy: the only future source of hydrogen? International Journal of Hydrogen Energy 2007;32:1616–24.

[30] Moriarty P, Honnery D. What energy levels can the Earth sustain? Energy Policy 2009;37:2469–74.

[31] Bagla P. Thorium seen as nuclear's new frontier. Science 2015;350:726–7.

[32] Dittmar M. Nuclear Energy: status and future limitations energy 2012;37:35–40.

[33] Clery D. New review slams fusion project's management. Science 2014;343:957–8.

[34] Froggatt A, Schneider M. Nuclear power versus renewable energy: a trend analysis. Proceedings of the IEEE 2015;103(4):487–90.

[35] International Atomic Energy Association (IAEA). Energy, electricity and nuclear power estimates for the period up to 2050. Vienna: IAEA; 2012.

[36] Hall CAS, Lambert JG, Balogh SB. EROI of different fuels and the implications for society. Energy Policy 2014;64:141–52.

[37] van Vuuren DP, Stehfest E, Elzen MG, Kram T, Vliet JV, Deetman S, et al. RCP2.6: exploring the possibility to keep global mean temperature increase below 2°C. Climatic Change 2011;109:95–116.

[38] de Castro C, Mediavilla M, Miguel LJ, Frechoso F. Global solar electric potential: a review of their technical and sustainable limits. Renewable and Sustainable Energy Reviews 2013;28:824—35.

[39] Hafele W. Energy in a finite world: a global systems analysis. Cambridge (MA): Ballinger; 1981.

[40] Makarieva AM, Gorshkov VG, Li B-L. Energy budget of the biosphere and civilization: rethinking environmental security of global renewable and nonrenewable resources. Ecological Complexity 2008;5:281—8.

[41] Trenberth KE. An imperative for climate change planning: tracking Earth's global energy. Current Opinion in Environmental Sustainability 2009;1:19—27.

[42] de Vries BJM, van Vuuren DP, Hoogwijk MM. Renewable energy sources: their global potential for the first-half of the 21st century at a global level: an integrated approach. Energy Policy 2007;35:2590—610.

[43] Lu X, McElroy MB, Kiviluoma J. Global potential for wind-generated electricity. Proceedings of the National Academy of Sciences of the United States of America 2009;106:10933—8.

[44] Jacobson MZ, Delucchi MA. Providing all global energy with wind, water and solar power, part 1: technologies, energy resources, quantities and areas of infrastructure, and materials. Energy Policy 2011;39:1154—69.

[45] Moriarty P, Honnery D. What is the global potential for renewable energy? Renewable and Sustainable Energy Reviews 2012;16:244—52.

[46] Moriarty P, Honnery D. Can renewable energy power the future? Energy Policy 2016;93:3—7. http://dx.doi.org/10.1016/j.enpol.2016.02.051.

[47] Field CB, Campbell JE, Lobell DB. Biomass energy: the scale of the potential resource. Trends in Ecology and Evolution 2008;23(2):65—72.

[48] de Castro C, Mediavilla M, Miguel LJ, Frechoso F. Global wind power potential: physical and technological limits. Energy Policy 2011;39:6677—82.

[49] Moriarty P, Honnery D. Preparing for a low-energy future. Futures 2012;44:883—92.

[50] Electricite de France (EdF). Worldwide electricity production from renewable energy sources. Fifteenth Inventory 2013. http://www.energies-renouvelables.org/observ-er/html/inventaire/pdf/15e-inventaire-Chap01-Eng.pdf.

[51] Kleidon A. The climate sensitivity to human appropriation of vegetation productivity and its thermodynamic characterization. Global and Planetary Change 2006;54:109—27.

[52] World Energy Council (WEC). World energy resources: 2013 survey. London: WEC; 2013 [Also earlier surveys].

[53] Moriarty P, Honnery D. Is there an optimum level for renewable energy? Energy Policy 2011;39:2748—53.

[54] Chatzivasileiadis S, Ernst D, Andersson G. The global grid. Renewable Energy 2013;57:372—83.

[55] Seboldt W. Space- and Earth-based solar power for the growing energy needs of future generations. Acta Astronautica 2004;55:389—99.

[56] Pickard WF. Smart grids versus the Achilles' Heel of renewable energy: can the needed storage infrastructure be constructed before the fossil fuel runs out? Proceedings of the IEEE 2014;102(7):1094—105.

[57] Lemmon JP. Reimagine fuel cells. Nature 2015;525:447—9.

[58] Steinke F, Wolfrum P, Hoffmann C. Grid vs. storage in a 100% renewable Europe. Renewable Energy 2013;50:826—32.

[59] Buijs P, Bekaert D, Cole S, van Hertem D, Belmans R. Transmission investment problems in Europe: going beyond standard solutions. Energy Policy 2011;39:1794—801.

[60] Coady D, Parry I, Sears S, Shang B. How large are global energy subsidies? IMF Working Paper WP/15/105. 2015. Available at: https://www.imf.org/external/pubs/ft/wp/2015/wp15105.pdf.

[61] Darmstadter J. The International Monetary Fund confronts global energy subsidies. Environment: Science and Policy for Sustainable Development 2013;55(5):41—4.

[62] Delucchi MA, Murphy JJ. US military expenditures to protect the use of Persian Gulf oil for motor vehicles. Energy Policy 2008;36:2253—64.

[63] Interagency Working Group on Social Cost of Carbon. Technical update of the social cost of carbon for regulatory impact analysis under executive order 12866. Washington D.C.: US Government; 2013. Available at: http://go.nature.com/vzpkkb.

[64] Revesz RL, Howard PH, Arrow K, Goulder LH, Kopp RE, Livermore MA, et al. Improve economic models of climate change. Nature 2014;508:173—5.

[65] Ackerman F, Stanton EA. Climate risks and carbon prices: revising the social cost of carbon. Economics 2012;6(10). http://dx.doi.org/10.5018/economics-ejournal.ja.2012-10.

[66] Aldy JE. Pricing climate risk mitigation. Nature Climate Change 2015;5:396—8.

[67] Moore FC, Diaz DB. Temperature impacts on economic growth warrant stringent mitigation policy. Nature Climate Change 2015;5:127—31.

[68] Ward P. Unfair aid: the subsidies keeping nuclear energy afloat across the globe. Nuclear Monitor 2005:630—1. http://citeseerx.ist.psu.edu/viewdoc/download?doi=10.1.1.363.641&rep=rep1&type=pdf.

[69] Edenhofer O. King coal and the queen of subsidies. Science 2015;349(6254):1286—7.

[70] Moriarty P, Honnery D. Future cities in a warming world. Futures 2015;66:45—53.

[71] Hajat A, Hsia C, O'Neill MS. Socioeconomic disparities and air pollution exposure: a global review. Current Environmental Health Reports 2015;2:440—50.

[72] Unruh GC. Escaping carbon lock-in. Energy Policy 2002;30:317—25.

[73] Davis SJ, Caldeira K, Matthews HD. Future CO_2 emissions and climate change from existing energy infrastructure. Science 2010;329:1330—3.

[74] Koseoglu NM, van den Bergh JCJM, Lacerda JS. Allocating subsidies to R&D or to market applications of renewable energy? Balance and geographical relevance. Energy for Sustainable Development 2013;17:536—45.

[75] Newell P. The political economy of carbon markets: the CDM and other stories. Climate Policy 2012;12:135—9.

[76] Buigues G, Zamora I, Mazón AJ, Valverde V, Pérez FJ. Sea energy conversion: problems and possibilities. 2006. 8 pp. http://icrepq.com/icrepq06/242-buigues.pdf.

[77] Vidal O, Goffé B, Arndt N. Metals for a low-carbon society. Nature Geoscience 2013;6:894—6.

[78] Moriarty P, Honnery D. Reliance on technical solutions to environmental problems: caution is needed. Environmental Science and Technology 2015;49:5255—6.

[79] Soloveichik GL. Flow batteries: current status and trends. Chemical Reviews 2015;115:11533—58.

[80] Lienau PJ, Ross H. Final report: low-temperature resource assessment program. 1996. http://geoheat.oit.edu/pdf/db7.pdf.

CHAPTER TWO

Environmental Impact Assessment of Different Renewable Energy Resources: A Recent Development

M.M. Rahman, S. Salehin, S.S.U. Ahmed and A.K.M. Sadrul Islam
Islamic University of Technology, Gazipur, Bangladesh

2.1 INTRODUCTION

Energy is the prerequisite for sustainable development of the modern civilization. Global energy demand is expected to grow by 36% for the year 2011−30 with a share of 88% from the nonrenewable (e.g., oil, natural gas, and coal) energy resources [1]. According to 2014 data, in the United States, the electricity sector and the transportation sector consumed 39% and 27% of the total energy, respectively, and most of the energy comes from petroleum with a share of 35% and natural gas with a share of 28%, respectively [2]. The electricity sector was the major greenhouse gas (GHG)−emitting sector followed by the petroleum and natural gas sector, and the petroleum refinery sector, respectively [3]. In 2014, world's electricity production was 22,433 terawatt-hours (TWh) and the shares of coal, natural gas, hydroelectric, nuclear, and liquid fuel were 39%, 22%, 17%, 11%, and 5%, respectively [4]. Fossil fuel−based energy sources are depleting, and combustion of fossil fuels releases harmful GHGs into the atmosphere resulting in global warming and ozone layer depletion. Global CO_2 emission was increased by 78% from the year 1981−2011 and will increase by 85% from year 2000−30 [1]. Different policy regulations have come into play to reduce GHG emissions by encouraging the use of more and more alternative sources, such as renewable energy sources. Renewable energy sources—solar, wind and hydro power, geothermal, biomass, and so on—are now considered as sustainable alternatives. In 2014, renewable sources accounted for about 24% of the world's total energy generation [4], and by the year 2070, the share will be increased to 60%.

People already know the environmental impacts of fossil fuel−based electricity generation. However, there is limited number of studies to present the environmental impacts of electricity generation from renewable sources. Although the renewable energy sources have no/very little operational GHG emissions, a large amount of energy is required to manufacture the parts of the renewable energy systems, which will ultimately produce

Clean Energy for Sustainable Development
ISBN 978-0-12-805423-9, http://dx.doi.org/10.1016/B978-0-12-805423-9.00002-8

GHG emissions. Hence, there is a need of clear understanding of how much cleaner these sources are compared to conventional sources of energy generation.

2.1.1 Life Cycle Assessment

Life cycle assessment (LCA) is a tool to quantify the energy usage and environmental impacts associated with all the stages of a product or system throughout its whole lifetime. Fig. 2.1 represents the framework of LCA. There are four stages of LCA—goal and scope definition, inventory analysis, impact assessment, and interpretation.

For conducting an LCA, it is very important to define the goal and scope of the study, functional unit, and system boundary, which is the first stage of the LCA. The second stage is inventory analysis, which involves quantifying the materials and resources flowing throughout the lifetime of a product or a system. Impact assessment involves categorizing and aggregating the resource consumption and emissions for different environmental problems, such as global warming potential (GWP), land use, water use, acidification, ozone layer depletion, and so on. The function of interpretation stage is to make conclusions that are consistent with the defined goal and scope. With the findings from the interpretation phase, the improvement potential can be found to lower the environmental impacts.

LCA has become an extremely useful method to assess the environmental impacts of renewable energy technologies. It is a common understanding that power generation from renewables is free of GHG emissions. There is no/very little operation emissions associated with these technologies as they are free from fossil fuel use. Manufacturing and transportation of different parts of the systems, installation, decommissioning, and recycling involve energy use that ultimately results in GHG emissions. In this chapter, an extensive review of LCA of different renewable energy technologies—solar photovoltaic (PV), wind, biomass, biogas, hydro, and geothermal—has been presented along with the LCA of biomass for biodiesel production. At the end of this chapter, some LCA results of fossil fuel—based power generation technologies—diesel, natural gas, coal, and so on—have been presented for comparison purpose.

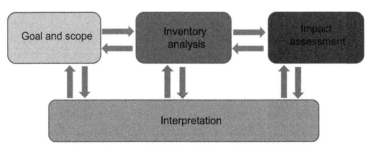

Figure 2.1 Life cycle assessment framework. *Adapted from ISO 14040: Environmental management— life cycle assessment—principles and framework. 2006.*

2.2 LIFE CYCLE ASSESSMENT OF SOLAR PHOTOVOLTAIC SYSTEM

2.2.1 Methodology

Solar PV cells convert the sunlight into direct current (DC). Semiconducting materials are used to manufacture PV modules. Photons from the sunlight reach the surface of the semiconducting materials and trigger electrons to produce electricity. There are different types of PV modules—monocrystalline silicon (mono-Si), multicrystalline silicon (multi-Si), amorphous silicon (a-Si), CIS thin film (CIS), and CdTe thin film (CdTe). The sun is a huge source of energy; the PV system is one of the most widely used technologies to harness energy from the sun. A solar PV system is thought of as a clean technology for energy generation. However, manufacturing of solar PV modules involves energy consumption. And some people think that energy consumption in manufacturing PV modules is larger than the energy production by the modules throughout its entire lifetime [6]. Hence, there is a necessity of transparently quantifying the energy use and the resulting GHG emission of a solar PV system used for electricity generation. Two different indicators—energy payback time (EPBT) and GHG emission—have been used to check the performance of solar PV systems. EPBT of a PV system can be defined as the time (number of years) required producing the same amount of energy that is consumed throughout its lifetime. Energy is utilized at different stages of a PV system. EPBT can be calculated using Eq. (2.1), where E_{input} (MJ) is the total amount of primary energy required throughout its lifetime to produce the PV modules, battery, inverter, supporting structure, cable, transportation, installation, maintenance, and decommissioning of the system, and E_{output} (MJ) is the amount of energy produced by the system in a year [6]:

$$\text{EPBT} = \frac{E_{input}}{E_{output}} \tag{2.1}$$

GHG emission (g-CO_2eq/kWh) can be calculated using Eq. (2.2), where GHG_{system} (g-CO_2eq) is the total amount of GHG emission throughout the entire life of the system (emission from manufacturing different components of PV system, installation, transportation, decommissioning, recycling, and so on) and E_{total} (kWh) is the total electricity produced throughout the lifetime of the system [6]:

$$\text{GHG emission} = \frac{GHG_{system}}{E_{total}} \tag{2.2}$$

The lifetime of a solar PV system usually varies from 20 to 30 years [7]. The EPBT and GHG emission depend on the energy consumption to manufacture various components, and electricity production, which are ultimately dependent on various factors—module type, conversion efficiency, manufacturing process, type of supporting structure, installation location (ground-mounted or rooftop), location, and so on. Table 2.1 represents the location, module efficiency, and lifetime of various types of solar PV modules considered in different studies in the existing literature.

Various studies have been conducted to quantify the energy consumption to manufacture solar modules. Mono-Si PV modules have the highest conversion efficacy, but consume much more energy than the other types [6]. The main stages to manufacture Si-based PV modules (mono-Si, multi-Si, and a-Si) are quartz reduction, metallurgical-grade silicon purification, electronic silicon or solar-grade silicon production, mono-Si or multi-Si crystallization, wafer sawing, and cell production [27]. Due to the higher cost and higher energy intensity of Si-based modules, people have started to manufacture thin-film solar modules that use less material and energy.

Table 2.1 Location, Module Efficiency, and Lifetime of Different Types of Photovoltaic Modules

Module Type	Location	Efficiency (%)	Lifetime (years)	References
Mono-Si	UK	12	20	[8]
Mono-Si	Japan	12.2	20	[9]
Mono-Si	South European	13.7	30	[10]
Mono-Si	South European	14	30	[11]
Mono-Si	Switzerland	14	30	[12]
Multi-Si	South European	13	25	[13]
Multi-Si	Japan	11.6	20	[9]
Multi-Si	South European	13	30	[14]
Multi-Si	Gobi Desert of China	12.8	30	[15]
Multi-Si	Italy	10.7	30	[16]
Multi-Si	South European	13.2	30	[11]
Multi-Si	USA	12.9	20	[17]
Multi-Si	Switzerland	13.2	30	[12]
a-Si	USA	5	25	[18]
a-Si	Northwestern European	6	—	[19]
a-Si	South European	7	30	[14]
a-Si	Switzerland	6.5	30	[12]
a-Si	USA	6.3	20	[17]
CdTe	Northwestern European	6	—	[19]
CdTe	Japan	10.3	20	[20]
CdTe	USA	9	30	[21]
CdTe	South European	9	30	[11]
CdTe	Switzerland	7.1	30	[12]
CdTe	South European	9	20	[22]
CdTe	South European	10.9	—	[23]
CdTe	Europe	10.9	30	[24]
CIS	Switzerland	10.7	30	[12]
CIS	South European	11	20	[22]
CIS	China	11	30	[25]
CIS	China	11	—	[26]

Table 2.2 shows the energy consumption to manufacture mono-Si, multi-Si, a-Si, CIS, and CdTe modules.

The variation of energy requirements is due to the variation in assumptions, energy mix, and manufacturing processes used in different studies. Most of the LCA studies found in the public domain are based on the LCA of mono-Si or multi-Si modules for power generation.

An LCA of a 4.2 kW_p standalone solar PV system was conducted in Spain by García-Valverde et al. [28]. The system considered in the study was a rooftop mono-Si solar PV system. Fig. 2.2 depicts the system boundary used in the study by García-Valverde et al. [28]. Transportation, supporting structure, and copper cables were also kept within the system boundary.

A similar kind of study was conducted by Kannan et al. [29] for Singapore. The authors considered a 2.7 kW_p grid-connected mono-Si solar PV system. On the other hand, Sumper et al. [30] assessed a 200 kW rooftop PV system with polycrystalline silicon modules for Catalonia, Spain. The same methodology (according to LCA framework) was used in these studies. There is very limited data for the energy consumption in manufacturing the balance-of-system (BOS), such as battery, inverter, charge controller, cable, supporting structure, and other accessories. The energy consumptions for various components used in these studies are represented in Table 2.3.

Transportation and installation usually require very little amount of energy. It is very difficult to trace the amount of energy that is consumed for the installation purpose. Transportation energy requirement can be calculated using the transportation distance and

Table 2.2 The Range of Energy Requirements (MJ/m^2) to Manufacture Various Solar Photovoltaic Modules

Module Type	Mono-Si	Multi-Si	a-Si	CIS	CdTe
Energy requirement (MJ/m^2)	2860—5253	2699—5150	710—1990	1069—1684	790—1803

Adapted from Peng J, Lu L, Yang H. Review on life cycle assessment of energy payback and greenhouse gas emission of solar photovoltaic systems. Renewable and Sustainable Energy Reviews 2013;19:255—74.

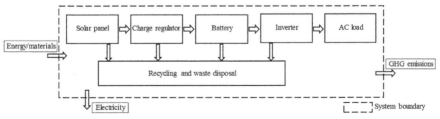

Figure 2.2 Life cycle assessment system boundary. *Adapted from García-Valverde R, Miguel C, Martínez-Béjar R, Urbina A. Life cycle assessment study of a 4.2 kW p stand-alone photovoltaic system. Solar Energy 2009;83:1434—45.*

Table 2.3 The Energy Consumption for Various Components Used in the Earlier Studies in Literature

Component	García-Valverde et al. [28]	Kannan et al. [29]	Sumper et al. [30]
Photovoltaic module	1.583 MWh$_{th}$/m^{2a}	16 MWh$_{th}$/kW$_p$	4.59×10^6 MJ
Al module frame	41.7 kWh$_{th}$/kg^b, 2.08 kWh$_{th}$/kg^c	Value taken from GEMIS^g	—
Charge regulator	277 kWh$_{th}$/kW$_{el}$	—	10.96×10^4 MJ^h
Inverter	277 kWh$_{th}$/kW$_{el}$	0.17 MWh$_{el}$/kW$_p$	3.02×10^4 MJ
Lead-acid battery	331 kWh$_{th}$/kWh^b, 242 kWh$_{th}$/kWh^{c,d}	—	—
Supporting structure	9.72 kWh$_{th}$/kg^b, 2.5 kWh$_{th}$/kg^{c,e}	Value taken from GEMIS^g	—
Cables	19.44 kWh$_{th}$/kg^b, 13.9 kWh$_{th}$/kg^{c,f}	—	—

[a]Frameless module was considered. Ten percent of the module weight was considered as the aluminum frame. Recycling for aluminum frame was considered as 35% for Spain [28].
[b]New materials.
[c]Recycled materials.
[d]50% of the lead-acid batteries are recycled.
[e]90% of the steel are recycled.
[f]43% of the copper cables are recycled.
[g]GEMIS, 2002. Global emission model for integrated systems, GEMIS 4.1 Database (September 2002), Oko-Institut Darmstadt, Germany.
[h]Energy requirements for BOS excluding the inverter.

the energy intensity (MJ/t-km) of transportation medium. The energy requirement of the system must be amortized over the lifetime of the individual components. The electricity production can be found experimentally over the total lifetime (i.e., 20 or 30 years) of the PV plant. Also the amount of electricity generation can be calculated using Eq. (2.3), where El (kWh) is the amount of electricity produced throughout the plant's life, E (kWh/m^2/year) is the amount of solar radiation received by the selected location, η_p and η_e are the efficiencies of the PV modules and inverter, respectively, A (m^2) is the total area covered by the modules, and L (year) is the total lifetime of the PV system:

$$El = E \times \eta_p \times \eta_e \times A \times L \qquad (2.3)$$

The energy consumption is multiplied with the emission factors to get GHG emissions. Thermal energy can be converted to electrical energy with a thermoelectric conversion efficiency of 35% [28]. Table 2.4 represents the emission factors (EF) used in the study by García-Valverde et al. [28].

2.2.2 Life Cycle Assessment Results of Solar Photovoltaic System

The energy consumption, EPBT, and GHG emission are very much specific to the location under study. There is a wide range of these performance indicators observed among the studies in the public domain. The EPBT and GHG emission obtained from different studies are furnished in Table 2.5.

Table 2.4 Emission Factors for the Elements Used in the Solar Photovoltaic System

Component	Production From New Materials	Production From Recycled Materials	Recycling Process
Multi-Si module	93.6 g-CO_2/kWh$_{th}$	—	—
Al frame	14.6 kg-CO_2/kg	0.73 kg-CO_2/kg	0.73 kg-CO_2/kg
Charge regulator	93.6 g-CO_2/kWh$_{th}$	—	—
Inverter	93.6 g-CO_2/kWh$_{th}$	—	—
Lead-acid battery	93.6 g-CO_2/kWh$_{th}$	93.6 g-CO_2/kWh$_{th}$	0.16 g-CO_2/kg
Supporting structure	2.82 kg-CO_2/kg	0.45 kg-CO_2/kg	0.45 kg-CO_2/kg
Cable	5.57 kg-CO_2/kg	3.98 kg-CO_2/kg	3.98 kg-CO_2/kg

Adapted from García-Valverde R, Miguel C, Martínez-Béjar R, Urbina A. Life cycle assessment study of a 4.2 kW p stand-alone photovoltaic system. Solar Energy 2009;83:1434−45.

Table 2.5 Life Cycle Assessment Results Obtained From Different Studies

Module Type	Location	EPBT (years)	GHG Emissions (g-CO_2eq/kWh)	References
Mono-Si	UK	7.4−12.1	—	[8]
Mono-Si	Japan	8.9	61	[9]
Mono-Si	South European	2.6	41	[10]
Mono-Si	South European	2.1	35	[11]
Mono-Si	Switzerland	3.3	—	[12]
Multi-Si	South European	2.7	—	[13]
Multi-Si	Japan	2.4	20	[9]
Multi-Si	South European	3.2	60	[14]
Multi-Si	Gobi Desert of China	1.7	12	[15]
Multi-Si	Italy	3.3	—	[16]
Multi-Si	South European	1.9	32	[11]
Multi-Si	USA	2.1	72.4	[17]
Multi-Si	Switzerland	2.9	—	[12]
a-Si	USA	3	—	[18]
a-Si	Northwestern European	3.2	—	[19]
a-Si	South European	2.7	50	[14]
a-Si	Switzerland	3.1	—	[12]
a-Si	USA	3.2	34.3	[17]
CdTe	Northwestern European	3.2	—	[19]
CdTe	Japan	1.7	14	[20]
CdTe	USA	1.2	23.6	[21]
CdTe	South European	1.1	25	[11]
CdTe	Switzerland	2.5	—	[12]
CdTe	South European	1.5	48	[22]
CdTe	South European	0.79	18	[23]
CdTe	Europe	0.7−1.1	19−30	[24]
CIS	Switzerland	2.9	—	[12]
CIS	South European	2.8	95	[22]
CIS	China	1.6	10.5	[25]
CIS	China	1.8	46	[26]

Adapted from Peng J, Lu L, Yang H. Review on life cycle assessment of energy payback and greenhouse gas emission of solar photovoltaic systems. Renewable and Sustainable Energy Reviews 2013;19:255−74.

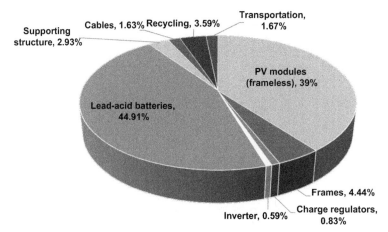

Figure 2.3 Breakdown of CO_2 emission in a 4.2 kW$_p$ stand-alone photovoltaic (PV) facility in the southeast of Spain. *Adapted from García-Valverde R, Miguel C, Martínez-Béjar R, Urbina A. Life cycle assessment study of a 4.2 kW p stand-alone photovoltaic system. Solar Energy 2009;83:1434–45.*

Most of the emissions come from manufacturing of lead–acid batteries (45%), followed by PV modules (39%). The breakdown of CO_2 emission is depicted in Fig. 2.3. Transportation, recycling, cables, and supporting structure together contribute about 10% CO_2 emission.

The reported values of EPBT and GHG emissions (see Table 2.5) of different PV systems vary significantly among the studies due to the variation of the influencing factors—PV module type, cell manufacturing technologies, installation methods, locations, weather conditions, and so on. Therefore, it is very important to conduct a location-specific LCA of a solar PV system.

2.3 LIFE CYCLE ASSESSMENT OF WIND ENERGY SYSTEM

2.3.1 Methodology

In the present world, wind energy is becoming more and more popular for power generation in the areas where wind speed is sufficient for electricity generation. Several studies conducted LCAs of different capacity wind turbines around the globe. Tremeac and Meunier [31] assessed the environmental impacts of 4.5 MW and 250 W wind turbines. All the life cycle stages—manufacturing, transports, installation, maintenance, disassembly, and disposal—were considered in the study. The system boundary used for the study is depicted in Fig. 2.4. EPBT and CO_2 emissions were calculated, and it was found that wind energy can provide excellent environmental solution.

The functional unit used in most of the studies is kWh of electricity. GHG emissions are presented in the unit of g-CO_2eq/kWh$_{el}$. The main components of wind turbine

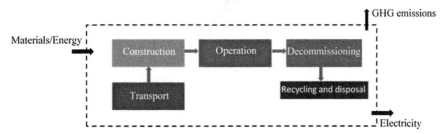

Figure 2.4 Life cycle assessment system boundary for wind turbine power generation. *Adapted from Tremeac B, Meunier F. Life cycle analysis of 4.5 MW and 250 W wind turbines. Renewable and Sustainable Energy Reviews 2009;13:2104—10.*

systems are rotor, nacelle, tower, foundation, and grid connection cables. Guezuraga et al. [32] conducted an LCA of two different types of wind turbines with a lifetime of 20 years. The authors reported the material requirements for different parts of a wind turbine system. Table 2.6 represents the material requirements for a 1.8 MW gearless and a 2 MW geared wind turbines, respectively.

The recycling and waste disposal rates of different materials are furnished in Table 2.7. The recycling rates are very high for steel, cast iron, and copper that lead to lesser energy consumption.

The operational phase hardly requires any energy consumption. The energy requirements in the transportation of wind turbine parts to the wind site can be calculated using the transportation distance and energy intensity of modes of transportation used. The energy consumption in transportation of raw materials to the manufacturing plants can be ignored as it is very difficult to trace [33].

A wind turbine is used to convert the kinetic energy of wind into mechanical power. The rotating shaft is coupled with the generator, which converts the mechanical power into electrical power. The electricity provided by a wind turbine can be AC or DC. The amount of effective mechanical power can be estimated using Eq. (2.4), where P (W) is the mechanical power generated, c_p (%) is the capacity factor, ρ (kg/m^3) is the density of air, A (m^2) swept area of the blades, and V (m/s) is the average wind velocity:

$$P = \frac{1}{2} c_p \rho A V^3 \qquad (2.4)$$

The electrical energy delivered by the turbine throughout the lifetime can be found from the power developed, generator efficiency, and operation hours of the turbine per year, and the lifetime of the wind turbine (i.e., 20 years).

2.3.2 Life Cycle Assessment Results of Power Generation From Wind

The primary energy consumption and emission factors for manufacturing various materials used in wind turbines and for recycling and landfilling are presented in

Table 2.6 Material Requirements for the 1.8 MW Gearless and 2 MW Geared Turbines

Material	1.8 MW Gearless Turbine		2 MW Geared Turbine[a]	
	Mass (tonnes)	Wt. (%)	Mass (tonnes)	Wt. (%)
Stainless steel	178.4	29.9	296.4	19.3
Cast iron	44.10	5.9	39.35	2.6
Copper	9.90	1.6	2.40	0.2
Epoxy	4.80	1.8	10	0.6
Plastic	1.85	0.3	2.40	0.2
Fiberglass	10.20	2.6	24.30	1.6
Reinforced concrete	360	57.9	1164	75.6

[a]The 2 MW geared turbine is 2.5 times heavier than the 1.8 MW gearless turbine.
Adapted from Guezuraga B, Zauner R, Pölz W. Life cycle assessment of two different 2 MW class wind turbines. Renewable Energy 2012;37:37—44.

Table 2.7 Recycling and Waste Disposal Rates of Different Materials

Material	Type of Dismantling
Stainless steel	90% recycle, 10% landfill
Cast iron	90% recycle, 10% landfill
Copper	90% recycle, 10% landfill
Epoxy	100% incinerated
Plastic	100% incinerated
Fiberglass	100% incinerated
Concrete	100% landfill

Adapted from Guezuraga B, Zauner R, Pölz W. Life cycle assessment of two different 2 MW class wind turbines. Renewable Energy 2012;37:37—44.

Table 2.8. According to Guezuraga et al. [32], most of the energy is consumed in the manufacturing phase, which is 84.4%, followed by transportation (7%) and maintenance (4.3%), respectively. Fig. 2.5 represents the breakdown of energy consumption in different stages of a wind turbine power plant.

It was observed that material manufacturing and transportation are the unit processes that mostly affect the life cycle energy and resulting GHG emissions. Table 2.9 shows a country-specific overview of energy and CO_2 analysis of wind turbines.

The reported values of energy intensity and emissions (see Table 2.9) of different capacity wind turbines vary significantly among the studies due to the variation of different influencing factors—turbine manufacturing technologies, wind speed, energy mix of the location, and so on. Hence, it is very important to conduct a location-specific LCA of a wind turbine electricity generation system.

Table 2.8 Embodied Energy and Emission Factors for Different Materials, Recycling, and Landfilling

Material	Energy Requirements (GJ/tonnes)	GHG Emissions (kg/tonnes)		
		CO_2	CH_4	N_2O
Material Production				
Steel	34	2473	0.04	0.07
Stainless steel	53	3275	0.04	0.07
Rebar steel	34.26	2163.83	0.10	0.07
Glass	8.70	566	0.04	0.01
Epoxy	45.70	3941	0.04	0.12
Polyester	45.70	3941	0.08	0.12
Copper	78.20	6536	0.16	0.19
Aluminum	39.15	3433.50	0.07	0.11
Concrete	0.81	119.02	0.03	8.7E-5
Material Recycling		*Kg-CO_2eq/tonnes*		
Steel	9.70	1819		
Aluminum	16.80	738		
Copper	6.40	3431		
Landfilling operations	0.04	0.90		

Adapted from Kabir MR, Rooke B, Dassanayake GM, Fleck BA. Comparative life cycle energy, emission, and economic analysis of 100 kW nameplate wind power generation. Renewable Energy 2012;37:133–41.

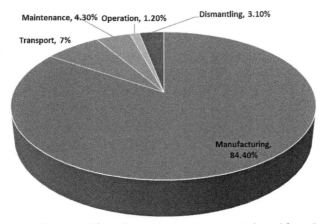

Figure 2.5 Breakdown of the total life cycle energy requirements. *Adapted from Guezuraga B, Zauner R, Pölz W. Life cycle assessment of two different 2 MW class wind turbines. Renewable Energy 2012;37:37–44.*

Table 2.9 The Energy Intensity and Emissions of Different Wind Turbine Plants Around the Globe

Power Rating (kW)	Location	Energy Intensity (kWh/kWh)	Emissions (g-CO$_2$/kWh)	References
30	Denmark	0.1	—	[34]
100	Japan	0.456	123.7	[35]
500	Brazil	0.069	—	[36]
1500	India	0.032	—	[37]
6600	UK	—	25	[38]
100	Japan	0.16	39.4	[39]
300	Japan	—	29.5	[40]
3	USA	1.016	—	[41]
10 × 500[a]	Denmark	—	16.5	[42]
18 × 500[b]		—	9.7	[42]
30−800	Switzerland	—	11	[43]
1.8 and 2	Austria	—	9	[32][c]
5, 20, and 100	Canada	—	(42.7, 25.1, and 17.8)[d]	[33][e]

[a]Offshore.
[b]Onshore.
[c]This study found the energy payback time as 7 months.
[d]The unit of these values is g-CO$_2$eq/kWh.
[e]This study found the energy payback time as 1.4, 0.8, and 0.6 years for 5, 20, and 100 kW, respectively.

2.4 LIFE CYCLE ASSESSMENT OF BIOFUELS

Biofuels are one of the oldest fuels in place. From time immemorial, wood and straw have been used as fuel for various purposes, which suited the time. With the industrial revolution, however, the need for concentrated source of energy pushed biofuels on the margin and welcomed the use of fossil fuels. In the late 20th and 21st centuries, however, the problems of global warming and other environmental abuse have ushered a renewed interest in renewable energy and thus in biofuels as well. A proof of that is, since 2000s there has been a marked increase in the production of bioethanol. It rose from 16.9 billion liters in 2000 to 72 billion liters in 2009 [44]. In Poland, it is expected that there will be a sharp rise in production of bioelectricity from 2010 to 2030. Electricity production from solid biomass will increase from 5.5 to 11.1 TWh/year in Poland [45]. Although biomass has a distinct advantage of being deemed as carbon neutral, it comes with caveats, such as "food versus fuel" debate and consequences related to land use [46]. To identify savings in energy and emissions from biofuel production, its utilization, and its corresponding environmental effects, a thorough evaluation of the corresponding life cycle is to be carried out carefully. LCA is an effective tool for this as it can unravel and quantify the potential environmental impacts and evaluate the inputs and outputs [5].

It should be noted that a striking feature of LCA for biofuel is that it gives differing results, and thus a range of results is obtained for even the same fuel as illustrated in

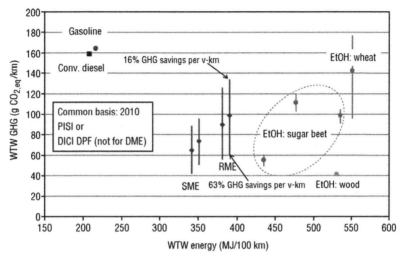

Figure 2.6 Well to wheel (WTW) energy requirements and greenhouse gas emissions for conventional biofuel pathways compared with gasoline and diesel pathways showing the range of life cycle assessment results of biofuels [49].

Fig. 2.6. These differences can be attributed to different types of feedstock sources, conversion technologies, end-use technologies, system boundaries, and reference energy system. Also, region plays a significant part in LCA and so does the development of fuel conversion technology [47,48]. Despite this, it can be said with confidence that most biofuel-based LCA shows a reduction of GHG emission when used as a substitute or in combination with fossil fuel in transportation sector [49–51]. In the environmental aspect, such as land use or eutrophication, however, majority of studies shows that biofuel have negative impacts when aimed to reduce the GHGs [49,52,53]. Thus, making a decision regarding the use of biofuels is sometimes as black and white as deemed.

This section of the review seeks to discuss briefly the findings regarding the first- and second-generation biofuels and their implications. The aim of this section is to summarize the key LCA issues influencing outcomes for bioenergy and to provide an overview of the GHG and energy balances of the most relevant bioenergy chains in comparison with fossil fuels.

2.4.1 Biomass Source

When including both the first- and the second-generation biofuels, the source of biomass covers a wide range. A basic definition for biomass is that it is renewable organic matter that includes plant and animal products and excretions, food processing and forestry by-product, and urban waste [54]. The first-generation biofuels are derived from the parts that are or can also be used as food materials for humans, while the second generation uses lignocellulosic parts of the biomass. Hence bioethanol derived from sugarcane or rapeseed is first-generation biofuel, while from *Jatropha* or wood would be second generation. In LCA study, the biomass supply plays a key role since the source has a big impact on the LCA outcome. The biomass supply can be divided into two main categories: residues and energy crop.

Biomass residues and waste are not specifically produced as energy resource, but are by-products from agriculture, forestry, households, and so on. Most of these biomass and residues are inevitable in any economic activity or industrial process [55]. Due to this aspect, its use in making biofuels usually does not adversely affect the environment; although there is some backlash, since there is already a system established in nature and displacing something would hamper it. For example, the removal of agriculture by-products hampers the carbon cycle by reducing the carbon storage in the soil. Not to mention the use of such residue could potentially invite a more thorough use of the parent mechanism and thus could cause some permanent damage to the ecosystem, such as deforestation.

Energy crops, on the other hand, are cultivated with the intent to provide a feedstock for biofuel development [56]. Those feedstocks generated from agricultural activities and forest log can be included in this as well [57]. To avoid excess environmental burden related to agricultural chain, the types of energy crops grown are suggested to use high-yielding species [57], which would require minimal maintenance [58] and can survive in marginal or degraded lands. While it is difficult to find a crop that meets all these criteria, perennial C4 grasses, such as *Miscanthus* and switchgrass (*Panicum virgatum* L.), are particularly promising [59]. Excessive cultivation of such crops can lead to some problems such as deforestation, directly for its cultivation or indirectly by displacing a nonenergy crop, and also eutrophication due to the use of fertilizers and pesticides. However, cultivation of energy crops also has the added benefit of providing certain ecosystem services, such as C-sequestration, increase in biodiversity, salinity mitigation, and enhancement of soil and water quality. It should be noted that to be accepted as an energy crop, it must fall within the parameters of sustainable agriculture.

A typical form of biofuel source is manure and if not treated well, which may be the case in developing countries, the effect is quite damaging to the environment. In China, it is estimated that poultry and livestock manure reached about 3.97×10^9 tons in 2007, majority of which was drained into the river resulting in water pollution and GHG emissions. In countries with these problems, biogas production from manures is a sound solution both for the environment and for electricity production [60].

2.4.2 Methodology

LCA is a structured approach to analyze a system and thus it follows a particular strand of approach. It seeks to incorporate the environmental and economic impacts of all the stages in a production chain and removes ambiguities thereby giving a holistic picture of a system. This section will give a general idea of the methodology of LCA and will give an idea of how biofuels fit into this methodology.

2.4.2.1 Goal, System Boundaries, and Functional Unit

The first step involves in defining the goal and scope, which defines the purpose, audiences, and system boundaries. Since this process will be a controller for the rest to come, it has to be very specific in detailing what to include and what to exclude thereby creating an analysis system boundary. It can simply include the GHG emission, and thus the study will be centered around the direct and indirect emissions during the formation of the biofuel. It can also exclude the GHG and concentrate on the environmental impacts in which case the LCA would collect, assess, and interpret the data obtained regarding issues, such as eutrophication, acidification, biodiversity change, and so on.

The functional unit is defined as the quantification of the identified performance characteristics of the products [5]. It gives a reference to which the input and output data are normalized and harmonizes the establishment of the inventory. This also provides a means for comparison among different LCA studies provided that the system boundaries are similar. It should be noted that different functional units for the same LCA will give different outcomes, and hence it is imperative to choose the one that satisfies the goals of LCA the best. For example, to study the effect biofuels have on the transportation sector, the most comprehensive functional unit is to record the effect (e.g., GHG emission) by vehicle-km basis [49]. If the goal is to ensure the optimum use of land while using energy crops as the biomass supply, a useful way would be to represent the results on a per hectare basis [61]. In order to be independent from the kind of feedstock, the results should be expressed in per unit output, for example, kWh, and should be expressed in per unit input, for example, kg, in order to be independent of the conversion process.

To illustrate the method described, here is an example where both the goal scope and the functional unit are clearly specified. It is taken from the study that develops an LCA for the generation of bioelectricity [62] and is as follows: "The aim of this study is to evaluate the contribution to CO_2 emission reduction that can be achieved by means of using biomass for energy production, in comparison with conventional fuel use. The functional unit is the produced energy unit (1 MJ), to which inventory data and results are referred to."

2.4.2.2 Life Cycle Inventory Analysis

In this process, for each unit of input and output of energy, mass flows and emission data are collected, validated, and categorized. The system boundary plays a very important role in this part since it dictates what data to incorporate and what to reject. Something indirectly important to the LCA might not even be applicable. For example, a study by

Wiloso et al. [63] documents that out of 25 studies that use enzymatic action in processing lignocellulosic biomass, only 15 incorporated enzyme production in their inventory analysis. To further illustrate the life cycle inventory (LCI) analysis, Fig. 2.7 adapted from a study by Corti and Lombardi [62] is presented.

2.4.2.3 Allocation Methods

Allocation is a process of attributing environmental burden of multifunctional process to only those functions that are associated with it [64]. It is, however, to be avoided if possible either through the division of the whole process into subprocesses related to coproducts or by expanding the system boundaries (substitution approach) to include the additional functions related to them [5]. If not avoided, then the question of which method to use and what numerical values to be used is raised. And depending on the method of allocation to the coproduct or the source technology, the LCA values can change significantly. For example, in a case study where bioethanol is produced from wheat, the output differs significantly as shown in Fig. 2.8.

In Fig. 2.8, different models represent different means by which the power was obtained for the conversion process in the LCA. The description of each model is given in Table 2.10.

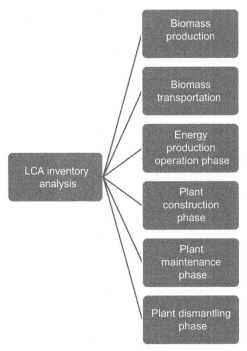

Figure 2.7 An example of life cycle inventory.

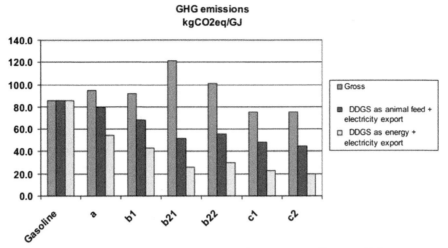

Figure 2.8 Variations in CO_2 output due to source allocation [65].

Table 2.10 Model Description Used in the Study Conducted by Punter et al. [65]

Model	Description
a	Conventional natural gas—fired steam boiler + imported electricity
b1	Conventional natural gas—fired steam boiler + backpressure steam turbo-generator
b21	Natural gas—fired gas turbine + unfired HRSG + backpressure steam turbo-generator
b22	Natural gas—fired gas turbine + cofired HRSG + backpressure steam turbo-generator
c1	Straw—fired steam boiler + backpressure steam turbo-generator
c2	Straw—fired steam boiler + backpressure and condensing steam turbo-generator

Table 2.11 adapted from a study by Singh et al. [66] shows different allocation methods for the production of bioethanol.

2.4.2.4 Impact Assessment

Impact assessment creates a connection between the product or process and its potential impacts on human health, environment, and source depletion. The impact assessment requires categorizing the effect of the biofuel production. It mainly looks into whether or not systems give surplus energy, followed by concern on global warming, eutrophication, acidification, water and land use, and so on. For example, it has been suggested that biofuels based on Jatropha are appropriate for small-scale, community-based production aimed at local use [52,67]. In the study by Corti and Lombardi [62], the category for impact assessment is kg of equivalent CO_2 per functional unit (MJ). This is because as the goal stated the aim was to evaluate the contribution of CO_2 emission

Table 2.11 Biomass and Its Reported Allocation Methods

Biomass	Allocation Method
Maize (grain)	Displacement, replacement, system expansion, Economy value energy content of outputs, mass, subdivision
Maize (stover)	System expansion, substitution, mass
Cellulose	System expansion, displacement
Sugarbeet and wheat (grain)	System expansion, mass, energy, market value
Sugarcane	None

reduction during the harvesting of bioenergy. Thus the goal and scope, and functional unit matter when it comes to defining the impact category.

2.4.3 Energy Balance, Greenhouse Gas Emission, and Environmental Concerns

2.4.3.1 Energy Balance

Usually, net energy value (NEV) is used in studying the issue of energy surplus. It is basically an efficiency term calculated by taking the difference between the usable energy produced from a biofuel crop and the amount of energy required in the production of that fuel for energy. A negative NEV indicates energy loss; that is, more energy is required to produce the biofuel than the amount of energy that can be used for fuel. A positive NEV is an estimate of the energy gained for fuel use in the production process indicating surplus of energy.

However, this categorization has an obvious flaw; that is, NEV calculation is too simplistic and gives a raw energy output, but not all forms of energy are the same. Different forms incur different costs and benefits. In reality, the service gained from fuel energy matters mostly. Therefore, it might be more accurate to compare biofuel energy balance directly to the fossil fuel energy equivalent that can be displaced. This is generally reported as a ratio of the amount of energy produced to the amount of fossil fuel energy required to produce it. This new term is called fuel energy ratio (FER). An FER < 1 indicates net energy loss, while an FER > 1 indicates a surplus of energy. Table 2.12 adapted from a study by Davis et al. [68] shows the FER for different crops.

2.4.3.2 Greenhouse Gas Emission

Since biomass as fuel is relatively recent when compared to fossil fuels, which are locked below the earth for thousands of years, burning the former is considered to be carbon neutral. This comes from the assumption that it burns off the same amount of carbon dioxide that it consumed throughout its life, which is relatively short. This entails that bioenergy has an almost closed CO_2 cycle. This though sounds perfect presents us with a hidden caveat that there are GHG emissions in its life cycle largely from the production

Table 2.12 Estimated Fuel Energy Ratio
(FER) Values of Some Biofuel Crops

Biofuel Crop	FER
Corn	1.95
	1.76
	1.67
	1.64
	1.62
	1.60
	1.52
	1.51
	1.39
	1.34
	1.32
	1.28
	1.27
	1.25
	1.22
	1.21
	1.08
	0.99
	0.95
	0.92
	0.8
	0.78
	0.69
Lignocellulosic crops (generalized)	5.6
	4.3
	3.51
	2.62
	2.19
	1.8
Miscanthus (combustion)	1.16
Miscanthus (gasification)	0.99
Switchgrass	4.43
	0.44

stages where fossil fuel inputs are required to produce and harvest the feedstocks, in processing and handling the biomass, in converting the biomass to fuel, and in transporting the feedstocks and biofuels.

The most uncertain greenhouse gas in the biofuel life cycle is nitrous oxide (N_2O). It evolves from the decomposition of organic matter and from the application of fertilizers [69]. Since fertilization rates are higher for annual crops than for perennial energy crops,

N_2O emission is higher in those. Crops grown in high rainfall environments or under flood irrigation have particularly high N_2O emissions, as denitrification, the major process leading to N_2O production, is favored under moist soil conditions where oxygen availability is low. As the emission of N_2O lacks a point source, the estimation of its emission is very difficult [70]. This uncertainty is even more magnified since the GWP of the N_2O is very high around 298 times as that of CO_2 [49].

Methane (CH_4) is the last major emission in the life cycle after CO_2 and N_2O. This is released by anaerobic decomposition of organic feedstocks or by reducing the oxidation of the soil thereby reducing the methane content in it while releasing it into the atmosphere. Its GWP is 23 times as that of CO_2, thus considerably lower than that of N_2O.

The most pertinent GHGs mentioned earlier; that is, CO_2, CH_4, and N_2O, are called direct GHGs since they impact the climate directly. Other gases that are emitted throughout the life cycle are carbon monoxide (CO), nonmethane organic compound (NMOC), and ozone (O_3) [49]. In order to take a holistic look into the GHG emissions, both the emissions need to be accounted for and all of the emissions are to be converted to the carbon dioxide equivalents. Taking all of the direct and indirect emissions, sometimes it may so happen that a particular brand of biofuel emits more GHGs than the conventional fuel. Thus a blanket statement regarding the GHG emission is not possible. Fig. 2.9 shows the GHG emission in CO_2 equivalents for different biofuels. It has been adapted from the study by Fritsche and Hennenberg [71], and it gives the maximum and minimum values for all the biofuels given. From Fig. 2.9, it can be noted that some fuels have higher GHG emission than conventional diesel and gasoline.

The study by Corti and Lombardi [62] gives comparative results of the equivalent CO_2 emission per MJ of electricity produced for both biofuels-based and coal-based electricity production. The results are adapted to graphical format (Fig. 2.10 for biomass-based electricity and Fig. 2.11 for coal-based electricity) for comparative purpose.

2.4.3.3 Land Use and Other Environmental Issues

The land use for biomass can be divided into direct land use and indirect land use. Direct land use is when biofuel feedstock is cultivated in a land already in use for something. Thus, if there was a forest or any other kind of agriculture, such as rice, wheat, and so on, and it was displaced to grow sugarcane or sunflower for biofuel, then that would be direct land use. Such land use can bring issues, such as change in indigenous carbon cycle. It necessarily is not bad since whether or not the soil carbon content increases depends on the biomass feedstock crop and the previous crop/plantation.

Indirect land use is the term used to describe the phenomena that occurs when the biofuel feedstock occupies a land already in use forcing the previous plantations to occupy another land. This could happen if the production demand for the previous land

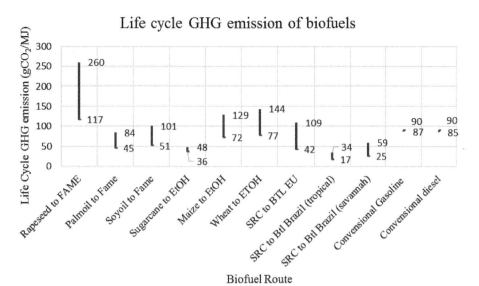

Figure 2.9 Life cycle GHG emissions of different biofuels and conventional gasoline and diesel including indirect land use change. *Adapted from Fritsche UR, Hennenberg K. The "iLUC Factor" as a means to hedge risks of GHG emissions from indirect land-use change associated with bioenergy feedstock provision. In: Background paper for the EEA expert meeting in Copenhagen, 2008.*

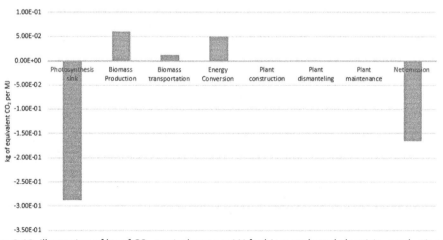

Figure 2.10 Illustration of kg of CO_2 equivalents per MJ for biomass-based electricity production [62].

use still exists; for example, in the case of crops such as rice or wheat, and thus these are cultivated in another land which may prompt unfavorable land use change [71].

Excessive land use causes little pertinent damage to the environment since the land is used to grow crops and that comes with a baggage of other needs, such as fertilizers and

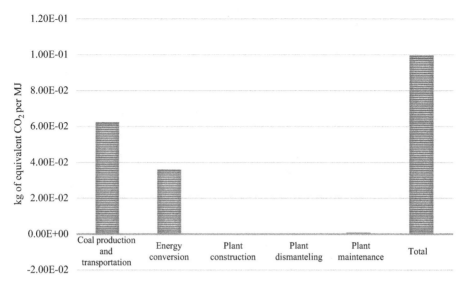

Figure 2.11 Illustration of kg of CO_2 equivalent per MJ for coal-based electricity production [62].

pesticides. Use of such may increase the risk of eutrophication and acidification if it is washed away into local water bodies or seeps into ground water. Although in the case of biogas production, which is mainly a product of anaerobic action on manure, in some cases it can elevate the environmental condition if the manure is released into the environment untreated otherwise [60]. One way to note the damage is to define the impact categories meticulously, a categorization is Eco-Indicator 99 methodology which includes the impact categories into three types of damage: "Damage to human health," which includes the following impacts: carcinogenesis, organic respiratory effects, inorganic respiratory effects, climate change, ionizing radiation, and reduction of the ozone layer; "damage to ecosystem quality," which includes ecotoxicity, acidification/eutrophication, and land use; and "resource damage," which includes minerals and fossil fuels [72].

In the study conducted to analyze the production of biofuels from different vegetable oils [72], it is noted that the production of soybean has an impact of 70% in the category of carcinogens, 34.3% in the category of respiratory inorganics, 55% in the category of acidification and eutrophication, and 35.5% in the category of land use. If the impact categories are normalized to show how much each factor, noted in Eco-Indicator 99, is effected compared to each other, the result showed in Fig. 2.12 is obtained.

Thus, from the study it can be readily concluded that apart from climate change, the culturing of biofuel feedstocks and making biofuels have adverse effect on the environment. Thus making biofuels more ubiquitous is a double-edged sword. One needs to make a balance between the environmental degradation in comparison to the

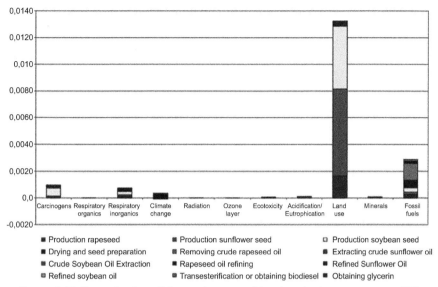

Figure 2.12 Normalization of the environmental burdens by impact category [72].

net reduction of GHGs into the atmosphere. There is no universal right choice regarding what to do, rather it is always case specific.

2.4.4 Reasons for Uncertainties in Biofuel Life Cycle Assessment

It should be obvious by now that LCA of biofuels results in somewhat ambiguous results, and sometimes one data contradicts the other. It is best illustrated in Table 2.12 where corn gives both FER > 1 and FER < 1. Also it can be seen in Fig. 2.9 where some fuel range shows the emission can be both lower and greater than conventional diesel and gasoline. This feature of LCA results partly from the myriads of un-certainties that creep into the analysis throughout its working and partly from the fact that there are wide range of plausible values for key input parameters with values often dependent on local condition. Same feed materials and output can have more than one path to follow. For example, a study by Wiloso et al. [63] notes that using lignocellulose as feed to obtain bioethanol can be done in two ways. The widespread one is to hydrolyze the biomass feedstock and then ferment it; however, it can also be done by gasifying the feed to form syngas and then either fermenting it or using a catalyst to convert it to ethanol. Now even with the fact that the product and the feed were the same, the path undertaken was different and as shown before, the path taken to produce will have an effect on the final outcome of the LCA. Also land use and technology use differ from place to place and time to time and thus LCA outputs vary from time and place.

Keeping these reasons aside, Larson provides four basic reasons why there are such major uncertainties [49] (All of these reasons are discussed in brief in this discussion and hence not repeated):

1. the climate-active species included in the calculation of equivalent GHG emissions (Section 2.4.3.2);
2. assumptions around N_2O emissions (Section 2.4.3.2);
3. the allocation method used for coproduct credits (Section 2.4.2.3);
4. soil carbon dynamics (Section 2.4.3.3).

From this study, it should be obvious that biofuels are not a door to utopia. It has its fair share of problems, which can be a major issue if not regulated. Biofuels are shown to lower the overall GHG into the atmosphere if the right one is chosen. However, in doing so it introduces major environmental degradation. Thus, to make a choice to move toward biofuels must be an educated one, which accounts for the consequence and is prepared to offset it thereby making the fuel truly sustainable.

2.5 LIFE CYCLE ASSESSMENT OF BIOGAS

Biogas is a fuel gas produced from biomass and/or from the biodegradable fraction of wastes, which can be purified to natural gas quality, to be used as biofuel. The gas consists mostly of CH_4 and CO_2 and is produced from the bacterial anaerobic action on the feedstock. Fig. 2.13 shows the typical composition of biogas.

Electricity from biogas concept is on the rise especially for European countries as the GHG emission needs to be curbed to favorable level. In Poland it has been estimated that electricity from biogas is expected to rise from 0.4 to 6.6 TWh/year from

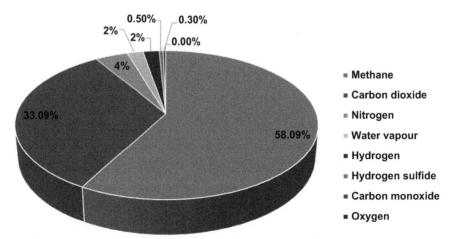

Figure 2.13 Composition of biogas. *Adapted from A Biogas Road Map for Europe. European Biomass Association 2010, https://www.americanbiogascouncil.org/pdf/euBiogasRoadmap.pdf.*

2010 to 2030. Also, bioelectricity provides a huge economical potential as it has been estimated that it can provide almost 30% of energy demand in China [45]. Also the quantity of small-scale biogas digesters has increased from about 1.8×10^9 m^3 in 1996 to 1.0×10^{10} m^3 in 2007, while the number of large- and medium-scale biogas projects has increased from about 1.2×10^{11} m^3 in 1996 to 6.0×10^{12} m^3 in 2007.

2.5.1 Feedstock and Environmental Effects

The feedstock of biogas is not limited to a single biomass source. However, the waste and manure is the most obvious source for biogas. Thus, it is sometimes a surprise to know that even the herbaceous materials can and do contribute to the formation of biofuels. Fig. 2.14 shows the list of sources and its use along with the percentage of methane that is formed from those sources.

The environmental effects of biomass use have already been briefed in Section 2.4.3.3; however, in the case of biogas since waste and animal discharge plays a big part, there is an added benefit. The environmental advantages of using sewage-derived biogas relate to the reduction of problematic sludge by about 50% (dry mass basis). Also, biogas that has methane, carbon dioxide, and hydrogen sulfide, which is toxic, is naturally released from wastes in landfills and its oxidation is necessary for prohibiting the release of methane and volatile organic compounds to the atmosphere thus improving local air quality [45].

2.5.2 Greenhouse Gas Emission and Electricity Production

The toxic emission of biogas combustion is relatively cleaner than petrol and diesel. This makes it a very attractive fuel. The relative reduction of toxic emission, which includes GHG, in comparison to petrol and diesel is given in Fig. 2.15.

However, the other side of the story is hinted in the study by Ishikawa et al. [75]. It documents the total equivalents of CO_2 emission in kg for a biogas plant in Betsukai, Hokkaido. The CO_2 emission for each item is given in Table 2.13.

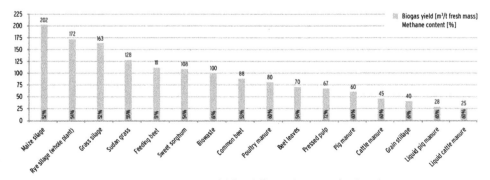

Figure 2.14 Biogas yield for different biomass feedstock.

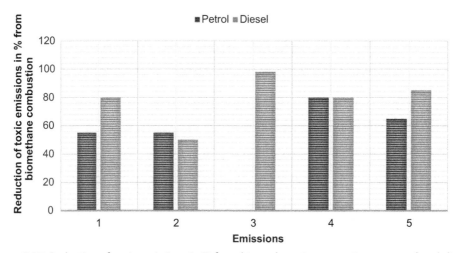

Figure 2.15 Reduction of toxic emissions in % from biomethane in comparison to petrol and diesel. *Adapted from Rutz D, Janssen R. Biofuel technology handbook. Munich (Germany): WIP Renewable Energies; 2007.*

Table 2.13 CO_2 Emissions of Biogas Plant in Hokkaido [75]

Items	CO_2 Emissions (kg)
Initial energy investment	2,589,000
Operating energy	78,000
Maintenance energy	—
Total	2,667,000

Budzianowski [45] notes and compares three ways of producing electricity from biomethane. First one is combustion of biomethane with air, the second one involves oxyfuel combustion of biomethane where the components other than oxygen in air are stripped out and pure oxygen is used instead of air for combustion. The third one is called oxy-reforming fuel cell (ORFC) where CH_4 is split into CO and H_2 in the oxygen stream atmosphere. The energy required for this splitting is obtained from the exothermic reaction as CO is converted to CO_2. The H_2 on the other hand is used in the H_2 fuel cells for production of electricity. The result after the comparison of the three mentioned method to produce electricity is given in Table 2.14.

Thus, it can be seen that electricity generation from biogas though viable has its issues as well. The environmental issues discussed in Section 2.4.3.3 still stands. It is a better form of fuel than the fossil fuels in use; however, there are some issues with GHG during the formation and operation of biogas plants. As stated before there is no choice that can lead to utopia. Every technology comes with its disadvantage. It is thus wise to know what they are to make prudent judgment regarding adapting any technology.

Table 2.14 Comparison Among Different Ways of Producing Electricity From Biogas

	Combustion	Oxyfuel Combustion	Oxy-Reforming Fuel Cell
Net electricity output (kJ/mol CH_4)	431.2	411.2	450.2
Net efficiency (%)	53	51	56

Adapted from Budzianowski WM. Can 'negative net CO_2 emissions' from decarbonised biogas-to-electricity contribute to solving Poland's carbon capture and sequestration dilemmas? Energy 2011;36:6318—25.

2.6 LIFE CYCLE ASSESSMENT OF HYDROPOWER PLANTS

Hydropower is the major renewable electricity generation technology being used in 159 countries having advantages such as high level of reliability, high efficiency, proven technology, very low operating and maintenance cost, flexibility, and large storage capacity. It contributes to more than 16% of worldwide electricity generation and about 85% of global renewable electricity. International Energy Agency (IEA) roadmap on hydropower predicts a global capacity of 2000 GW resulting in over 7000 TWh of electricity by the year 2050. China, Brazil, Canada, and the United States together produce half the world's hydropower (Table 2.15). In 2010, 36 countries generated more than 50% of their total electricity from hydropower [76] (Table 2.16).

Hondo [40] has conducted a life cycle GHG emission analysis for a hydropower plant in Japan having a gross output of 10 MW with a capacity factor of 45%. The plant lifetime was considered to be 30 years. The plant analyzed was a run-of-the-river type with a small reservoir. The constituents of the plant are a small concrete dam (2000 m^3 volume), a penstock (9000 m), a pressure pipe (490 m), and a powerhouse. The maximum intake to the powerhouse was 4.8 m^3/s. Table 2.17 shows the life cycle GHG emission factors (LCEs) and their breakdowns for the hydropower plant studied.

Table 2.15 Top 10 Hydropower Producers in 2010 [76]

Country	Hydroelectricity (TWh)	Share of Electricity Generation (%)
China	694	14.8
Brazil	403	80.2
Canada	376	62.0
United States	328	7.6
Russia	165	15.7
India	132	13.1
Norway	122	95.3
Japan	85	7.8
Venezuela	84	68
Sweden	67	42.2

Table 2.16 Countries With More Than Half of Their Electricity Generation From Hydropower in 2010

Share of Hydropower	Countries
~100%	Albania, DR of Congo, Mozambique, Nepal, Paraguay, Tajikistan, Zambia
>90%	Norway
>80%	Brazil, Ethiopia, Georgia, Kyrgyzstan, Namibia
>70%	Angola, Columbia, Costa Rica, Ghana, Myanmar, Venezuela
>60%	Austria, Cameroon, Canada, Congo, Iceland, Latvia, Peru, Tanzania, Togo
>50%	Croatia, Ecuador, Gabon, DPR of Korea, New Zealand, Switzerland, Uruguay, Zimbabwe

Adapted from International Energy Agency. Technology roadmap: hydropower, https://www.iea.org/publications/freepublications/publication/2012_Hydropower_Roadmap.pdf.

Table 2.17 Life Cycle GHG Emission Factors for Hydropower Plant Studied by Hondo [40]

	g-CO_2/kWh	Share (%)
Construction	**9.3**	**82.8**
Machinery	0.9	8.0
Dam	0.5	4.5
Penstock	4.5	39.8
Other foundations	2.4	21.0
Site construction	1.1	9.6
Operation	**1.9**	**17.2**
Total	11.3	100.00

The results suggest that 82.8% of CO_2 is emitted during construction as compared to 17.2% CO_2 emission during operation. The LCEs for hydropower plant depend prominently on the assumption of lifetime and capacity factor [40]. Tables 2.18 and 2.19 show the effect of lifetime and capacity factor on LCE (g-CO_2/kWh) for hydropower plant.

Pascale et al. [77] have conducted an LCA study of a 3 kW run-of-river community hydroelectric system located in Huai Kra Thing (HKT) village in rural Thailand. They have modeled the construction, operation, and the end-of-life phases of the hydropower

Table 2.18 Effect of Lifetime on Life Cycle GHG Emission Factor for Hydropower Plant Studied by Hondo

Lifetime (years)	10	20	30	50	100
g-CO_2/kWh	30	16	11	8	5

Adapted from Hondo H. Life cycle GHG emission analysis of power generation systems: Japanese case. Energy 2005;30:2042−56.

Table 2.19 Effect of Capacity Factor on Life Cycle GHG Emission Factor for Hydropower Plant Studied by Hondo

Capacity factor	−10 pt	−5 pt	Reference	+5 pt	+10 pt
g-CO_2/kWh	14	13	11	10	9

Adapted from Hondo H. Life cycle GHG emission analysis of power generation systems: Japanese case. Energy 2005;30:2042–56.

plant over a period of 20 years. The model includes all the relevant equipment, materials, and transportation; 1 kWh of electrical energy has been considered as functional unit in the study. Fig. 2.16 shows the scope and system boundary of the study. Table 2.20 shows the results of the study in terms of GWP in g-CO_2/kWh [77].

Gallagher et al. [78] have calculated the environmental impacts of three (50–650 kW) run-of-river hydropower projects in the United Kingdom using the LCA tool. The GHG emissions from the projects to generate electricity ranged from 5.5 to 8.9 g-CO_2eq/kWh, which is very low as compared to 403 g-CO_2eq/kWh for UK marginal grid electricity. The system boundary considered for the study included raw material extraction, processing, transport, and all installation and grid connection operations. The functional unit was 1 kWh of electricity generated and the lifespan has been considered to be 50 years. Fig. 2.17 shows the system boundary used in the study. Table 2.21 shows the descriptions of three run-of-river hydropower plant case studies and their environmental impacts.

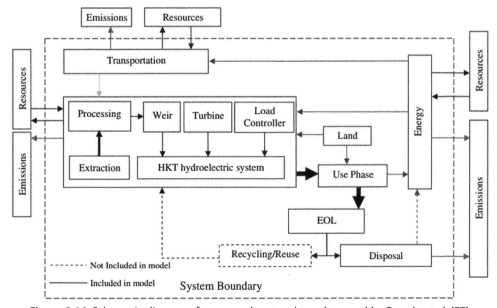

Figure 2.16 Schematic diagram of scope and system boundary used by Pascale et al. [77].

Table 2.20 Life Cycle Assessment Results for Different Components of the Hydropower Plant

	Weir, Intake, Canal and Forebay	Penstock	Powerhouse, Turbine, and Outflow	Transmission Line	Control House and Control and Conditioning Equipment	Distribution	3 kW Hydropower Scheme Total
g-CO$_2$/kWh	3.7	9.8	9.0	14.7	2.7	12.9	52.7

Adapted from Pascale A, Urmee T, Moore A. Life cycle assessment of a community hydroelectric power system in rural Thailand. Renewable Energy 2011;36:2799—808.

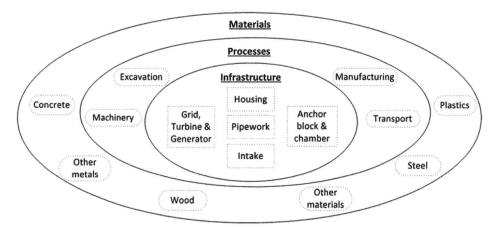

Figure 2.17 Key materials, processes, and infrastructure considered within the system boundaries for run-of-river hydropower projects [78].

Table 2.21 Description and Environmental Impacts of Three Run-of-River HP Case Studies

Parameter	Hydropower Project 1	Hydropower Project 2	Hydropower Project 3
Location	North Wales	North Wales	North England
Net head	175 m	128 m	105 m
Flow	~450 L/s	~100 L/s	~90 L/s
Design capacity	650 kW	100 kW	50 kW
Annual output	1.8—2.1 GWh	0.4—0.5 GWh	0.2—0.3 GWh
g-CO$_2$/kWh	5.46	7.39	8.93

Adapted from Gallagher J, Styles D, McNabola A, Williams AP. Current and future environmental balance of small-scale run-of-river hydropower. Environmental Science & Technology 2015;49:6344—51.

Suwanit and Gheewala [79] have studied the LCA of five mini-hydropower plants in Thailand. The functional unit has been considered as MWh of electricity and lifespan of the plants has been considered as 50 years. The design capacities of the five power plants considered in this study are Mae Thoei (2.25 MW), Mae Pai (1.25 MW × 2), Mae Ya (1.15 MW), Nam San (3 MW × 2), and Nam Man (5.1 MW), having a net efficiency ranging between 40% and 50%. Table 2.22 shows the overall description of the

Table 2.22 Overall Description of the Mini-Hydropower Plants Studied

Description of the Study Site	Nam Man	Nam San	Mae Pai	Mae Thoei	Mae Ya
Project Description					
Geographic location	Dan Sai, Loei province	Phu Rua, Loei province	Pai, Mae Hong Son province	Om Koi, Chiang Mai province	Jom Thong, Chiang Mai province
Installed capacity	5.1 MW	3 MW × 2	1.25 MW × 2	2.25 MW	1.15 MW
Proximity to population served	1558 households	453 households	6 villages	940 households	190 households
Condition for electricity use	Local electricity grid and supply electricity for main transmission line				
Design of the system	Run-of-river (extra, tunnel 2.45 × 2.45 × 1800 m)	Run-of-river (extra, tunnel 2.45 × 2.45 × 2400 m)	Run-of-river	Run-of-river	Run-of-river
Local river condition	The flow of rivers changes following seasons—having a rapid flow for 4 months in rainy season, medium flow for 4 months, and low flow for 4 months, electricity is generated for only 10 months with varying capacity; for the calculations, annual electricity production data are used from the actual records.				
Project area (ha)	7.3	9.6	23	12	6.4
Project Design					
Design flow rate (m³/s)	6.0	4.36	1.39	2	1.73
Water head (m)	127	95	106.7	137.1	98.1
Turbine type	43 in. Twin Jet Turgo	43 in. Twin Jet Turgo	22.5 in. Twin Jet Turgo	22.5 in. Twin Jet Turgo	22.5 in. Twin Jet Turgo
Generator type	Synchronous	Synchronous	Synchronous	Synchronous	Induction
Weir	Mass concrete, 4 m high and 35.5 m long	Mass concrete, 4 m high and 55 m long	Mass concrete, 3.5 m high and 21.5 m long	Mass concrete, 2 m high and 18 m long	Mass concrete, 3.6 m high and 46 m long
Penstock or pressure pipe line	Steel, 1.51 m diameter and 304 m long	Steel, 1.82 m diameter and 250 m long	Steel, 1.15 m diameter and 182 m long	Steel, 1 m diameter and 404 m long	Steel, 0.9 m diameter and 360 m long
Water gate and screen	17 sets	19 sets	15 sets	14 sets	13 sets

Adapted from Suwanit W, Gheewala SH. Life cycle assessment of mini-hydropower plants in Thailand. The International Journal of Life Cycle Assessment 2011;16:849—58.

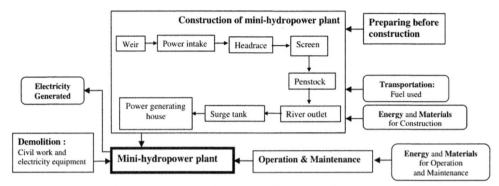

Figure 2.18 Life cycle inventory for the mini-hydropower plants studied by Suwanit and Gheewala [79].

mini-hydropower plants studied. The LCI has been taken for five stages: (1) before construction, (2) construction of the hydropower plant, (3) transportation, (4) operation and maintenance, and (5) demolition of the plants. Fig. 2.18 shows the LCI of the mini-hydropower plants.

For the mini-hydropower plants, the major contributors for global warming are construction at 60% (48–72%), transportation at 32% (18–50%), and operation and maintenance accounting 8%. The significant emission is CO_2 contributing more than 83–88% of GWP, CO contributing 10–12%, N_2O about 2–3%, and CH_4 accounting 2–5% (from the cast iron production, cement production, and transportation by truck). The related activities are combustion of diesel oil used for construction equipment and transportation, electricity used for construction equipment, activities in the construction period, and operation of mini-hydropower plants [79]. Table 2.23 shows the life cycle environmental impact potentials of five mini-hydropower plants studied.

Zhang et al. [80] have compared the carbon footprints of two types of hydropower schemes: comparing earth-rockfill dams (ECRDs) and concrete gravity dams (CGDs) for Nuozhadu power station in China as a case study. This power station is the largest of its kind in Asia and the third largest in the world, having a 5.85 GW rated capacity. To compare two different schemes, ECRD and CGD systems were designed separately for the same plant in the planning phase. The model with the ECRD system is constituted of a clay-core rockfill dam with a 258 m height, an open crest spillway, spillway tunnels, bank protection, water diversion and power generation system, and diversion construction.

Table 2.23 Life Cycle Environmental Impact Potentials of Five Mini-Hydropower Plants [79]

Power plant location	Nam Man	Nam San	Mae Pai	Mae Thoei	Mae Ya	Average
kg-CO_2eq	11.01	23.01	16.28	22.71	16.49	17.62

The model of the CGD system is composed of a CGD with a 265 m height, plunge pool and subsidiary dam, bank protection, water diversion and power generation system, and diversion construction. The study considered 44 years of time span of which 14 years is the lifespan of the construction phase and 30 years is the lifespan of the plant; 1 kWh of electricity has been considered as the functional unit. The total carbon footprint throughout the power plant life cycle was assessed by amassing the emissions from the material production, transportation, construction, and operation and maintenance stages. For the 44-year time period, the total carbon footprint for the ECRD system is 8.8 million tons of CO_{2e}, while that of the CGD is 11.69 million tons of CO_{2e}. The ECRD system reduces the total CF by about 24.7% compared with the CGD system [80].

Varun et al. [81] have presented life cycle GHG emission correlations for small hydro power schemed in India. They have presented the data for 145 small plants of three types—run-of river, canal-based, and dam-toe. The rated power capacity for these plants varies from 50 kW to 16 MW with head ranging from 1.97 to 427.5 m. The GHG emission for these power plants ranges from 11.34 to 74.87 kg-CO_2eq/kWh. As a case study, they have shown the calculation for Karmi-III micro hydropower project (50 kW capacity with 55 m head) in Uttarakhand, India. The GHG emissions for the total electricity generated over the lifetime of 30 years have been found as 74.87 g-CO_{2eq}/kWh_e.

Hertwich [82] has studied the biogenic GHG emissions from hydropower in tropical region using LCA. A hydropower plant with installed capacity of 250 MW in Balbina reservoir at Brazilian amazon emits 8.5 kg CO_2/kWh. Another plant, Petit Saut power station in French Guyana, which produces 560 GWh/year emits 1.55 kg CO_2/kWh. For upstream Nam Leuk reservoir in Laos, the emissions are in the order of 0.05−0.1 kg CO_2/kWh.

2.7 LIFE CYCLE ASSESSMENT OF GEOTHERMAL POWER PLANTS

Geothermal energy has a vital role to play in meeting goals in energy security, economic development, and mitigating climate change, since geothermal technologies use renewable energy resources to generate electricity and heating and cooling while emitting very low levels of GHG. This energy is stored in rock and is trapped in vapor/liquids, for example, water or brines that can be used to generate electricity and for providing heating. Electricity generation usually requires geothermal resource's temperature of over 100°C. According to the roadmap by International Energy Agency, geothermal electricity generation has the potential to reach 1400 TWh/year, which is around 3.5% of global electricity production by 2050, reducing almost 800 megatonnes (Mt) of CO_2 emissions per year [83]. The quantity of gases and metals contained within the geothermal fluids depends on the depth and location of the geothermal reservoir, characteristics of the electricity generation systems and the abatement systems [84].

Bayer et al. [85] reviewed the direct environmental impacts of geothermal power plants in terms of land use, geological hazards, waste heat, atmospheric emissions, solid waste, emissions to soil and water, water use and consumptions, impact on biodiversity, noise, and social impact. Armannsson et al. [86] reported that direct carbon dioxide (CO_2) emissions from geothermal power plants extend to a broad range originating from degassing magma, infrequently from decomposition of organic sediments and metamorphic decarbonization. Bertani and Thain [87] have conducted a global survey for International Geothermal Association for a large number of geothermal power plants (85% of 2001 geothermal capacity of 6.65 GW) and found the CO_2 emission ranging from 4 to 740 g/kWh. Fridleifsson [88] has reported this value to be 3—380 g/kWh.

According to DiPippo [89], the range is 50—80 g-CO_2/kWh, whereas Kagel et al. [90] have reported the emission to be 44 g/kWh. Bloomfield et al. [91] have provided a value of 91 g/kWh of CO_2, which is the same as the weighted average value for geothermal power plants in the United States. For New Zealand, Rule et al. [92] have reported a range of 30—570 g/kWh. Armannsson et al. [86] have reported the CO_2 emission value for three plants in Iceland. The values are 152, 181, and 26 g-CO_2/kWh for Krafla, Svartsengi, and Nesjavellir, respectively, for the year 2000. The US department of Energy reported dry-steam plants at the Geysers (California) to produce about 41 g/kWh and flash plants to generate about 28 g/kWh [93].

Bravi and Basosi [84] have conducted environmental impact study for four geothermal power plants in Mount Amiata area, Italy from environmental perspective. There is 1 unit in the Bagnore site, which has an area of 5 km^2 having 7 production wells and 4 injection wells, 3 units in Piancastagnaio site covering an area of 25 km^2 having 19 production wells and 11 injection wells. The descriptions of the sites are given in Table 2.24. In particular, the authors have analyzed the emissions of noncondensable gases from geothermal fluids in 2002—09. The production time of the selected

Table 2.24 Description of Four Geothermal Power Plants Used by Bravi and Basosi

Units	Bagnore 3	Piancastagnaio 3	Piancastagnaio 4	Piancastagnaio 5
Province	Grosseto	Siena	Siena	Siena
Acronym	BG3	PC3	PC4	PC5
Installed capacity, MWe	20	20	20	20
Type of unit	**Single Flash**	Steam with entrained water separated at wellhead		
Well depth, km		**From 2 to 4**		
Temperature, °C		Between 300 and 350		
Pressure, bar		Around 200		
Annual energy produced, GWh/year (2008)	169.7	160.4	139.1	145.3

Adapted from Bravi M, Basosi R. Environmental impact of electricity from selected geothermal power plants in Italy. Journal of Cleaner Production 2014;66:301—8.

geothermal power plants was considered by studying the yield of the emission materials from the chimneys. However, the authors did not consider the consumption of resources associated with drilling, construction, and operation of the wells, and the supplementary materials needed for the construction and operation of plants since the effect of plant construction is dispersed over the assumed 25 years of plant operation and only accounts for an insignificant amount of total foreground and background emissions.

The system boundary in this study includes the production period of the plants, disregarding the drilling, construction, and decommissioning periods. The foreground emissions into the environment were accounted for evaluating the potential impact of electricity production from geothermal power plants. The authors reasoned that due to the dilution of construction phase emissions, the conclusion of the study was not affected by excluding certain emissions. The functional unit of the study has been considered as 1 MWh electric energy production from a geothermal power plant [84].

The geothermal electricity production units in the Mount Amiata region discharge noncondensable products, that is, CO_2, H_2S, NH_3, and CH_4. Out of the emitted products, CO_2 is the main gas from the geothermal field having actual range from 245 to 779 kg/MWh with a weighted average of 497 kg/MWh. The range of NH_3 emissions is between 0.086 and 28.94 kg/MWh with a weighted average of 6.54 kg/MWh. NH_3 emissions per MWh in the geothermal field of Bagnore are about 4 times higher than those recorded in the units of Piancastagnaio. H_2S has a mean range of 3.24 kg/MWh, with values varying between 0.4 and 11.4 kg/MWh. Like the ammonia emission, the average values of H_2S in Piancastagnaio are four times higher than those of the geothermal fields of Bagnore. These values are related to the characteristics of the geothermal fluid available in the sites. The GWP average value is 693 kg-CO_2eq/MWh, with values ranging between 380 and 1045 kg/MWh [84].

Hondo [40] has conducted a life cycle GHG emission analysis for a geothermal power plant (double flash type) in Japan having a gross output of 55 MW with a capacity factor of 60%. The plant lifetime was considered to be 30 years. Installation of plants and drilling of production wells and exploration wells were considered in the study. The depth of 5 exploration wells was assumed to be 1500 m whereas the depth of 14 production wells and 7 reinjection wells were assumed to be 1000 m. The drilling failure was also considered for the analysis. While in the operation, each year an additional production well was drilled, and an additional reinjection well was drilled every two years.

Table 2.25 shows the LCEs and their breakdowns for the geothermal power plant studied [40]. The results suggest that 64.7% of CO_2 is emitted during operation as compared to 35.3% CO_2 emission during construction. This is obvious since significant CO_2 emissions take place while digging additional wells and with manufacturing and replacing hot water heat exchanger pipes. The LCEs for geothermal power plant depend prominently on the assumption of lifetime and capacity factor. Tables 2.26 and 2.27 show

Table 2.25 Life Cycle GHG Emission Factor for Geothermal Power Plant Studied by Hondo [40]

	g-CO$_2$/kWh	Share (%)
Construction	**5.3**	**35.3**
Foundations	2.0	13.2
Machinery	3.2	21.2
Exploration	0.1	0.9
Operation	**9.7**	**64.7**
Drilling of additional wells	2.9	19.6
General maintenance	2.3	15.1
Exchange of equipment	4.5	30.0
Total	15.0	100.00

Table 2.26 Effect of Lifetime on Life Cycle GHG Emission Factor for Geothermal Power Plant Studied by Hondo

Lifetime (years)	10	20	30	50	100
g-CO$_2$/kWh	26	18	15	13	11

Adapted from Hondo H. Life cycle GHG emission analysis of power generation systems: Japanese case. Energy 2005;30:2042−56.

Table 2.27 Effect of Capacity Factor on Life Cycle GHG Emission Factor for Geothermal Power Plant Studied by Hondo

Capacity Factor	−10 pt	−5 pt	Reference	+5 pt	+10 pt
g-CO$_2$/kWh	18	16	15	14	13

Adapted from Hondo H. Life cycle GHG emission analysis of power generation systems: Japanese case. Energy 2005;30:2042−56.

the effect of lifetime and capacity factor on LCE (g-CO$_2$/kWh) for geothermal power plant.

Karlsdottir et al. [94] have conducted an LCA of combined heat and power production from Hellisheidi geothermal power plant in Iceland. This combined heat and power (CHP) plant is located at Hengill geothermal area close to Reykjavik, the capital of Island. As of February 2009, the power generation capacity was 213 MW. When completed, the Hellisheidi plant will have estimated production capacity of 300 MW of electricity and 400 MW of thermal energy. The plant is double flash type with high- and low-pressure turbines and separators. For the LCA model, a steady production of 213.6 MW of electricity and 121 MW heat is used. Figs. 2.19 and 2.20 show the schematic diagram and flow diagram for LCA analysis of the Hellisheidi CHP plant.

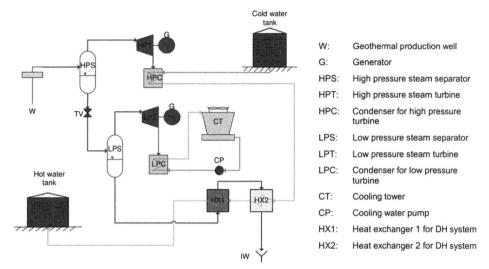

Figure 2.19 Schematic diagram of the Hellisheidi geothermal combined heat and power plant [94].

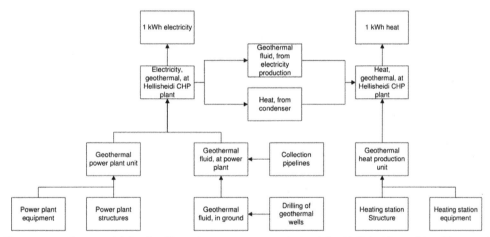

Figure 2.20 Flow model for the life cycle assessment analysis of the Hellisheidi combined heat and power (CHP) plant [94].

The study has analyzed two energy performance indicators (the primary energy efficiency and the CO_2 emissions) for the electricity and heat production from Hellisheidi CHP plant. The functional unit of the study is chosen to be MWh of electricity or heat produced in this plant. Regarding the system boundary, the process included in the study are the operation and construction of the plant. The decommissioning/demolition of the plant is disregarded due to insufficient data along with energy and materials flow for maintenance. The project life has been considered to be 30 years. Table 2.28 shows the

Table 2.28 CO_2 Emissions From Electricity Generation From Geothermal Energy

Source of Electricity	kg-CO_2/MWh
Electricity from Hellisheidi geothermal power plant	29
Electricity from Hellisheidi geothermal power plant, with reinjection	29
Electricity from Hellisheidi combined heat and power plant	29

Adapted from Karlsdottir MR, Palsson OP, Palsson H. LCA of combined heat and power production at hellisheiði geothermal power plant with focus on primary energy efficiency. Power 2010;2:16.

Table 2.29 CO_2 Emissions (g-CO_2/kWh) From Conventional Electricity Generation Systems

Conventional Energy Source	Emission of CO_2 (g-CO_2/kWh)
Hard coal	660–1050
Lignite	800–1300
Natural gas	380–1000
Oil	530–900
Nuclear power	3–35

Adapted from Turconi R, Boldrin A, Astrup T. Life cycle assessment (LCA) of electricity generation technologies: overview, comparability and limitations. Renewable and Sustainable Energy Reviews 2013;28:555–65.

results obtained from the LCA study. The origins of CO_2 emission are geothermal fluids (87.5%), geothermal well drilling (8%), power plants and components (4%), and collection lines (0.5%). The emission of CO_2 is the same for all three cases of electricity production as reinjection and utilization of waste stream do not have substantial effects on the total emissions [94].

2.8 COMPARISON WITH CONVENTIONAL SYSTEMS

To get a clear picture, it is important to compare the environmental impacts of electricity generation from renewable resources to the conventional sources of electricity (e.g., coal, natural gas, oil, and nuclear). This will help in understanding how much cleaner a renewable technology is compared to conventional systems. For the conventional electricity generation systems, most of the GHG emissions come from the combustion of fossil fuels. GHG emissions from material manufacturing contribute insignificantly to the total emissions. Table 2.29 represents a comparison of life cycle CO_2 emissions for various conventional fuels for electricity generation.

Table 2.29 shows that coal power plants are the most GHG intensive. However, over the last decade cleaner technologies for coal extraction and combustion have come into play, which can reduce GHG emissions significantly. Nuclear power plants are the least GHG intensive, which are comparable to hydro and wind power plants. But disposal of radioactive materials involves higher damage to the surroundings.

2.9 CONCLUSIONS

Due to the depletion of fossil fuel sources and global warming, renewable energy technologies are becoming more popular. Worldwide, the investment on renewable energy has been increased substantially. Generally, the electricity generation from renewable resources is costlier than electricity generation from the conventional sources. The selection of a power generation system does not only depend on the cost parameters but also on the environmental impacts of the system. The quantification of GHG emissions from renewable power generation technologies is very important for decision making toward sustainability. The LCA tool is the most widely used tool to quantify GHG emissions of a system. In this chapter, a thorough review on LCA of different renewable power generation technologies is conducted. The environmental footprints of different conventional electricity generation systems are also presented. This chapter indicates that GHG emissions from renewable power generation technologies are very less compared to conventional sources of power generation. The emission of GHGs can further be reduced by increasing the efficiency of renewable power generation technologies. This chapter shows a wide range of results for all the technologies mainly due to the variation in system boundaries, assumptions, and data quality. So it is very important to conduct a location-specific LCA of power generation technologies. Although energy can be produced with lower emissions, the problem of renewable energy systems is that they are not capable of generating power the whole day around-the-clock. For example, when there is no sunlight and wind, PV system and wind turbines cannot generate electricity. If it is possible to design a mixed system (i.e., solar PV, wind, and diesel generator), the demand for electricity can be met in an environmentally friendly manner.

REFERENCES

[1] Islam MT, Shahir SA, Uddin TMI, Saifullah AZA. Current energy scenario and future prospect of renewable energy in Bangladesh. Renewable and Sustainable Energy Reviews 2014;39:1074—88.
[2] Enegy Information Administration [cited 2016 March 07], https://www.eia.gov/totalenergy/data/monthly/pdf/flow/css_2014_energy.pdf.
[3] Rahman MM, Canter C, Kumar A. Well-to-wheel life cycle assessment of transportation fuels derived from different North American conventional crudes. Applied Energy 2015;156:159—73.
[4] The shift project data portal [cited 2016 March 07], http://www.tsp-data-portal.org/Breakdown-of-Electricity-Generation-by-Energy-Source#tspQvChart.
[5] ISO 14040: Environmental management—life cycle assessment—principles and framework. 2006.
[6] Peng J, Lu L, Yang H. Review on life cycle assessment of energy payback and greenhouse gas emission of solar photovoltaic systems. Renewable and Sustainable Energy Reviews 2013;19:255—74.
[7] Sherwani AF, Usmani JA, Varun. Life cycle assessment of solar PV based electricity generation systems: a review. Renewable and Sustainable Energy Reviews 2010;14:540—4.
[8] Wilson R, Young A. The embodied energy payback period of photovoltaic installations applied to buildings in the UK. Building and Environment 1996;31:299—305.
[9] Kato K, Murata A, Sakuta K. Energy pay-back time and life-cycle CO_2 emission of residential PV power system with silicon PV module. Progress in Photovoltaics: Research and Applications 1998;6:105—15.

[10] Alsema E, de Wild-Scholten M. The real environmental impacts of crystalline silicon PV modules: an analysis based on up-to-date manufacturers data. In: Presented at the 20th European photovoltaic solar energy conference; 2005. p. 10.

[11] Alsema E, de Wild-Scholten M, Fthenakis V. Environmental impacts of PV electricity generation-a critical comparison of energy supply options. In: 21st European photovoltaic solar energy conference, Dresden, Germany; 2006.

[12] Jungbluth N, Dones R, Frischknecht R. Life cycle assessment of photovoltaics; update of the ecoinvent database. MRS Proceedings: Cambridge Univ Press; 2007. p. 1041-R01-R03.

[13] Phylipsen G, Alsema E. Environmental life-cycle assessment of multicrystalline silicon solar cell modules: department of Science, Technology and Society. Utrecht University; 1995.

[14] Alsema E. Energy pay-back time and CO_2 emissions of PV systems. Progress in Photovoltaics: Research and Applications 2000;8:17—25.

[15] Ito M, Kato K, Sugihara H, Kichimi T, Song J, Kurokawa K. A preliminary study on potential for very large-scale photovoltaic power generation (VLS-PV) system in the Gobi desert from economic and environmental viewpoints. Solar Energy Materials and Solar Cells 2003;75:507—17.

[16] Battisti R, Corrado A. Evaluation of technical improvements of photovoltaic systems through life cycle assessment methodology. Energy 2005;30:952—67.

[17] Pacca S, Sivaraman D, Keoleian GA. Parameters affecting the life cycle performance of PV technologies and systems. Energy Policy 2007;35:3316—26.

[18] Lewis GM, Keoleian GA. Life cycle design of amorphous silicon photovoltaic modules. US Environmental Protection Agency, National Risk Management Research Laboratory; 1997.

[19] Alsema E. Energy requirements of thin-film solar cell modules—a review. Renewable and Sustainable Energy Reviews 1998;2:387—415.

[20] Kato K, Hibino T, Komoto K, Ihara S, Yamamoto S, Fujihara H. A life-cycle analysis on thin-film CdS/CdTe PV modules. Solar Energy Materials and Solar Cells 2001;67:279—87.

[21] Fthenakis V, Kim HC. Energy use and greenhouse gas emissions in the life cycle of CdTe photovoltaics. MRS Proceedings: Cambridge Univ Press; 2005. p. 0895-G03-G06.

[22] Raugei M, Bargigli S, Ulgiati S. Life cycle assessment and energy pay-back time of advanced photovoltaic modules: CdTe and CIS compared to poly-Si. Energy 2007;32:1310—8.

[23] Fthenakis V, Kim H, Held M, Raugei M, Krones J. Update of PV energy payback times and life-cycle greenhouse gas emissions. In: 24th European photovoltaic solar energy conference, 21—29 September 2009, Hamburg, Germany; 2009.

[24] Held M, Ilg R. Update of environmental indicators and energy payback time of CdTe PV systems in Europe. Progress in Photovoltaics: Research and Applications 2011;19:614—26.

[25] Ito M, Kato K, Komoto K, Kichimi T, Kurokawa K. A comparative study on cost and life-cycle analysis for 100 MW very large-scale PV (VLS-PV) systems in deserts using m-Si, a-Si, CdTe, and CIS modules. Progress in Photovoltaics: Research and Applications 2008;16:17—30.

[26] Ito M, Komoto K, Kurokawa K. Life-cycle analyses of very-large scale PV systems using six types of PV modules. Current Applied Physics 2010;10:S271-S3.

[27] Jungbluth N. Life cycle assessment of crystalline photovoltaics in the Swiss ecoinvent database. Progress in Photovoltaics: Research and Applications 2005;13:429—46.

[28] García-Valverde R, Miguel C, Martínez-Béjar R, Urbina A. Life cycle assessment study of a 4.2 kW p stand-alone photovoltaic system. Solar Energy 2009;83:1434—45.

[29] Kannan R, Leong K, Osman R, Ho H, Tso C. Life cycle assessment study of solar PV systems: an example of a 2.7 kW p distributed solar PV system in Singapore. Solar Energy 2006;80:555—63.

[30] Sumper A, Robledo-García M, Villafáfila-Robles R, Bergas-Jané J, Andrés-Peiró J. Life-cycle assessment of a photovoltaic system in Catalonia (Spain). Renewable and Sustainable Energy Reviews 2011;15:3888—96.

[31] Tremeac B, Meunier F. Life cycle analysis of 4.5 MW and 250 W wind turbines. Renewable and Sustainable Energy Reviews 2009;13:2104—10.

[32] Guezuraga B, Zauner R, Pölz W. Life cycle assessment of two different 2 MW class wind turbines. Renewable Energy 2012;37:37—44.

[33] Kabir MR, Rooke B, Dassanayake GM, Fleck BA. Comparative life cycle energy, emission, and economic analysis of 100 kW nameplate wind power generation. Renewable Energy 2012;37:133—41.

[34] Krohn S. The energy balance of modern wind turbines. Wind Power Note 1997;16:1—16.

[35] Uchiyama Y. Life cycle analysis of photovoltaic cell and wind power plants. Assessment of greenhouse gas emissions from the full energy chain of solar and wind power and other energy sources. Working Material Vienna (Austria): IAEA; 1996.

[36] Lenzen M, Wachsmann U. Wind turbines in Brazil and Germany: an example of geographical variability in life-cycle assessment. Applied Energy 2004;77:119—30.

[37] Gürzenich D, Mathur J, Bansal NK, Wagner H-J. Cumulative energy demand for selected renewable energy technologies. The International Journal of Life Cycle Assessment 1999;4:143—9.

[38] Proops JL, Gay PW, Speck S, Schröder T. The lifetime pollution implications of various types of electricity generation. An input-output analysis. Energy Policy 1996;24:229—37.

[39] Nomura N, Inaba A, Tonooka Y, Akai M. Life-cycle emission of oxidic gases from power-generation systems. Applied Energy 2001;68:215—27.

[40] Hondo H. Life cycle GHG emission analysis of power generation systems: Japanese case. Energy 2005;30:2042—56.

[41] Haack BN. Net energy analysis of small wind energy conversion systems. Applied Energy 1981;9:193—200.

[42] Schleisner L. Life cycle assessment of a wind farm and related externalities. Renewable Energy 2000;20:279—88.

[43] Jungbluth N, Bauer C, Dones R, Frischknecht R. Life cycle assessment for emerging technologies: case studies for photovoltaic and wind power (11 pp). The International Journal of Life Cycle Assessment 2005;10:24—34.

[44] Sorda G, Banse M, Kemfert C. An overview of biofuel policies across the world. Energy Policy 2010;38:6977—88.

[45] Budzianowski WM. Can 'negative net CO_2 emissions' from decarbonised biogas-to-electricity contribute to solving Poland's carbon capture and sequestration dilemmas? Energy 2011;36:6318—25.

[46] De Oliveira MED, Vaughan BE, Rykiel EJ. Ethanol as fuel: energy, carbon dioxide balances, and ecological footprint. BioScience 2005;55:593—602.

[47] Börjesson P. Good or bad bioethanol from a greenhouse gas perspective — what determines this? Applied Energy 2009;86:589—94.

[48] Nguyen TLT, Gheewala SH, Garivait S. Full chain energy analysis of fuel ethanol from cane molasses in Thailand. Applied Energy 2008;85:722—34.

[49] Larson EA. Review of LCA studies on liquid biofuel systems for the transport sector. In: GEF. STAP workshop on liquid biofuels for the transport sector New Delhi, India; 2005.

[50] Kim S, Dale BE. Allocation procedure in ethanol production system from corn grain I. system expansion. The International Journal of Life Cycle Assessment 2002;7:237—43.

[51] Von Blottnitz H, Curran MA. A review of assessments conducted on bio-ethanol as a transportation fuel from a net energy, greenhouse gas, and environmental life cycle perspective. Journal of Cleaner Production 2007;15:607—19.

[52] Achten WM, Akinnifesi FK, Maes W, Trabucco A, Aerts R, Mathijs E, et al. Jatropha integrated agroforestry systems: biodiesel pathways towards sustainable rural development. In: Jatropha curcas as a premier biofuel: cost, growing and management; 2010. p. 85—102.

[53] Zah R, Böni H, Gauch M, Hischier R, Lehmann M, Wäger P. Life cycle assessment of energy products: environmental impact assessment of biofuels (for the Federal Office for Energy BFE, the Federal Office for the Environment BFE and the Federal Office for Agriculture BLW). 2007.

[54] Lal R. World crop residues production and implications of its use as a biofuel. Environment International 2005;31:575—84.

[55] Hoogwijk M, Faaij A, van den Broek R, Berndes G, Gielen D, Turkenburg W. Exploration of the ranges of the global potential of biomass for energy. Biomass and Bioenergy 2003;25:119—33.

[56] Spatari S, Zhang Y, MacLean HL. Life cycle assessment of switchgrass-and corn stover-derived ethanol-fueled automobiles. Environmental Science & Technology 2005;39:9750—8.

[57] Cherubini F, Strømman AH. Life cycle assessment of bioenergy systems: state of the art and future challenges. Bioresource Technology 2011;102:437−51.

[58] Luo L, van der Voet E, Huppes G, De Haes HAU. Allocation issues in LCA methodology: a case study of corn stover-based fuel ethanol. The International Journal of Life Cycle Assessment 2009;14:529−39.

[59] Venturi P, Venturi G. Analysis of energy comparison for crops in European agricultural systems. Biomass and Bioenergy 2003;25:235−55.

[60] Wang X, Chen Y, Sui P, Gao W, Qin F, Wu X, et al. Efficiency and sustainability analysis of biogas and electricity production from a large-scale biogas project in China: an emergy evaluation based on LCA. Journal of Cleaner Production 2014;65:234−45.

[61] Schlamadinger B, Edwards R, Byrne K, Cowie A, Faaij A, Green C, et al. Optimizing the greenhouse gas benefits of bioenergy systems. In: Proceedings of the 14th European biomass conference; 2005. p. 17−21.

[62] Corti A, Lombardi L. Biomass integrated gasification combined cycle with reduced CO_2 emissions: performance analysis and life cycle assessment (LCA). Energy 2004;29:2109−24.

[63] Wiloso EI, Heijungs R, de Snoo GR. LCA of second generation bioethanol: a review and some issues to be resolved for good LCA practice. Renewable and Sustainable Energy Reviews 2012;16:5295−308.

[64] Azapagic A, Clift R. Allocation of environmental burdens in co-product systems: product-related burdens (Part 1). The International Journal of Life Cycle Assessment 1999;4:357−69.

[65] Punter G, Rickeard D, Larivé J, Edwards R, Mortimer N, Horne R, et al. Well-to-wheel evaluation for production of ethanol from wheat. Report by the LowCVP Fuels Working Group, WTW Sub-Group; 2004.

[66] Singh A, Pant D, Korres NE, Nizami A-S, Prasad S, Murphy JD. Key issues in life cycle assessment of ethanol production from lignocellulosic biomass: challenges and perspectives. Bioresource Technology 2010;101:5003−12.

[67] Achten WM, Maes W, Aerts R, Verchot L, Trabucco A, Mathijs E, et al. Jatropha: from global hype to local opportunity. Journal of Arid Environments 2010;74:164−5.

[68] Davis SC, Anderson-Teixeira KJ, DeLucia EH. Life-cycle analysis and the ecology of biofuels. Trends in Plant Science 2009;14:140−6.

[69] Stehfest E, Bouwman L. N2O and NO emission from agricultural fields and soils under natural vegetation: summarizing available measurement data and modeling of global annual emissions. Nutrient Cycling in Agroecosystems 2006;74:207−28.

[70] Wrage N, Van Groenigen J, Oenema O, Baggs E. A novel dual-isotope labelling method for distinguishing between soil sources of N_2O. Rapid Communications in Mass Spectrometry 2005;19:3298−306.

[71] Fritsche UR, Hennenberg K. The "iLUC Factor" as a means to hedge risks of GHG emissions from indirect land-use change associated with bioenergy feedstock provision. In: Background paper for the EEA expert meeting in Copenhagen, 2008.

[72] Requena JS, Guimaraes AC, Alpera SQ, Gangas ER, Hernandez-Navarro S, Gracia LN, et al. Life Cycle Assessment (LCA) of the biofuel production process from sunflower oil, rapeseed oil and soybean oil. Fuel Processing Technology 2011;92:190−9.

[73] A Biogas Road Map for Europe. European Biomass Association 2010, https://www.americanbiogas council.org/pdf/euBiogasRoadmap.pdf.

[74] Rutz D, Janssen R. Biofuel technology handbook. Munich (Germany): WIP Renewable Energies; 2007.

[75] Ishikawa S, Hoshiba S, Hinata T, Hishinuma T, Morita S. Evaluation of a biogas plant from life cycle assessment (LCA). International Congress Series. Elsevier; 2006. p. 230−3.

[76] International Energy Agency. Technology roadmap: hydropower, https://www.iea.org/publications/ freepublications/publication/2012_Hydropower_Roadmap.pdf.

[77] Pascale A, Urmee T, Moore A. Life cycle assessment of a community hydroelectric power system in rural Thailand. Renewable Energy 2011;36:2799−808.

[78] Gallagher J, Styles D, McNabola A, Williams AP. Current and future environmental balance of small-scale run-of-river hydropower. Environmental Science & Technology 2015;49:6344—51.

[79] Suwanit W, Gheewala SH. Life cycle assessment of mini-hydropower plants in Thailand. The International Journal of Life Cycle Assessment 2011;16:849—58.

[80] Zhang S, Pang B, Zhang Z. Carbon footprint analysis of two different types of hydropower schemes: comparing earth-rockfill dams and concrete gravity dams using hybrid life cycle assessment. Journal of Cleaner Production 2015;103:854—62.

[81] Varun PR, Bhat IK. Life cycle greenhouse gas emissions estimation for small hydropower schemes in India. Energy 2012;44:498—508.

[82] Hertwich EG. Addressing biogenic greenhouse gas emissions from hydropower in LCA. Environmental Science & Technology 2013;47:9604—11.

[83] International Energy Agency. Technology roadmap: geothermal heat and power. 2011. http://www.iea.org/publications/freepublications/publication/Geothermal_roadmap.pdf.

[84] Bravi M, Basosi R. Environmental impact of electricity from selected geothermal power plants in Italy. Journal of Cleaner Production 2014;66:301—8.

[85] Bayer P, Rybach L, Blum P, Brauchler R. Review on life cycle environmental effects of geothermal power generation. Renewable and Sustainable Energy Reviews 2013;26:446—63.

[86] Ármannsson H, Fridriksson T, Kristjánsson BR. CO_2 emissions from geothermal power plants and natural geothermal activity in Iceland. Geothermics 2005;34:286—96.

[87] Bertani R, Thain I. Geothermal power generating plant CO_2 emission survey. IGA News 2002;49:1—3.

[88] Fridleifsson IB. Geothermal energy for the benefit of the people. Renewable and Sustainable Energy Reviews 2001;5:299—312.

[89] DiPippo R. Geothermal power plants: principles, applications, case studies and environmental impact. Butterworth-Heinemann; 2012.

[90] Kagel A, Bates D, Gawell K. A guide to geothermal energy and the environment. Washington, DC: Geothermal Energy Association; 2005.

[91] Bloomfield K, Moore JN, Neilson RM. Geothermal energy reduces greenhouse gases. Bulletin Geothermal Resources Council 2003;32:77—9.

[92] Rule BM, Worth ZJ, Boyle CA. Comparison of life cycle carbon dioxide emissions and embodied energy in four renewable electricity generation technologies in New Zealand. Environmental Science & Technology 2009;43:6406—13.

[93] US DOE 2012 [cited 2016 March 07], http://energy.gov/eere/geothermal/geothermal-energy-us-department-energy.

[94] Karlsdottir MR, Palsson OP, Palsson H. LCA of combined heat and power production at hellisheiði geothermal power plant with focus on primary energy efficiency. Power 2010;2:16.

[95] Turconi R, Boldrin A, Astrup T. Life cycle assessment (LCA) of electricity generation technologies: overview, comparability and limitations. Renewable and Sustainable Energy Reviews 2013;28:555—65.

CHAPTER THREE

Clean and Sustainable Energy Technologies

M. Arshad

University of Veterinary and Animal Sciences Lahore, Lahore, Pakistan

3.1 INTRODUCTION

Consistent and secure energy resources are mandatory for our mobility, prosperity, and daily comfort in modern way of life. Current energy means have been divided into three broad classes: the first is derived from fossil fuels, the second is all the renewable resources, and the third one is energy taken from nuclear resource [1]. The world's energy future is anticipating renewable energy (RE) resources for the reason that optimum implications of such resources curtail environmental impacts and generate lesser wastes [2]. The energy sources that may be used as RE sources are solar, wind, biomass, and hydro energy sources [3]. Around the globe, renewable resources are frequently available naturally. As of mid-2016, about 14% of world's energy requirement is being met from these resources [4].

The RE resources presented in Table 3.1 emit fewer pollutants as compared to fossil fuels obeying the principles of sustainability.

The various RE policies, lessening the cost of numerous RE technologies, fluctuation in the fossil fuel prices, and rising energy demands have fortified the ongoing intensification in the use of RE (Table 3.2).

Generation of power through hydropower stations, various modern biomass opportunities, photovoltaic (PV) system, and wind turbines will upsurge in future and will increase the share of these technologies in hybrid systems that combine multiple technologies.

Table 3.1 Key Renewable Energy Resources and Their Usage Forms [5]

Energy Source	Energy Conversion and Usage Options
Hydropower	Power generation
Modern biomass	Heat and power generation, pyrolysis, gasification, digestion
Solar	Solar home system, solar dryers, solar cookers, direct solar photovoltaics, thermal power generation, water heaters
Wind	Power generation, wind generators, windmills, water pumps

Clean Energy for Sustainable Development
ISBN 978-0-12-805423-9, http://dx.doi.org/10.1016/B978-0-12-805423-9.00003-X

Table 3.2 Current and Projected Global Renewable Energy Usage by Category [6]

	2010	2020	2035
Bioenergy	0331	0696	1487
Hydro	3431	4513	5677
Wind	0342	1272	2681
Solar PV	0032	0332	0846
Concentrating solar power	2	50	278
Share of total production	10%	12%	14%

3.2 BIOMASS

This chapter evaluates biomass as a substitute source of fossil fuels for energy supply. Since human's dawn, biomass has fulfilled the world's energy needs and has provided fuel [7]. The industrialization had taken off as the consumption of fossil fuels started [8], and now their utilization has reached quickly at the top.

The term biomass is applied to biological materials originated from plant life together with algae derived through photosynthesis. Carbohydrates are produced as initial building blocks from the photosynthetic process occurring among CO_2, water, and solar rays [9]. Usually biomass is reaped to use it as feed, food, fiber, and as structural materials [10]. The remaining is left in the growth zones for natural decay and later on may be well used as fossil fuels. On the other hand, with the help of novel techniques, biomass and other wastes may be transformed into useful synthetic fuels [11].

3.2.1 Classification of Biomass Materials

European Commission categorized a number of biomass resources into products and byproducts with remnants from crop growing (agriculture), forestry, and linked industries, in addition to the decomposable portions of agricultural industries and urbanite waste [11].

Uses and purposes of the biomass resources are usually basics of their classification [12]. Table 3.3 gives a thorough classification of biomass resources and examples of each kind.

According to different varieties, biomass is grouped into four main types [13]:

* woody plants,
* herbaceous plants and grasses,
* aquatic plants,
* manures.

3.2.2 Processing of Biomass

Based on the processing techniques, biomass can be more categorized into those having high-moisture ratios and the ones with low-moisture content. Most of the commercial

Table 3.3 Sorting of Biomass Resources Presenting Their Commencement

Mode of Life	Plant/Animal Source	Class of Biomaterials	Major Representative
Terrestrial	Plants	Carbohydrate	Sugar cane, corn, sweet sorghum
		Starch	Maize, cassava, sweet potato
		Cellulosic materials	Tropical grasses, poplar, sycamore
			Forestry
		Hydrocarbon	Eucalyptus, green coral
		Lipids	Oil palm rapeseed sunflower
		Cellulose	Wheat bran, straw
			Vegetable residues, processing residues
			Farm residues
			Secondary forest
			Woodland remnants
			Crippled material in plants
	Fisheries/animal husbandry	Proteinaceous	Jettisoned and dead fish
		Organic matter	Animal manure
		Proteinaceous	Animal slaughtering waste
	Humans	Organic matter	Municipal and pulp sludge
		Organic matter	Family garbage, feces
Aquatic	Fresh water	Cellulose	Water hyacinth
	Ocean	Cellulose	Large kelp
	Microorganism	Cellulose, lipids, carbohydrates	Green algae, photosynthetic bacteria

research achievements have focused the lesser moisture—containing plants, such as woody plants and herbaceous species.

Wet processing techniques based on biochemical processes, such as fermentation, are much appropriate for aquatic materials and manures that naturally have elevated levels of moisture. Techniques such as gasification, pyrolysis, or combustion are more economically right to dry biomass such as wood chips. Wet handling techniques are employed where moisture contents of the materials are so high that the energy required for drying would be extremely high as compared to the energy content of the product formed. Other than moisture contents, ash, alkali, and trace component contents are considerable factors in consideration of suitable processing technique (Table 3.4).

Lot of work has been performed to understand the methane fermentation process to explore the biochemistry and microbiology of the organisms involved. Now the biomass conversion into fuels has been advanced. Complex biomolecules of the biomass are decomposed to lower molecular weight molecules, which are further transformed into methane and CO_2. If the fermentable biomass is frequently available, the anaerobic digestion process can be operated on a large scale for a long period just keeping the important fermentation parameters within acceptable range. Other than lignin and keratins that have low biodegradability, nearly all types of biomass can be processed.

Table 3.4 Major Pathways of Biomass Processing Showing Key Issues and Advantages With Current Advances [14—42]

Processing Pathways	Key Issues	Major Advantages With Current Improvements
Gasification	Production of tar is problematic [14].	According to Ref. [15], adsorption as well as catalytic transformation engagement with char-based adsorbents/catalysts can eliminate tar successfully. The ignition engine has been designed that can be driven on impure syngas and can tolerate tar issues [16]. Integrated gasification with gas cleaning and conditioning was proposed as a better option [17].
Pyrolysis	High content of O_2 and H_2O being there lowers the quality [18]. Oil obtained is below par for direct blending with fossil fuels [19]. Chemistry of the product is yet to be explored [20].	Rapid pyrolysis at high temperatures 300—500°C in the presence of catalyst can result into fuels, which have the oxygen removed [21]. Biomass torrefaction was upgraded to reduce functional groups having oxygen [22].
Hydrothermal liquefaction	In effect, solvents with suitable catalysts to reduce the number of products is still to be searched. Biomass-processing cost through liquefaction is very high [23]. As water is the processing medium, large amount of water is required [24].	About 80% energy from biomass is recovered to fuel through hydrothermal liquefaction, which is excellent as compared to other biomass-processing pathways [25]. The product almost resembles petroleum crude other than high nitrogen and oxygen ratios. Homogenous alkaline catalysts in solution form can reduce the nitrogen and oxygen ratios in the final product [26—28].
Enzymatic hydrolysis	Enzymes required for pretreatment are very costly [29].	Biomass hydrolysis is performed at pH 4.8 and 45—50°C temperature, then the applicable enzymes cost can be reduced [30]. Applications of genetically reconstructed microbes that can produce ethanol from xylose and other pentose directly. Optimization of enzyme application can significantly improve ethanol production efficiency reducing the production cost [31].

Method		
Dilute acid hydrolysis (DAH)	Retrieval of sugar is low. Production of furfural and related compounds inhibits the fermentation of sugars to ethanol [32].	Decreasing the feedstock size can improve the recovery of sugar and pretreatment cost [33].
Concentrated acid hydrolysis (CAH)	Recycling of acid is a difficult task [34]. Corrosion problems. Ca[OH]$_2$ is added to counteract the acid, so calcium sulfate originates [35]. Its disposal is an additional task.	Major benefits of concentrated acid hydrolysis are better sugar recovery, minimum concentration of inhibitors [32].
Ionic liquids (ILs)	To recover the ionic liquids is a difficult task as both sugars and ionic liquids have comparable solubility [36]. Further, ionic liquids inhibit the fermentation process also [37]. Another issue with ionic liquids is their high cost [38].	Most of the ionic liquids are environment friendly [38]. Having H^+ and hydrogen sulfate $[HSO_4]^-$ anion are comparable in effectiveness and process cost with other pretreatment methods.
Mechanical extraction	Extra heating with high temperature ends in a low nutritional value cake and lower quality oil.	Simple process. High-skilled supervisors are not required. Provides high protein cake [39].
Chemical extraction	The operative design must be improved considering the mass transfer kinetics [40].	The process is environment and human friendly. Properties of CO_2 may be adjusted to improve selectivity [41].
Transesterification of vegetable oils	High viscosity issues. Low heating values. Commercialization can be done, but cost is very high [42].	It is decomposable, recyclable, and nontoxic.

3.2.3 Conclusion

Energy generation from biomass is the global move to reduce the environmental impact of fossil fuel. Energy produced from nonfood feed sources can practically substitute the power generated from fossil fuels. The use of indigenous sources can also increase energy security with the mitigation of global warming. During 2000s, the energy generation from biomass has demonstrated a fast growth, as many countries look after it well.

Energy yielded through biochemical and thermochemical routes is suitable to fulfill the current needs. The thermochemical processes transform biomass into useable energy within less time as compared to biochemical methods that proceed up to many hours for transformation. Now research has been focused on modeling transformational pathway for their optimization to increase the performance and decrease the production cost. Eventually, an approach considering all issues is expected for better generation of bioenergy utilizing the indigenous biomass.

3.3 SOLAR POWER

There have been continuous efforts to explore alternative ways to replace the fossil fuel and to meet the globally cumulated need for energy due to rapidly increasing population and intensifying demand from developing countries. The challenge has to be replied with a low-cost solution employing raw materials available in abundance. Clearly the sun is the ultimate focal point for unpolluted and inexpensive energy, exploited by nature to support virtually whole life on earth; it can offer a fully developed solution for the energy crisis [43]. Therefore, solar cells can be taken as a major RE resource once their production cost is reduced to a reasonable level, similar to other available energy resources. Accordingly, fixing the energy from the solar system with PV equipment seems to be a sensible huge scales response to the current energy issue [44].

Various methods are available to harness the energy of solar radiation from the sun. Active solar heating, passive solar heating, and solar engines for electricity generation are included. For small-scale heating, such as at the domestic level, active solar energy system is utilized that can reduce electrical consumption [45].

Passive efficiency of housing and other buildings can be improved through passive solar heating systems. In this technique, the equipment that can consume the energy of solar radiation to heat a building is employed. It may take in conservatory, Trombe wall, and direct gain−type applications [46].

Solar heat engines are meant for electricity generation. By and large, reflective glasses are fixed to direct the solar radiations over a water source or some other fluid, steam is generated through evaporation. The steam is employed to run a turbine for power generation [47].

3.3.1 Application and Advantages of Solar Energy

Production of electricity using solar energy to replace fossil fuels has been increased globally as it is clearly environment friendly as compared to all the other energy sources. The natural resources are not used up; neither CO_2 nor other gaseous and solid waste products are released [48].

Following are the major advantages:

- zero greenhouse gases (GHGs; CO_2, NO_x) discharge;
- no release of toxic gases (SO_2, particulates);
- reparation of barren land;
- reduced cost of transmission lines from electricity grids;
- security of energy supply and diversification with national energy independence;
- speeding up of rural electrification.

Due to increased global climate change, pressure of intensifying energy consumption rate, and international arrangements to diminish the GHGs release, it is being thought that how to access solar energy. For this, governments worldwide are launching their national objectives for the provision of electricity from RE and are hence trying to set up the various solar energy policies in different countries [49].

Moreover, solar energy—based electricity generation is getting pace in every corner of the world. Solar electricity is normally generated from two methods: first is PV and the other is concentrated solar power (CSP).

3.3.2 Solar Photovoltaics

Solar PV units are solid-state semiconductor equipment combined of many elements, such as cells; mechanical; and electrical mountings, having the ability to transform solar energy into electricity [50].

When photons of the solar light smash the cells surface, these are absorbed and pair of electrons and holes is generated. The generated electrons and holes rush toward the n-type side and p-type side. As the two sides of the PV cell are attached through its load, an electric current is produced and it flows as long as solar system is available to hit the cell. PV power generation systems are built on batteries, inverters, chargers, discharge controllers, and solar tracking control systems, other than solar cells. Constituents of PV sheets are monocrystalline silicon, polycrystalline silicon, microcrystalline silicon, copper indium selenide, and cadmium telluride [51]. The C—Si technology is globally getting almost 87% of the total PV sales in the year 2010 [52]. Leading producer in PV cell is China, whereas European countries are leaders by the installation capacities of PV power outputs of 39 GW by the end of 2011 [53]. Although CSP plants keep high capacity to add for future energy needs, in the year 2012, about 98% of solar plants installed were based on PV systems [54]. It is the world's rapid-growing energy technology, as PV production has doubled every 2 years, since 2002 [55].

Up-to-date solar PV systems possess abilities to generate $10-60$ MW and can now be functional up to 10 years at 90% and for up to 25 years at 80% of its rated power capacity [56]. The leading PV manufacturers include First Solar, Suntech Power, Sharp, Q-Cells, Yingly Green Energy, JA Solar, Kyosera, Trina Solar, Sunpower, and Gintech [57].

3.3.3 Solar Thermal Application

Thermal solar energy is the most commonly available source that can be utilized for cooking, water heating, crop drying, and so on [58]. Solar cooking is the utmost direct and useful application of energy from the sun [59,60,61].

Benefits and disadvantages of solar ovens were compared with traditional firewood and electric stoves [62]. The payback period of a common hot box—type solar oven, even if used $6-8$ months a year, is around $12-14$ months, about 16.8 million tons of firewood can be saved and the emission of 38.4 million tons of CO_2 per year can also be prevented.

According to Ref. [63], solar water heating system of 100 L/day volume installed at home can alleviate around 1237 kg of CO_2 emissions in a year. Solar-drying technology offers an alternative, which can process the vegetables and fruits in clean, hygienic, and sanitary conditions with zero energy costs. It saves energy and time, occupies less area, and improves the product quality of heliostat field collectors [64].

3.3.4 Concentrated Solar Power

In CSP, solar radiations are concentrated to generate steam to drive a conventional turbine or engine for the production of electricity. The major difference from solar PV is that the heat may be kept, commonly through using molten salts or oil as the liquid medium in the solar receiver, and electricity can be generated outside of solar light hours [65]. Furthermore, solar thermal technology offers the ability to match increased supply during periods of intense summer radiations with peak demand associated with space cooling requirements.

Solar thermal technology is commonly used for hot water systems. Solar thermal electricity, also known as concentrating solar power, is typically designed for large-scale power generation. Solar thermal technologies can also operate in hybrid systems with fossil fuel power plants, and, with appropriate storage, have the potential to provide base load electricity generation. Solar thermal technologies can also potentially provide electricity to remote townships and mining centers where the cost of alternative electricity sources is high [66,67].

3.3.5 Conclusion

As fossil fuel supplies are expected to be less available, more expensive, and of increasing environmental concern in the coming century, increasing dependence on energy conservation and alternative energy sources is expected. The most obvious alternative energy source is the sun.

The solar-based energy-generating system is rapidly growing worldwide. To keep its growth up, fresh improvements in material utilization, alternate designs, and consistency of production technologies are highly needed. Key attraction for international funding to sponsor PV energy systems is its competence to keep up a clean energy source. It can also help to improve basic living standard. Slowly but surely, solar energy system is being employed in programs that develop education, water supply, and healthcare.

3.4 WIND POWER

Wind power is the second largest, developed, and commercially utilized RE technology applied for electricity generation, which is achieved on an average annual growth of 28% during the period 2001−11 [68,69], and its average installed capacity has doubled every 3 years.

Wind energy can be transformed into convenient forms: through wind turbines to generate electricity, by wind mills for mechanical power and wind pumps can pump water or drainage, or sails to propel ships [70]. Humans have been using wind power since almost 3000 years ago, but up to the early 20th century, it was just used to provide mechanical power to pump water or to grind grain. Fossil fuels replaced wind energy at the start of the industrialization era [71]. Electricity is produced from the wind through utilization of the kinetic energy that the air possesses. The kinetic energy of the air is firstly transformed to mechanical energy and then to electrical energy. The challenge for the modern industry is to design cost effective wind turbines and power plants to do these energy-form transformations. Available kinetic energy in the wind can be extracted up to 40−50% only. Therefore design of the wind turbines must be improved to maximize the energy captured. Since mid-1960s different onshore wind turbine configurations with horizontal and vertical axes have been investigated. The horizontal axis design came to dominate with time, even though configurations varied, especially the number of blades and blades' orientation. Wind power plants, sometimes named as wind farms (5−300 MW in size), are created by installing together the many wind turbines [72,73].

In 2007, wind-generated electricity fulfilled above 1% of the global demand [74]. As the growth continued in 2008, a further 27 GW of capacity was commissioned [75]. It has been predicted that installed capacity will increase fivefold over the next 10-year period [76].

Now the wind energy technology is mature enough and marketable as the price of wind power is generally reasonable compared to other types of power generation. Emission avoided by this technology ranges from 391 to 828 g of CO_2/kWh [77].

It is worth mentioning that almost 80% of the worldwide wind capacity is installed in Germany, USA, Denmark, India, and Spain. Hence, most of the knowledge and experience of wind energy recline in these five countries only.

3.4.1 World Wind Energy Scenario

Potential of onshore wind power is very high, that is, $20,000 \times 10^9$ to $50,000 \times 10^9$ kWh per annum as compared to current total world electricity consumption of $15,000 \times 10^9$ kWh. The economic potential depends upon factors such as average wind speed, statistical wind speed distribution, turbulence intensities, and the cost of wind turbine systems. The aggregate global wind energy size has been grown to 46,048 MW.

The five major countries with the highest total installed wind power capacity are Germany; 16,500 MW, Spain; 8000 MW, the United States; 6800 MW, Denmark; 3121 MW, and India; 2800 MW, nearly 80% of total wind energy installed worldwide. Other countries, such as Italy, the Netherlands, Japan, and the United Kingdom, are above or near the 1000 MW mark.

3.4.2 Problems Associated With Wind Turbines

Wind turbine components are subjected to various problems. Some methods used for reducing failure of wind turbine components has been reviewed in this chapter.

References	Suggestions
[78]	Discussed the fatigue issue and their remedy.
[79]	Stated that the fatigue-specific failure mechanism depends on material or structural defect.
[80]	Designed a new analytical model against corrosion fatigue.
[81]	Studied the structural dynamic characteristics of rotor blades to avoid sympathetic vibration problem.
[82]	Proposed a model to avoid ice deposits on wind turbines.
[83]	Discussed the environmental impact of wind power system.
[84]	Applied multilayered metallic coating against fatigue cracks.
[85]	Used asbestos-free friction-lining material.
[86]	Discussed the problems faced at wind farms and how to tackle these.
[87]	Given details about downwind turbine noise issues.
[88]	Presented a solution for the uncertainties in the system load.

3.4.3 Conclusion

Advancement in technology has made outstanding developments in designs of wind turbines. Different factors, for instance, choice of site; elevation level; selection of wind generators; speed of wind; and wind power potential, have been well-thought-out for development of model wind turbines. Vibration issue of wind turbines with lifetime prediction of wind turbine blades has been well studied. Now, after this improved technology, wind turbine has been designed for optimum power production at lesser cost, and the wind turbine technology has a bright future globally.

3.5 HYDROPOWER

Hydropower is the energy resulting from tidal energy possessed by flowing water due to height difference and flow speed. Energy possessed by moving water can generate electricity through turbines [89]. It is prophesied that the electricity generation from renewable sources will be shared majorly by hydropower. Hydropower sources provide 90% of RE and above 16% of total electricity globally [90], without emitting GHGs.

First, at the world summit on sustainable development in Johannesburg [91], and for a second time at the third world water forum in Kyoto (2003), the delegates of over 170 nations declared hydropower a RE source unanimously [92].

Hydro-based electricity is now being generated in over 150 countries. Present worldwide hydroelectricity installed capacity is about 970 GW [93]. Global hydropower production remained about 3500 TWh in the year 2011. Nearly 50% of global hydropower is produced in just three countries, United States, China, and Canada, collectively [94].

Hydropower projects can be designed at wide range and in several types to outfit specific requirements of particular site conditions. Hydropower neither consumes nor pollutes the water, to produce electricity, but it lets go this vital source to be accessible for other usages. The incomes made by sales of power can fund other arrangements crucial for humans, such as drinking water supply systems, irrigation structures for agriculture, navigation organization, and tourism. Every form of life on earth needs water. Unluckily, its distribution is uneven; some portions of the world are susceptible to drought, while in others parts, floods are the major cause of loss of lives and property [95].

Water has been always collected and stored in dams and reservoirs, through the history, to meet human needs [96].

3.5.1 Main Attributes of Hydropower As Renewable Energy Source

A major source of renewable energy:

Kinetic energy of moving water is utilized to get hydroelectricity, with no depletion of sources; so all kinds of hydropower ventures, minor or major, run-of-river or storage, fulfill the definition of RE.

Backbone of other renewables energy resources:

Hydropower projects with storage facilities provide an extraordinary operational flexibility that these can better bear out immediate changes in electricity demand.

This ability makes hydro energy very capable and cost-effective technology to support the placement of intermittent RE sources, such as wind and solar power [97].

Energy security and price constancy:

The river water is a local resource, so it is free from world market instabilities [98].

Storage of fresh water:

Lakes for hydroelectricity generation gather the water of rain fall, also serve as a source of drinking and irrigation. Further, aquifers cannot be depleted.

Electric grid stability:

The management of electric grids depends on fast, flexible generation sources to meet peak power demands, maintaining system voltage level, and quickly restoring the service after a blackout [99].

Helpful in climate change scenario:

As the life cycle of hydropower releases minimum GHGs, it can help to slow global warming. Currently, hydroelectricity evades burning of 4.4 million barrels of oil equivalent daily [100].

Improvement of air quality:

No air pollutants are generated and substitute fossil-fired generation, thus decreasing acid rain and smog chances.

Contribution to development:

Hydropower facilities bring electricity, roads, industry, and commerce to communities, thereby developing the economy, improving access to health and education, and enhancing the quality of life.

Clean and affordable energy for present and future:

Easy to upgrade and fit in the latest innovations. Minimum operational and maintenance costs.

A tool for sustainable development:

Hydroelectricity projects are economically viable, environment friendly and socially responsible, and with the ability to serve future generations.

3.5.2 Conclusion

Although the hydropower is a site-specific technology, it is a more concentrated energy resource than others. The energy available is readily predictable and continuously available on demand with no environmental impact. Moreover, it is highly cost-effective, reliable, and environmentally sound means of providing power. Globally, there are many hilly areas of the world where grid electricity will perhaps not reach, but those regions have enough hydropower resources to fulfill local needs. To unfold the potential, it requires significant efforts and resources to be allocated for technology transfer.

3.6 FUTURE PROSPECTS AND CHALLENGES FOR RENEWABLE ENERGY TECHNOLOGIES

RE systems are rapidly growing worldwide. To keep the growth rate up, they need novel improvements in the materials used, better designs, and highly reliable and productive technologies.

Presently, RE production systems market is being run by subsidies and tax exceptions. The key attraction is the competence of RE technologies to favor cleaner energy production sources. Following are the key areas to be addressed for promotion of these clean technologies.

The initial installation cost of RE systems are very much high, so the major challenge faced by such technologies is their costs. To apply RE system at a massive scale, technology must be cost-effective as compared to fossil fuel.

Improvement in manufacturing technologies and reduction of waste products (e.g., in biomass treatment) are required.

Power generation from RE sources (other than hydro) generates power in an intermittent way, so such technologies are not a good choice for a continuous load requirement. Therefore, these must be operated in conjunction with the utility grid or some kind of energy storage in order to achieve the required continuity in power supply.

These energy technologies produce no air or water pollution and do not emit any GHGs, but do have some indirect impacts on the environment.

ACKNOWLEDGMENT

The author is highly thankful to Ms. Sidra Jamil, Lecturer, English (Jhang-Campus) University of Veterinary and Animal Sciences Lahore Pakistan.

REFERENCES

[1] Demirbas A. Recent advances in biomass conversion technologies. Energy Education Science and Technology 2000;6:19—40.
[2] Panwar NL, Kaushik SC, Surendra K. Role of renewable energy sources in environmental protection: a review. Renewable and Sustainable Energy Reviews 2011;15(3):1513—24.
[3] Rathore NS, Panwar NL. Renewable energy sources for sustainable development. New Delhi (India): New India Publishing Agency; 2007.
[4] Sahoo SK. Renewable and sustainable energy reviews solar photovoltaic energy progress in India: a review. Renewable and Sustainable Energy Reviews 2016;59:927—39.
[5] Demirbas A. Ethanol from cellulosic biomass resources. Energy Sources 2005;27(4):327—37.
[6] World Energy Outlook; [cited on 11 May 2016]. Available from: http://www.worldenergyoutlook.org.
[7] Berndesa G, Hoogwijkb M, Broek R. The contribution of biomass in the future global energy supply: a review of 17 studies. Biomass and Bioenergy 2003;25:1—28.
[8] Pandey A, Bhaskar T, Stocker M, Sukumaran RK. In recent advances in thermo-chemical conversion of biomass edition. 2015.
[9] Long H, Li X, Wang H, Jia J. Biomass resources and their bioenergy potential estimation: a review. Renewable and Sustainable Energy Reviews 2013;26:344—52.

[10] Solangi KH, Islam MR, Saidura R, Rahimb NA, Fayazb H. A review on global solar energy policy. Renewable and Sustainable Energy Reviews 2011;15:2149—63.

[11] Clifton J, Boruff BJ. Assessing the potential for concentrated solar power development in rural Australia. Energy Policy 2010;38(9):5272—80.

[12] ABARE. Australian energy statistics: Canberra. August 2009.

[13] McKendry P. Energy production from biomass [part 1]: overview of biomass. Bioresource Technology 2002;83:37—46.

[14] Narvaez I, Orio A, Aznar MP, Corella J. Biomass gasification with air in an atmospheric bubbling fluidized bed. Effect of six operational variables on the quality of the produced raw gas. Industrial Engineering and Chemical Research 1996;35:2110—20.

[15] Shen Y. Chars as carbonaceous adsorbents/catalysts for tar elimination during biomass pyrolysis or gasification. Renewable and Sustainable Energy Reviews 2015;43:281—95.

[16] Bhaduri S, Contino F, Jeanmart H, Breuer E. The effects of biomass syngas composition, moisture, tar loading and operating conditions on the combustion of a tar-tolerant HCCI [Homogeneous Charge Compression Ignition] engine. Energy 2015;87:289—302.

[17] Heidenreich S, Foscolo PU. New concepts in biomass gasification. Progress in Energy Combustion Science 2015;46:72—95.

[18] Sharma A, Pareek V, Zhang D. Biomass pyrolysis—A review of modelling, process parameters and catalytic studies. Renewable and Sustainable Energy Reviews 2015;50:1081—96.

[19] Zacher AH, Olarte MV, Santosa DM, Elliott DC, Jones SB, Elliott, et al. A review and perspective of recent bio-oil hydrotreating research. Green Chemistry 2014;16:491—515.

[20] Mettler MS, Vlachos DG, Dauenhauer PJ. Top ten fundamental challenges of biomass pyrolysis for biofuels. Energy Environment and Science 2012;5:7797—809.

[21] Kantarelis E. Steam pyrolysis of biomass for production of liquid feedstock [doctoral dissertation]. KTH-Royal Institute of Technology; 2014.

[22] Chen D, Zheng Z, Fu K, Zeng Z, Wang J, Lu M. Torrefaction of biomass stalk and its effect on the yield and quality of pyrolysis products. Fuel 2015;159:27—32.

[23] Huang HJ, Yuan XZ. Recent progress in the direct liquefaction of typical biomass. Progress in Energy Combustion Science 2015;49:59—80.

[24] Elliott DC, Biller P, Ross AB, Schmidt AJ, Jones SB. Hydrothermal liquefaction of biomass: developments from batch to continuous process. Bioresource Technology 2015;178:147—56.

[25] Toor SS, Rosendahl L, Rudolf A. Hydrothermal liquefaction of biomass: a review of subcritical water technologies. Energy 2011;36:2328—42.

[26] Shakya R, Whelen J, Adhikari S, Mahadevan R, Neupane S. Effect of temperature and Na2CO3 catalyst on hydrothermal liquefaction of algae. Algal Research 2015;12:80—90.

[27] Guo Y, Yeh T, Song W, Xu D, Wang S. A review of bio-oil production from hydrothermal liquefaction of algae. Renewable and Sustainable Energy Reviews 2015;48:776—90.

[28] Karagoz S, Bhaskar T, Muto A, Sakata Y, Oshiki T, Kishimoto T. Low-temperature catalytic hydrothermal treatment of wood biomass: analysis of liquid products. Chemical Engineering Journal 2005;108:127—37.

[29] Binder JB, Raines RT. Fermentable sugars by chemical hydrolysis of biomass. Proceedings of the National Academy of Sciences 2010;107:4516—21.

[30] Duff SJB, Murray WD. Bioconversion of forest products industry waste cellulosics to fuel ethanol. Bioresource Technology 1996;55:1—33.

[31] Sun Y, Cheng J. Hydrolysis of lignocellulosic materials for ethanol production: a review. Bioresource Technology 2002;83:1—11.

[32] Moe ST, Janga KK, Hertzberg T, Hagg MB, Oyaas K, Dyrset N. Saccharification of lignocellulosic biomass for biofuel and biorefinery applications-A renaissance for the concentrated acid hydrolysis? Energy Procedia 2012;20:50—8.

[33] Badger PC. Ethanol from cellulose: a general review in Trends in new crops and new uses. In Janick J. Whipkey A, edition. Alexandria VA, ASHS Press,, 2002:17—21.

[34] Demirbas A. Ethanol from cellulosic biomass resources. International Journal of Green Energy 2004;1:79—87.

[35] Demirbas A. Progress and recent trends in biofuels. Progress in Energy Combustion 2007;33:1—18.

[36] Sen SM, Binder JB, Raines RT, Maravelias CT. Conversion of biomass to sugars via ionic liquid hydrolysis: process synthesis and economic evaluation. Biofuels Bioproduct & Biorefinery 2012;6:444—52.

[37] Sanderson K. A chewy problem, the inedible parts of plants are feeding the next generation of biofuels. But extracting the energy-containing molecules is a challenging task. Nature 2011; 474:12—4.

[38] George A, Brandt A, Tran K, Zahari SMSNS, Klein-Marcuschamer D, Sun N, et al. Design of low-cost ionic liquids for ligoncellulosic biomass pretreatment. Green Chemistry 2015;17:1728—34.

[39] Singh J, Bargale PC. Development of a small capacity double stage compression screw press for oil expression. Journal of Food Engineering 2000;43:75—82.

[40] Patel RN, Bandyopadhyay S, Ganesh A. A simple model for super critical fluid extraction of bio oils from biomass. Energy Conversion Management 2011;52:652—7.

[41] de Melo MMR, Silvestre AJD, Silva CM. Supercritical fluid extraction of vegetable matrices: applications, trends and future perspectives of a convincing green technology. Journal Supercritical Fluids 2014;92:115—76.

[42] Al-Zuhair S. Production of biodiesel: possibilities and challenges. Bioproducts and Biorefining 2007;1(1):57—66.

[43] Kuwahata R, Monroy CR. Market stimulation of renewable-based power generation in Australia. Renewable and Sustainable Energy Reviews 2011;15:534—43.

[44] Clean Energy Council. Clean energy fact sheets all about solar photovoltaic. Melbourne: Clean Energy Council; 2008.

[45] Currie MJ, Mapel JK, Heidel TD, Goffri S, Baldo MA. High-efficiency organic solar concentrators for photo voltaics. Science 2008;321:226—8.

[46] Geoscience Australia. Renewable energy power stations. 2009. http://www.ga.gov.au/renewable/operating/operating_renewable.xls.

[47] Australian Bureau of Agriculture and Resources Economics, Energy in Australia, Canberra Department of Resources Energy and Tourism: 2009.

[48] Chambouleyron I. Photo voltaics in the developing world. Renewable Energy 1996;21:385—94.

[49] Asim N, Sopian K, Ahmadi S, Saeedfar K, Alghoul MA, Saadatian O, et al. A review on the role of materials science in solar cells. Renewable and Sustainable Energy Reviews 2012;16:5834—47.

[50] Yusaf T, Goh S, Borserio JA. Potential of renewable energy alternatives in Australia. Renewable and Sustainable Energy Reviews 2011;15:2214—21.

[51] Mathews JA. Carbon negative biofuels. Journal of Energy Policy 2007;36:940—5.

[52] Bhutto AW, Bazmi AA, Zahedi G. Greener energy: issues and challenges for Pakistan- solar energy prospective. Renewable and Sustainable Energy Reviews 2012;16:2762—80.

[53] Mekhilef S, Sadur R, Safari A. A review on solar energy use in industries. Renewable and Sustainable Energy Reviews 2011;15:1777—90.

[54] Kalogirou AS. Solar thermal collectors and applications. Progress in Energy Combustion Science 2004;30:231—95.

[55] Kumar A, Kandpal TC. CO_2 emissions mitigation potential of some renewable energy technologies in India. Energy Sources Part A 2007;29(13):1203—14.

[56] Jyotirmay M, Kumar NK, Wagner HJ. Energy and environmental correlation for renewable energy systems in India. Energy Source Part A 2002;24:19—26.

[57] Nandwani SS. Solar cookers cheap technology with high ecological benefits. Ecological Economics 1996;17(2):73—81.

[58] Biermann E, Grupp M, Palmer R. Solar cooker acceptance in South Africa: results of a comparative field-test. Solar Energy 1999;66:401—7.

[59] Tucker M. Can solar cooking save the forests? Ecological Economics 1999;31:77—89.

[60] Wentzel M, Pouris A. The development impact of solar cookers: a review of solar cooking impact research in South Africa. Energy Policy 2007;35:1909—19.

[61] Thirugnanasambandam M, Iniyan S, Goic R. A review of solar thermal technologies. Renewable and Sustainable Energy Reviews 2010;14:312—22.

[62] Parida B, Iniyan S, Goic R. A review of solar photo voltaic technologies. Renewable and Sustainable Energy Reviews 2011;15:1625—36.

[63] Devabhaktuni V, Alam M, Shekara S, Reddy S, Robert D, Green II , et al. Solar energy: trends and enabling technologies. Renewable and Sustainable Energy Reviews 2013;19:555—64.

[64] Kropp R. Solar expected to maintain its status as the world's fastest-growing energy technology. 2009. http://www.socialfunds.com/news/article.cgi/2639.htmls.

[65] Razykov TM, Ferekide CS, Morel D, Stefanakos E, Ullal HS, Upadhyaya HM. Solar photo voltaic electricity: current status and future prospects. Solar Energy 2011;85:1580—608.

[66] Pavlović TM, Radonjić IS, Milosavljević DD, Pantić LS. A review of concentrating solar power plants in the world and their potential use in Serbia. Renewable and Sustainable Energy Reviews 2012;16:3891—902.

[67] Tyagi VV, Rahim NAA, Rahim NA, Selvaraj JAL. Progress in solar PV technology: research and achievement. Renewable and Sustainable Energy Reviews 2013;20:443—61.

[68] Milford E. Record growth for wind: what comes next? World Renewable Energy 2008:37—47.

[69] de Vries E. The DEWI report: wind energy study 2008. World Renewable Energy 2008: 93—101.

[70] Blanco MI. The economics of wind energy. Renewable and Sustainable Energy Reviews 2009;13:1372—82.

[71] European Wind Energy Association. Wind energy. The facts. A guide to the technology, economics and future of wind power. London: Earth Scan; 2009. p. 568.

[72] Kaygusuz K. Wind power for a clean and sustainable energy future. Energy Source. Part B 2009;4:122—32.

[73] Islam MR, Mekhilef S, Said R. Progress and recent trends of wind energy technology. Renewable and Sustainable Energy Reviews 2013;21:456—68.

[74] Eltamaly AM. Design and implementation of wind energy system in Saudi Arabia. Renewable Energy 2013;60:42—52.

[75] Ackermann T, Soder L. Wind energy technology and current status: a review. Renewable and Sustainable Energy Reviews 2000;4:315—74.

[76] de Vries E. 40,000 MW by 2020: building offshore wind in Europe. World Renewable Energy 2008:36—47.

[77] Herbert GMJ, Iniyan S, Rajayapandian S, Sreevalsan E. Parato analysis on various problems of grid connected wind energy conversion system. Windpro Journal 2005;89:25—30.

[78] Blom AF, Svenkvist P, Sven-erik T. Fatigue design of large wind energy conservation systems and operational experience from the Swedish prototypes. Journal of Wind Engineering and Industrial Aerodynamics 1990;34(1):45—76.

[79] Dasgupta A, Pechat M. Material failure mechanisms and damage models. IEEE Transactions on Reliability 1991;40(5):531—6.

[80] Ramsamooj DV, Shugar TA. Modeling of corrosion fatigue in metals in an aggressive environment. International Journal of Fatigue 2001;23:301—9.

[81] Zhiquan YE, Haomin MA, Nengsheng B, Yan C, Kang D. Structure dynamic analysis of a horizontal axis wind turbine system using a modal analysis method. Wind Engineering 2001;25(4):237—48.

[82] Makkonen L, Laaakso T, Marjaniemi M, Karen JF. Modeling and prevention of ice accretion on wind turbine. Wind Engineering 2001;25:3—21.

[83] Nair CV. Wind power-the near term commercial renewable energy source. Australian Science 1995;16:25—6.

[84] Stoudt MR, Ricker ER, Cammarata RC. The influence of a multilayered metallic coating on fatigue cracks nucleation. International Journal of Fatigue 2001;23:215—23.

[85] Chand N, Hashmi SAR, Lomash S, Naik A. Development of asbestos free brake pad. IE [I] J-MC 2004;85:13—6.

[86] Lyntte R. Status of the U.S. wind power industry. Journal of Wind Engineering and Industrial Aerodynamics 1988;27:327—36.

[87] Jacksan J. Going commercial with a down wind turbine. Windpro Journal 2004;55:15—6.

[88] Ekanayake JB, Jenkins N. Harmonic issues of the application of an advanced static var compensator of a wind farm. Wind Energy 1997;21:215−26.

[89] Yuksel I. Development of hydropower: a case study in developing countries. Energy Sources Part B 2009;1:100−10.

[90] Kaygusuz K. Sustainable development of hydropower. Energy Sources 2002;24:803−15.

[91] IHA, International Hydropower Association. The role of hydropower in sustainable development. IHA White Paper; February 2003. p. 1−140.

[92] Seifried D, Witzel W. Renewable energy—the facts. 1st ed. London; Washington (DC): Earth Scan Publishing for a Sustainable Future; 2010. p. 114−20.

[93] Dincer I. Renewable energy and sustainable development: a crucial review. Renewable and Sustainable Energy Reviews 2000;4:157−75.

[94] Yuksel I. Hydropower in Turkey for a clean and sustainable energy future. Renewable and Sustainable Energy Reviews 2008;12:1622−40.

[95] Ardizzon G, Cavazzini G, Pavesi G. A new generation of small hydro and pumped hydro power plants: advances and future challenges. Renewable and Sustainable Energy Reviews 2014; 31:746−61.

[96] Darmawi R, Sipahutar S, Bernas M, Imanuddin MS. Renewable energy and hydro power utilization tendency worldwide. Renewable and Sustainable Energy Reviews 2013;17:213−5.

[97] Deane JP, Gallacho BPO, McKeogh EJ. Techno-economic review of existing and new pumped hydro energy storage plant. Renewable and Sustainable Energy Reviews 2010;14:1293−302.

[98] Shadman F, Sadeghipour S, Moghavvemi M, Saidur R. Drought and energy security in key ASEAN countries. Renewable and Sustainable Energy Reviews 2016;53:50−8.

[99] Paish O. Small hydro power: technology and current status. Renewable and Sustainable Energy Reviews 2002;6:537−56.

[100] Raadala HL, Gagnonb L, Modahla IS, Hanssena OJ. Life cycle greenhouse gas [GHG] emissions from the generation of wind and hydro power. Renewable and Sustainable Energy Reviews 2011;15:3417−22.

CHAPTER FOUR

Bioenergy With Carbon Capture and Storage (BECCS): Future Prospects of Carbon-Negative Technologies

M.A. Quader[1] and S. Ahmed[2]
[1]University of Malaya, Kuala Lumpur, Malaysia
[2]Islamic University of Technology, Dhaka, Bangladesh

4.1 INTRODUCTION

Carbon dioxide (CO_2) is one of the most important contributors for the increase of the greenhouse effect. The IPCC (Intergovernmental Panel on Climate Change) Fifth Assessment Report (AR5) showed that in order to limit the long-term global temperature increase to 2°C above preindustrial levels and avoid dangerous climate change consequences, a radiative forcing of below 3 W/m^2 is required around the end of the century [1]. However, it seems increasingly likely that we will overshoot this limit. The IEA (International Energy Agency), in its World Energy Outlook, announces that "the door to 2°C is closing."

In order to meet both environmental and economic constraints, there must be a comprehensive mitigation portfolio that includes multiple options. This would, for example, mean measures that improve efficiency, favor energy conservation, renewable energy, and enhancement of carbon sinks, as well as CCS (carbon capture and storage). Therefore, it may become necessary to develop technologies that capture emissions from the atmosphere (negative CO_2 emissions technologies). By capturing CO_2 from the air (directly or indirectly), CO_2 emissions can be sequestered and the stock of atmospheric CO_2 reduced to correct the overshoot. This technique could also be used to offset additional anthropogenic emissions from sectors where emission reductions are difficult or uneconomical. A range of negative emissions options have been identified that directly remove CO_2 from the atmosphere called direct air capture technologies such as artificial trees and lime-soda process) and some options remove emissions indirectly, for example, augmented ocean disposal processes, biochar, and BECCS (bioenergy with carbon capture and storage). Given that the cost of direct air-capture technologies is still very uncertain and high, BECCS appears to be the negative CO_2 emissions technology with the most immediate potential to reduce emissions. This chapter describes a new opportunity for CO_2 abatements: geological storage of CO_2 from biomass, or BECCS.

Clean Energy for Sustainable Development
ISBN 978-0-12-805423-9, http://dx.doi.org/10.1016/B978-0-12-805423-9.00004-1

4.1.1 What Is BECCS?

BECCS is a technology that integrates biomass systems with geological carbon storage. During combustion, fermentation, putrefaction, biodegradation, and other biological processes, large amounts of CO_2 are emitted from trees, plants, and agricultural crops. These processes are, for example, found in biomass-fueled power plants, pulp and paper industries, steel plants, ethanol plants, and biogas plants.

As biomass grows, CO_2 is absorbed from the atmosphere. Through photosynthesis, carbon is incorporated into plant fibers, while oxygen from the decomposed CO_2 molecule is set free. The energy for the process comes from the sun that induces photosynthesis. When biomass is broken down through combustion or any other natural process, the carbon atoms that the plant was composed of are released. Together with the oxygen in the air, they form CO_2. In this way, large amounts of biogenic CO_2, obtained though natural biodegradation processes, are released back into the atmosphere. The CO_2 molecules are then split again through the growth of new biomass, which is captured in the next generation of plants. When applying BECCS, the CO_2 previously tied up in biomass is captured from the atmosphere, and the gas flow is diverted to the bedrock for permanent storage. In this way, BECCS systems create a flow of CO_2 from the atmosphere into the underground (see Fig. 4.1).

The BECCS technology was first mentioned in scientific publications in the 1990s. Since then, the BECCS technology has been discussed as a variant of the CCS technology that is applied to fossil sources. Most interest has been directed toward the fact that BECCS provides an opportunity to create permanent negative carbon emissions, that is, the removal of CO_2 from the atmosphere. Since BECCS is a new and complex technology, it has come to be known by different names depending on the author and context. The IPCC uses the acronym "BECCS" to describe the technology in its fourth assessment report from 2007. Other authors use the abbreviations "BECS,"

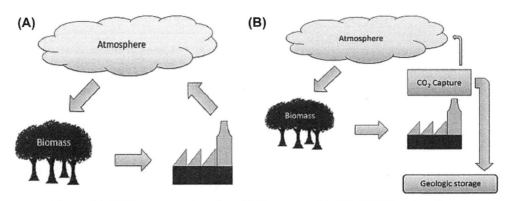

Figure 4.1 (A) Bioenergy carbon flow. (B) Bioenergy with CCS (BECCS) carbon flow.

"biomass–based CCS," "BCCS," and "biotic CCS." This chapter uses the acronym BECCS, as applied by the IPCC, throughout [2].

4.1.2 Negative Emissions With BECCS

For an overview of the main flows of carbon and CO_2 in different energy systems, see Fig. 4.2. Please note that in addition to these main system flows, the support systems for construction, fuel extraction, and transportation have been included. In other words, it is needed to consider the fact that all systems currently involve certain fossil emissions at some part of the production chain. Even the design and installation of wind turbines involve carbon emissions, though the quantities are relatively small.

The radical difference between negative carbon emissions and other energy systems becomes evident when looking at Fig. 4.2. Fossil fuels increase the amount of CO_2 in the atmosphere in absolute terms. As fossil coal and oil, which are not part of the natural carbon cycle, are extracted and combusted, CO_2 is added to the atmosphere. Fossil fuels with CCS also increase the amount of CO_2, but not as much as without CCS. Renewable energy generated by wind, solar, geothermic, and hydroelectric power plants affects the carbon cycle to a very limited extent, once in operation. Bioenergy emits as much carbon as the biomass previously captured. BECCS, however, only emits parts of the previously captured CO_2, and the rest is permanently removed from the atmosphere.

Biomass use for energy production in processes such as combustion and gasification, and its use to produce biofuels such as bioethanol, results in emissions of CO_2. This CO_2 produced during combustion is about the same quantity consumed during biomass growth; therefore emissions from biomass combustion are considered to be CO_2 neutral.

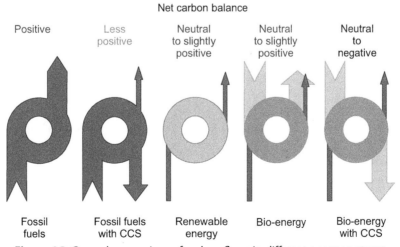

Figure 4.2 General comparison of carbon flows in different energy systems.

Capture and long-term storage of these CO_2 emissions would effectively result in net removal of atmospheric CO_2. Biomass with CCS is potentially one of the few options for "negative emission."

4.2 CARBON-NEGATIVE TECHNOLOGIES

4.2.1 Artificial Photosynthesis

Artificial photosynthesis, which is described as "Chemistry's Greatest Challenge," attempts to replicate the natural processes of photosynthesis, and, at least in near future, the goal of artificial photosynthesis is to use sunlight energy to make high-energy chemicals to store energy. A challenge in artificial photosynthesis is to use cheap and environmentally friendly compounds. Many components currently proposed for use in artificial photosynthetic systems are expensive, toxic, inefficient, or nondurable.

An "artificial tree" is a device that mimics the processes used by biological plant life to absorb CO_2 from the atmosphere. In nature, plants combine CO_2 from the atmosphere with water from their sap chemically, forming various hydro and oxy-hydrocarbons. However, in the case of artificial trees, the output from the "tree" is a stream of essentially pure CO_2 at high pressure, ready for sequestration. Klaus Lackner, a physicist at Columbia University, is working on a filter that can do just that. Lackner envisions artificial trees small enough to fit in a shipping container but large enough to capture a ton of CO a day [3]. Lackner's trees are essentially passive devices (i.e., no energy input required for the capture of CO_2) that present a large surface area of CO_2-absorbing material to the atmosphere—akin to the leaves of natural trees. Wind is used to drive a current of CO_2-laden air across an absorbent surface so that mass transfer of CO_2 to the absorbent takes place. the sorbent, over time, becomes saturated with CO_2 and must be regenerated (Fig. 4.3; Scheme 4.1).

Lackner developed an absorbent that can be regenerated by simple rehydration; soaking the saturated sorbent with water results in it releasing a portion of the CO_2 chemically bound to it. This process must be done in a sealed chamber held at reduced pressure. After regeneration, the sorbent can be reexposed to the air where it first dries, and then absorbs another tranche of CO_2 from the atmosphere. It is claimed that this absorption/stripping cycle can be repeated many thousands of times without degradation of the sorbent, and experiments have confirmed this on laboratory scale. All that remains is to dehydrate and compress the CO_2 released in the regeneration chamber ready for transport to the sequestration site.

A feature of Lackner's trees, therefore, is that the only significant energy requirement is the electricity needed to drive the gas compressors. Some heat input is required in the regeneration process, but this could be supplied from heat recovery in the CO_2 compression process. However, due to the dehydration step, a process that contributes to the overall energy balance of the system, the devices require a significant (but not

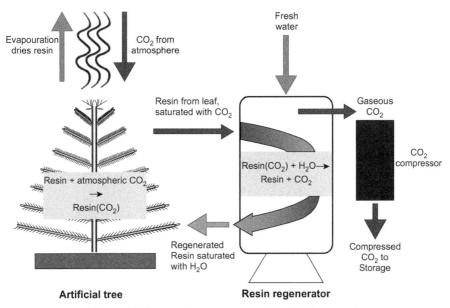

Figure 4.3 Proposed arrangement of artificial trees [2].

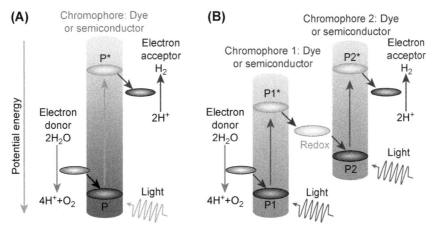

Scheme 4.1 An artificial photosynthetic system: single-step reactions (A), and two-step (Z-scheme) reactions (B). *P*, chromophore of a single-step reaction system; *P**, excited state of P; *P1*, the first chromophore of a two-step reaction system; *P1**, excited state of P1; *P2*, second chromophore of a two-step reaction system; *P2**, excited state of P2. Adapted from Tachibana Y, Vayssieres L, Durrant JR. Artificial photosynthesis for solar water-splitting. Nature Photonics 2012;6(8):511—518.

quantifiable based on the present literature) amount of water, which may limit the application of artificial trees to nonarid regions.

However, there are a number of studies to design different systems using chemicals and methods that may not have much to do with how natural organisms perform it. All researchers must be encouraged, but the point is that learning from the natural systems makes sense since these have been doing it successfully for millions of years (Scheme 4.1). Artificial photosynthesis is, as discussed by Collings and Critchley, an umbrella term that includes reactions from water splitting, CO_2, and N_2 reduction, to engineered bacteria. In this issue, we collected new progresses in this field from different scientists in different countries [5].

4.3 CARBON-NEGATIVE BIOFUELS

4.3.1 Solid Biofuels

4.3.1.1 Biomass

Biomass is one of the most important renewable energy sources in the near future. It has the potential benefits of decreasing pollutant generation and being CO_2 neutral. Compared to other sources of energy, biomass offers some unique advantage with respect to the environment since it is "carbon neutral." Biomass is biological material derived from living, or recently living organisms. In the context of biomass for energy, this is often used to mean plant-based material, but biomass can equally apply to both animal- and vegetable-derived material. It is carbon based and is composed of a mixture of organic molecules containing hydrogen, usually including atoms of oxygen, often nitrogen, and also small quantities of other atoms, including alkali, alkaline earth, and heavy metals.

4.3.1.1.1 Classification of Biomass

Fig. 4.4 shows the different sources of biomass production that are discussed in the following sections.

4.3.1.1.1.1 Energy Crops   The energy crops are grown specifically for use as fuel and offer high output per hectare with low inputs. These crops are usually low cost and need low maintenance. The energy crops mainly consist of herbaceous energy crops, woody energy crops, agricultural crops, and aquatic crops [6].

Herbaceous energy crops are perennials that are harvested annually. It takes 2−3 years to reach in complete production. These crops include grasses such as switchgrass, miscanthus, bamboo, sweet sorghum, tall fescue, kochia, wheatgrass, reed canary grass, coastal Bermuda grass, alfalfa hay, thimothy grass, and others. The Biowert, Germany uses meadow grass to manufacture green electricity and innovative materials such as plastics, insulation materials, and fertilizers.

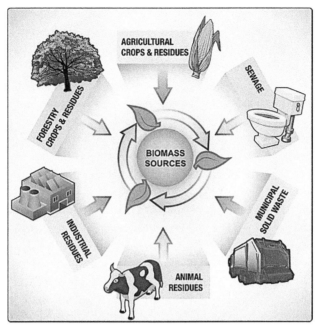

Figure 4.4 Different sources of biomass.

Woody energy crops are fast-growing hardwood trees that are harvested within 5—8 years of plantation. These crops include hybrid poplar, hybrid willow, silver maple, eastern cottonwood, green ash, black walnut, sweetgum, sycamore, and so on. For the manufacturing of paper and pulp the short rotation woody energy crops are traditionally used. On the other hand, *agricultural crops* such as oil crops (e.g., jatropha, oilseed rape, linseed, field mustard, sunflower, castor oil, olive, palm, coconut, and groundnut), cereals (e.g., barley, wheat, oats, maize, and rye), and sugar and starchy crops (e.g., sweet sorghum, potato, sugar beet, and sugarcane) are generally grown to produce vegetable oils, sugars, and extractives. These crops have potentials to produce plastics, chemicals, and products as well. *Aquatic crops* include several varieties of aquatic biomass, for instance, algae, giant kelp, other seaweed, marine microflora, and so on.

The energy crops are extensively grown for production of biofuels, for example, sugarcane in Brazil for ethanol, maize in the United States for ethanol, and oilseed rape in Europe for biodiesel.

4.3.1.1.1.2 Agricultural Residues and Waste
Agricultural residues mainly comprise of stalks and leaves that are generally not harvested from fields for commercial use. Sugar cane bagasse, corn stover (stalks, leaves, husks, and cobs), wheat straw, rice straw, rice hulls, nut hulls, barley straw, sweet sorghum bagasse, olive stones, and so on are some of the examples of agricultural residues. With vast areas of corn cultivated worldwide, corn

stover is expected to be a major feedstock for biorefinery. The use of agricultural residues for biorefinery is beneficial as it eliminates the need of sacrificing arable lands [7].

4.3.1.1.1.3 Forestry Waste and Residues

The forestry waste and residues are referred to the biomass that is usually not harvested from logging sites in commercial hardwood and softwood stands. The forestry residues also include biomass resulting from forest management operations (thinning of young stands and removal of dead and dying trees). Utilization of this biomass for biorefinery near its source is highly desirable to avoid expensive transportation. However, limited accessibility to dense forests largely increases operation costs for logging/collection activities [8].

4.3.1.1.1.4 Industrial and Municipal Wastes

These include municipal solid waste (MSW), sewage sludge, and industrial waste. Residential, commercial, and institutional postconsumer waste usually contains good amounts of plant-derived organic materials that can be used as potential source of biomass. The waste paper, cardboard, wood waste, and yard waste are examples of MSW. The waste product generated during wood pulping, called black liquor, is an example of industrial waste.

4.3.1.2 Liquid Biofuels

4.3.1.2.1 First Generation

First-generation fuels allude to the biofuel a product of sugar, starch, vegetable oil, or animal fats utilizing conventional technology. These fluid biofuels include the accessible fuels such as pure plant oil from oil-yielding crops, biodiesel from esterification of immaculate plant oil or waste vegetable oils, bioethanol from sugar or starch crops maturation, and ethanol derivate ETBE (i.e., the t-butyl ether of ethanol) [9].

4.3.1.2.1.1 Vegetable Oil

Vegetable oil can be used for either food or fuel. The potential to run engines on straight run vegetable oils dates back to the 19th century, notably to attempts by the famous German inventor, Rudolph Diesel leading to the successful development of his Diesel engine in 1895. In most cases, vegetable oil is utilized to fabricate biodiesel, which is good with most diesel motors when mixed with conventional diesel fuel.

4.3.1.2.1.2 Biodiesel

Biodiesel refers to a variety of ester-based fuels (fatty esters) generally defined as the monoalkyl esters made from several types of vegetable oils, such as soybean, canola, or hemp oil, or sometimes from animal fats through a simple transesterification process. When oils are mixed with methanol and sodium hydroxide, biodiesel and glycerol are produced by the chemical reaction. First part of glycerol is produced for every 10 parts biodiesel. Biodiesel can be successfully used in any diesel engine when it is mixed with mineral diesel in a neat composition.

4.3.1.2.1.3 Bioalcohol   Ethanol, propanol, and butanol are commonly biologically produced alcohols that are manufactured by the action of microcosm and enzymes through sugar stretches or cellulose (which is more difficult). It is likely that butanol is able to produce enough energy to be burnt "straight" in the existing gasoline engines since it is less corrosive and water soluble than that of ethanol which could be distributed through the existing system.

4.3.1.2.2 Second Generation

The second-generation biofuel technologies have been developed to overcome some important limitations of the first-generation biofuel, notably their use as food. Biomass of trees as second-generation fuels is said to contain more carbohydrate and the raw material for biofuel than that of food crops. Genetic modification (GM) engineering is constantly used to attempt to lessen the level of lignin in trees and change the structure of the hemicelluloses. Cellulosic ethanol is taken from nonfood crops or inedible waste products that have less impact on food such as switch gases, sawdust, rice hulls, paper pulp, and wood chips. Lignocelluloses are the "woody" structural material of plants. The greenhouse gas emissions savings for lignocellulosic ethanol are greater than those obtained by the first-generation biofuels.

4.3.1.2.3 Third Generation
4.3.1.2.3.1 Microalgae Drive Biofuels   Algae are the fastest growing organisms in the world. Microalgae are known for faster growth rates than terrestrial crops. The per unit area yield of oil from algae is estimated to be from 18,927 to 75,708 liters per acre, per year; this is 7−31 times greater than terrestrial crops although there are claims of higher yields of up to 100,000 liters per hectare per year. Studies show that algae can produce up to 60% of their biomass in the form of oil because the cells grow in aqueous suspension where they have more efficient access to water, CO_2, and dissolved nutrients. Many fuel–grade products could be gained from algae [9]. Different types of biofuel, such as biodiesel, hydrogen, methanol, and ethanol, are produced from different types of microalgae (Table 4.1). Microalgae can provide several types of renewable fuels described in the following sections:

4.3.1.2.3.2 Biodiesel   Biodiesel, typically produced from oil plants including food crops, has received a lot of concerns about the sustainability of this practice. Microalgae as an alternative to conventional crops, such as sunflower and rapeseed, can produce more oil and consume less space. Microalgal biodiesel contains no sulfur and can replace diesel in today's cars with little or no modifications of vehicle engines, while, on the other hand, the use of it can decrease the emissions of particulate matters, CO, hydrocarbons, and SO_x. Compared to biodiesel derived from land-based crops, producing a substantial amount of biodiesel from microalgae has been considered as the most efficient way to make biodiesel sustainable [10].

Table 4.1 Different Types of Biofuel Production From Different Microalgae [11,12]

Microalgae	Algae Type	Biofuel	Productivity of Biofuel
Arthrospira maxima	Green	Hydrogen, biodiesel	40–69%
Chlamydomonas reinhardtii	Green	Hydrogen	2.5 mL/h/11.73 g/L
Chlorella	Green	Biodiesel	
Chlorella biomass	Green	Ethanol	22.6 g/L
Chlorella minutissima	Green	Methanol	
Chlorella protothecoides	Green	Biodiesel	15.5 g/L
Chlorella regularis	Green	Ethanol	
Chlorella vulgaris	Green	Ethanol	
Chlorococcum humicola	Green	Ethanol	7.2 g/L or 10 g/L
Chlorococcum infusionum	Green	Ethanol	0.26 g ethanol/g algae
Chlorococum sp.	Blue-green	Biodiesel	10.0 g/L
Chlorococum sp.	*Blue-green*	Ethanol	3.83 g/L
Dunaliella sp.	Green	Ethanol	11.0 mg/g
Haematococcus pluvialis	Red	Biodiesel	420 GJ/ha/yr
Neochlorosis oleabundans	Green	Biodiesel	56.0 g/g
Platymonas subcordiformis	Green	Hydrogen	
Scenedesmus obliquus	Green	Methanol, hydrogen	
Spirogyra	Green	Ethanol	
Spirulina platensis	Green	Hydrogen	1.8 μmol/mg
S. platensis UTEX 1926	Blue-green	Methane	0.40 m^3/kg
Spirulina Leb 18	Blue-green	Methane	0.79 g/L

4.3.1.2.3.3 Biohydrogen   Microalgae can also directly use sunlight and water to generate biohydrogen in the absence of oxygen in a closed culture system. During photolysis, microalgae can split two water molecules via photosynthesis to form one oxygen molecule and four hydrogen ions which can be converted into two hydrogen molecules by hydrogenase enzyme. Usually hydrogen is generated from algae through three methods: (1) Biochemical processes—A microscopic green algae (known as *Chlamydomonas reinhardtii*, or pond scum) split water into hydrogen and oxygen under controlled conditions. Under these conditions, enzymes in the cell act as catalysts to split the water molecules. A recent breakthrough in controlling the algae's hydrogen yield has prompted interest in commercial-scale H_2 production from algae. (2) Gasification of algal biomass—During gasification, biomass is converted into a gaseous mixture comprising primarily of hydrogen and carbon monoxide, by applying heat under pressure in the presence of steam and a controlled amount of oxygen. A number of methods are available for the separation of H_2 from syngas. (3) Steam reformation of methane—Fermentation of algal biomass produces methane. The traditional steam reformation (SMR) techniques can be used to derive hydrogen from methane. Steam reforming is a method that produces hydrogen commercially, besides being used in the industrial synthesis of ammonia. It is also the cheapest method.

Table 4.2 Bioethanol Production Yields From Different Microalgae Species [13,14]

Feedstock	Pretreatment	Yield (g Ethanol/g Substrate)
Kappaphycus alvarezii	Sulfuric acid	0.457
Gracilaria verrucosa	Sulfuric acid and enzymatic	0.430
Saccharomyces cerevisiae	Enzymatic	0.259
Chlorococum humicolo	Sulfuric acid	0.520
Chlorella vulgaris	Acid and enzymatic	0.400
Chlorococum sp.	Supercritical CO_2	0.383
Chlorococum infusionum	Alkaline	0.260
Gelidium amansii	Sulfuric acid	0.888
Chlamydomonas reinhardtii	Enzymatic	0.240

4.3.1.2.3.4 Bioethanol   Bioethanol made from food crops is viewed as "first-generation" biofuel, which competes with animal feed and human food for the source materials (Table 4.2). To minimize the adverse impacts, manufacturing "second-generation" bioethanol from nonfood lignocellulosic plant materials has been explored. Indeed, lignocellulosic materials are widely available: forest slashes, crop residues, yard trimmings, food processing waste, and municipal organic refuses can be the feedstock for bioethanol. Lignocellulosic bioethanol production involves three categories of costs: the costs of feedstock, the costs of sugar preparation, and the costs of ethanol production. Among these three categories, conversion of cellulosic components into fermentable sugars is the major technological and economical bottleneck. Research is focusing on development of cost-effective techniques for extracting simple sugars from lignocellulosic biomass. Two feedstock treatment technologies have been proposed: acid hydrolysis and enzymatic hydrolysis. Bioethanol is ethanol produced from vegetative biomass through fermentation, in which the following biochemical reactions are involved:

$$(C_6H_{10}O_5)n(starch, cellulose, sugar) + nH_2O \rightarrow nC6H_{12}O_6(glucose, fructose)$$

$$(C_5H_8O_4)n(hemicellulose) + nH_2O \rightarrow nC_5H_{10}O_5(xylose, mannose, arabinose, and soon)$$

$$C6H_{12}O_6 \rightarrow 2CH_3CH_2OH(ethanol) + 2CO_2$$

$$C_5H_{10}O_5 \rightarrow 5CH_3CH_2OH(ethanol) + 5CO_2$$

4.3.1.2.3.5 Biobutanol   The green waste leftover from the algae oil extraction can be used for the production of butanol. This fuel has an energy density similar to gasoline and greater than that of either ethanol or methanol. It can be used in most gasoline engines in place of gasoline with no modifications. In several tests, butanol consumption is similar to that of gasoline and when blended with gasoline, it provides better performance and corrosion resistance thanethanol or E85.

4.3.1.2.3.6 Jet Fuel   Commercial application of algae—derived jet fuel was further buttressed when on January 8, 2009, Continental Airlines ran the first test for the first flight of an algae-fueled jet. The test was done using a twin-engine commercial jet consuming a 50/50 blend of biofuel and normal aircraft fuel. A series of tests executed at 38,000 ft. (11.6 km), including a mid-flight engine shutdown, showed that no modification to the engine was required. The fuel was praised for having a low flash point and sufficiently low freezing point issues that have been problematic for other biofuels.

However, the extraction or production of different kinds of biofuel is mostly dependent on feedstock availability and the technological option [15]. The application of microalgae biomass can displace the use of fossil fuel and, consequently, this can lead to reduction of CO_2 emissions. Fig. 4. 5 shows the paths for the different energy products that can be produced by microalgae.

4.3.1.3 Gaseous Biofuels

4.3.1.3.1 Biogas

Biogas, which is generally referring to gas from anaerobic digestion (AD) units, is a promising means of addressing global energy needs and providing multiple environmental benefits, as shown in Table 4.3. Natural gas, consisting of 95% methane (CH_4) and 5% ethane (C_2H_6), propane (C_3H_8), butane (C_4H_{10}), nitrogen (N_2), and carbon dioxide (CO_2), is a gaseous fossil fuel formed from buried plants and animals that

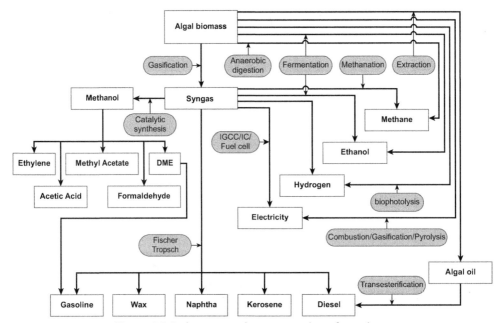

Figure 4.5 Paths to several energy products from algae.

Table 4.3 Biogas Environmental Benefits Analysis [16,17]

Biogas	Corresponding Contents
Green energy production	Electricity
	Heat
	Vehicle fuel
	Trigeneration
Organic waste disposal	Agricultural residues
	Industrial wastes
	Municipal solid wastes
	Household wastes
	Organic waste mixtures
Environmental protection	Pathogen reduction through sanitation
	Less nuisance from insect flies
	Air and water pollution reduction
	Eutrophication and acidification
	Reduction
	Forest vegetation conservation
	Replacing inorganic fertilizer
Biogas linked agrosystem	Livestock—biogas—fruit system
	Pig—biogas—vegetable greenhouse system
	Biogas—livestock and poultry farms system
GHG emission reduction	Substituting conventional energy sources

experienced great heat and pressure over thousands of years. The energy content of natural gas is 38.2 MJ/m^3 (1027 Btu/ft.^{3}) or 53.5 MJ/kg^3 [the density of natural gas is 0.717 kg/m^3 at standard temperature (0°C) and pressure (1 atm)].

Biogas is a renewable gaseous fuel alternative to natural gas. It is generated by AD of organic wastes. Raw biogas consists of 60—65% of methane (CH_4), 30—35% of CO_2, and small percentages of water vapor, H_2, and H_2S. After purification to remove CO_2, H_2S, and other impurities, the upgraded, pipeline-quality biogas (now named biomethane) is used as a natural gas substitute [18].

4.3.1.3.2 Syngas

Syngas is another gaseous biofuel produced from gasification or pyrolysis of plant materials. Chemically, syngas consists of 30—60% CO, 25—30% H_2, 5—15% CO_2, 0—5% CH_4, and lesser portions of water vapor, H_2S, COS, NH_3, and others, depending on the feedstock types and production conditions.

Syngas is commercially produced by gasification process. In the operation, carbon-rich materials, such as coal, petroleum, natural gas, and dry plant biomass, are rapidly heated to above 700°C in the high-temperature (e.g., 1200°C) combustion chamber of a

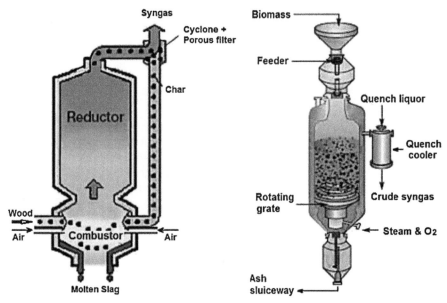

Figure 4.6 The structure of a Mitsubishi Heavy Industries gasifier (left) and a Lurgi dry-ash gasifier (right) for syngas.

gasifier and partially burned in the presence of controlled air flow to yield syngas (Fig. 4.6). In the gasifier, wood biomass experiences three thermal transformation phases: dehydration, pyrolysis, and partial oxidation. In the initial dehydration phase, air-dry biomass swiftly loses its moisture before its temperature reaches 200 C. Pyrolysis begins as the temperature increases, and biomass is converted to char and vapor. In the presence of O_2, char is partially oxidized to generate CO and CO_2, while the vapor is combusted to CO_2 and H_2O. As the hot char particulates, CO, CO_2, and H_2O rise in the combustion chamber, further reactions occur, that is, char is oxidized by CO_2 to yield CO, or by H_2O to yield CO and H_2, and CO reacts with H_2O to produce CO_2 and H_2. The mixture of CO, H_2, and CO_2 is then recovered as syngas (Fig. 4.6) [18]. The major reactions can be simplified as follows:

$$Wood \rightarrow char + vapor$$

$$Char + O_2 \rightarrow CO + CO_2$$

$$Vapor + O_2 \rightarrow CO_2 + H_2O$$

$$Char + CO_2 \rightarrow CO$$

$$Char + H_2O \rightarrow CO + H_2$$

$$CO + H_2O \rightarrow CO_2 + H_2$$

4.4 BIOFUEL CONVERSION TECHNOLOGIES

As there is wide diversity in the characteristics and properties of these different classes of material, and their various subgroups, there is also a wide range of conversion technologies to make optimum use of them, which include both thermal and chemical conversion technologies.

4.4.1 Conventional Combustion

Combustion is the process by which flammable materials are allowed to burn in the presence of air or oxygen with the release of heat. The basic process is oxidation. When the flammable fuel material is a form of biomass the oxidation is of predominantly carbon (C) and hydrogen (H) in the cellulose, hemicellulose, lignin, and other molecules present to form carbon dioxide (CO_2) and water (H_2O). Fire—the combustion of organic matter—is prompt oxidation of biocarbon compounds by oxygen at high temperature and can be simply described as:

$$C_6H_{10}O_5 + 6O_2 \rightarrow 6CO_2 + 5H_2O + heat + light$$

4.4.2 Thermochemical Conversion

These are processes in which heat is the dominant mechanism to convert the biomass into another chemical form. The basic alternatives are separated principally by the extent to which the to proceed (Fig. 4.7).

4.4.2.1 Gasification

Gasification is considered as a key technology for the use of biomass. Biomass gasification is a complex thermochemical process in which biomass is converted to synthetic gas (syngas) under substoichiometric conditions. The syngas could be then used as a fuel in internal combustion engines, gas turbines, or fuel cells for the production of heat, mechanical energy, or power, or as a feedstock for the synthesis of liquid fuels and chemicals. The results of Herzog and Golomb [19] showed that the overall energy efficiency for a wood gasification process targeting the production of synthetic natural gas (SNG) in SCWG is 70%. The overall energy efficiency is given in equation as:

$$\varepsilon = \frac{\Delta h^0_{SNG} \dot{m}^-_{SNG} + \dot{E}^- + \dot{Q}^-}{\Delta h^0_{biomass} \dot{m}^+_{biomass,daf} + \dot{E}^+}$$

where Δh^0_{SNG} refers to the lower heating value (LHV), m refers to the mass flow, E refers to the mechanical or electrical power, Q refers to the heat flow, and *daf* refers to the dry ash−free basis. The superscripts − and + refers to the flows leaving the system and flows entering the system, respectively. The same net efficiency for an SNG production process

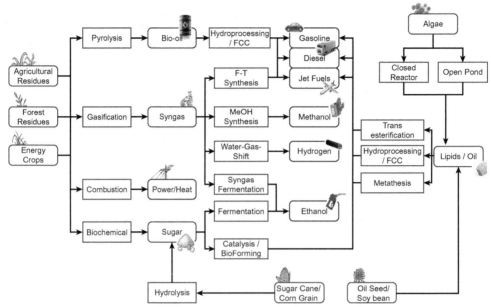

Figure 4.7 Selected conversion pathways from terrestrial and aquatic biomass to intermediates and to final biofuel products.

from wood were found to be 54.3% for an entrained-flow gasifier operating at 3 MPa, 58.1% for a circulating fluidized bed operating at 1 MPa, and 66.8% for an allothermal gasifier operating at 0.1 MPa on LHV basis.

As far as gasification is concerned, the gasifiers can be broadly cataloged into fixed-bed, fluidized-bed, and entrained-flow reactors (Table 4.4). For power generation, the purpose of biomass gasification is to produce a combustible producer gas to run the engine, which rotates the generator shaft. However, the engines have some specific requirements for accepting fuel gas. For instance, the producer gas must have a certain percentage of burnable gas (>20% CO and >10% H_2), a minimum amount of tar content (100 mg/Nm^3), and be completely free of dust and other poisonous gases (NH_3, SO_2, and so on). To satisfy the requirement of product gas, a comprehensive research has been done in the last couple of decades. Those researches mainly focused on the development of different types of reactors. The entire reactor systems can be classified into two categories: (1) updraft gasifier and (2) downdraft gasifier.

4.4.2.1.1 Updraft Gasification

Updraft gasification is basically a counter-current gasification system where the air and other gasifying agents are injected from the bottom, while the biomass enters from the top and moves downward under the force of gravity. Updraft gasifier can be classified as

Table 4.4 Effect of the Key Process Parameters on Biomass Gasification

		Effect on Gasification	Recommended Values

Gasifying agent	Air	• Poor gas product quality (N_2-rich); heating value: 4—7 MJ/m3; used for heat and power purposes.
		• The amount of air in the gasifier (in equivalence ratio) needs to be within a certain range to achieve an optimal gasification process.
		  • Small amounts of air result in significant amounts of tar and low carbon conversion.
		  • Large amounts of air decrease the heating value, the energy content of the product gas, and the H_2 content.
			Equivalence ratio: 0.2—0.4

	Oxygen	• High gas product quality (N_2-free); heating value: 10—18 MJ/m3; biofuel production purposes. • The effect of equivalence ratio on the gasification process is in line with that observed in the presence of air.	Equivalence ratio: 0.2—0.4

	Steam	• High gas product quality (H_2-rich). Heating value: 10—16 MJ/m3
		• The amount of steam in the gasifier (in steam to biomass ratio) needs to be within a certain range to achieve an optimal gasification process.
		  • Small amounts of steam limit steam reforming and water—gas shift reactions, resulting in small amounts of syngas.
		  • Large amounts of steam enhance the decomposition of tar and the production of syngas. However, it decreases the gas yield, energy conversion, and carbon conversion.
			Steam to biomass ratio: 0.5—0.7

(Continued)

Table 4.4 Effect of the Key Process Parameters on Biomass Gasification—cont'd

		Effect on Gasification	Recommended Values
Temperature		• An increase in temperature often enhances the H_2 content, the fuel gas production, and the carbon conversion whereas the CH_4 content and the heating value are hindered. • The influence of temperature on CO and CO_2 formation are not straightforward due to the complexity of the reactions in which both species are involved.	Fluidized bed: 800—900°C Entrained flow: 1300—1500°C
Pressure		• Pressurized systems allow the reduction of the reactor size. • An increase in pressure results in higher amounts of H_2, CO_2, CH_4, and carbon conversion and a decrease in CO attributed to the water—gas shift constant enhancement and an increase in secondary pyrolysis reactions.	Pressure: 20—50 bar
Bed material	Natural catalysts	• Dolomite reduces the tar content effectively and increases the gas product and the H_2 content. However, it has a low attrition-resistance. • Olivine also increases the gas product and H_2 content and in contrast to dolomite is more attrition-resistant. However, it is less active in tar removal.	
	Synthetic catalysts	• Alkali-based catalysts enhance gasification rates. They might agglomerate at high temperatures and are difficult to recover. • Nickel-based catalysts are highly active for steam reforming of hydrocarbons and tars at relatively low temperatures, resulting in large amounts of syngas. These catalysts deactivate rapidly due to sintering and coke deposition.	Addition of nobel metals as catalyst promoters

(Continued)

Table 4.4 Effect of the Key Process Parameters on Biomass Gasification—cont'd

		Effect on Gasification	Recommended Values
Biomass feedstock	Type	• Noble metal—based catalysts show high resistance to coke deposition. These catalysts are very expensive and are available in limited amounts. • Among all the chemical and physical properties of a certain type of biomass, the moisture content is the factor that influences carbon conversion, cold gas efficiency, and heating value to a larger extent. The possible catalytic effect of ash on steam gasification of char makes the ash content and composition other important fuel characteristics to consider in biomass gasification.	
	Particle size	• Smaller particles improve the gas quality (larger amounts of H_2, CO, CO_2, and CH_4 and lower amounts of char and heavy tars), fuel conversion, heating value, and cold gas efficiency due to a more effective mass and heat transfer.	

updraft fixed-bed gasifiers, fluidized-bed gasifiers, and circulating fluidized-bed gasifiers. The operating principle of this type of gasifier is shown in Fig. 4.8A. Feedstock material is first introduced into the drying zone at the top, followed by the pyrolysis and reduction zone, and finally the unconverted solid passes through the combustion zone. In the combustion zone, solid charcoal is combusted producing heat, which effectively transfers to the solid particles during counter-current flow of the rising gas and descending solids. In this gasification system, the contamination of substantial amount of tars is the major problem of updraft gasifiers. If the gas is to be utilized for turbines or internal combustion engines for electricity generation or mechanical power, it must go through a series of filtering and cleaning devices in order to reduce the tar content to an acceptable range. The intensive cleaning process adds considerably higher investment cost and reduces the overall efficiency of the whole process. Therefore, the application of updraft gasification is not suitable for internal combustion engines.

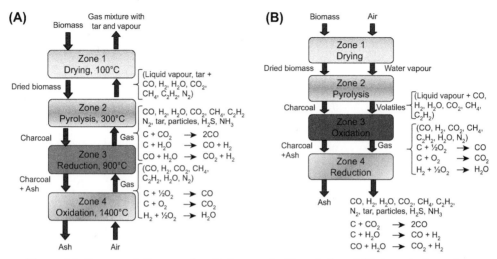

Figure 4.8 Conceptual diagram of multiple steps in (A) updraft and (B) downdraft gasifier.

4.4.2.1.2 Downdraft Gasification

The downdraft gasifier features a co-current flow of air needed for gasification, where product gases and solids flow downward. The operating principle of this gasifier, as shown in Fig. 4.8B, is such that the biomass and air are fed from the top, and are first introduced into the drying zone, followed by the pyrolysis, oxidation, and reduction zones, and finally the product gas is drawn out from the bottom, through the reduction zone. In this system, the gas is quite clean from the downdraft gasifier, it is suitable for internal combustion engines and turbines for electricity generation; however, because the gas leaves the gasifier at a relatively high temperature, it needs to be cooled down before downstream application.

4.4.2.1.2.1 Operating Variables

1. Temperature: In the gasification of biomass, temperature is one of the most important parameters that can control the gas composition, tar concentration, reaction rate, ash build-up, and so on. Therefore, it needs to be highly controlled. Low-temperature (700°C−1000°C) gasification is attributed to high tar content and low CO and H_2 content in the product gas. On the other hand, high-temperature (1200°C−1600°C) gasification leads to a desired high yield of CO and H_2, while reducing the tar content.
2. Pressure.
3. Gasifying agent.

4. Air—fuel ratio (AFR) and equivalence ratio (ER): The mass ratio of air to fuel in any combustion unit is defined as AFR. The ER can be defined as the ratio between the AFR of the gasification process and the AFR for complete combustion. The mathematical representations of AFR and ER are as follows:

$$\text{Air} - \text{fuel ratio} = \frac{\text{mol of air}}{\text{mol of fuel}}$$

$$\text{ER} = \frac{\text{actual air} - \text{fuel ratio}}{\text{air} - \text{fuel ratio for complete combustion}}$$

4.4.2.1.3 Plasma Gasification

In a plasma, gas molecules are ionized by electric discharges. A plasma is highly reactive due to the electrons, ions, and the high energy density in the gas. To generate a plasma, direct current (DC) discharge, alternating current (AC) discharge, radio frequency (RF) induction discharge, or microwave discharge are typically used. Thermal and cold plasmas can be distinguished. Cold plasmas are generated at vacuum pressure, whereas thermal plasmas are achieved at atmospheric pressure. As plasma gases argon, nitrogen, hydrogen, water vapor, or gas mixtures are used, the temperatures of thermal plasmas can be 5000K and higher.

Even if the main application for plasma gasification is waste treatment, the use of plasma gasification has gained attraction for syngas production from biomass. Gasification of injected biomass in the plasma takes place within milliseconds without any intermediate reaction at the very high temperatures. Advantages of plasma gasification are syngas with high hydrogen and CO content, low CO_2 content, low tar concentration, high heating value, useable for wet biomass, such as sewage sludge, and no influence of particle size and structure of the biomass. Disadvantages are the high electricity consumption to operate the plasma, high investment costs, and lower efficiencies.

4.4.2.1.4 Supercritical Water Gasification

The supercritical water gasification (SCWG) process is an alternative to both conventional gasification as well as the AD processes for conversion of wet biomass. This process does not require drying and the process takes place at much shorter residence times; a few minutes at most. SCWG is therefore considered to be a promising technology for the efficient conversion of wet biomass into a product gas that after upgrading can be used as substitute natural gas. Water in its supercritical condition above its critical point of P = 22.12 MPa and T = 374.12°C (see Fig. 4.9) has unique properties as solvent and as reactant. Solubility of organic materials and gases is significantly increased and materials that are insoluble in water or water vapor can be dissolved whereas solubility of inorganic material is decreased.

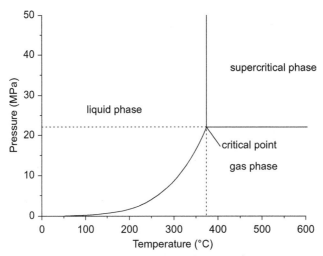

Figure 4.9 Schematic phase diagram of water.

Gasification of biomass is mainly influenced by the density, viscosity, and dielectric constant of water. Above the critical point, physical properties of water drastically change and water behaves as a homogeneous fluid phase. In its supercritical state, water has a gas–like viscosity and liquid–like density, two properties which enhance mass transfer and solvation properties, respectively. Main advantages of biomass gasification in supercritical water are: wet biomass can be treated without predrying; even liquid biomass waste can be treated, for example, olive mill wastewater production of hydrogen–rich

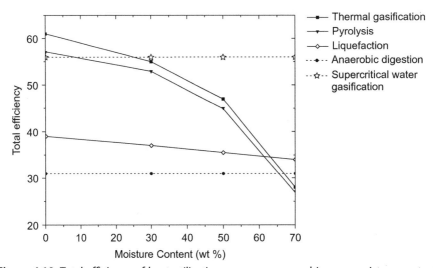

Figure 4.10 Total efficiency of heat utilization processes versus biomass moisture content.

gas, high gasification efficiency, and low tar formation. Main disadvantages are: high investment costs due to the need of special corrosion— and high pressure— and high temperature—resistant materials, and high-energy need to heat up the water to the re-action temperature [20]. Fig. 4.10 shows the comparison of SCWG with the other biomass conversion routes on heat utilization efficiency basis.

4.4.2.2 Pyrolysis

Pyrolysis is a thermochemical decomposition process in which organic material such as biomass is converted into a carbon-rich solid and a volatile matter by heating in the absence of oxygen. The solid product of this process is known as the biochar or char, and is generally high in carbon content. The volatile fraction of this process is partly condensed to a liquid fraction called tar or bio-oil along with a mixture of the noncondensable gases. The bio-oil is stored and further used for energy production. The gases can be utilized for providing heat energy to the pyrolysis reactor. The pyrolysis products are formed from both primary decomposition of the solid biomass as well as secondary reactions of volatile condensable organic products into low-molecular weight gases, secondary tar, and char. One of the significant benefits of pyrolysis is that it can be conducted at temperatures lower (normally in range of 673—973K) than those required in gasification (>973K) and combustion (>1173K) processes. The pressure requirement is also much lower in pyrolysis process (0.1—0.2 MPa) as compared to hydrothermal liquefaction (10—25 MPa) of biomass to generate bio-oil. The operation of a pyrolysis process depends on several parameters such as the feedstock and reactor conditions that lead to formation of products in different proportions as shown in Fig. 4.11.

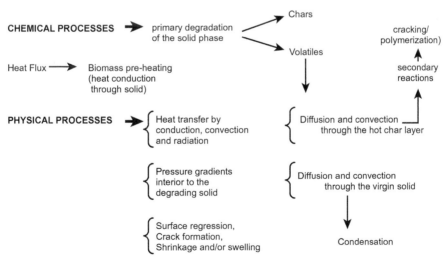

Figure 4.11 Schematic diagram of chemical and physical processes inside biomass particle during pyrolysis [20].

CFD models have been used to study the biomass pyrolysis in various reactor configurations for analyzing the effect of process parameters on the reactor hydrodynamics and the product yields. It is a powerful tool for analyzing the reactor performance for different kind of processes. Multiphase CFD modeling of pyrolysis process is essentially a combination of Eulerian and Lagrangian approaches. These modeling approaches are further subdivided into different classes, such as Euler—Euler and Euler—Lagrangian. Table 4.5 describes applicability of these models at different scales.

4.4.3 Applications of Thermal Conversion

4.4.3.1 Combined Heat and Power

Combined heat and power (CHP), production or cogeneration, is the simultaneous generation of electricity and heat. Trigeneration is a further extension to include a refrigeration process for air conditioning as well. Despite issues related to the use of biomass in energy production, using it in a CHP plant is the most efficient way to produce both heat and electricity from biomass. There are also several combustion technologies that have been developed and optimized for the use of biomass, depending predominantly on the scale required. Although larger-scale biomass systems can make use of well-established technologies from conventional electricity generation, below about 1MWe most technologies can currently be considered to be under development. However, a growing number of successful projects are demonstrating the capabilities of biomass CHP. There are mainly two types: (1) large-scale CHP generation and (2) small-scale CHP generation.

Applications of CHP: CHP is most suitable when there is year round demand for heat to balance the demand for electricity, but is useful:

Table 4.5 Description of Different CFD Models [22]

Name	Gas Phase	Solid Phase	Gas—Solid Coupling	Scale
Discrete bubble model (Model 1)	Lagrangian	Eulerian	Drag closures for bubbles	Industrial (10 m)
Two-fluid model (Model 2)	Eulerian	Eulerian	Gas—solid drag closures	Engineering (1 m)
Unresolved discrete particle model (Model 3)	Eulerian (unresolved)	Lagrangian	Gas-particle drag closures	Laboratory (0.1 m)
Resolved discrete particle model (Model 4)	Eulerian (resolved)	Lagrangian	Boundary condition at particle surface	Laboratory (0.01 m)
Molecular dynamics (Model 5)	Lagrangian	Lagrangian	Elastic collisions at particle surface	Mesoscopic (<0.001 m)

1. when there is a requirement for space heating or process heat close to the generator;
2. to provide low-temperature (up to 90°C) hot water heating for local district schemes;
3. for applications that require (low-grade) process heat, especially those that can supply their own fuel (i.e., sawmills and wood process industry which use heat for timber drying and steaming);
4. at sites such as hospitals, leisure centers, greenhouses, and retirement complexes which have a year round heat demand;
5. to provide steam for other industrial applications;
6. where there is a requirement for environmentally responsible disposal of waste (i.e., sewage sludge, clinical waste, or agricultural residue) and where transport costs for disposal are high;
7. to power an absorption refrigerator to provide cooling in summer, giving trigeneration.

4.4.3.1.1 Co-Firing

Biomass co-firing is regarded as one of the attractive short-term options for biomass in the power-generation industry. It is defined as the simultaneous blending and combustion of biomass with other fuels, such as coal and/or natural gas, in a boiler in order to generate electricity. Solid biomass co-firing is the combustion of solid biomass fuels, such as wood chips and pellets, in coal-fired power plants. Gas biomass co-firing is the simultaneous firing of gasified biomass with natural gas or pulverized coal (PC) in gas power plants in a technique usually referred to as indirect co-firing [23]. There are three types of co-firing: direct co-firing, indirect co-firing, and parallel co-firing.

4.4.3.1.1.1 Direct Co-firing   Direct co-firing is a simple approach and the most common and least expensive method of co-firing biomass with coal in a boiler, usually a

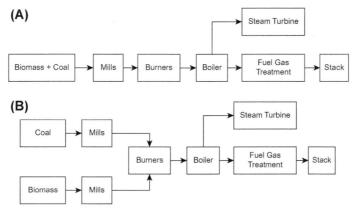

Figure 4.12 Direct biomass co-firing technologies: (A) Mixing biomass with coal. (B) Separate biomass feeding arrangement.

PC boiler. As shown in Fig. 4.1, in direct co-firing technology biomass is fed directly into the furnace after either being milled together with the base fuel (Fig. 4.12A) or being milled separately (Fig. 4.12B) The fuel mixture is then burned in the burner. The co-firing rate is usually in the range of 3–5%. This rate may rise to 20% when cyclone boilers are used, although the best results are achieved with PC boilers.

4.4.3.1.1.2 Indirect Co-Firing

Indirect co-firing technology allows biomass to be co-fired in an oil- or gas-fired system. It exists in two forms, gasification-based co-firing and pyrolyzation-based co-firing. In gasification-based co-firing, the biomass feedstock is fed into a gasifier at the early stages of the process to produce syngas which is rich in CO, CO_2, H_2, H_2O, N_2, CH_4, and some light hydrocarbons. This syngas is then fired together with either natural gas or gasified coal in a dedicated gas burner. The other kind of indirect co-firing is based on pyrolysis, where the biomass fuel undergoes a destructive distillation process to produce a liquid fuel, such as bio-oil, as well as solid char, and then the bio-oil is co-fired with a base fuel such as natural gas in a power station. An illustration of indirect biomass co-firing is shown in Fig. 4.13.

4.4.3.1.1.3 Parallel Co-Firing

In parallel biomass co-firing technology, as shown in Fig. 4.14, biomass preprocessing, feeding, and combustion activities are carried out in separate, dedicated biomass burners. Parallel co-firing involves the installation of a completely separate external biomass-fired boiler in order to produce steam used to generate electricity in the power plant. Instead of using high-pressure steam from the main boiler, the low-pressure steam generated in the biomass boiler is used to meet the process demands of the coal-fired power plant. Parallel co-firing offers more opportunity

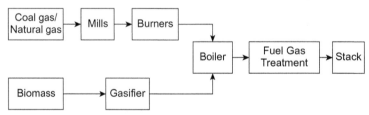

Figure 4.13 Indirect biomass co-firing technologies.

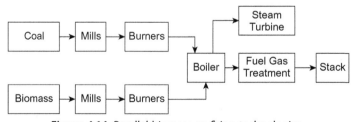

Figure 4.14 Parallel biomass co-firing technologies.

for higher percentages of biomass fuels to be used in the boiler. This technology also offers lower operational risk and greater reliability due to the availability of separate and dedicated biomass burners running in parallel to the existing boiler unit.

4.4.4 Chemical Conversion

A range of chemical processes may be used to convert biomass into other forms, such as to produce a fuel that is more conveniently used, transported, or stored, or to exploit some property of the process itself.

4.4.4.1 Anaerobic Digestion

AD is a biological treatment performed in the absence of oxygen to stabilize organic matter while producing biogas, a mixture formed mainly of methane and carbon dioxide. AD is the process whereby bacteria break down organic material in the absence of air, yielding a biogas containing methane. The products of this process are:

- biogas [principally methane (CH_4) and carbon dioxide (CO_2)];
- a solid residue (fiber or digestate) that is similar, but not identical, to compost;
- a liquid liquor that can be used as a fertilizer.

There are two basic AD processes, which take place over different temperature ranges: (1) mesophilic digestion takes place between 20°C and 40°C and can take a month or two to complete and (2) thermophilic digestion takes place from 50 to 65°C and is faster, but the bacteria are more sensitive.

Biofuel production from microalgal feedstock has several benefits. It is anticipated that the incorporation of AD in microalgae biofuel production and biorefinery processes will increase the cost effectiveness of the production methods, helping it to become economically feasible and environmentally sustainable. Fig. 4.15 illustrates the conceptual implementation of AD into algal production processes. Three pathways have been defined. Pathway 1 shows the direct AD after the biomass harvest and concentration step.

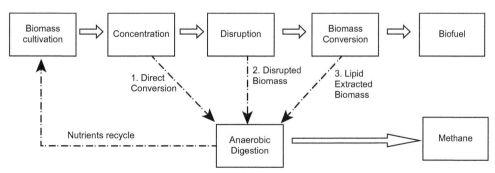

Figure 4.15 Conceptual visualization of anaerobic digestion incorporation into algal biofuel production [23].

Pathway 2 could be utilized in a wastewater process where the cell wall is degradable by bacterial activity within the digester. The second pathway illustrates the AD of biomass after cell wall disruption prior to conversion. Pathway 3 is the traditional biodiesel practice where lipid is extracted and residual algal biomass is converted to biogas by AD and methane fermentation [24].

4.4.4.2 Transesterification

Transesterification is the conversion of a carboxylic acid ester into a different carboxylic acid ester. The most common method of transesterification is the reaction of the ester with an alcohol in the presence of an acid catalyst. For biodiesel production the relevant lipids from microalgae oil are the nonpolar lipids triacylglycerols (TAGs) and free fatty acids (FFAs) while acid, alkali, or enzymatic catalysis (Fig. 4.16; Table 4.6) can be used

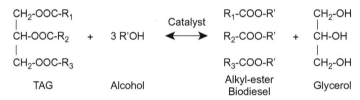

Figure 4.16 Transesterification reaction for biodiesel production.

Table 4.6 Application of Transesterification Technologies

Type of Transesterification	Advantages	Disadvantages
Chemical catalysis	1. Reaction condition can be well controlled 2. Large-scale production 3. The cost of the production process is cheap 4. The methanol produced in the process can be recycled 5. High conversion of the production	1. Reaction temperature is relative high and the process is complex 2. The later disposal process is complex 3. The process needs much energy 4. Needs an installation for methanol recycling 5. The waste water pollutes the environment
Enzymatic catalysis	1. Moderate reaction condition 2. The small amount of methanol required in the reaction 3. Have no pollution to natural environment	1. Limitation of enzyme in the conversion of short chain of fatty acids 2. Chemicals exist in the process of production are poisonous to enzyme
Supercritical fluid techniques	1. Easy to be controlled 2. It is safe and fast 3. Friendly to environment	1. High temperature and high pressure in the reaction condition leads to high cost of production and wastes energy

for the transesterification reaction of lipids with an alcohol (methanol or ethanol) to form fatty acid (m)ethyl esters.

4.5 CO_2 CAPTURE AND STORAGE

4.5.1 CO_2 Capture Technique

To reduce the effects caused by this environmental problem, several technologies were studied to capture CO_2 from large emission source points: (1) absorption; (2) adsorption; (3) gas-separation membranes; and (4) cryogenic distillation. Absorption of CO_2 is effected using various chemical agents such as monoethanolamine (MEA), solid adsorbents like activated carbon, or zeolite 5A. Membranes and cryogenic fractionation have also been employed for the removal of CO_2. The chemical methods of CO_2 separation are highly energy-intensive and expensive. Conventional carbon capture technologies (largely using chemical methods) have a capture efficiency of 85—95%. It has been reported that 3.7 GJ of energy/tonne of CO_2 absorbed is required during the regeneration of MEA, which corresponds to around 370 kg of extra CO_2 emitted if this energy input comes from a fossil fuel such as coal. The resulting streams with high CO_2 concentrations are transported and stored in geological formations. However, these methodologies, known as CCS technologies, are considered as short-term solutions, as there are still concerns about the environmental sustainability of these processes. Fig. 4.17 shows the different types of conventional CO_2 capture technologies.

However, a potential and promising biological approach, microalgae-based CO_2 fixation and energy/resource utilization, has received significant attention over the last two decades due to its techno-economic feasibility and environmental friendliness. In essentials, the microalgae essentially biologically fix and store CO_2 via photosynthesis, which can convert water and CO_2 into organic compounds without secondary pollution. Microalgal-CO_2 fixation features potential advantages over other CCS approaches, such as a wide distribution, high photosynthesis rate, good environmental adaptability, and easy operability. Additionally, the microalgal biomass can be harvested after CO_2 fixation to produce microalgal biofuel that can be utilized as a renewable or sustainable energy source (Fig. 4.18).

4.5.2 Biosynthesis of Lipid in Microalgae

Microalgae are photosynthetic microorganisms with simple growing requirements (light, sugars, CO_2, nitrogen, phosphorous, and potassium) that can produce lipids in large amounts over short periods of time. Microalgae transform the solar energy into the carbon storage products, leads to lipid accumulation, including TAG (triacylglycerols), which then can be transformed into biodiesel, bioethanol, and biomethanol. Most microalgae species produce lipids, carotenoids, antioxidants, fatty acids, enzyme

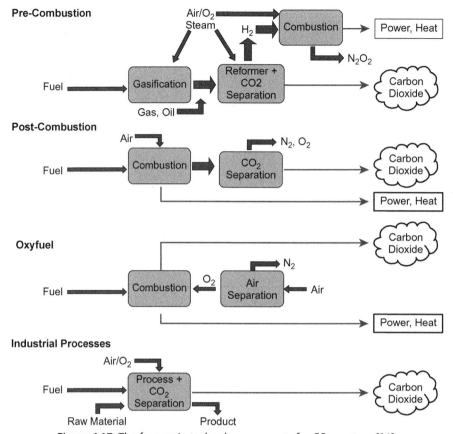

Figure 4.17 The four main technology concepts for CO_2 capture [21].

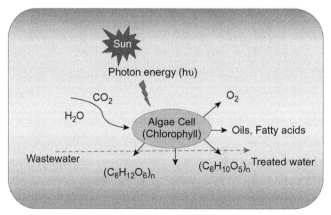

Figure 4.18 Microalgae species uses light energy (e.g., from the sun) to produce chemical energy by photosynthesis during natural growth cycle.

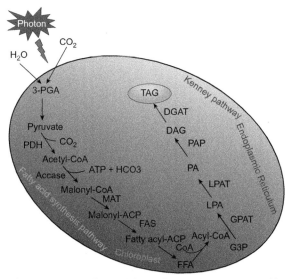

Figure 4.19 Pathway of the synthesis of triacylglycerols (TAGs), as storage of lipid as chemical energy by microalgae [20].

polymers, peptides, toxins, and sterols. A diverse range of microalgae species uses light energy (e.g., solar energy) to produce chemical energy by photosynthesis with the natural growth cycle of just few days (Fig. 4.19).

4.6 BIOLOGICAL CO$_2$ FIXATION

A promising technology is the biological capture of CO$_2$ using microalgae. These microorganisms can fix CO$_2$ using solar energy with efficiency 10 times greater than that of terrestrial plants. Moreover, the capture process using microalgae has the following advantages: (1) being an environmentally sustainable method; (2) using directly the solar energy; and (3) coproducing high added value materials based on biomass, such as human food, animal feed mainly for aquaculture, cosmetics, medical drugs, fertilizers, biomolecules for specific applications, and biofuels. Microalgae can typically be used to capture CO$_2$ from three sources: (1) atmospheric CO$_2$, (2) CO$_2$ emission from power plants and industrial processes, and (3) CO$_2$ from soluble carbonate.

Furthermore, four applications are achieved by using microalgae biomass production as a CO$_2$ reduction strategy: (1) production of biofuels, (2) enhancement of the economic yield of the CCS through production of commodities or by-products from flue gases, (3) utilization of bacteria—microalgae consortiums to reduce the energy required for aeration in wastewater treatment plants, and (4) utilization of microalgae to reduce the total CO$_2$ emissions released by wastewater treatment plants.

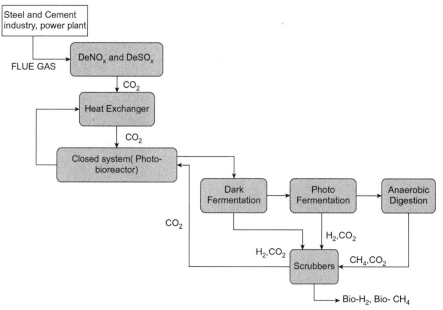

Figure 4.20 Microalgae CO_2 removal using industrial flue gases [19].

4.6.1 Microalgae Removal of CO_2 From Industrial Flue Gas

CO_2 capture from flue gas emissions from power plants that burn fossil fuels achieves better recovery due to the higher CO_2 concentration of up to 20%. Since microalgae CO_2-fixation involves photoautotrophic growth of cells, CO_2 fixation capability of specific species should positively correlate with their cell growth rate and light utilization efficiency (Fig. 4.20). The advantages of using microalgae to capture CO_2 from coal combustion flue gas are:

1. High-purity CO_2 gas is not required for algal culture. Flue gas containing varying amounts of CO_2 can be fed directly to the microalgal culture. This simplifies CO_2 separation from flue gas significantly.
2. Some combustion products, such as NO_x or SO_x, can be effectively used as nutrients for microalgae. This could potentially negate the use of flue gas scrubbing systems for power plants.
3. The microalgae could yield high−value commercial products. The sale of these high-value products could offset the capital and operating costs of the process.
4. The envisioned process is a renewable cycle with minimal negative impacts on the environment.

4.6.1.1 Power Plant

Flue gases from coal power plants can be a potential CO_2 source for the production of microalgal biomass. Microalgae can utilize CO_2 with the help of solar energy, 10 times more efficiently than terrestrial plants. Microalgae can be grown in saline conditions or wastewater throughout the year. Flue gases are generally dominated by N_2 (72–74%), CO_2 (4.8–26.9%), H_2O (9–13.8%), and O_2 (0.7–15%). However, they also contain smaller quantities of NO (59–1500 mg/Nm^3), NO_2 (2–75 mg/Nm^3), SO_2 (20–1400 mg/Nm^3), SO_3 (0–32 mg/Nm^3), CxYx (0.008–0.4%), CO (100–11,250 mg/Nm^3), particulate matter (2000–15,000 mg/Nm^3), and heavy metals (2.2 mg/Nm^3). Typically, flue gases are treated for the removal of particulate matter, heavy metals, and NO_x and SO_x to comply with the regulations on effluent discharge and air quality.

4.6.1.2 Cement Industry

The cement industry is one of the major CO_2-producing sectors being responsible for about 8% of global emissions. A production of 193×10^6 metric tons of CO_2 by the cement industry is reported, considering only member states of the European Union and Norway. Hasanbeigi et al. [25] reviewed 18 technologies for the reduction of CO_2 emissions by cement industry. They classified algal biomass utilization as an emerging technology in demo stage. Only a few studies regarding flue gas usage from cement industry have been developed. Borkenstein et al. [26] evaluated the air lift cultivation of *Chlorella emersonii* using flue gas derived from a cement plant. Pure CO_2 injection was used as a control and 5.5 L photobioreactors with controlled pH were used. After 30 days of cultivation, the flue gas had no visible adverse effects compared with the control reactors. The control essay (pure CO_2) resulted in a biomass yield of 2 g/L, CO_2 fixation of 3.25 g/L, and growth rate of 0.1/day, meanwhile the flue gases reactors resulted in very similar parameters with 2.06 g/L in biomass yield, 3.38 g/L in CO_2 fixation, and a growth rate of 0.13/day. Although there was no accumulation of flue gas residues in the culture media, the lead concentration in the microalgae biomass was three times higher with the flue gases. Therefore, lead accumulation and its effect on the downstream processing for biofuels' production have to be investigated.

Lara-Gil et al. [27a] performed toxicity tests of a simulated cement industry flue gas in cultures of *Desmodesmus abundans* and *Scenedesmus* sp. The results suggest that nitrite and sulfite are not toxic for the tested microalgae at the maximum concentrations of 1067 ppm and 254 ppm, respectively, differing from bisulfate where concentrations above 39 ppm were toxic. Studies related to flue gas from different industries can be considered useful despite slight changes in flue gas composition.

4.7 MICROALGAE CULTIVATION TECHNOLOGY

Microalgae with the composition of $CH_{1.7}O_{0.4}N_{0.15}P_{0.0094}$ are simple photosynthetic organisms living in aquatic environments, where they can convert CO_2 and H_2O to biomass using sunlight. Factors influencing the microalgae growth include: abiotic factors, such as light intensity and quantity, temperature, O_2, CO_2, pH, salinity, and nutrients (N, P, K, etc.); biotic factors, such as bacteria, fungi, viruses, and competition for abiotic matters by other microalgae species; and operational factors, such as mixing and stirring degree, width and depth, dilution rate, harvest frequency, and addition of bicarbonate. The cultivation technologies being pursued to produce microalgae for biofuel generation mainly include open ponds, photobioreactors, and fermenters.

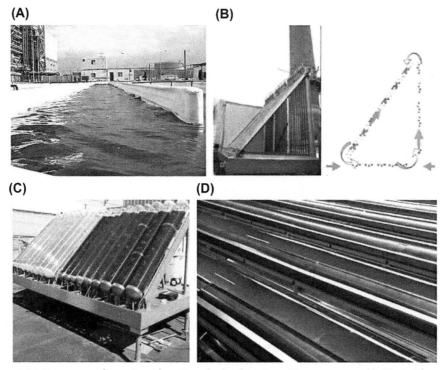

Figure 4.21 Reactor configurations for microalgal cultivation: (A) raceway pond; (B) air-lift reactor; (C) bubble column reactor; and (D) horizontal tubular reactor. (A) From Seambiotic; (B) from Green Fuel Technologies—MIT [27b]; (C) from Green Fuel Technologies; and (D) from http://www.algaelink.com.

4.7.1 Bioreactors

Microalgae can grow either in open ponds or closed systems (photobioreactors). Fig. 4.21 shows images of the most common bioreactor configurations. Table 4.1 makes a comparison between the open and closed bioreactors concerning the production of microalgae. The production in open ponds depends on the local climate due to the lack of control in this type of bioreactors. The contamination by predators is an important drawback of this cultivation system. Thus, high production rates in open ponds are achieved with algal strains resistant to severe culture environment; for instance, the *Dunaliella*, *Spirulina*, and *Chlorella* spp. are cultivated in high salinity, alkalinity, and nutrition, respectively. Besides the technological simplicity, the production in open systems is not cheap due to the downstream processing costs.

4.7.1.1 Open Systems

The production in open ponds depends on the local climate due to the lack of control in this type of bioreactors. The contamination by predators is an important drawback of this cultivation system. Thus, high production rates in open ponds are achieved with algal strains resistant to severe culture environment; for instance, the *Dunaliella*, *Spirulina*, and *Chlorella* spp. are cultivated in high salinity, alkalinity and nutrition, respectively. Besides the technological simplicity, the production in open systems is not cheap due to the downstream processing costs.

4.7.1.1.1 Limitations of Open Pond Systems

The major limitations in open systems include the following: (1) poor light utilization by the cells, (2) significant evaporative losses, (3) limited diffusion of CO_2 from the atmosphere, and (4) requirement of large areas of land.

. In addition, contamination is another major problem of open systems with large-scale microalgal production. Unwanted algae, mold, fungi, yeast, and bacteria are the common biological contaminants often found in these open systems.

4.7.1.2 Closed System

Closed systems, mainly known as photobioreactors, can address some of the problems associated with open pond systems. The major advantages of the closed systems are as follows: (1) minimization of water evaporation and (2) reduction of the growth of competitive algal weeds, predators, and pathogens that may kill the desired microalgae. It is important to acknowledge that although photobioreactors significantly reduce the growth of competitive algal weeds, they cannot completely eliminate the growth of contaminants. A detailed comparison of different closed photobioreactor systems and their biomass productivities are reported in Tables 4.7 and 4.8.

Table 4.7 Characteristics' Comparison of Open Ponds, Photobioreactors, and Fermenters [30–33]

Parameter	Open Systems (Raceway Ponds)	Closed Systems (Photobioreactors)	Fermenter
Land requirement	High	Variable	Low
Water loss	Very high, may also cause salt precipitation	Low and it may be high if water spray is used for cooling	Low
Hydrodynamic stress on algae	Very low	Low–high	Unknown
Gas transfer control	Low	High	High
CO_2 loss	High, depending on pond depth	Low	No CO_2 is required
O_2 inhibition	Usually low enough because of continuous spontaneous outgassing	High (O_2 must be removed to prevent photosynthesis inhibition)	O_2 supply should be sufficient
Temperature	Highly variable	Cooling often required	Should be controlled to some special level
Startup	6–8 weeks	2–4 weeks	2–4 weeks
Construction costs	High—US $100,000 per hectare	Very high—US $1,000,000 per hectare: PBR plus supporting systems	Low
Operating costs	Low—paddle wheel, CO_2 addition	Very high—CO_2 addition, pH control, oxygen removal, cooling, cleaning, maintenance	Very high—oxygen addition, cleaning, sterilization, maintenance
Limiting factor for growth	Light	Light	O_2
Control over parameters	Low	Medium	Very high
Technology base	Readily available	Under development	Readily available
Risk of pollution	High	Medium	Low
Pollution control	Difficult	Easy	Easy
Species control	Difficult	Easy	Easy
Weather dependence	High—light intensity, temperature, rainfall	Medium—light intensity, cooling required	Low
Maintenance	Easy	Hard	Hard

(Continued)

Table 4.7 Characteristics' Comparison of Open Ponds, Photobioreactors, and Fermenters [30–33]—cont'd

Parameter	Open Systems (Raceway Ponds)	Closed Systems (Photobioreactors)	Fermenter
Ease of cleaning	Easy	Hard	Hard
Susceptibility to overheating	Low	High	Unknown
Susceptibility to excessive O_2 levels	Low	High	Unknown

Table 4.8 Comparison of Different Closed Photobioreactor Systems [29]

Closed System Photobioreactor Type	Light Source	Capacity (L)	Algal Strain	Biomass Conc. (g/L)	Biomass Conc. (g/m³ day)
Tubular	Artificial	5.5	*Spirulina platensis*	0.62	
	Sun	200	*Phaeodactylum tricornutum*	1.19	
	Sun	75	*Phaeodactylum tricornutum*	1.38	
	Sun	10,000	*Spirulina*		25
Airlift	Artificial	3	*Haematococcus pluvialis*	4.09	
	Artificial	170	*Chaetoceros*	0.80	
	Artificial		Nannochloropsis		32.5−95.3
	Artificial		Chlorella		109−264
Bubble column	Artificial	170	*Chaetoceros*	3.31	
	Artificial	1.9	*Phaeodactylum*	−	
	Artificial	4.5	*Monoraphidium*		23
	Artificial	1.8	Cyanobium sp.	0.071	
	Artificial	3.5	Spirulina	4.13	
	Sun	64	*Monodus*	0.03−0.20	
	Artificial	1.8	*Sc. obliquus*	2.12	
	Artificial	1.8	*Chlorella vulgaris*	1.41	
Flat plate	Artificial	3.4	*Dunaliella*	1.5	
	Sun	5	*Phaeodactylum*	1.38	

4.7.1.2.1 Tubular Photobioreactor

Tubular photobioreactors are made with transparent materials and are placed in outdoor facilities under sunlight irradiation (Fig. 4.22A). A gas exchange vessel where air, CO_2,

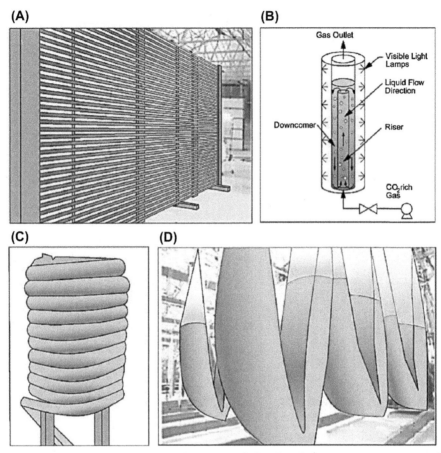

Figure 4.22 Closed cultivation system: (A) horizontal tubular photobioreactor at Varion Aqua Solution Ltd., United Kingdom; (B) bubble column air-lift photobioreactor, BBSRC, United Kingdom; (C) helical–tubular photobioreactor; and (D) large-scale plastic bag photobioreactors [28].

and nutrients are added and O_2 is removed is connected to the main reactor. One of the basic characteristics of these cultivation vessel designs is their large surface area per unit volume. This is done to maximize exposure of the microalgae to sunlight. Tube sizes are generally less than 10 cm in diameter to secure sunlight penetration. In a typical tubular microalgae culture system, the medium is circulated through the tubes, where it is exposed to sunlight for photosynthesis. The medium is circulated back to a reservoir with the help of a mechanical pump or an airlift pump. The pump also helps to maintain a highly turbulent flow within the reactor, preventing the algal biomass from settling. A fraction of the algae is usually harvested after it circulates through the solar collection tubes, making the system a continuous operation. Until today, most of the tubular photobioreactors, studied in presence of artificial light, have been developed at small/laboratory scale (0–20 L capacities). There is, in this respect, a limited number of studies reporting data for large-scale closed photobioreactors. James and Al-Khars studied the growth and the productivity of *Chlorella* and *Nannochloropsis* in a translucent vertical airlift photobioreactor. This is a bubble column unit with good light penetration. Its implementation at full commercial scale-up still seems challenging Fig. 4.22B.

In some photobioreactors, the tubes are coiled spirals forming helical–tubular photobioreactors (Fig. 4.22C). Usually, these types of reactors are suitable for the culture of microalgal species in the presence of sunlight. Despite this, these systems sometimes require artificial illumination as well as natural light to enhance the microalgae growth. However, the introduction of artificial light adds to production costs, making the helical–tubular bioreactor only adequate for the manufacturing of high value added products. Another category of closed systems is the airlift photobioreactor. In this reactor, liquid motion is characterized by large circulatory currents in a heterogeneous flow regime.

4.7.1.2.2 Plastic Bag Photobioreactor

There are studies which suggest that microalgae can be produced in transparent polyethylene bags, as shown in Fig. 4.22D. Generally, these bags are either hung or placed in a cage under the sunlight irradiation. In such arrangements, the algae cultures are mixed with air at the bottom of the bags. Transparent polyethylene sleeves sealed at the bottom in a conical shape which are used to prevent cell settling are also widely used. Using 50 L polyethylene bag cultures operated as turbidostats, Trotta obtained yields of $20–30 \text{ g/m}^3$ day for Tetraselmis.

4.7.1.2.3 Airlift Photobioreactor

In airlift photobioreactors (Fig. 4.22B), the fluid volume of the vessel is divided into two interconnected zones using a baffle or a draft tube. Liquid movement is characterized by large circulatory currents in the heterogeneous flow regime. Airlift photobioreactors are sometimes difficult to scale up given their complex flow pattern.

4.7.1.2.4 Flat-Plate Photobioreactor

Vertical plate photobioreactors mixed by air bubbling seem to be even better than bubble columns in terms of productivity and ease of operation. Flat-plate photobioreactors allow the following: (1) large irradiated surface area, (2) suitable for outdoor cultures, (3) good for algae immobilization, and (4) good biomass productivities. These photobioreactors are relatively cheap and easy to clean.

4.7.2 Key Growth Parameters

The research on CO_2 removal by microalgae covers two fields: (1) the CO_2 capture from flue gases (10−20% CO_2) and (2) the CO_2 capture from closed spaces (less than 1% CO_2). The process variables that could influence the success of cultivation are carbon sources, nitrogen sources, the light distribution and saturation, temperature, pH, salinity, nutrient qualitative and quantitative profiles, dissolved oxygen concentration, and the presence of toxic elements (heavy metals). Several environmental (location of the cultivation system, rainfall, solar radiation, and so on), engineering (pond depth, CO_2 delivery system, methods of mixing, power consumption, and so on), and biological (light, pH, oxygen accumulation, salinity, algal predators, and so on) parameters affect the biomass productivity in the open pond system.

4.7.2.1 Characterizing Parameters

Various species of microalgae have been cultivated for biofuel production. The National Renewable Energy Laboratory (NREL) has identified over 3000 microalgal strains, which makes the selection of a suitable strain not an easy task. The selection process requires the evaluation of various parameters, including lipid productivity, CO_2 fixation rate, and adaptability to grow in harsh conditions, such as high temperature and salinity.

4.7.2.2 Microalgal Harvesting Test

The equation for the harvesting efficiency is shown in the following equation:

$$\text{Harvesting efficiency (\%)} = \frac{OD_i - OD_f}{OD_i} \times 100 \text{ or } \frac{N_i - N_f}{N_i} \times 100$$

where OD_i and OD_f are the optical density (OD) of the initial microalgal culture and the supernatant liquid after microalgal harvesting, respectively. N_i and N_f are the sum of the cell numbers between 1.97 and 4.977 μm of the initial medium and supernatant liquids after microalgal harvesting.

4.7.2.3 Kinetic Study

To design and scale up a cultivation system, the development of a kinetic model that adequately describes cells growth is essential. Microbial cell growth kinetics studies are commonly carried out in batch cultures. CO_2 and HCO_3 are the potential inorganic carbon sources for photosynthesis in microalgae. The SGR (specific growth rate) to describe the microalgae biomass kinetics is:

$$K = \frac{(\ln X_t - \ln X_0)}{t - t_0}$$

Table 4.9 Fixation of CO_2 by Different Microalgae

Microalgae Species	CO_2 Fixation Rate ($g/m^3/h$), or Removal Efficiency (%)
Chlorogleopsis sp.	0.8–1.9
Chlorella sp.	16–58%
Chlorella sp. *NCTU-2*	63%
Chlorella sp.	10–50%
Chlorella vulgaris	128 and 141
Chlorella vulgaris	80–260
Chlorella vulgaris	148
Euglena gracilis	3.1
Porphyridium sp.,	3–18
S. platensis	38.3–60

where K is the SGR (day^{-1}), and X_t and X_0 are the biomass concentration (based on dry cellular weight) (g/L) at culture time t and t_0, respectively.

The BPR (biomass production rate), that is, the linear growth rate or average growth rate, was used to estimate the microalgae biomass productivity according to the following equation:

$$P = \frac{(X_t - X_0)}{(t - t_0)}$$

where P is the BPR (g/L/day).

4.7.2.4 CO$_2$ Fixation Rates

CFR (CO_2 fixation rate) could be estimated according to the biomass production rate and carbon content in microalgae biomass (Table 4.9). Some investigations statistically estimated it via an approximate molecular formula of microalgae biomass with $CO_{0.48}H_{1.83}N_{0.11}P_{0.01}$ (C, O, H, N, and P represent the elements carbon, oxygen, hydrogen, nitrogen, and phosphorus, respectively). However, the elemental composition is usually varied with environmental conditions, culture processes, and in particular with microalgae species. To characterize the CFR for most unicellular microalgae, this work uses the following equation for calculation:

$$R_{CO_2} = P_M \times C \times \frac{M_{CO_2}}{M_C}$$

where R_{CO_2} is the CO_2 fixation rate, C is the determined carbon content of the biomass; M_{CO_2} and M_C are CO_2 and elemental carbon molecular weights, respectively (Table 4.10).

Table 4.10 Comparison of the Growth Characteristics and CO_2 Fixation Performance of Microalgae Strains Under Different CO_2 Concentrations, Temperature, and NO_x/SO_x Contents

Microalgae Species	CO_2 (%)	Temperature (°C)	NO_x/SO_x (mg/L)	Biomass Productivity (mg/L/day)	CO_2 Consumption Rate (mg/L/day)
Nannochloris sp.	15	25	0/50	350	658[a]
Nannochloropsis sp.	15	25	0/50	300	564[a]
Chlorella sp.	20	35	60/20	950	1790[a]
Chlorella sp.	20	40	N.S.	700	1316[a]
Chlorella sp.	50	25	N.S.	386	725
Chlorella sp.	15	25	0/60	1000	1880
Chlorella sp.	50	25	N.S.	500	940[a]
Chlorogleopsis sp.	5	50	N.S.	40	20.45
Chlorococcum littorale	50	22	N.S.	44	82

N.S., not specified.
[a]Calculated from the biomass productivity according to the following equation: CO_2 fixation rate $(P_{CO_2}) = 1.88 \times$ biomass productivity (mg/L/day), which is derived from the typical molecular formula of microalgal biomass, $CO_{0.48} H_{1.83} N_{0.11} P_{0.01}$.

4.8 MICROALGAE HYBRID TECHNOLOGIES

4.8.1 Algal Biorefinery Concept

The concept of biorefining is similar to the petroleum refineries in which multiple fuels and chemicals are derived using crude oil as the starting material. Similarly, biorefining is sustainable biomass processing to obtain energy, biofuels, and high-value products through processes and equipment for biomass transformation. A more specific and comprehensive definition of a biorefinery has been given by IEA Bioenergy Task 42 documents which states, "the sustainable processing of biomass into a spectrum of marketable products and energy." In a broad definition, biorefineries convert all kinds of biomass (all organic residues, energy crops, and aquatic biomass) into numerous products (fuels, chemicals, power and heat, materials, and food). Algae can easily be part of this concept because each strain produces certain amount of lipids, carbohydrates, or proteins that can be used as biomass in different processes. The biorefinery concept has been identified as the most promising way to create a biomass-based industry. There are four main types of biorefineries: biosyngas-based refinery, pyrolysis-based refinery, hydrothermal upgrading—based refinery, and fermentation-based refinery. Biorefinery includes fractionation for separation of primary refinery products. The main goal of the biorefinery is to integrate the production of higher value chemicals and commodities, as

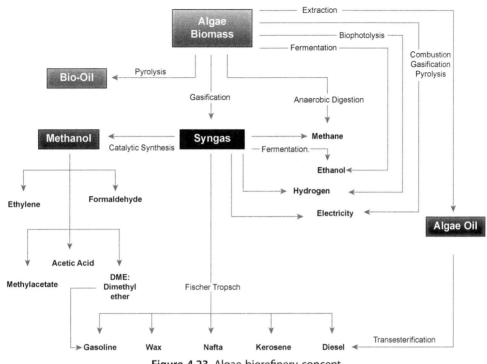

Figure 4.23 Algae biorefinery concept.

well as fuels and energy, and to optimize the use of resources, maximize profitability and benefits, and minimize wastes [34].

Microalgae are considered to be futuristic raw material for establishing a biorefinery because of their potential to produce multiple products. A new biorefinery-based integrated industrial ecology encompasses the different value chain of products, coproducts, and services from the biorefinery industries. Cross-feeding of products, coproducts, and power of the algal biofuel industry into the allied industries is desirable for improving resource management and minimization of the ecological footprint of the entire system. The biomass, after the oil has been extracted from it, can be used as animal feed, converted to fertilizer, and used for power generation. The power generated can then be put back to produce more biomass. The CO_2 released by the power-generation plant can be used again for the production of algal biomass, thus reducing CO_2 in the atmosphere. Selected species of microalgae (freshwater algae, saltwater algae, and cyanobacteria) were used as a substrate for fermentative biogas production in a combined biorefinery. Anaerobic fermentation has been considered as the final step in future microalgae-based biorefinery concept (Fig. 4.23) [35].

Table 4.11 Removal of Inorganic and Organic Pollutant From Wastewater by Different Algae

Microalgae Species	Pollution Control
Anabaena, Oscilatoria, Spirulina, S. platensis	NO_3^-, NO_2^-, NH_4^+, PO_4^{3-}
Anabaena sp.	2,4,6-Trinitrotoluene
Ankistrodesmus sp., *Scenedesmus* sp., *Microactinium* sp., *Pediastrum* sp.,	CO_2
Chlamydomonas reinhardtii	Hg (II), Cd(II), Pb(II)
Chlorella sp.	Boron
Chlorella miniata	Tributyltin (TBT)
Chlorella vulgaris, Chlorella sp.	Tributyltin (TBT)
Chlorella vulgaris	Azo compounds
Chlorella vulgaris	NH_4^+, PO_4^{3-}
Chlorella spp.	P
Chlorella vulgaris	Cd, Zn,
Chlorella vulgaris, Scenedesmus rubescens	N and P
Chlorella salina	Co, Zn, Mn
Coelastrum proboscideum	Pb
Isochrysis galbana	NH_4^+
Ochromonas danica	Phenols
Oedogonium hatei	Ni
Oedogonium sp., *Nostoc* sp.	Pb
Oscillatoria sp. *H1*	Cd(II)
Phormidium bigranulatum	Pb(II), Cu(II), Cd(II)
Phormidium laminosum	Cu(II), Fe(II), Ni(II), Zn(II)
Scenedesmus quadricauda	Cu(II), Zn(II), Ni(II)
Spirulina platensis	Cr(VI)
Streptomyces viridochromogenes, Chlorella regularis	U
Ulva lactuca	Pb (II), Cd (II)
Undaria pinnatifida	Ni, Cu

4.8.2 Wastewater Treatment

Photosynthetic microorganisms, such as microalgae, can use pollutants as nutrients (N, P, and K) and grow in accordance with environmental conditions, such as light, temperature (generally $20-30°C$), pH (around 7.0), salinity, and CO_2 content. On the other hand ecofriendly pollutant removal is a major issue in current-day research. Many researchers consider microalgae as green technological medium for pollutant removal from wastewater. Removal of organic and inorganic pollutants (NO_3^-, NO_2^-, NH_4^+, PO_4^{3-}, CO_2, Cd, Zn, Ni, Co, Mn, Cu, Cr, U, Hg(II), Cd(II), Pb(II), B, TBT (tributyltin), phenols, and Azo compounds) from wastewater by different algae is shown in Table 4.11.

There are several reasons for the cultivation of microalgae in wastewater, such as: (1) cost-effective treatment, (2) low-energy requirement, (3) reduction in sludge formation, and (4) production of algal biomass for biofuel production. Microalgae are efficient to remove different types of pollutants and toxic chemicals, such as nitrogen, phosphorous, potassium, nitrite, silica, iron, magnesium, and other chemicals from municipal and industrial wastewater. In addition, microalgae have high capacity to accumulate heavy metals (selenium, chromium, lead, etc.), metalloids (arsenic), and organic toxic compounds (hydrocarbons) to form microalgae biomass which subsequently can be used for biofuel production. The Chlorella spp. has diverse range of different pollutants compared to other microalgae. Other several algae such as *Ourococcus multisporus*, *Nitzschia* cf. *pusilla*, *Chlamydomonas mexicana*, *Scenedesmus obliquus*, *Chlorella vulgaris*, and *Micractinium reisseri* were efficient to remove nitrogen, phosphorus, and inorganic carbon. The highest achieved capacity *C. mexicana* for removal of nitrogen, phosphorus, and inorganic carbon were 62%, 28%, and 29%, respectively. Simultaneously, the lipid productivity and lipid content were reported to be 0.31 ± 0.03 g/L and $33\% \pm 3\%$, respectively. Using microalgae for combined renewable energy production along with efficient wastewater treatment systems at a low cost offers an innovative promising direction for an integral approach to water and energy problems and climate change mitigation.

4.9 THE ECONOMIC POTENTIAL FOR BECCS

According to climate change mitigation scenario modeling, BECCS is a cost-effective technology for reducing the concentration of CO_2 in the atmosphere and for meeting ambitious climate targets. For ambitious CO_2 levels such as 350 ppm and

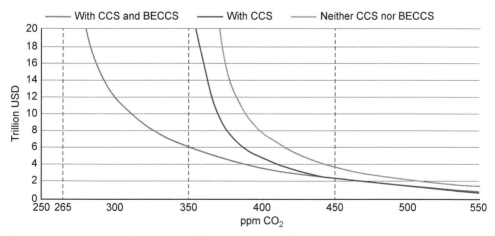

Figure 4.24 Cost of reaching various CO_2 concentration targets depending on mitigation portfolio [35].

below, alternative options are to be considered inadequate or too expensive. It may be necessary to reach these levels in order to avoid severe climate change. It is worth noting that according to the scientific studies refered earlier, the BECCS technology also reduces the cost of less-ambitious climate targets, if included in the total portfolio for climate mitigation measures, see Fig. 4.24. With delayed policy decisions for climate change mitigation,

BECCS may be needed to reach higher stabilization levels such as 400 and 450 ppm in an economically attainable way.

The International Energy Agency has published a report on the role of CCS and BECCS in the global energy portfolio, using their BLUE map scenario. The report shows that BECCS has a very important role to play, if we want to meet the 450 ppm emission target. Using technical, physical, and economic constraints in the optimization model, BECCS is shown to have a profound overall impact. It was found that CCS applied to biomass has more potential than all other industrial applications combined.

4.10 DISCUSSION AND CHALLENGES FOR BECCS

The deployment of large-scale bioenergy faces biophysical, technical, economic, and social challenges, and CCS is yet to be implemented widely. Four major

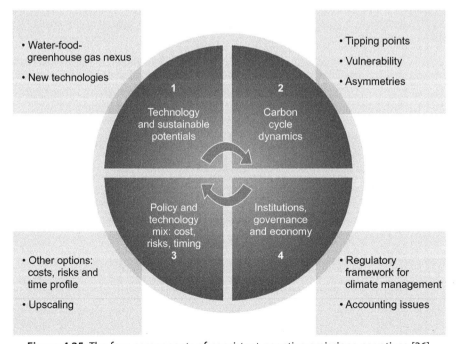

Figure 4.25 The four components of consistent negative emissions narratives [36].

uncertainties need to be resolved: (1) the physical constraints on BECCS, including sustainability of large-scale deployment relative to other land and biomass needs, such as food security and biodiversity conservation, and the presence of safe, long-term storage capacity for carbon; (2) the response of natural land and ocean carbon sinks to negative emissions; (3) the costs and financing of an untested technology; and (4) socio-institutional barriers, such as public acceptance of new technologies and the related deployment policies [36].

A consistent narrative of negative emissions management therefore has four components (Fig. 4.25) relating to the key uncertainties. The first component refers to technological aspects: with BECCS being the negative emissions technology most widely applied by IPCC-integrated assessment models (IAMs), the implied heavy demands for sustainable biomass availability are suggested to be at least 100 EJ/year and up to more than 300 EJ/year of equivalent primary energy by 2050. Also, CO_2 storage potential in geological layers (aquifers, depleted fossil carbon reservoirs) and other resources, such as water and fertilizer, in the face of increasing food demand will need to be addressed. Bioenergy and water recycling with solar-powered distillation, algae grown offshore and fertilized with previously captured CO_2, and other innovations are among possible technologies enabling negative emissions to be achieved with lower pressure on land biomass production. However, such technologies require significant new research and development.

The second component in Fig. 4.25 describes carbon cycle uncertainties and dynamics in the Earth system. If negative emission options such as BECCS are used only after significant climate change, then the response of the global carbon cycle can make the necessary amount of negative emissions even larger than for a scenario where the future CO_2 trajectory is contained below 430—480 ppm. This could occur through decreasing terrestrial and ocean sink efficiencies due to climate change, and net releases of CO_2 by the land and ocean reservoirs due to CO_2 removal over several decades.

The third component acknowledges that negative emissions will be part of a wider mitigation effort and their deployment will depend on the cost, risks, and timing profile of other options. The spectrum ranges from more established mitigation technologies for which it might then be too late to solar-radiation management geoengineering options, which are quicker and cheaper to ramp up, but which embody a much larger scale of mostly unknown risks and are not able to deal with other consequences of increased CO_2 concentrations such as ocean acidification. This emphasizes that we are not in a position to discard the negative emissions option easily, despite the above challenges. The fact that negative emissions solutions such as BECCS will require time to achieve sufficient scale confirms that the future option space depends strongly on today's decisions.

The final component is concerned with institutional and policy challenges. CO_2 removal will be expensive and contentious, whereas emissions will remain cheap in the absence of strong climate policies. Therefore, any CO_2 removal strategy requires an extraordinary global regulatory framework taking into account national economic conditions. In the absence of a global climate agreement requiring stringent mitigation efforts and given the asymmetric distribution of mitigation potentials, negative emissions could help to offset emissions from countries that might not participate in reduction efforts or have less capacity to do so. This could open new perspectives on global climate management. Rigorous monitoring, reporting, and verification will be needed to facilitate these options.

4.11 CONCLUSIONS

BECCS is a carbon reduction technology offering permanent net removal of CO_2 from the atmosphere. This has been termed "negative carbon emissions," and offers a significant advantage over other mitigation alternatives. BECCS is able to do this because it uses biomass that has removed atmospheric carbon while it was growing, and then stores the carbon emissions resulting from combustion permanently underground.

It has been suggested that BECCS could be applied to a wide range of biomass-related technologies, and may also be attractive from a relative cost perspective. To date, however, the combination of bioenergy and CCS has not been fully recognized or realized. It is important to keep in mind that the possible contribution of BECCS depends heavily on the potential and societal acceptance of bioenergy on one hand, and the deployment of capture and storage technologies on the other. Although there may be significant potential for this technology, uncertainties and concerns remain regarding technology development, carbon-negative life cycle assessment, food security, and biodiversity.

REFERENCES

[1] Stocker TF. Climate change 2013: the physical science basis: working group I contribution to the fifth assessment report of the intergovernmental panel on climate change. Cambridge University Press; 2014.
[2] Gough C, Upham P. Biomass energy with carbon capture and storage (BECCS or Bio-CCS). Greenhouse Gases: Science and Technology 2011;1(4):324—34.
[3] Lackner KS. Carbonate chemistry for sequestering fossil carbon. Annual Review of Energy and the Environment 2002;27(1):193—232.
[4] Tachibana Y, Vayssieres L, Durrant JR. Artificial photosynthesis for solar water-splitting. Nature Photonics 2012;6(8):511—8.
[5] Collings AF, Critchley C. Artificial photosynthesis: from basic biology to industrial application. John Wiley & Sons; 2007.
[6] Maity SK. Opportunities, recent trends and challenges of integrated biorefinery: Part I. Renewable and Sustainable Energy Reviews 2015;43:1427—45.
[7] Akhtar J, Amin NAS. A review on process conditions for optimum bio-oil yield in hydrothermal liquefaction of biomass. Renewable and Sustainable Energy Reviews 2011;15(3):1615—24.
[8] Carriquiry MA, Du X, Timilsina GR. Second generation biofuels: economics and policies. Energy Policy 2011;39(7):4222—34.
[9] Ullah K, et al. Assessing the potential of algal biomass opportunities for bioenergy industry: a review. Fuel 2015;143:414—23.

[10] Huang G, et al. Biodiesel production by microalgal biotechnology. Applied Energy 2010;87(1):38—46.

[11] Munasinghe PC, Khanal SK. Biomass-derived syngas fermentation into biofuels: opportunities and challenges. Bioresource Technology 2010;101(13):5013—22.

[12] Lundquist TJ, et al. A realistic technology and engineering assessment of algae biofuel production. Berkeley, California: Energy Biosciences Institute; 2010. p. 1.

[13] Hargreaves PI, et al. Production of ethanol 3G from *Kappaphycus alvarezii*: evaluation of different process strategies. Bioresource Technology 2013;134:257—63.

[14] Zhu L, et al. Microalgal biofuels: flexible bioenergies for sustainable development. Renewable and Sustainable Energy Reviews 2014;30:1035—46.

[15] Maity JP, et al. Microalgae for third generation biofuel production, mitigation of greenhouse gas emissions and wastewater treatment: present and future perspectives — a mini review. Energy 2014;78:104—13.

[16] Rehl T, Müller J. Life cycle assessment of biogas digestate processing technologies. Resources, Conservation and Recycling 2011;56(1):92—104.

[17] Tambone F, et al. Assessing amendment and fertilizing properties of digestates from anaerobic digestion through a comparative study with digested sludge and compost. Chemosphere 2010;81(5):577—83.

[18] Guo M, Song W, Buhain J. Bioenergy and biofuels: history, status, and perspective. Renewable and Sustainable Energy Reviews 2015;42:712—25.

[19] Herzog H, Golomb D. Carbon capture and storage from fossil fuel use. Encyclopedia of Energy 2004;1:1—11.

[20] Heidenreich S, Foscolo PU. New concepts in biomass gasification. Progress in Energy and Combustion Science 2015;46:72—95.

[21] Di Blasi C. Modeling and simulation of combustion processes of charring and non-charring solid fuels. Progress in Energy and Combustion Science 1993;19(1):71—104.

[22] Van der Hoef M, et al. Numerical simulation of dense gas-solid fluidized beds: a multiscale modeling strategy. Annual Review of Fluid Mechanics 2008;40:47—70.

[23] Agbor E, Zhang X, Kumar A. A review of biomass co-firing in North America. Renewable and Sustainable Energy Reviews 2014;40:930—43.

[24] Ward A, Lewis D, Green F. Anaerobic digestion of algae biomass: a review. Algal Research 2014;5:204—14.

[25] Hasanbeigi A, Price L, Lin E. Emerging energy-efficiency and CO_2 emission-reduction technologies for cement and concrete production: a technical review. Renewable and Sustainable Energy Reviews 2012;16(8):6220—38.

[26] Borkenstein C, et al. Cultivation of *Chlorella emersonii* with flue gas derived from a cement plant. Journal of Applied Phycology 2011;23(1):131—5.

[27a] Lara-Gil JA, Álvarez MM, Pacheco A. Toxicity of flue gas components from cement plants in microalgae CO_2 mitigation systems. Journal of Applied Phycology 2014;26(1):357—68.
[27b] Vunjak-Novakovic G, et al. Air-lift bioreactors for algal growth on flue gas: mathematical modeling and pilot-plant studies. Industrial & Engineering Chemistry Research 2005;44(16):6154—63.

[28] Zhu L, et al. Recycling harvest water to cultivate *Chlorella zofingiensis* under nutrient limitation for biodiesel production. Bioresource Technology 2013;144:14—20.

[29] Razzak SA, et al. Integrated CO_2 capture, wastewater treatment and biofuel production by microalgae culturing—a review. Renewable and Sustainable Energy Reviews 2013;27:622—53.

[30] Pulz O. Photobioreactors: production systems for phototrophic microorganisms. Applied Microbiology and Biotechnology 2001;57(3):287—93.

[31] Carlsson AS, Bowles DJ. Micro-and macro-algae: utility for industrial applications: outputs from the EPOBIO project. CPL Press; September 2007.

[32] Alabi AO, et al. Microalgae technologies & processes for biofuels-bioenergy production in British Columbia: current technology, suitability & barriers to implementation: final report. British Columbia Innovation Council; 2009.

[33] Mata TM, Martins AA, Caetano NS. Microalgae for biodiesel production and other applications: a review. Renewable and Sustainable Energy Reviews 2010;14(1):217—32.

[34] Cuellar-Bermudez SP, et al. Photosynthetic bioenergy utilizing CO_2: an approach on flue gases utilization for third generation biofuels. Journal of Cleaner Production 2015;98:53−65.

[35] Trivedi J, et al. Algae based biorefinery—how to make sense? Renewable and Sustainable Energy Reviews 2015;47:295−307.

[36] Fuss S, et al. Betting on negative emissions. Nature Climate Change 2014;4(10):850−3.

Solar Energy Applications

CHAPTER FIVE

Solar Kilns: A Green Technology for the Australian Agricultural and Forest Industries

M. Hasan and T.A.G. Langrish
The University of Sydney, Sydney, NSW, Australia

5.1 INTRODUCTION

Solar drying of agricultural products, such as fruits and vegetables, and forest products, e.g., timber, has been internationally demonstrated to be a promising technology to improve the end-use quality; reduce greenhouse gas (GHG) emissions from process industries; improve the shelf life; reduce operational, storage, and transportation costs; encapsulate the original flavors and nutritional values of fruits and vegetables, as mentioned by [1–3]. A comparative performance study, with different drying materials, e.g., banana, chili, and coffee, between a greenhouse solar dryer and open sun drying was carried out by [1]. It was experimentally found that, for all the three drying materials, the solar dryer required a lower drying time compared with the open-sun drying process. It was also found that the solar dryer produced better-quality dried product while protecting the drying materials from insects, animals, and rain than sun drying.

A comprehensive review on the development of solar drying applications has been presented by [3]. The study also discussed the positive role of solar energy for drying systems in the context of economic, environmental, and political aspects. Drying is an energy-intensive process, which makes wood products more stable for use in furniture, construction, and joinery, and in the case of agricultural products, allows them to be stored safely for a specific period of time. The largest part of the required energy in the drying process is used in the form of thermal energy, which can be provided from various sources, including electricity, gas, fossil fuel, and the sun. Although this application of solar energy has existed since ancient times, it has not yet been commercialized widely, mainly due to the lack of understanding about solar kiln technology and its sustainability benefits over a wider spectrum, such as in the agricultural and forest industries. The rapid depletion of fossil fuels, coupled with the consequences of increasing the consumption of conventional fuel, has prompted governments, industries, and researchers to consider solar energy as a clean

Clean Energy for Sustainable Development
ISBN 978-0-12-805423-9, http://dx.doi.org/10.1016/B978-0-12-805423-9.00005-3

and sustainable source of heat energy for drying processes. The development of solar kiln technology has the potential to enable the forest and industrial sectors to modify their energy requirements, improve sustainability, and increase the overall profitability.

Traditionally, designing and choosing a particular solar kiln has been based on small-scale testing to assess the kiln performance. However, this experimental procedure poses practical difficulties due to the extremely large number of variables that must be considered and the significant time required, combined with the practical difficulties in repeating specific testing environments (natural climatic situations), as mentioned by [4] and [5]. There is no or little information in the literature about the life cycle energy effectiveness of solar kilns for wood drying. Several experimental and modeling approaches [6—9] have been adopted to assess the drying behaviors of wood and agricultural products during solar drying and the performances of the kilns themselves. Despite the abundance of work regarding the performance evaluation of drying facilities for the agricultural and forest products, most methods still have a number of limitations. For example, the optimization study of solar kilns for the drying of forest products (i.e., timber) by [6] was based on a parametric study of the thermal performance and did not consider the embodied energy (EE) requirement. Another example is the experimental study, as carried out by [8], for drying agricultural products, e.g., apple and carrot. The study considered only the energy required to evaporate moisture from the drying materials over a single drying cycle, and no recommendation was made regarding the energy requirements over the whole service life of the dryer. In general, the key limitations of the methods for performance evaluation of solar drying systems to date are that: (1) the EE requirement has been ignored, (2) the long-term (over the whole service life) energy gains or losses have not been considered, and (3) the methods have only been suitable either for a specific type of solar kiln or for a specific location. Taking all these aspects into consideration, a new method for life cycle (LC) performance evaluation of this green technology (i.e., solar kilns) has been introduced in this chapter It should be noted here that since a large number of modeling and simulation outputs for each drying system are needed as the inputs of the methodology presented in this chapter and since the modeling and simulation approach may vary significantly from system to system, a comparative study of different drying systems (e.g., microwave drying, convective drying, intermittent drying, and solar drying) has been beyond the scope of this single chapter. Consequently, the objectives of this chapter have been: (1) to provide information about the significance and the scope of solar drying in Australian context, (2) to introduce an innovative approach for evaluating the life cycle performance of solar kilns, and (3) to demonstrate its application to two case-study solar kilns in the context of hardwood drying.

5.2 SIGNIFICANCE AND SCOPE OF SOLAR DRYING IN AUSTRALIA

Australia has a strong primary industry sector, while its vegetable-growing industry contributed around $3.7 billion to the gross value of agricultural production in 2013–14, which was a 12% increase from 2011 to 2012 [10]. However, higher yields during the growing seasons have traditionally caused a large influx of those fruits and vegetables into the domestic market of Australia, leading to reduced prices and large amounts of waste per year. An analysis of household food waste by [11] mentioned that Australian households throw out more than $5 billion worth of food each year, of which the fruits and vegetables were valued at $1.1 billion. Although 2 million people in Australia still rely on food relief, as mentioned in a Commonwealth Scientific and Industrial Research Organisation (CSIRO) magazine (ECOS), total food wastage is currently costing Australians up to $10 billion each year. In addition to the direct financial costs of this waste, the environmental impact associated with excessive GHG emissions and water use is substantial.

Within the wider agricultural industry, the hardwood products industry is also a growing area of Australia's primary and secondary industry sectors, with hardwood plantations increasing sixfold over the period between 1994 and 2010 [12]. Drying of hardwood is an energy-intensive process for reducing the moisture contents (MCs) down to a level at which the wood is immediately suitable for normal services, such as furniture, joinery, and construction materials, as mentioned by [13] and [14]. The better productivity and quality of solar-assisted kiln-dried timber (e.g., [15–17]), together with the low operating and EE costs (e.g., Refs. [18] and [19]), have motivated the application of solar kilns for wood drying to ensure that the final kiln-dried products remain competitive in the market.

The strong primary industry sector, with significant spoilage of fruits and vegetables, low export volumes, and the need for high-quality timber products, together with the availability of solar radiation [20], is likely to make Australia a favorable place for the development of solar drying technology, especially for industries involved in timber, fruit, and vegetable processing, as evident from [21] and [22]. However, there are only two or three small companies in Australia that manufacture solar dryers for timber and fruit processing, indicating that still there exists a significant opportunity to develop this business, especially if Australia is to decarbonize its economy at the required rate and compete in ever-increasing international trade, as mentioned by Refs. [23] and [24].

5.2.1 Australian Horticultural Industry Outlook

Australia's horticultural industry consists of a wide range of products, including fruits and vegetables. In 2012–13, the production of tomatoes, bananas, and grapes increased by 23%, 16%, and 6%, respectively, since 2011–12 [12]. Australia's horticultural industry was the nation's third largest agricultural industry in 2011–12 based on the gross value of

Table 5.1 Import and Export Values (Millions of Dollars) for Horticultural Commodities [26]

Commodity	Imports		Exports	
	2010–11	2011–12	2010–11	2011–12
Fruits[a]	1022 (56%)	1194 (57%)	651 (59%)	734 (59%)
Vegetables[a]	786 (44%)	908 (43%)	460 (41%)	505 (41%)
Total	1808	**2102**	1111	**1239**

[a]Includes fresh and processed products.

production [25]. To show the overall domestic and international markets in the years of 2010–11 and 2011–12 for fruits and vegetables in Australia, the import and export values for these commodities have been given in Table 5.1.

Despite strong agricultural growth in 2011–12 [12], Australia exported $1.239 billion worth of total fruits and vegetables (i.e., fresh plus processed products) and imported $2.102 billion worth of these materials, resulting in an overall trade deficit of $863 million, as shown in Table 5.1. Although Australia had a trade surplus (i.e., exports exceed imports values) in fresh vegetables [25], the trade deficit was mainly due to large imports of processed fruits and vegetables [26]. This issue has also been addressed in the Australian 44th parliamentary library briefing book by [27], mentioning that "the Australian processed food sector faces a difficult future, as increased imports erode the sector's domestic market share." It was mentioned in [25] that produce was imported into Australia out of season or during periods of domestic shortage due to production failures, an inability to produce the commodity, or production shortfalls relative to demand.

The total amount of food wastage in Australia is valued at $10 billion each year [28], with 2 million people still relying on food relief, and opportunities may be opened up for exporting processed products to new markets in Asia or the Middle East. Therefore preservation of surplus produce during peak production periods can diversify the opportunities for generating income and employment by small and marginal farmers and small and medium-scale horticultural industries in Australia. This situation, coupled with increased fuel and labor costs, is likely to motivate the extension of solar-drying facilities for fruits and vegetables in Australia to ensure food security for the entire population. This preservation is likely to maximize the potential capacity of the primary production sector, and stabilize food supplies in the domestic and international markets throughout the year at stable and reasonably remunerative prices.

5.2.2 Solar Drying—Hardwood Industry Perspective

Every year in Australia thousands of tons of wood is harvested for use in construction, flooring, and furniture [29]. The timber used for all these products must be dried to a certain MC to prevent it from warping and deforming in service. Solar drying of hardwood timber reduces the drying costs and improves the quality of the end-use products, as evident from [4,15,16]. Due to the significant production of native and plantation hardwood sawlogs across the Australian states and territories, there is a substantial potential for solar kilns in hardwood-drying processes in Australia. Table 5.2 summarizes the production of sawn hardwoods, by state and territory, throughout Australia and the potential of solar kilns to reduce the processing costs and GHG emissions associated with the value-addition processes (drying) for these logs. The total production of sawn hardwoods has been extracted from Australian forest and wood products statistics.

In Table 5.2, the estimated drying energy has been calculated as the minimum energy required for drying hardwood from an average initial MC of 50% to an average final MC of 15%. The approach for estimating the drying costs and the GHG emissions associated with the hardwood-drying processes was similar to that used for the fruits and vegetables. Table 5.2 shows that the production of sawn hardwood in New South Wales (including Australian Capital Territory) was 324 kilotons (kt) in 2012–13, representing the highest sawn hardwood production (31% of the total) among the states/territories in Australia. In the same period of time, approximately 25%, 18%, 15%, and 11% of the total sawn hardwood production in Australia were generated in the states of Victoria, Tasmania, Western Australia, and Queensland, respectively, as shown in Table 5.2. A total of 1030 kt sawn hardwood was produced in Australia over the period 2012–13. Table 5.2 shows that, if all of this hardwood is processed using solar kilns, an approximate saving, compared with electrically heated kilns, of AU$ 128 million and 123 kt in processing costs and GHG emissions, respectively, per year could be obtained. These values of savings in the processing costs and GHG emissions were likely to be AU$ 31 million and

Table 5.2 Production of Sawn Hardwood, by States/Territories, in Australia, and the Potential Savings in Drying Costs and GHG Emissions (Base Year 2012–13)

States	Quantity[b] (kt/Year)	Energy Saving[c] (TJ/Year)	Cost Saving[d] (Million AU$/Year)		GHG Savings[e] (kt CO_2/Year)	
			Electricity	Gas	Electricity	Gas
New South Wales[a]	324 (31%)	171	41	10	39	24
Victoria	260 (25%)	136	32	8	31	19
Tasmania	185 (18%)	97	23	6	22	14
Western Australia	141 (15%)	74	17	4	17	10
Queensland	119 (11%)	62	15	3	14	9
Total	**1030**	**540**	**128**	**31**	**123**	**76**

[a]Includes Australian Capital Territory.
[b]Quantities are based on ABARES survey data (Refs. [69] and [30]).
[c]Estimated energy required to dry from an average initial moisture content (MC) of 50% to an average final MC of 15%.
[d]Based on local utility price of electricity and gas (Refs. [70] and [71]).
[e]Based on Australian National Greenhouse Accounts [72].

76 kt, respectively, when compared with gas-fired solar kilns, as shown in Table 5.2. The potential savings in these parameters are even more significant when the solar drying of softwood, especially radiata pine, is taken into consideration.

5.3 SIGNIFICANCE AND BENEFITS OF SOLAR KILNS

Although drying can be accomplished by numerous ways, such as by using chemical desiccants, by freezing using liquid and solid, and (mechanically) by using compression and centrifugal forces, the most common drying method is thermal drying, which involves simultaneous heat and mass transfer processes [3]. Thermal drying is used in numerous industrial (e.g., drying of bricks, leather, wood and timber, textile, sewage sludge, tea, and dairy) and agricultural (e.g., paddy, oil seed, carrot, herb and spices, and vegetables) processes. The energy consumption in such processes is high and depends mostly on both the materials to be dried and the technology used for the process. The implementation of the fossil-fuel-based drying systems, such as microwave drying, freeze drying, and conventional kilns, results in high energy consumption and financial burdens [31]. In the following sections, the technical, economic, environmental, and political aspects of solar kilns will be described.

5.3.1 Technical Aspects

A solar kiln is a piece of equipment where solar energy is used in the form of thermal energy for drying materials. The lightweight and simple design, together with the simple operational procedure, of solar kilns means fewer resources and less supervision are required in manufacturing and maintaining them than conventional kilns. Unlike some solar energy capturing systems that require resource-intensive photovoltaic solar panels, solar kiln technology uses solar energy through a simple combination of layered plastic, a ventilation chamber, and an air circulation system. This natural system of heat generation means that this equipment produces few by-products other than heat. Also, this technology does not need boilers and wood-waste burners to provide heat, so there are less pollutants and environmental hazards. The low infrastructure requirements for greenhouse-type solar kilns, as described in this chapter, mean that the kilns do not have to be installed around a high-energy supply center and/or static boiler system, resulting in flexibility with choosing the location of the equipment.

Also, there has been some evidence, including [6,15], and [4], indicating that kiln drying of materials (wood) improves the productivity and quality of the end-use timber, when compared with the open-air drying systems. This better productivity and quality of kiln-dried timber, together with the rising price of conventional fuel (fossil fuel), have driven the use of solar energy for drying processes to ensure that the final kiln-dried products are sustainably processed and are economically competitive in the market.

5.3.2 Economic Aspects

Ref. [32] conducted an economic analysis, where the costs of solar-dried products were compared with those of products dried using other drying systems that used fossil fuels. The analysis was carried out by three methods, namely, annualized cost, present cost of annual savings, and present worth of cumulative savings. In the first method, the cost of drying 1 kg of pineapple was found to be Indian rupees (Rs.) 11 (US$ 1 = Rs. 45) and Rs. 19.73 for the solar and electric dryers, respectively. By the second method, it was found that the investment on the solar dryer was Rs. 550,000, whereas the saving was nearly 17 million rupees over an assumed solar dryer life span of 20 years. The payback period of the solar dryer was also found to be only 0.54 year, which is equivalent to 191 drying days (a short time compared with the lifetime of 20 years). Another economic analysis of solar dryers for crop drying was performed by Ref. [33]. In this study, the estimated total cost of the solar dryer was found to be Rs. 3,200,000, whereas the saving was found to be Rs. 1,512,000 per year. Consequently, the payback period was determined to be approximately 2 years, which shows a good return in using the solar energy for drying applications.

5.3.3 Environmental Aspects

Undoubtedly, technologies that are based on solar energy can play a significant role in reducing GHG emissions and can reduce global warming by decreasing the dependence of industrial and agricultural processes on fossil fuels. Solar drying technology involves processes that are relatively clean and sanitary, and conforms to national and international standards without high operating costs for energy [34]. Overall, solar kilns can save energy and time, occupy less area, improve product quality, and most importantly, solar technology protects the environment [34,35].

The production of carbon dioxide (CO_2) in a drying system was estimated by Ref. [36]. They studied a drying system whose electricity energy consumption was 100 kWh/day on the basis of 25 days per month and 11 months per year operation. Under these conditions, the estimated CO_2 emission was approximately 14.77 tons per year. In another study [37] on solar crop drying and CO_2 emission potential, the estimate was that a 1 m^2 area of solar irradiation can prevent the production of 463 kg of CO_2 over the determined life cycle. The mitigation of CO_2 emissions due to the implementation of solar drying was also investigated by Ref. [38]. They suggested that solar drying for cash crops, such as tobacco, tea, small cardamom, coriander, seeds, and onion flakes, is one of the possible areas for the immediate intervention of solar energy that can reduce CO_2 emissions.

5.3.4 Political Aspects

Due to the obvious importance of sustainability and protecting the environment, governments and industries from around the world are continuously finding ways to mitigate

GHG emissions, as mentioned by many workers, including [34,39—41]. Governments from around the world have started to address this issue by establishing energy policies in various forms, including legislation, international treaties, and incentives to investors [3]. Government policies can considerably mitigate the effects of global warming and energy crisis due to energy resource efficiency, as mentioned by Refs. [40] and [41]. Policies, such as feed-in-tariff (FIT), renewable portfolio standard, and incentives have already been (and are being) implemented by many governments across the globe [37,40], which may further facilitate the use of solar energy in drying systems. The solar energy policies have been reviewed by Ref. [40] in some countries, including the United States, Canada, Germany, Spain, France, China, Pakistan, Australia, and Malaysia. They indicated that, depending on the situation, each country creates energy policies to encourage industries, as well as the society, to use solar energy as a clean, green, and low-cost energy as much as possible.

Most of the policies aim to facilitate the use of various electricity generation mixes, reduce state reliance on fossil fuel, increase renewable energy deployment, reduce carbon emission, or a combination of any of these. In late 2015, the federal government of United States enacted a 5-year income tax credit for solar and wind power producers, phasing out from 30% for 2016—19 to 10% in 2020 [42]. It is expected to add $38 billion of investment for 20 GW of solar energy [43]. The Canadian government has implemented new policies on taxation, energy, trade, infrastructure and research development, and commercialization for the manufacturing and processing sectors to build a foundation for sustainable and long-term economic growth [40]. Germany implements two policies, namely, FIT and incentives. The mechanism of FIT is that a constant reward is provided for energy production, whereas incentives and beneficial credit terms provide additional support [41]. In Australia, issues such as FIT, subsidies, and emission reduction are considered to promote renewable energy commercialization [44]. Because of the large amounts of domestic reserves of coal in Australia, which makes electricity generation relatively inexpensive, the level of GHG emissions per capita is extremely high in Australia compared with other industrialized countries [45]. However, Australian policymakers have established a mechanism to increase the proportion of emission-free renewable energy as well as to prevent the production of more GHGs [44].

From the foregoing discussion, it is clear that the solar kiln technology (i.e., application of solar energy for industrial and agricultural drying systems) may have significant potential to reduce drying costs, to reduce GHG emissions, and, in general, to improve the living quality of the whole community.

5.4 PERFORMANCE EVALUATION AND SELECTION OF SOLAR KILNS—AN INNOVATIVE APPROACH

Solar drying technology can be used for processing in diverse industry sectors, including food and dairy, wood, paper, biomass, pharmaceuticals, textiles, and mineral concentrates, each having its own unique characteristics. The performance of a given solar drying facility also depends on a large number of variables, as mentioned before. This situation creates challenges for the performance evaluation and hence the selection of a particular solar kiln. A fundamental characteristic of most energy-intensive projects is that the energy costs and benefits, with which they are associated, are spread out over time. However, society is very interested in the timing of energy costs and benefits—usually it prefers to receive benefits as early as possible and pay for costs as late as possible. It is, therefore, important that the valuation of energy costs and benefits takes into account the time value of energy. Traditionally, manufacturers, investors, and decision makers choose projects based on their life cycle economic performance. However, the economic performance is strongly influenced by a volatile parameter, i.e., "the time value of money." The life cycle viability of a process cannot be assessed in economic terms alone. This approach needs to be developed when directly analyzing energy-intensive technologies/utilities, such as solar dryers, because it measures cash flows rather than energy flows [46]. The monetary value of energy does not always represent its true value to society because of subsidies, inaccurate pricing techniques and policies, and accounting confusion caused by inflation, as mentioned by Refs. [47] and [48]. Most of the previous studies, including Refs. [49—51], regarding the overall energy benefits and energy costs associated with a particular process/system/facility, have been carried out either on a financial basis alone or did not account for the time value of energy when energy gains or losses occurred at different times in the future.

In this chapter, the technique of discounted net energy analysis (DNEA) has been adopted as part of a methodology for assessing the overall energy effectiveness of solar kilns. This method accounts for the time value of future energy flows that occur at different times and involves the assessment (through an appropriate numerical model) of the annual, on-going operating energy requirements, and the assessment of the EE requirements [through an appropriate life cycle assessment (LCA) model] requirements. This method is a new approach to assess the performance of existing energy-intensive systems, such as solar kilns for drying processes, over the lifetime of the system.

Several other energy analysts and ecologists [52—54] also emphasized the need to use a net energy analysis (NEA) approach in their respective studies. It was mentioned in Ref. [47] that the pervasiveness of energy, along with its uniqueness, makes energy the ideal commodity for a standard measure of value. In one study [52] it was stated that "The true value of energy is the net energy." A group of scientists from Stanford University said in a recent press that "Net energy analysis should become a standard policy tool." Their ideas about the NEA and energy return on investment have been published in Nature Climate Change [55]. However, it was not clear from those studies whether and how (no or little explanation of calculation procedure) the time-value of energy was considered for assessing the performance of a project/system. There appears, therefore, to be a significant gap in the literature regarding taking the time value of energy into account for these life cycle energy studies.

In the following sections, a DNEA will be described in detail for the two greenhouse-type solar kilns as the case study in the context of hardwood drying, taking this time value of energy into consideration as part of a net energy analysis.

5.4.1 Basic Streams of Energy for a Typical Solar Kiln

Fig. 5.1 shows the key components of the overall energy flows associated with a representative solar kiln for wood drying. The energy consumptions in a solar kiln include the energy for the fan motors, the energy for the loading and unloading system, and the energy embodied in the kiln structure. Solar energy, which is captured by the kiln absorbers, is partly used for drying the timber, whereas the rest is lost through conduction, convection, and radiation heat-transfer processes.

The relative proportion of the energy used for drying and the ratio of the energy losses to the incoming solar energy, both depend upon the energy effectiveness associated with the particular kiln design. The better the kiln design, the larger the proportion of energy used for drying and the lower the energy losses from the kiln.

5.4.2 A Brief Description of the Case Study Solar Kilns

Two greenhouse-type wood-drying solar kilns of different designs, namely, the Oxford kiln and the Boral kiln, have been used for the analysis in this chapter. The overall dimensions and the key components for the Oxford and Boral kilns are shown in Figs. 5.2 and 5.3, respectively.

The Oxford kiln consists of a wooden frame on which polythene glazing is placed to cover the timber stack. Fig. 5.2 shows that the size of wood stack is 4.88 m long, 2 m wide, and 1.83 m high, which leads to an overall kiln capacity of about 10 m^3 of timber. Two painted matt-black corrugated-iron solar absorbers rest on wooden frames and are used to capture solar energy to provide heat to dry the wood in the stack. The north

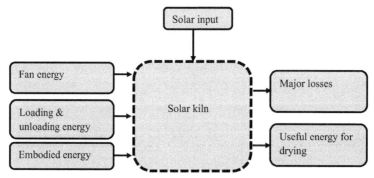

Figure 5.1 Basic energy streams around a solar kiln.

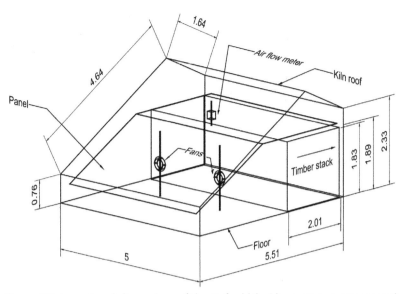

Figure 5.2 Layout and dimensions of the Oxford kiln (dimensions are in meters).

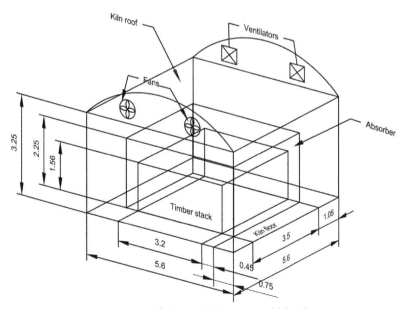

Figure 5.3 Layout and dimensions of the scaled-down Boral kiln (dimensions are in meters).

panel is placed parallel to the inclined north roof, whereas the south panel is horizontal. Two axial fans driven by 0.25-hp electric motors are used to circulate air over the absorber surfaces and through the wood stack.

The dimensions of the original Boral kiln, which was manufactured in Western Australia by Advanced Environmental Structures Pty Ltd., have been scaled down to give the same stack capacity as the Oxford kiln. The scaling procedure has been described in Ref. [18]. The resultant dimensions, as shown in Fig. 5.3, show that the size of the scaled-down Boral kiln is 5.6 m long, 5.6 m wide, and 3.25 m high, to give a net kiln capacity of approximately 10 m^3. This equipment is also a greenhouse-type solar kiln, which has a rectangular stack chamber within a tunnel-like greenhouse structure. The glazing plastic materials used to cover the wood stack are placed on a large, curved aluminum frame and are always kept inflated by a small fan. As in the Oxford kiln, two axial fans were assumed to be installed in the scaled-down Boral kiln to circulate air over the absorbing surfaces and through the stack of timber.

5.4.3 Life Cycle Energy Assessment

There is little information in the literature about the life cycle energy requirements of solar kilns for drying materials. The total life cycle energy use in solar kilns is the sum of the EEs and the total on-going operational energies (OEs) consumed over the whole operational life. EE is the total amount of energy consumed during the extraction of raw materials, including the transport, manufacturing, construction, use (maintenance and renovation), and disposal phases. This energy is analogous to the capital cost in financial analysis. By contrast, OE (analogous to operating costs in financial analysis) is the energy required to operate (or generated by operating) the built facility (kiln) in terms of energy-transfer processes, such as space conditioning, lighting, and operating other appliances.

To carry out a DNEA for solar kilns, it is necessary to calculate all the energy streams (direct or indirect) used to construct, operate, and maintain the wood-drying solar kilns over the whole analysis period. Assessing the operational-energy flows means assessing the annual, on-going operating energy (costs) throughputs, whereas the EE assessment means assessing the capital energy (cost) requirements. Obtaining appropriate estimates for these energy flows require using a mathematical model for solar kilns (combined drying and energy flow models), an LCA model for the kiln construction materials, and a manual calculation in an Excel spreadsheet, depending upon the component of energy being calculated. The two previously described [18] greenhouse-type wood-drying solar kilns, namely, the Oxford kiln and the Boral kiln, have been used for the analysis in this chapter.

5.4.4 Key Features of the Operational Energy Analysis Model

The energy use for drying timber (i.e., drying energy), the energy losses from the kiln, and the solar input into the kiln have been predicted in this chapter by using a previously developed numerical simulation [17] for a solar-kiln model. This particular model was chosen mainly due to two reasons: (1) the basic solar kiln model was experimentally validated by Ref. [7] for a specific design, site, and climatic conditions, and (2) the modified simulation procedure for the model, as developed by Ref. [17], is capable of not only robustly solving the model equations, accommodating a wide range of climatic and geographical conditions, but also simulating kilns with different designs. A detailed description of these operational parameters for the two kilns, combined with the explanation of why those results were obtained, was given in Ref. [18]. Although it is not the purpose of this chapter to repeat them here, the key features of the model, as used in this work, are given in Table 5.3.

The meteorological and geographical data of Brisbane (latitude 27.46°S) in Australia were used in the simulation, as shown in Table 5.3. This particular location was chosen for the current study because [17] found that Brisbane was the most favorable location for solar wood drying out of three representative locations (Sydney, Melbourne, and Brisbane) in Australia. For a fair performance comparison, the reference functional unit for the assessment of OE was defined as the drying of 10 m^3 of hardwood timber from a specific species (Eucalyptus), typical board and sticker thicknesses (0.25 and 0.2 m, respectively) from an average initial MC of 53% to the final average MC of 15% (dry basis)—a typical final MC level for hardwoods (10−20%). The fan energy has been estimated on the basis of 12 h per day operation for two 0.25-hp electric fans, as mentioned by Ref. [56]. The energy consumption of the loading and unloading system has been estimated for a typical electric powered forklift on the basis of the average motor power of 7.25 kW [61] and an estimated loading and unloading time [58].

5.4.5 Key Features of the Embodied Energy Analysis Model

The EEs associated with the construction and maintenance periods for the two solar kilns were calculated by using an LCA model, which was constructed in "SimaPro"

Table 5.3 Key Features of the Drying and Energy Flow Model

Parameters	Value/Method	Remarks/References
Climatic and geographical conditions	Brisbane, Australia in the year 2013	Better performance compared with other locations [17]
Functional unit	Drying of 10 m^3 of Eucalyptus from 53% moisture content (MC) to 15% (dry basis)	Typical average MC levels [18]
Fan energy	Two 0.25-hp fans with 12 h per day operation	Ref. [56]
Loading/unloading energy	Electric powered forklift with 7.25 KW motor power	Refs. [57] and [58]

software in accordance with ISO 14044 guidelines on LCA. Although the details of the model for the assessment of the EE were given in one of the authors' previous study [19], and it is not the purpose of this chapter to repeat them all here, the key features of the model, as used in this study, are given in the following paragraphs.

The EE assessment was done, as far as possible, from an Australian perspective. The inventory data were sourced either from the Australian Life Cycle Inventory (AusLCI) library or the ecoinvent library, whereas the model was analyzed by using the Australian Impact Method with Normalization including Cumulative Energy Demand (AIM-CED). AusLCI and AIM-CED were chosen for the purpose of EE assessment because they are complementary and include Australian region-specific data [59–61]. The functional unit for assessing the EE was the construction and the subsequent use over a 20-year service life of two different solar kilns (Oxford and Boral) for wood drying with the same timber load capacity of 10 m^3. The system boundary for this model and the quantities of the key materials for the construction of the two solar kilns were described in Ref. [19] and have been adopted in this study.

5.4.6 Methods for Discounted Net Energy Analysis

The standard approach that has been used here to valuing future energy flows occurring at different times is based on the principle that every year further into the future in which energy is produced, the less valuable it may be considered to society today. This principle can be expressed in the following way:

$$PVE = FVE * (1 + d)^{-n} \tag{5.1}$$

Here, FVE is the future value of energy, PVE is the present value of energy, d is the minimum attractive rate of return (MARR), and n is the time (years) into the kiln service life when the FVE is produced or consumed. The value of "d" adjusts the future energy value to its corresponding present value.

To carry out a discounted life cycle energy flow analysis for the two solar kilns considered here, it was necessary to predict all the future incoming and outgoing energy flows associated with the construction, operation, and maintenance of these wood-drying kilns over the entire service life. All these future energy inflows and outflows were adjusted to their relative present values by taking account of the time at which they occur (Eq. 5.1). The net present values for the stream of energy benefits (or costs) were calculated to be the sum of all the annual energy benefits (or costs), with each annual benefit (or cost) discounted by an appropriate discount rate (d) to convert it into present value terms. These calculations were carried out using Eqs. (5.2) and (5.3).

$$NPEPV = \sum_{n=0}^{N} FEPV_n * (1 + d)^{-n} \tag{5.2}$$

$$\text{NPECV} = \sum_{n=0}^{N} \text{FECV}_n * (1 + d)^{-n} \qquad (5.3)$$

Here, FEPV and FECV are the future energy production and consumption values, respectively. Similarly, NPEPV and NPECV are the net present energy production and consumption values, respectively. These net present energy values have been used in this study as the basis for decision making in ranking the two solar kilns. Unlike a life cycle analysis based around costs, instead of focusing on the greatest net cash flow, this approach focuses on producing the greatest net present energy. The kiln that produces the higher net energy, or achieves the most energy utilized while consuming the least energy or losing the least incoming energy, has been considered to be the better or more energy efficient kiln. The present value of net energy (PVNE) is the difference between the NPECV and NPEPV, i.e., NPEPV−NPECV, which gives the amount of presently valued energy available for the drying process during the kiln's service life. Another important factor that can be used to determine the relative value of a kiln is the net energy benefit to cost ratio (NEBCR), i.e., NPEPV/NPECV, which describes how much presently valued energy is produced relative to the consumed energy over the lifetime of the kiln. A kiln that returns an NEBCR equal to or greater than unity means that, over its life, it not only produces more energy than it consumes, but also at a rate at or above the MARR, which is the lowest rate of return that investors are willing to accept before they invest.

As mentioned before, d in Eq. (5.1) is the MARR, which is the minimum rate of return that companies/clients are willing to accept on their investment. This MARR is normally considered to be the rate for government bonds, because buying government bonds is a secured (relatively risk free) investment. The recommendation made by the NSW Treasury [62] has been adopted in this study to use a discount rate of 7% (with sensitivity tests using 4% and 10%).

5.5 RESULTS AND DISCUSSION

The two greenhouse-type solar kilns for hardwood drying were analyzed in terms of the life cycle discounted energy analysis to determine their relative performance over a service life of 20 years. The key results and the associated discussion have been provided in the following sections.

5.5.1 Discounted Energy Analysis Results

The main discounted performance parameters per cubic meter of dried timber for the two solar kilns over a lifetime of 20 years are shown in Fig. 5.4.

Fig. 5.4 shows that the PVNE and the present value of the drying energy (PVDE) per cubic meter of dried timber were larger for the Oxford kiln (657 and 575 MJ,

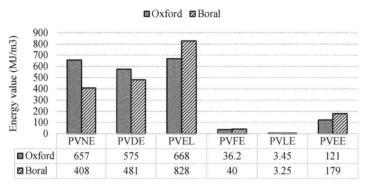

Figure 5.4 Present values of future energy flows per cubic meter of dried timber for the Oxford and Boral kilns over a service life of 20 years.

respectively) than for the Boral kiln (408 and 481 MJ, respectively). On the other hand, the present values of EE (PVEE) and energy losses (PVEL) were smaller for the Oxford kiln (124 and 786 MJ, respectively) than for the Boral kiln (186 and 974 MJ, respectively). These values indicate that not only is more energy consumed in constructing and maintaining the Boral kiln, but the Boral design also loses more energy than the Oxford kiln, which would otherwise be used for drying timber. The larger PVDE for the Oxford kiln compared with the Boral kiln implies that a greater amount of input solar energy is used for drying purposes in the Oxford kiln.

The energy consumption values (EEs, fan energies, loading/unloading energies), together with the energy loss values, were subtracted from the incoming solar energy (energy benefits) when the net values of energy were calculated. When mutually exclusive projects are considered, as in this chapter, the project that maximizes the PVNE should be chosen, as suggested in many energy and economic analysis guidebooks/ studies, including [63,46], and [65]. Thus the larger PVNE for the Oxford kiln indicates that the Oxford kiln is more desirable than the Boral kiln over the analysis period of 20 years.

5.5.2 Complementary Parameters

In addition to the PVNE, a number of other decision parameters may be used for evaluating a particular energy-producing system relative to the other available alternative systems. These decision parameters may include the NEBCR, internal rate of return (IRR), net present energy to EE ratio (NPEEE), and discounted energy payback period (DEPBP). The values of these factors for the solar kilns being considered here are given in Table 5.4.

The NEBCR has been estimated as the ratio of the discounted recurrent net present energy value (net energy benefits) to the net present energy consumption value (net

Table 5.4 Values of Alternative Decision-Making Rules for the Two Solar Kilns

Items	Oxford	Boral	% Difference[a]
NEBCR	5.5	2.5	+54
IRR (%)	71	37	+47
NPEEE	5.44	2.28	+58
DEPBP (years)	2	3.5	−85

DEPBP, discounted energy payback period; *IRR,* internal rate of return; *NEBCR,* net energy benefit to cost ratio; *NPEEE,* net present energy to embodied energy ratio.
[a]% change = [(Oxford − Boral)/Oxford * 100].

capital energy). When appropriately estimated, this ratio can generally assist with the selection of a particular kiln (project) without replacing the objective of maximizing the net present value. Table 5.4 shows that NEBCR for the Oxford kiln is 54% larger than that for the Boral kiln. However, the PVNE and NEBCR rules may rank mutually exclusive solar kilns differently because, in more general cases, NEBCR is biased toward small projects and must be used cautiously, as mentioned in a cost-benefit analysis book prepared by Ref. [64]. In this situation, the decision should be made depending on the PVNE, since the NEBCR complements the objective of maximizing the PVNE. From Fig. 5.4 and Table 5.4, it is seen that both the PVNE and NEBCR values are larger for the Oxford kiln than for the Boral kiln, which indicates that the Oxford kiln is likely to be superior to the Boral kiln over an analysis period of 20 years.

The IRR, which compares the discounted energy outflows (energy costs) and the discounted energy inflows (energy benefits), was estimated to be 47% higher for the Oxford kiln than for the Boral kiln, as shown in Table 5.4. When the IRR of a project exceeds the selected discount rate for estimating the present values of future energy streams, the project produces a positive PVNE and is normally considered to be effective. The IRR, like NEBCR, can be misleading, especially when ranking alternative drying systems with different lifetimes. In this case, where the nature and timing of the energy inflows and outflows are similar for the two kilns, and both the kilns are analyzed over the same service life (20 years), the IRR decision rule is consistent with the PVNE decision rule.

Table 5.4 shows that the NPEEE and the DEPBP values for the Oxford kiln are 58% larger and 85% lower, respectively, than the corresponding values for the Boral kiln. These values indicate that the Oxford kiln is likely to produce a greater positive quantity of net energy over the entire service life while consuming lower energy in its construction and maintenance (EE) than the Boral kiln. In addition, the Oxford kiln used a larger proportion of the incoming energy for the drying process. Reviewing the results given in Fig. 5.4 and Table 5.4 implies that the PVNE is the most reliable parameter in the decision-making context. Other alternative decision rules may reduce the frequency of inferior choices, but should be used only when they do not imply incorrect or misleading recommendations.

5.5.3 Comparative Social and Environmental Benefits

Application of solar energy in drying processes provides significant social and environmental benefits. The social and environmental benefits for the Oxford and Boral kilns are summarized in Table 5.5. It shows that, based on the DNEA, the savings in GHG emissions are 124 and 103 kg of CO_2-e per m^3 of dried timber per year for the Oxford and Boral kilns, respectively. It is also seen, from Table 5.5, that the average CO_2 emissions from a passenger car and from a standard house per year are 2632 and 5228 kg, respectively.

Although the CO_2 emissions from the cars have been collected from the National Average Carbon Emissions data [66], the average distance travelled by a passenger car per year in Australia has been assumed to be 14,000 km, as mentioned in the Survey of Motor Vehicle Use [67]. This survey was carried out by the Australian Bureau of Statistics over the period July 1, 2011, to June 30, 2012. The CO_2 emissions by a standard family house per year have been calculated from the Household Energy Consumption Survey data [68], which was 122.3 kWh per week or 17.47 kWh per day. The total production of sawn hardwoods in Australia in the year 2012–13 was 1030 kt, as mentioned in Table 5.2. Table 5.5 shows that, if 50% of this total swan hardwoods were dried by the solar kilns, the total savings in the GHG emissions would be 55 and 46 kt of CO_2-e per year for the Oxford and Boral kilns, respectively. These savings in GHG emissions for the Oxford and Boral kilns are equivalent to taking 20,896 and 17,525 passenger cars, respectively, off the Australian roads per year or taking 10,518 and 8822 standard family houses off the Australian electricity grid per year for the Oxford and

Table 5.5 Social and Environmental Benefits for the Oxford and Boral Kilns

| Items | Savings (kg/Unit/Year) | | Total Savings[d] | | % Difference |
	Oxford	Boral	Oxford	Boral	
CO_2 savings	124[a]	103[a]	55×10^6 (kg/year)	46×10^6 (kg/year)	
CO_2 emissions by passenger cars	2632[b]	2632[b]	20,896 (car/year)	17,525 (car/year)	+17
CO_2 emissions by standard houses	5228[c]	5228[c]	10,518 (houses/year)	8,822 (houses/year)	

[a]Equivalent CO_2 values obtained from energy flow simulation.
[b]Based on average distance traveled and national carbon emissions data (Refs. [66] and [67]).
[c]Based on household energy consumption survey [68].
[d]Based on 50% of total sawn hardwood production in Australia (Refs. [30] and [69]).

Boral kilns, respectively. This situation indicates that, although both the Oxford and Boral kilns have significant potential to give social and environmental benefits to the Australian community for drying processes, the Oxford kiln design is likely to give more of these benefits, by approximately 17%, than the Boral kiln.

5.5.4 Sensitivity Analysis

As mentioned before, the preferred way to deal with uncertainty in energy costs and benefits streams is to test the sensitivity of the project's PVNE against variations in the key parameters (here the discount rate, d). This analysis is a straightforward and rapid technique, and therefore is the generally recommended approach to gauge the robustness of the PVNE estimates. In this section, a sensitivity analysis involving changes to the discount rate used for the evaluation of the two solar kilns has been carried out to assess how they affect the key decision-making parameters. Considering a central discount rate of 7%, as used in the PVNE estimates of this study, the results associated with the sensitivity testing at 4% and 10% discount rates have been given in Table 5.6.

Table 5.6 shows that, irrespective of the discount rate, the Oxford kiln is the preferred choice in terms of the PVNE. The future energy streams increased in value with reduced discount rates and vice versa. The values of the NEBCR were changed from 4.9 to 2.27, at a low discount rate (4%), to 3.44 and 1.48, at a high discount rate, for the Oxford and Boral kilns, respectively. Similarly, the remaining parameters, as included in Table 5.6, were only slightly affected by the discount rate. The small changes in the values of these parameters with respect to the different discount rates were mainly because both the energy inflows (benefits) and outflows (costs) were adjusted to their respective present values with the same discount rate. It should be noted here that any variation made in the discount rate did not alter the ranking of the kilns—the Oxford kiln is better than the Boral kiln, as shown in Table 5.6. So, the general conclusions of this study are relatively insensitive to the uncertainties associated with the discount rate.

Table 5.6 Effects of the Discount Rate on the Decision-Making Parameters

Parameters	Discount Rate, d (4%)		Discount Rate, d (7%)		Discount Rate, d (10%)	
	Oxford	Boral	Oxford	Boral	Oxford	Boral
PVNE (GJ)	1515	870	1414	633	885	468
NEBCR	4.9	2.27	4.12	1.84	3.44	1.48
IRR (%)	71	37	71	37	71	37
DEPBP (years)	2	3.43	2	3.6	2.19	3.78

DEPBP, discounted energy payback period; *IRR*, internal rate of return; *NEBCR*, net energy benefit to cost ratio; *PVNE*, present value of net energy.

5.6 CONCLUSIONS

The potential opportunities for solar dryers in Australian horticulture and timber industries have been identified and discussed in this chapter In recent years, the demand for high-quality dried products has increased, which is associated with the increased energy uses of already energy-intensive unit operations. Since the energy costs or benefits that the solar kilns give are spread out over the entire service life, an innovative approach has been adopted here to assess the relative performance of the two designs of solar kilns. This analysis involved the use of a combined drying and energy flow model for solar kilns to predict the future flows of operating energy, and an LCA model to assess the EE over the assumed lifetime of the kilns. All the future streams of energy were translated to their corresponding present values and used in the DNEA to assess and evaluate the relative energy effectiveness for the two solar kilns. The general results indicated that the PVNE, the NEBCR, and the IRR were larger for the Oxford kiln than those for the Boral kiln, by approximately 37%, 55%, and 47%, respectively. However, the DEPBP and the PVEL were smaller for the Oxford kiln by 85% and 24%, respectively, than those for the Boral kiln. From an environmental point of view, the Oxford kiln was likely to have a higher GHG emission saving potential by 17% than the Boral kiln. A sensitivity analysis showed that the ranking of the kilns (Oxford is superior to Boral) was relatively insensitive, in terms of the key parameters including PVNE, NEBCR, IRR, and DEPBP, to the uncertainties associated with the discount rates. However, there was a significant difference in PVNE with a change in the discount rate used, due simply to the compounding effect of the discounting process. In general, the Oxford kiln had a higher PVNE than the Boral kiln and thus was suggested to be the preferred choice for hardwood-drying processes in Australia.

ACKNOWLEDGMENTS

The authors acknowledge the financial support of the School of Chemical and Biomolecular Engineering within the Faculty of Engineering and Information Technologies, The University of Sydney, Australia.

NOMENCLATURE

D	Discount rate, %
DEPBP	Discounted energy payback period, years
DNEA	Discounted net energy analysis
EE	Embodied energy, J
FECV	Future energy consumption value, J
FEPV	Future energy production value, J
FVE	Future value of energy, J
IRR	Internal rate of return, %
LCA	Life cycle assessment
MARR	Minimum attractive rate of return, %

MC	Moisture content, $kgkg^{-1}$
n	Time into the kiln service life, years
NEA	Net energy analysis
NEBCR	Net energy benefit cost ratio
NPECV	Net present energy consumption value, J
NPEPV	Net present energy production value, J
OE	Operational energy, J
PVDE	Present value of drying energy, J
PVE	Present value of energy, J
PVEE	Present value of the embodied energy, J
PVEL	Present value of the energy losses, J
PVFE	Present value of the fan energy, J
PVLE	Present value of the loading/unloading energy, J
PVNE	Present value of the net energy, J

REFERENCES

[1] Janjai S, et al. A large-scale solar greenhouse dryer using polycarbonate cover: modeling and testing in a tropical environment of Lao People's Democratic Republic. Renewable Energy 2011;36(3): 1053—62.

[2] Prakash O, Kumar A. Solar greenhouse drying: a review. Renewable & Sustainable Energy Reviews 2014;29:905—10.

[3] Pirasteh G, et al. A review on development of solar drying applications. Renewable & Sustainable Energy Reviews 2014;31:133—48.

[4] Langrish TAG, Keey RB, Kumar M. Improving the quality of timber from red beech (N. fusca) by intermittent drying. Drying Technology 1992;10(4):947—60.

[5] Thibeault F, et al. Numerical and experimental validation of thermo-hygro-mechanical behaviour of wood during drying process. International Communications in Heat and Mass Transfer 2010;37(7): 756—60.

[6] Khater HA, et al. Optimization of solar kiln for drying wood. Drying Technology 2004;22(4): 677—701.

[7] Haque MN, Langrish TAG. Mathematical modelling of solar kilns for drying timber: simulation and experimental validation. Drying Technology 2003;21(3):457—77.

[8] Romano G, Kocsis L, Farkas I. Analysis of energy and environmental parameters during solar cabinet drying of apple and carrot. Drying Technology 2009;27(4):574—9.

[9] Hasan M, Langrish TAG. Time-valued net energy analysis of solar kilns for wood drying: a solar thermal application. Energy 2016;96:415—26.

[10] Australian Bureau of Agricultural and Resource Economics and Sciences. Australian vegetable growing farms: an economic survey, 2012—13 and 2013—14. 2014. Available from: http://data.daff. gov.au/data/warehouse/9aab/9aabf/2014/avfesd9absf20141114/ AustVegGrwFrmEcoSurvey20141114_1.0.0.pdf.

[11] Baker D, Fear J, Denniss R. What a waste: an analysis of household expenditure on food. Policy brief. 2009.

[12] Australian Bureau of Statistics. Agricultural commodities, Australia. cat. no. 7121.0. Available from: http://www.abs.gov.au/ausstats/abs@.nsf/Latestproducts/7121.0Main%20Features72012-2013?open document&tabname=Summary&prodno=7121.0&issue=2012-2013&num=&view=; 2013.

[13] Falk B. Wood as a sustainable building material. Forest Product Journal 2009;59(9):6—12.

[14] Bentayeb F, Bekkioui N, Zeghmati B. Modelling and simulation of a wood solar dryer in a Moroccan climate. Renewable Energy 2008;33(3):501—6.

[15] Chadwick WB, Langrish TAG. A comparison of drying time and timber quality in the continuous and cyclic drying of Australian turpentine timber. Drying Technology 1996;14(3—4):895—906.

[16] Helwa NH, et al. Experimental evaluation of solar kiln for drying wood. Drying Technology 2004;22(4):703—17.

[17] Hasan M, Langrish TAG. Numerical simulation of a solar kiln design for drying timber with different geographical and climatic conditions in Australia. Drying Technology 2014;32(13):1632—9.

[18] Hasan M, Langrish TAG. Performance comparison of two solar kiln designs for wood drying using a numerical simulation. Drying Technology 2014;33(6):634—45.

[19] Hasan M, Langrish TAG. Embodied energy and carbon analysis of solar kilns for wood drying. Drying Technology 2015;33(8):973—85.

[20] Bureau of Resources and Energy Economics Australian Energy Resource Assessment. Second. Available from: http://www.ga.gov.au/webtemp/image_cache/GA21797.pdf; 2014.

[21] Fuller R. Solar drying in Australia: a historical review. In: Proceedings of the international solar energy society conference, Adelaide; 2001.

[22] Lovegrove K, Dennis M. Solar thermal energy systems in Australia. International Journal of Environmental Studies 2006;63(6):791—802.

[23] Odonoghue J, Fuller R. Experiences with the Australian version of the Hohenheim solar tunnel dryer. In: Australian solar council scientific conference. Australian Solar Council; 1999.

[24] Fuller RJ. Solar industrial process heating in Australia — past and current status. Renewable Energy 2011;36(1):216—21.

[25] Horticulture Australia Limited. Horticulture fact sheet. 2012. Available from: http://www.agriculture.gov.au/Style%20Library/Images/DAFF/__data/assets/pdffile/0020/2109206/Horticulture_Fact_Sheet.pdf.

[26] Australian Bureau of Agricultural and Resource Economics and Sciences. Agricultural commodity statistics. 2012. Available from: http://data.daff.gov.au/data/warehouse/agcstd9abcc002/agcstd9abcc0022012/ACS_2012_1.1.0.pdf.

[27] Dossor R. The future of the Australian processed food sector, in Parliamentary library briefing book. In: Heriot D, editor. Key issues for the 44th parliament. Parliament of Australia; 2013.

[28] OzHarvest. Australia takes part in global UN initiative to cut food waste, in ECOS. CSIRO; 2014.

[29] Burke R. Solar dryers. Australia Pty Ltd.; 2015. Available from: http://www.solardry.com.au/solardry/.

[30] Australian Bureau of Agricultural and Resource Economics and Sciences. Australian forest and wood products statistics. 2012. Available from: http://www.agriculture.gov.au/abares/forestsaustralia/australian-forest-and-wood-products-statistics.

[31] Burns K, Burke B. ABARES national wood processing survey, 2010—11. Canberra (ACT): Australian Bureau of Agricultural and Resource Economics; 2012.

[32] Department of Energy and Water Supply. Current electricity prices. 2015. Available from: https://www.dews.qld.gov.au/energy-water-home/electricity/prices/current-prices.

[33] Origin Energy. QLD small business energy price fact sheet. 2015. Available from: http://www.originenergy.com.au/content/dam/origin/business/Documents/energy-price-fact-sheets/qld/QLD_Natural%20Gas_Small%20Business_Australian%20Gas%20Networks%20Brisbane_Business%20eSaver.PDF.

[34] Department of the Environment. National greenhouse accounts factors. 2014. Available from: http://www.environment.gov.au/system/files/resources/b24f8db4-e55a-4deb-a0b3-32cf763a5dab/files/national-greenhouse-accounts-factors-2014.pdf.

[35] Celma AR, Cuadros F. Energy and exergy analyses of OMW solar drying process. Renewable Energy 2009;34(3):660—6.

[36] Sreekumar A. Techno-economic analysis of a roof-integrated solar air heating system for drying fruit and vegetables. Energy Conversion and Management 2010;51(11):2230—8.

[37] Hollick JC. Commercial scale solar drying. Renewable Energy 1999;16(1—4):714—9.

[38] Sharma A, Chen CR, Vu Lan N. Solar-energy drying systems: a review. Renewable & Sustainable Energy Reviews 2009;13(6):1185—210.

[39] Aravindh MA, Sreekumar A. Solar drying. A sustainable way of food processing. In: Sharma A, Kar KS, editors. Energy sustainability through green energy. New Delhi (India): Springer; 2015. p. 27—46.

[40] Piacentini RD, Mujumdar AS. Climate change and drying of agricultural products. Drying Technology 2009;27(5):629—35.

[41] Kumar A, Kandpal TC. Solar drying and CO_2 emissions mitigation: potential for selected cash crops in India. Solar Energy 2005;78(2):321—9.

[42] Ekechukwu OV, Norton B. Review of solar-energy drying systems II: an overview of solar drying technology. Energy Conversion and Management 1999;40(6):615—55.

[43] Abdelaziz EA, Saidur R, Mekhilef S. A review on energy saving strategies in industrial sector. Renewable and Sustainable Energy Reviews 2011;15(1):150—68.

[44] Solangi KH, et al. A review on global solar energy policy. Renewable and Sustainable Energy Reviews 2011;15(4):2149—63.

[45] Ekins P. Step changes for decarbonising the energy system: research needs for renewables, energy efficiency and nuclear power. Energy Policy 2004;32(17):1891—904.

[46] Fehrenbacher K. Solar, wind companies cheer as crucial tax credit nears extension, in Fortune. New York: Time Inc.; 2015.

[47] Randall T. What just happened in solar is a bigger deal than oil exports, in Bloomberg. New York: Bloomberg L.P.; 2015.

[48] The Parliament of Australia. Renewable energy (electricity) amendment act 2015. 2015 [Canberra ACT, Australia].

[49] Diesendorf M, Saddle H. Australia's polluting power: coal-fired electricity and its impact on global warming. World Wide Fund (WWF); 2003.

[50] Rock G. Using the time value of energy. 2006. Available from: http://www.hubbertpeak.com/netenergy/netenergyprofitratio_gregrock.pdf.

[51] Hannon B. Energy discounting. Technological Forecasting & Social Change 1982;21(4):281—300.

[52] Cleveland CJ. Net energy analysis. 2013. Available from: http://www.eoearth.org/view/article/154821/.

[53] Wang X, et al. Improving benefit-cost analysis to overcome financing difficulties in promoting energy-efficient renovation of existing residential buildings in China. Applied Energy 2015;141(0): 119—30.

[54] Trivedi P, et al. Energy return on investment for alternative jet fuels. Applied Energy 2015;141(0): 167—74.

[55] Buonomano A, et al. Energy and economic analysis of geothermal—solar trigeneration systems: a case study for a hotel building in Ischia. Applied Energy 2015;138(0):224—41.

[56] Cleveland CJ. Net energy from the extraction of oil and gas in the United States. Energy 2005;30(5):769—82.

[57] Dale M, Krumdieck S, Bodger P. Net energy yield from production of conventional oil. Energy Policy 2011;39(11):7095—102.

[58] Carbajales-Dale M, et al. A better currency for investing in a sustainable future. Nature Climate Change 2014;4(7):524—7.

[59] Carbajales-Dale M, Barnhart CJ, Benson SM. Can we afford storage? A dynamic net energy analysis of renewable electricity generation supported by energy storage. Energy and Environmental Science 2014;7(5):1538—44.

[60] Langrish TAG, Thompson FB, Plumptre RA. The measurement of energy flows around a solar kiln used for drying timber. Commonwealth Forestry Review 1993;72(2):95—104.

[61] Trucks CL. In: Trucks CL, editor. EP20K PAC - EP25K PAC, EP30K PAC - EP35K PAC, specifications; 2008. p. 2.

[62] Solar Kilns. General specifications. 2014. Available from: http://www.solarkilns.com/solar_kilns/features_and_specifications/specification-matrix.htm.

[63] Australian Life Cycle Assessment Society. Australian life cycle inventory (AusLCI). 2009. Available from: http://alcas.asn.au/AusLCI/.

[64] Tharumaharajah A, Grant T. Australian national life cycle inventory database: moving forward. In: 5th Australian life cycle assessment society conference, Melbourne, Australia; 2006.

[65] Newton PW, Hampson K, Drogemuller R. Technology, design and process innovation in the built environment. New York: Spon Press; 2009.

[66] Harrison M. Valuing the Future: the social discount rate in cost-benefit analysis. 2010. Available at: SSRN 1599963.

[67] Grant T, Peters G. Best practice guide to life cycle impact assessment in Australia. Australian Life Cycle Scociety 2008.

[68] Commonwealth of Australia. Handbook of cost-benefit analysis. The financial management guidance. 2006. Available from: http://www.finance.gov.au/finframework/docs/Handbook_of_CB_analysis.pdf.

[69] Heun MK, de Wit M. Energy return on (energy) invested (EROI), oil prices, and energy transitions. Energy Policy 2012;40:147—58.

[70] Australian Federal Chamber of Automotive Industries. National average carbon emission. 2011. Available from: http://www.fcai.com.au/environment/co2-emissions-and-climate-change.

[71] Australian Bureau of Statistics. Survey of motor vehicle use. 2012. Available from: http://www.abs.gov.au/ausstats/abs@.nsf/mf/9208.0/.

[72] Australian Bureau of Statistics. Household energy consumption survey (HECS). 2012. Available from: http://www.abs.gov.au/ausstats/abs@.nsf/Lookup/4670.0main+features100052012.

CHAPTER SIX

Small-Scale Dish-Mounted Solar Thermal Brayton Cycle

W.G. Le Roux and J.P. Meyer
University of Pretoria, Pretoria, South Africa

6.1 INTRODUCTION

Solar power generation holds endless opportunities for countries in solar-rich areas; however, more efficient and cost-effective small-scale solar-to-electricity technologies are required. The small-scale dish-mounted solar thermal Brayton cycle (STBC), shown in Fig. 6.1, using a recuperator and an off-the-shelf turbocharger has the potential for high energy conversion efficiency. The closed Brayton cycle was developed in the 1930s for power applications and was adapted to the design and development of STBCs for space power in the 1960s, with the success of lightweight and high-performance gas turbines for aircraft [1]. The STBC in the 1- to 20-kW range can be applied to generate power for small communities.

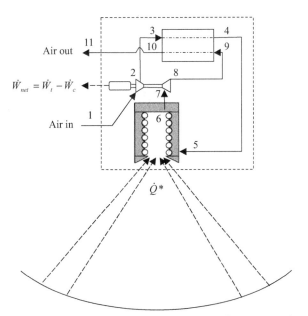

Figure 6.1 The open and direct solar thermal Brayton cycle.

Clean Energy for Sustainable Development
ISBN 978-0-12-805423-9, http://dx.doi.org/10.1016/B978-0-12-805423-9.00006-5

The parabolic dish reflects and concentrates the sun's rays onto the receiver aperture so that solar heat can be absorbed by the inner walls of the receiver. With reference to Fig. 6.1, the compressor (1—2) increases the air pressure before it is heated in the recuperator (3—4) and solar receiver (5—6). In the recuperator, hot turbine exhaust air (9—10) is used to preheat the compressed air. The compressed and heated air expands in the turbine (7—8), which produces rotational shaft power for the compressor and the electric load. The compressor, turbine, and generator are mounted on a single shaft and all spin at the same rate [2,3]. It is simple, robust, and easy to maintain. A recuperated STBC allows for lower compressor pressure ratios [4], higher efficiency [1], and a less complex solar receiver, which operates at lower pressure.

The open Brayton cycle uses air as working fluid, which makes this cycle very attractive for use in water-scarce countries. The hot exhaust air coming from the recuperator can be used for cogeneration, such as water heating. The use of cogeneration makes the cycle more efficient and highly competitive. Intercooling and reheating can also be applied to increase efficiency. The use of local small-scale power generation means that transmission lines do not have to cross vast stretches of land to bring electricity to isolated areas.

The STBC can be supplemented with natural gas as a hybrid system [5] for continuous operation when the sun is not shining. Storage systems such as packed rock bed thermal storage [6] can also be coupled to the cycle. Cameron et al. [7] showed how lithium fluoride could be used in the solar receiver of an 11-kW earth-orbiting STBC with recuperator to store heat for as much as 38 min during full power operation. The small-scale dish-mounted STBC also has an advantage in terms of bulk manufacturability and cost. Microturbines can be adapted as turbochargers from the vehicle market off the shelf [8], which allows for lower costs due to high production quantities [9].

According to Pietsch and Brandes [1], experimental testing of the STBC has proved high reliability and efficiencies above 30% with turbine inlet temperatures of between 1033K and 1144K. Dickey [10] also presented experimental test results of an STBC (20—100 kW), an initiative from HelioFocus Ltd. and Capstone Microturbine at the Weizmann Institute. A proprietary pressurized volumetric solar receiver was used in the experiment. A system efficiency of 11.76% was achieved with a solar receiver exit temperature of 871°C. The system generated 24.04 kW of electricity with the microturbine spinning at 96,000 rpm. Heller et al. [11] tested an open STBC without a recuperator and found that a volumetric pressurized receiver can produce air of 1000°C to drive a gas turbine. According to [11], the cost and performance of this cycle looks promising for future solar power generation. According to Chen et al. [12], the Brayton cycle is definitely worth studying when comparing its efficiency with those of other power cycles. Mills [9] predicted that small-scale Brayton microturbines might become more popular than Stirling engines due to high Stirling engine costs.

As shown earlier, the STBC has much potential and its efficiency can be improved using various methods. Another method to improve the efficiency of the cycle is through design optimization. When designing the STBC, there is a compromise between allowing effective heat transfer and keeping pressure losses in the components small. Heat losses and pressure losses in the components of the cycle decrease the net power output of the cycle. Limiting factors to the performance of the STBC include maximum receiver surface temperature and recuperator weight. The Brayton cycle and the optimization thereof have been studied by many authors; however, not many have studied the STBC with recuperator. Zhang et al. [13], for example, studied the performance of a closed STBC without a recuperator. Solar receivers and recuperators have been designed and optimized individually for the Brayton cycle and the STBC [14–16]; however, various studies such as [17] have emphasized the importance of the optimization of the global performance of a system, instead of optimizing components individually. To obtain the maximum net power output of the cycle, a combined effort of heat transfer, fluid mechanics, and thermodynamic thought is required. The method of entropy generation minimization combines these thoughts [18]. Jubeh [19] has done an exergy analysis of specifically the open STBC with recuperator.

Optimization of the geometry variables of cycle components using the method of total entropy generation minimization is considered to be a holistic optimization approach and has been applied to the STBC in recent work for maximum net power output [20–23]. In this chapter, the method is applied to a recuperated STBC with low-cost dish optics and low-cost tracking. Furthermore, three different off-the-shelf turbochargers from Garrett [8] are investigated for use in the STBC.

6.2 SOLAR COLLECTOR AND RECEIVER

Dish manufacturing and installation errors influence the position and shape of the focal point of the dish. Due to these errors and due to the sun's rays not being truly parallel, the reflected rays from the dish form an image of finite size centered on its focal point. The accuracy of the solar tracking system and the quality of the optics are important factors in the total cost of the small-scale dish-mounted STBC. Two-axis solar tracking is required to ensure that the sun's rays stay focused on the receiver aperture throughout a typical day. Typical solar tracking errors are between $0.1°$ and $2°$ [24]. The tracking accuracy is much dependent on sensor alignment, base-level alignment, momentum of the moving dish, and also drive nonuniformity and receiver alignment [4]. Error due to wind loading is also a measurable quantity [25]. For the dish, good reflectance and specular reflection of the entire terrestrial solar spectrum are important. Typical specularity errors range between 0 and 3.84 mrad, whereas typical slope errors are 1.75, 3, and 5 mrad [26]. The specular reflectance of any material is a function of time, regardless of the reflector [4].

According to Ref. [4], the focal length of a parabolic dish is defined by Eq. (6.1). The rim angle of a parabolic dish determines where its focal point is. A rim angle of $45°$ is paramount for concentrators with focal plane receivers [4].

$$f_c = \sqrt{\frac{(1 + \cos \psi_{rim})^2 \pi R^2}{4\pi \sin^2 \psi_{rim}}} \tag{6.1}$$

Solar receivers can be divided into tubular, volumetric, and particle receivers. A number of high-temperature and high-efficiency receivers are available from the literature, mostly for use in large-scale (10–200 MW) applications [27,28]. Typical receiver efficiencies and experimental data that have been obtained with pressurized volumetric receivers and tubular receivers are summarized in Ref. [24]. These receivers are mostly not optimized to perform well in a dish-mounted STBC with recuperator. A solar receiver might be designed for high efficiency, but if it is not optimized to achieve a common goal together with other components, it might not perform well in a recuperated Brayton cycle.

Most Brayton cycles are not self-sustaining at operating temperatures below $480°C$ [4]. For the STBC, the maximum receiver surface temperature and turbine inlet temperature are very important. The higher these temperatures, the better the Brayton cycle will perform, but more heat will be lost to the environment [21]. The maximum turbine inlet temperature of commercial off-the-shelf turbochargers is more or less $950°C$ [8,29], whereas for ceramic microturbines a temperature of $1170°C$ is envisaged [30]. The higher the turbine inlet temperature, the less material choices are available for the receiver.

6.3 THE TUBULAR OPEN-CAVITY RECEIVER

For simplicity and ease of optimization, a tubular open-cavity receiver [24] is considered in this work. Overall collector efficiencies of between 60% and 70% [31] are attainable with state-of-the-art open-cavity receivers operated in the temperature range of $500°C–900°C$ with an optimum area ratio of $0.0004 \le A' \le 0.0009$. The receiver efficiency is defined as shown in Eq. (6.2). Eq. (6.3) shows the overall efficiency of the STBC.

$$\eta_{rec} = \dot{Q}_{net}/\dot{Q}^* = \dot{m}c_{p0}(T_e - T_i)/\eta_{optical}\eta_{refl}\dot{Q}_{solar} \tag{6.2}$$

$$\eta_{STBC} = \eta_{col}\eta_{BC} = \eta_{refl}\eta_{REC}\eta_{BC} = \eta_{refl}\eta_{rec}\eta_{optical}\eta_{BC} \tag{6.3}$$

The open-cavity tubular solar receiver (see Fig. 6.2) consists of a coiled stainless steel tube through which pressurized air travels. The depth of the receiver is equal to $2a$. The receiver is covered with ceramic fiber insulation. Reflected solar beam irradiance gets

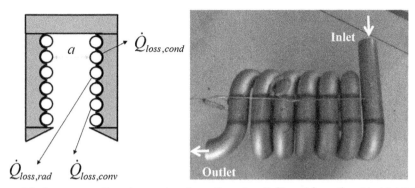

Figure 6.2 An open cavity solar receiver in section view (left) and from the side (right) [32].

absorbed at the inner walls of the cavity. The receiver has been tested at low temperatures in a low-cost solar dish with the use of weld-on thermocouples [32].

The heat loss from the receiver consists of convection, radiation, and conduction and can be modeled with the Koenig and Marvin heat loss model [33] presented by Harris and Lenz [31]. The heat loss per tube section is determined with Eq. (6.4). The larger the cavity aperture, the more heat can be lost, but more heat can also be intercepted.

$$\dot{Q}_{net,n} = \dot{Q}_{solar,n} - \dot{Q}_{loss,rad,n} - \dot{Q}_{loss,conv,n} - \dot{Q}_{loss,cond,n} \tag{6.4}$$

6.3.1 Conduction Heat Loss

The conduction heat loss rate through the insulation is calculated per tube section with Eq. (6.5) and Eq. (6.6) [24]. Eq. (6.6) was obtained by assuming an average wind speed of 2.5 m/s, an insulation thickness of $t_{ins} = 0.1$ m and an average insulation conductivity of 0.061 W/mK at 550°C [24,31]. The convection heat transfer coefficient on the outside of the insulation was determined by assuming a combination of natural convection and forced convection due to wind [24].

$$\dot{Q}_{loss,cond,n} = \frac{A_n\left(T_{s,n} - T_\infty\right)}{R_{cond}} = \frac{\left(T_{s,n} - T_\infty\right)}{\left(1/h_{out}A_n + t_{ins}/k_{ins}A_n\right)} \tag{6.5}$$

$$\left(1/h_{out} + t_{ins}/k_{ins}\right) \approx 1.77 \tag{6.6}$$

6.3.2 Radiation Heat Loss

The radiation heat loss rate from the receiver aperture is calculated per tube section in Eq. (6.7) [24]. The radiation heat loss and gain at each part of the inner wall is determined with the use of Eq. (6.8), where the view factors for the receiver are available from Ref. [24].

$$\dot{Q}_{loss,n,rad} = \varepsilon \sigma A_{ap} \left(T_{s,n}^4 - T_\infty^4 \right) \tag{6.7}$$

$$\dot{Q}_n = A_n \sum_{j=1}^{N} F_{n-j} \left(\varepsilon_n \sigma T_{s,n}^4 - \varepsilon_j \sigma T_{s,j}^4 \right) \tag{6.8}$$

6.3.3 Convection Heat Loss

The convection heat loss rate from the open cavity receiver is determined per tube section according to Eq. (6.9) [31], where the natural convection heat transfer coefficient is determined according to Ref. [24]. The natural convection heat transfer coefficient according to Ref. [24] is $h_{inner} = 2.75$ W/m^{2}K and a wind effect constant of $w = 2$ can be assumed.

$$\dot{Q}_{loss,conv,n} = w h_{inner} A_n \left(T_{s,n} - T_\infty \right) \tag{6.9}$$

6.3.4 Optimization and Efficiency

There are many different variables at play to model the efficiency of the receiver. These variables include concentrator shape, concentrator diameter, concentrator rim angle, concentrator reflectivity, concentrator optical error, tracking error, receiver aperture area, receiver material, receiver tube diameter, inlet temperature, and mass flow rate through the receiver.

The factors contributing to the temperature profile and net heat transfer rate on the receiver tube can be divided into two components: geometry dependent and temperature dependent. The geometry-dependent factors include the concentrator dish with its optics: tracking error, specularity error, slope error, reflectance, spillage, and shadowing. The effects of these factors can be found with the use of ray-tracing software. Computer software and algorithms as described by Ho [34] and Bode and Gauché [35] are available to compute the solar heat flux on a receiver as reflected from a reflector.

Regarding the temperature-dependent factors, a method to determine the receiver tube surface temperature and net heat transfer rate along the length of the receiver tube is described in Ref. [24], as shown in Eqs. (6.10) and (6.11). Eq. (6.10) is derived from Eq. (6.4).

$$\dot{Q}_{net,n} = \dot{Q}_{solar,n} - A_n \sum_{j=1}^{N} F_{n-j} \left(\varepsilon_n \sigma T_{s,n}^4 - \varepsilon_j \sigma T_{s,j}^4 \right)$$

$$\tag{6.10}$$

$$- A_n F_{n-\infty} \left(\varepsilon_n \sigma T_{s,n}^4 - \varepsilon_j \sigma T_\infty^4 \right) - h_n A_n \left(T_{s,n} - T_\infty \right) - \frac{A_n \left(T_{s,n} - T_\infty \right)}{R_{cond}}$$

$$\dot{Q}_{net,n} = \left(T_{s,n} - \sum_{i=1}^{n-1}\left(\frac{\dot{Q}_{net,i}}{\dot{m}c_{p0}}\right) - T_{in,0}\right) \bigg/ \left(\frac{1}{hA_n} + \frac{1}{2\dot{m}c_{p0}}\right) \qquad (6.11)$$

The heat loss terms in Eq. (6.10) can be linearized as shown in Eq. (6.12). By using Gaussian elimination, Eq. (6.11) and Eq. (6.12) can be solved simultaneously to determine the temperature profile $(T_{s,n})$ of the receiver tube and the net absorbed heat rate $(\dot{Q}_{net,n})$ at each tube section [24].

$$\dot{Q}_{net,n} = \dot{Q}_{solar,n} - A_n\varepsilon_n\sigma\left(m_1 T_{s,n} + c_1\right) + A_n \sum_{j=1}^{N} F_{n-j}\varepsilon_j\sigma\left(m_1 T_{s,j} + c_1\right)$$

$$\qquad (6.12)$$

$$- A_n\varepsilon_n\sigma F_{n-\infty}T_\infty^4 - A_n\left(m_2 T_{s,n} + c_2\right) - \frac{A_n}{R_{cond}}\left(T_{s,n} - T_\infty\right)$$

In Eq. (6.12), $\dot{Q}_{solar,n}$ can be determined with the use of a ray-tracing software such as SolTrace (see Fig. 6.3). SolTrace is a software tool developed at the National Renewable Energy Laboratory to model concentrating solar power optical systems and analyze their performance. The optical error of the dish is determined from the slope error and specularity error as shown in Eq. (6.13).

$$\omega_{optical} = \left(4\omega_{slope}^2 + \omega_{specularity}^2\right)^{\frac{1}{2}} \qquad (6.13)$$

6.4 RECUPERATOR

The addition of a recuperator in the cycle allows for a higher efficiency, a lower operating pressure, and a less complex receiver. The recuperator is used to preheat air going to the solar receiver by extracting heat from the turbine exhaust air (see Fig. 6.1). Heat exchangers are required to be efficient, safe, economical, simple, and convenient [36]. It is often beneficial for the cycle to have a large recuperator; however, the recuperator should be practical. Heat transfer and pressure losses as well as the optimization of cost, weight, and size should be considered when designing a heat exchanger [37]. The recuperator should have high effectiveness, compactness, 40,000-h operation life without maintenance, and low pressure loss (<5%) [29]. These criteria translate into a thin foil primary surface recuperator where flow passages are formed with stamping, folding, and welding side edges by an automated operation [29,38,39]. In solar applications, a compact counterflow recuperator [18,40,41] with multiple flow channels is often designed as integral to the microturbine. With the use of multiple flow channels, heat exchanger irreversibilities can be decreased by slowing down the fluid that is traveling through the heat exchanger [18].

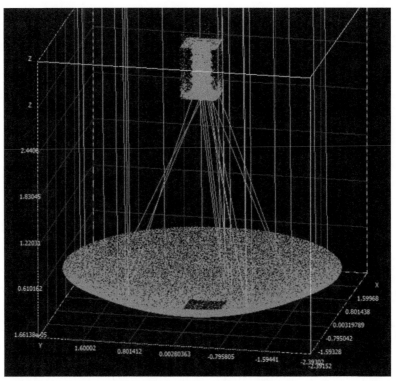

Figure 6.3 Example of a ray-trace analysis done in SolTrace for the open-cavity tubular receiver.

A counterflow plate-type recuperator is considered as shown in Fig. 6.4 [22]. The channels with length L_{reg} and aspect ratio a/b are shown. The recuperator effectiveness is modeled using an updated ε-NTU (effectiveness − number of transfer units) method [42]. This method takes the heat loss to the environment into consideration when calculating the recuperator efficiency, since the recuperator operates at a very high average temperature. According to Ref. [42], the hot side and cold side efficiencies can be calculated with Eq. (6.14) and Eq. (6.15) and the equations below.

$$\varepsilon_h = \left\{ \begin{array}{l} 1 - \Theta_{X=1}, Cr_h < 1 \\[1em] Cr_h(1 - \Theta_{X=1}), Cr_h > 1 \end{array} \right\} \tag{6.14}$$

$$\varepsilon_c = \left\{ \begin{array}{l} \dfrac{1 - \Theta_{X=0}}{Cr_h}, Cr_h < 1 \\[1em] 1 - \Theta_{X=0}, Cr_h > 1 \end{array} \right\} \tag{6.15}$$

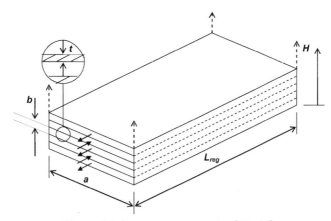

Figure 6.4 Recuperator geometry [20–22].

$$\Theta_{X=0} = \frac{\left(NTU_h(\chi_c + \chi_h) + \frac{Cr_h - 1}{Cr_h}\right)(Cr_h - 1) + (\chi_h + Cr_h\chi_c)\left(1 - e^{NTU_h(Cr_h - 1)}\right)}{(Cr_h - 1)\left(e^{NTU_h(Cr_h - 1)} - \frac{1}{Cr_h}\right)} \tag{6.16}$$

$$\Theta_{X=1} = NTU_h(\chi_c + \chi_h) + \frac{(\Theta_{X=0} - 1)}{Cr_h} + 1 \tag{6.17}$$

$$Cr_h = \frac{\dot{m}_h c_{p0,h}}{\dot{m}_c c_{p0,c}} \tag{6.18}$$

$$NTU_h = \frac{UA}{\dot{m}_h c_{p0,h}} \tag{6.19}$$

$$\chi_h = \frac{\dot{Q}_{loss,h}}{UA\left(T_{h,in} - T_{c,in}\right)} \tag{6.20}$$

$$\chi_c = \frac{\dot{Q}_{loss,c}}{UA\left(T_{h,in} - T_{c,in}\right)} \tag{6.21}$$

The heat loss rate from the hot side and cold side of the recuperator is calculated with Eq. (6.22) and Eq. (6.23) and the following equations.

$$\dot{Q}_{loss,h} = \frac{\dot{Q}_{loss,top,h}}{n} + \dot{Q}_{loss,side,h} \tag{6.22}$$

$$\dot{Q}_{loss,c} = \frac{\dot{Q}_{loss,bottom,c}}{n} + \dot{Q}_{loss,side,c} \tag{6.23}$$

$$\dot{Q}_{loss,top,h} = \frac{(T_9 + T_{10})/2 - T_\infty}{1/h_h aL + t_{ins}/k_{ins}aL + 1/h_{out}aL} \tag{6.24}$$

$$\dot{Q}_{loss,side,h} = \frac{(T_9 + T_{10})/2 - T_\infty}{1/h_h bL + t_{ins}/k_{ins}bL + 1/h_{out}bL} \tag{6.25}$$

$$\dot{Q}_{loss,bottom,c} = \frac{(T_3 + T_4)/2 - T_\infty}{1/h_c aL + t_{ins}/k_{ins}aL + 1/h_{out}aL} \tag{6.26}$$

$$\dot{Q}_{loss,side,c} = \frac{(T_3 + T_4)/2 - T_\infty}{1/h_c bL + t_{ins}/k_{ins}bL + 1/h_{out}bL} \tag{6.27}$$

6.5 TURBOCHARGER AS MICROTURBINE

The compressor isentropic efficiency, compressor corrected mass flow rate, compressor pressure ratio, and rotational speed are intrinsically coupled to each other and are available from the compressor map [8,43]. Compressor and turbine maps from standard off-the-shelf turbochargers from Garrett [8] are considered. The compressor isentropic efficiency and shaft speed is obtained with interpolation. The compressor should operate within its compressor map range, otherwise flow surge or choking can occur. According to Ref. [23], the turbine efficiency is determined by calculating the blade speed ratio [44−46] as shown in Eq. (6.28) and Eq. (6.29). The blade speed ratio is a function of the inlet enthalpy, pressure ratio, turbine wheel diameter, and rotational speed [23,45]. According to Guzzella and Onder [47], in automotive applications, typical values for the maximum turbine efficiency are $\eta_{t,max} \approx 0.65 - 0.75$. Note that the system mass flow rate is equal to the actual turbine mass flow rate and is calculated with Eq. (6.30) where P_7 is in pounds per square inch and T_7 is in degrees Fahrenheit, respectively [8].

$$BSR = \frac{\frac{2\pi N}{60}\left(\frac{D_t}{2}\right)}{\left[2h_{in}\left(1 - r_t^{\frac{1-k}{k}}\right)\right]^{\frac{1}{2}}} \tag{6.28}$$

$$\eta_t = \eta_{t,max}\left(1 - \left(\frac{BSR - 0.6}{0.6}\right)^2\right) \tag{6.29}$$

$$\dot{m}_t = \frac{\dot{m}_{tCF} \times P_7/14.7}{\sqrt{(T_7 + 460)/519}} \tag{6.30}$$

6.6 OPTIMIZATION AND METHODOLOGY

The method of entropy generation minimization is used to maximize the net power output of the STBC at steady state by optimizing the geometry variables of the receiver and recuperator [23]. When considering geometric optimization of components, in a system using a turbomachine, the compressor or turbine pressure ratio can be chosen as a parameter [48–50]. In this work, the turbine operating point (turbine corrected mass flow rate and turbine pressure ratio) is chosen. Note that the turbine corrected mass flow rate is a function of the turbine pressure ratio according to the turbine map.

$$\dot{W}_{net} = -T_\infty \dot{S}_{gen,int} + \left(1 - \frac{T_\infty}{T^*}\right)\dot{Q}^* + \dot{m}c_{p0}(T_1 - T_{11}) - \dot{m}T_\infty c_{p0} \ln\left(\frac{T_1}{T_{11}}\right) \tag{6.31}$$

$$\dot{S}_{gen,int} = \left[-\dot{m}c_{p0} \ln(T_1/T_2) + \dot{m}R \ln(P_1/P_2)\right]_{compressor}$$

$$+ \left[\dot{Q}_l/T_\infty + \dot{m}c_{p0} \ln(T_3/T_2) - \dot{m}R \ln(P_3/P_2)\right]_{duct23}$$

$$+ \left[\dot{m}c_{p0} \ln\left[\frac{T_{10}T_4}{T_9 T_3}\left(\frac{P_{10}P_4}{P_9 P_3}\right)^{-R/c_{p0}}\right] + \dot{Q}_l/T_\infty\right]_{recuperator}$$

$$+ \left[\dot{Q}_l/T_\infty + \dot{m}c_{p0} \ln(T_5/T_4) - \dot{m}R \ln(P_5/P_4)\right]_{duct45} \tag{6.32}$$

$$+ \left[-\dot{Q}^*/T^* + \dot{Q}_{loss}/T_\infty + \dot{m}c_{p0} \ln(T_6/T_5) - \dot{m}R \ln(P_6/P_5)\right]_{receiver}$$

$$+ \left[\dot{Q}_l/T_\infty + \dot{m}c_{p0} \ln(T_7/T_6) - \dot{m}R \ln(P_7/P_6)\right]_{duct67}$$

$$+ \left[-\dot{m}c_{p0} \ln(T_7/T_8) + \dot{m}R \ln(P_7/P_8)\right]_{turbine}$$

$$+ \left[\dot{Q}_l/T_\infty + \dot{m}c_{p0} \ln(T_9/T_8) - \dot{m}R \ln(P_9/P_8)\right]_{duct89}$$

The finite heat transfer and pressure drop in the compressor, turbine, recuperator, receiver, and other tubes are identified as entropy generation mechanisms. When doing an exergy analysis of the system at steady state and assuming $V_1 = V_{11}$ and $Z_1 = Z_{11}$ (see Fig. 6.1), the objective function is assembled as shown in Eq. (6.31). The function to be maximized (the objective function), is $\dot{W}_{net}$ (the net power output). Eq. (6.32) shows the total entropy generation rate in terms of the temperatures and pressures at different locations in the cycle (with reference to Fig. 6.1). The entropy generation rate of each component is added as shown in Eq. (6.32). T^* is the apparent sun's temperature as an exergy source as described in Ref. [23]. Note that $\dot{Q}^* - \dot{Q}_{loss} = \dot{Q}_{net}$. For validation, the net power output can also be calculated with the first law of thermodynamics using a control volume around the microturbine and also around the system.

As optimization, different combinations of three turbochargers by Garrett [8], three different receiver tube diameters, and 625 differently sized recuperators are used as parameters and variables to determine the net power output of the system. MATLAB [51] is used to determine the variables that will give the best results. The recuperator variables are the width of the recuperator channel, a, the height of a recuperator channel, b, the length of the recuperator, L, and the number of flow channels in one direction, n. The maximum receiver surface temperature is constrained to 1200K. The recuperator total plate mass is restricted to 300, 400, or 500 kg. The temperatures and pressures at every point of the STBC are found with iteration and by modeling every component [23] and using isentropic efficiencies of the compressor and turbine as well as recuperator and receiver efficiencies.

6.6.1 Assumptions

It is assumed that the receiver and recuperator are made from stainless steel. The pressure drop through the receiver tube and other tubes in the cycle is calculated with the friction factor using the Colebrook equation [52] for rough stainless steel. The pressure drop through the recuperator is calculated similarly or by using the friction factor for fully developed laminar flow, depending on the Reynolds number. It is assumed that $T_8 = T_9$ and $P_8 = P_9$ since the recuperator and microturbine are close to each other. It is further assumed that $T_2 = T_3$ and $P_2 = P_3$ since the recuperator and microturbine are close to each other. Also note that $T_1 = 300K$ and $P_1 = P_{10} = P_{11} = 86$ kPa. It is assumed that the thickness of the material between the hot and cold stream, t, in the recuperator, is 1 mm.

6.7 RESULTS

With the use of SolTrace, Fig. 6.5 is obtained as an initial result, showing the effect of the aperture size and the optical error on the receiver efficiency [24]. An average receiver surface temperature of 1150 K was assumed to estimate the heat loss. A pillbox sunshape parameter of 4.65 mrad was assumed in the SolTrace analysis and a parabolic dish rim angle of 45°. The reflectance of the receiver tube was assumed to be 15% (oxidized stainless steel) and the emissivity of the receiver material as 0.7. It was mentioned previously that the dish optics and tracking errors are important factors

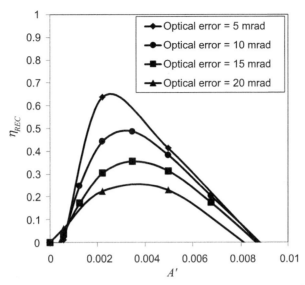

Figure 6.5 Overall receiver efficiency for a tracking error of 1° with receiver surface emissivity of 0.7 [24].

regarding the total cost of the system. A solar tracking error of 1° was used in the analysis. A tracking error of 1° indicates rather poor tracking and thus a low-cost tracking system. Note that an optimum area ratio exists for each optical error in Fig. 6.5. If an optical error of 10 mrad is assumed, the optimum area ratio is $A' \approx 0.0035$ [24].

In the following results, a solar dish with diameter of 4.8 m was considered and a solar direct normal irradiance (DNI) of 1000 W/m^2 was assumed. For a fixed area ratio of $A' = 0.0035$, the available solar power at the aperture of a receiver with $a = 0.25$ m (aperture area of 0.25 m × 0.25 m) is shown in Fig. 6.6 for different tracking errors and optical errors. Fig. 6.6 shows that the available solar power decreases as the optical error and tracking error increase. However, at a tracking error of 3°, the available solar power increases as the optical error increases, since the focal image increases as the optical error increases. Note that the available solar power stays almost constant when the tracking error is between 0° and 1° and the optical error below 10 mrad.

Fig. 6.7 shows the average net heat transfer rate at the cavity receiver inner wall. The net heat transfer rate is the available solar power minus the heat loss rate from the cavity receiver to the environment. Note that a dish reflectivity of 85% was assumed. Fig. 6.7 also shows that an optical error of 10 mrad or less with a tracking error of 1° or less is required to have an acceptable net heat transfer rate. A tracking error of 2° would thus not be acceptable for the collector.

Eq. (6.31) was used as an objective function in MATLAB to determine the maximum net power output of the cycle by optimizing the receiver tube diameter and counterflow

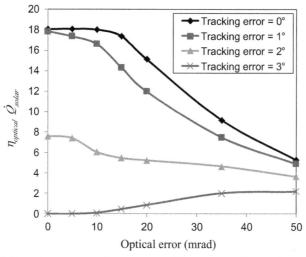

Figure 6.6 Available solar power at the aperture of the open-cavity receiver with $a = 0.25$ m.

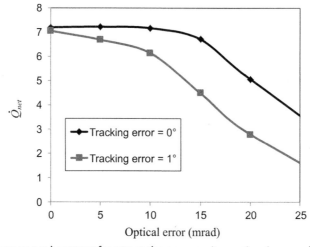

Figure 6.7 The average net heat transfer rate at the open-cavity receiver inner walls with $a = 0.25$ m.

plate-type recuperator geometries. A 4.8-m-diameter parabolic dish, rim angle of $45°$, $1°$ tracking error, 10 mrad optical error, $a = 0.25$, solar receiver emissivity of 0.7, a solar DNI of 1000 W/m^2, and a dish reflectivity of 85% was used to generate the results. It was also assumed that all tubing in the cycle has 10 mm insulation thickness and conduction heat transfer coefficient of 0.18 W/mK. For the open-cavity tubular receiver, efficiencies of between 43% and 70% were obtained with mass flow rates of between 0.06

and 0.08 kg/s, tube diameters of between 0.05 and 0.0889 m, and air inlet temperatures of between 900K and 1070K [24]. In Figs. 6.8–6.10 the maximum net power output of the STBC at steady state was found as a function of turbine pressure ratio. It is shown that the maximum net power output of the system can be found when a large receiver tube diameter is used, for all three of the microturbines. Note that each data point represents a maximum, which is achieved using a unique recuperator as shown in Table 6.1 for the GT2860RS turbocharger with a receiver tube diameter of 0.0833 m (compare with Fig. 6.9). From Table 6.1, a recuperator with $a = 0.225$ m, $b = 2.25$ mm, $L = 1.5$ m, and $n = 45$ was the best-performing recuperator since it gave the highest maximum net power output of 1.77 kW (Table 6.1). Note that in these results the recuperator mass was restricted to 500 kg and the receiver surface temperature restricted to 1200K.

The GT1241 provides the highest maximum power output of 2.11 kW and a conversion efficiency of 12% (Fig. 6.10). This efficiency compares well with photovoltaic panels. It should be noted that this efficiency can be improved in many ways. Note that in this chapter, low-cost tracking and solar optics were considered, which decrease the efficiency of the receiver. In this work, only standard off-the-shelf turbochargers were considered, which are not necessarily optimized to perform as a power-generating unit but rather for performance in a vehicle. However, the low cost of these units justifies the research into its application in this field. Further investigation into the correct matching of turbocharger compressors and turbines to improve efficiency should be done. The overall efficiency of the cycle can also be improved with the use of cogeneration. It should also be noted that higher efficiencies have already been obtained with state-of-the-art systems, as mentioned in Section 6.1.

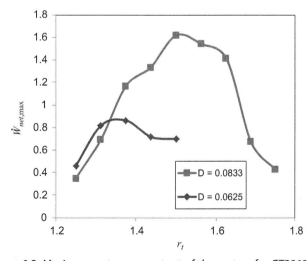

Figure 6.8 Maximum net power output of the system for GT2860R(2).

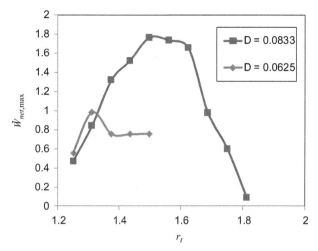

Figure 6.9 Maximum net power output of the system for GT2860RS.

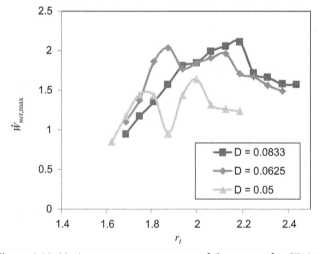

Figure 6.10 Maximum net power output of the system for GT1241.

For larger receiver tube diameters, a higher maximum net power output can be achieved at high turbine pressure ratios. Note that in Ref. [24], it was found that the highest second law efficiencies for the receiver were achieved when the tube diameter and inlet temperature were large and the mass flow rate small. This combination of variables would also create a very high surface temperature. When the surface temperature is then restricted to 1200K, a larger mass flow rate and lower inlet temperature can still provide high second law efficiencies, when a large tube diameter is chosen.

Table 6.1 Optimum Recuperator Geometries, Maximum Net Power Output, Maximum Receiver Surface Temperature, and Recuperator Mass for GT2860RS With Receiver Tube Diameter $D = 0.0833$ m

r_t	a (m)	b (mm)	L (m)	n	$\dot{W}_{net,max}$ (W)	$T_{s,max}$ (K)	Mass (kg)
1.25	0.15	4.5	1.5	22	470	1186	167
1.313	0.37	4.5	1.5	15	843	1197	273
1.375	0.37	3.7	1.5	22	1319	1195	409
1.438	0.45	3.0	1.5	22	1524	1180	489
1.5	0.22	2.2	1.5	45	1767	1127	491
1.563	0.45	2.2	1.5	22	1736	1053	488
1.625	0.22	2.2	1.5	45	1655	1000	491
1.688	0.22	2.2	1.5	45	979	940	491
1.75	0.22	2.2	1.5	45	597	885	491
1.813	0.22	2.2	1.5	45	90	830	491

The larger the mass of the recuperator, the higher the net power output of the system as shown in Fig. 6.11. When the recuperator mass is constrained to 400 kg, a recuperator with $a = 150$ mm, $b = 2.25$ mm, $L = 1.5$ m, and $n = 45$ was found to be the recuperator with the most common optimum dimensions. When the recuperator mass is constrained to 300 kg, a recuperator with $a = 150$ mm, $b = 2.25$ mm, $L = 1.5$ m, and $n = 37.5$ is best.

From Fig. 6.12 it is shown that it is optimum for the maximum receiver tube surface temperature to decrease with increasing turbine pressure ratio and with decreasing receiver tube diameter. At high turbine pressure ratios, the large receiver tube diameter

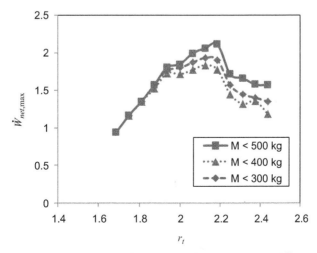

Figure 6.11 Maximum net power output of the system for GT1241 with different recuperator mass constraints.

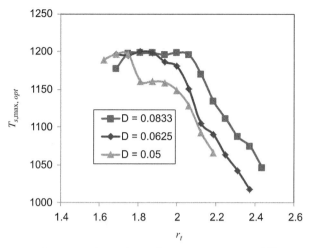

Figure 6.12 Optimum maximum receiver tube surface temperature at different operating temperatures for GT1241.

has a higher surface temperature and it allows for higher net power output of the small-scale STBC.

6.8 REMAINING CHALLENGES AND FUTURE POSSIBILITIES

The smaller a microturbine, the higher its operating speed. Off-the-shelf commercial turbochargers operate at speeds of 50,000–180,000 rpm [8]. A remaining challenge for the STBC using an off-the-shelf turbocharger is to connect the turbocharger's shaft to a high-speed generator that generates electric power of high and variable frequency before it is rectified to direct current [2]. Shiraishi and Ono [3] showed that a flexible coupling can be used to connect the turbocharger and generator. A two-shaft configuration can also be considered. According to Weston [53], the two-shaft arrangement allows for acceptable performance over a wider range of operating conditions. In the case where two-shaft technology is considered, the addition of reheating and intercooling, which is usually present on two-shaft configurations and which makes the cycle much more efficient, becomes more attractive. The possible commercial availability of high-speed gearboxes in future will invalidate the need for a high-speed generator and will allow for a low-speed generator to be coupled directly to the turbine shaft.

Future possibilities of the STBC include hybridization and thermal storage for continuous power generation, as was also mentioned in Section 6.1. Hot exhaust air leaving the STBC can be used to heat water or to run an absorption chiller. If the exhaust gas is not utilized, it is very important that the exhaust air is exposed of correctly and that it is not sucked in again at the compressor. Furthermore, commercial turbine operating temperatures might be increased in future with the use of ceramic materials, which will allow for higher STBC efficiencies.

6.9 CONCLUSION AND RECOMMENDATIONS

The small-scale dish-mounted STBC using a turbocharger as microturbine is a technology with a lot of potential to generate electricity from solar power. The chapter discussed the potential of the technology as well as the remaining challenges. One of the important advantages of the cycle is that it can be supplemented with natural gas as a hybrid system for continuous operation when solar radiation levels are low or during the night.

The cycle is faced with low compressor and turbine efficiencies, heat losses, and pressure losses, which decrease the net power output of the system. In this chapter, a recent development in the minimization of these losses was presented. The power output of the small-scale dish-mounted STBC with recuperator, off-the-shelf turbocharger, and low-cost optics was modeled and maximized using the method of total entropy generation minimization. A receiver and recuperator were optimized to perform in the cycle. Results showed that higher maximum net power output can be achieved at high turbine pressure ratios, when a large receiver tube diameter is used. It was found that a recuperator with $a = 225$ mm, $b = 2.25$ mm, $L = 1.5$ m, and $n = 45$ gives the best results for the setup with the recuperator mass constraint and receiver maximum surface temperature constraint. The larger the mass of the recuperator, the higher the power output of the system.

Results showed that solar-to-mechanical efficiencies of up to 12% could be achieved when using a standard off-the-shelf turbocharger and low-cost optics. The efficiency is low compared with other solar technologies; however, there are many ways in which the efficiency can be improved. Note that the current work only considered commercial turbochargers with an already-combined compressor and turbine. This combination is chosen for performance in a vehicle rather than for power generation. Further compressor and turbine matching should be performed to identify better combinations of these units for significant efficiency improvement. It should also be noted that a very simple solar receiver was used for the modeling. With the use of high absorptivity and low emissivity coatings, the efficiency of the receiver can be increased. It is recommended that high-temperature, low-emissivity receiver coatings and materials should be investigated further experimentally to improve efficiency. The efficiency can be further improved by having more precise tracking and dish optics and higher dish reflectivity. Note that, in this chapter, a receiver aperture area of 0.25 m × 0.25 m and a dish with poor tracking (1° error) and poor dish optics (10 mrad error) were assumed. According to Ref. [24], the receiver efficiency can be increased up to 71% when both the optical error and tracking error are changed to 5 mrad and $0°$, respectively, and the aperture area to $0.1 \text{ m} \times 0.1 \text{ m}$. By adding water heating using the hot exhaust air, the overall efficiency will be increased even further.

Furthermore, it is recommended that a cost-versus-efficiency study should be done regarding the small-scale dish-mounted STBC. With further research, the small-scale open STBC could become a competitive off-the-grid small-scale solar energy solution for small communities. It is recommended that the cycle should be further developed experimentally and investigated as a clean energy technology.

NOMENCLATURE

a	Receiver aperture side length or recuperator channel width, m
b	Recuperator channel height, m
A	Area, m^2
A'	Receiver aperture to dish area ratio
BSR	Blade speed ratio
c_1	Constant used in linear equation
c_2	Constant used in linear equation
c_{p0}	Constant pressure specific heat, J/kgK
Cr	Capacity ratio
D	Diameter, m
f_c	Focal length, m
F	View factor
h	Heat transfer coefficient, $\text{W/m}^2\text{K}$
h	Specific enthalpy, J/kg
H	Recuperator height, m
k	Thermal conductivity, W/mK
k	Gas constant
L	Length, m
m_1	Slope of linear equation
m_2	Slope of linear equation
$\dot{m}$	System mass flow rate, kg/s
M	Mass of recuperator, kg
n	Number of recuperator flow channels in one direction
N	Number of tube sections
N	Speed of microturbine shaft, rpm
NTU	Number of transfer units
P	Pressure, Pa
r	Pressure ratio
R	Gas constant, J/kgK
R	Dish radius, m
R	Thermal resistance, K/W
$\dot{Q}$	Heat transfer rate, W
$\dot{Q}*$	Available solar heat rate at receiver cavity, W

$\dot{Q}_{loss}$	Rate of heat loss, W
$\dot{Q}_{net}$	Net heat transfer rate, W
$\dot{Q}_{solar}$	Available solar heat rate on dish aperture, W
$\dot{S}_{gen}$	Entropy generation rate, W/K
t	Thickness, m
T	Temperature, K
$T^{\star}$	Apparent exergy-source sun temperature, K
U	Overall heat transfer coefficient, W/m^2K
w	Wind factor
$\dot{W}$	Power, W
V	Velocity, m/s
X	Dimensionless position
Z	Height, m

GREEK LETTERS

ε	Emissivity of receiver
ε	Recuperator effectiveness
η	Efficiency
Θ	Dimensionless temperature difference
σ	Stefan–Boltzmann constant, W/m^2K
χ	Dimensionless external heat load
ψ_{rim}	Rim angle
ω	Error, mrad

SUBSCRIPTS

0	Initial inlet to receiver
1–11	Refer to Fig. 6.1
ap	Aperture
BC	Brayton cycle
bottom	At the bottom
c	Compressor
c	Based on the cold side
CF	Corrected flow
col	Collector
cond	Due to conduction
conv	Due to convection
e	Exit
h	Based on the hot side
i	Inlet
in	At the inlet
ins	Insulation
int	Internal
inner	On the inside
l	Loss

max	Maximum
n	Tube section number
net	Net output
opt	Optimum
optical	Optical
out	On the outside of the insulation
rad	Due to radiation
rec	Receiver
refl	Due to dish reflectivity
REC	For the receiver including optical efficiency
reg	Recuperator
s	Surface
side	At the side
slope	Slope
solar	From solar as determined with SolTrace
specularity	Specularity
STBC	Solar thermal Brayton cycle
t	Turbine
top	At the top
∞	Environment

REFERENCES

[1] Pietsch A, Brandes DJ. Advanced solar Brayton space power systems. In: Intersociety energy conversion engineering conference, Los Alamitos, CA; 1989. p. 911−6.
[2] Willis HL, Scott WG. Distributed power generation. New York: CRC Press; 2000.
[3] Shiraishi K, Ono Y. Mitsubishi Heavy Industries, Ltd. Technical Review 2007;44(1).
[4] Stine BS, Harrigan RW. Solar energy fundamentals and design. New York: John Wiley & Sons; 1985.
[5] McDonald CF, Rodgers C. Journal of Engineering for Gas Turbines and Power 2002;124:835−44.
[6] Allen KG. Performance characteristics of packed bed thermal energy storage for solar thermal power plants [Thesis]. University of Stellenbosch; 2010.
[7] Cameron HM, Mueller LA, Namkoong D. Report NASA TM X-2552. Lewis Research Center; 1972.
[8] Garrett. Garrett by Honeywell: turbochargers, intercoolers, upgrades, wastegates, blow-off valves, turbo-tutorials. 2009. Available at: http://www.TurboByGarrett.com.
[9] Mills D. Advances in solar thermal electricity technology. Solar Energy 2004;76:9−31.
[10] Dickey B. Test results from a concentrated solar microturbine Brayton cycle integration. In: TurboExpo, GT2011, ASME conference proceedings. Vancouver (British Columbia, Canada): ASME; 2011.
[11] Heller P, Pfänder M, Denk T, Tellez F, Valverde A, Fernandez J, et al. Test and evaluation of a solar powered gas turbine system. Solar Energy 2006;80:1225−30.
[12] Chen L, Zhang W, Sun F. Power, efficiency, entropy-generation rate and ecological optimization for a class of generalized irreversible universal heat-engine cycles. Applied Energy 2007;84:512−25.
[13] Zhang Y, Lin B, Chen J. Optimum performance characteristics of an irreversible solar-driven Brayton heat engine at the maximum overall efficiency. Renewable Energy 2007;32:856−67.
[14] Stevens T, Baelmans M. Optimal pressure drop ratio for micro recuperators in small sized gas turbines. Applied Thermal Engineering 2008;28:2353−9.
[15] Neber M, Lee H. Design of a high temperature cavity receiver for residential scale concentrated solar power. Energy 2012;47:481−7.
[16] Hischier I, Hess D, Lipiński W, Modest M, Steinfeld A. Heat transfer analysis of a novel pressurized air receiver for concentrated solar power via combined cycles. Journal of Thermal Science and Engineering Applications 2009;1. 041002-1-6.

[17] Bejan A, Tsatsaronis G, Moran M. Thermal design and optimization. New York: John Wiley & Sons, Inc.; 1996.

[18] Bejan A. Entropy generation through heat and fluid flow. Colorado: John Wiley & Sons, Inc.; 1982.

[19] Jubeh NM. Exergy analysis and second law efficiency of a regenerative Brayton cycle with isothermal heat addition. Entropy 2005;7(3):172—87.

[20] Le Roux WG, Bello-Ochende T, Meyer JP. Operating conditions of an open and direct solar thermal Brayton cycle with optimised cavity receiver and recuperator. Energy 2011;36:6027—36.

[21] Le Roux WG, Bello-Ochende T, Meyer JP. Optimum performance of the small-scale open and direct solar thermal Brayton cycle at various environmental conditions and constraints. Energy 2012;46: 42—50.

[22] Le Roux WG, Bello-Ochende T, Meyer JP. Thermodynamic optimisation of the integrated design of a small-scale solar thermal Brayton cycle. International Journal of Energy Research 2012;36: 1088—104.

[23] Le Roux WG, Bello-Ochende T, Meyer JP. A review on the thermodynamic optimisation and modelling of the solar thermal Brayton cycle. Renewable and Sustainable Energy Reviews 2013;28:677—90.

[24] Le Roux WG, Bello-Ochende T, Meyer JP. The efficiency of an open-cavity solar receiver for a small-scale solar thermal Brayton cycle. Energy Conversion and Management 2014;84:457—70.

[25] Stafford B, Davis M, Chambers J, Martínez M, Sanchez D. Tracker accuracy: field experience, analysis, and correlation with meteorological conditions. In: Proceedings of the 34th photovoltaic specialists conference; 2009.

[26] Gee R, Brost R, Zhu G, Jorgensen G. An improved method for characterizing reflector specularity for parabolic trough concentrators. In: Proceedings of solar PACES international symposium of solar thermal, September, Perpignon, France; 2010. 0284.

[27] Ávila-Marín AL. Volumetric receivers in solar thermal power plants with central receiver system technology: a review. Solar Energy 2011;85:891—910.

[28] Ho CK, Iverson BD. Review of high-temperature central receiver designs for concentrating solar power. Renewable and Sustainable Energy Reviews 2014;29:835—46.

[29] Shah RK. Compact heat exchangers for micro-turbines. 2005. Available at: https://www.cso.nato.int/pubs/rdp.asp?RDP=RTO-EN-AVT-131.

[30] McDonald CF, Rodgers C. Small recuperated ceramic microturbine demonstrator concept. Applied Thermal Engineering 2008;28:60—74.

[31] Harris JA, Lenz TG. Thermal performance of solar concentrator/cavity receiver systems. Solar Energy 1985;34(2):135—42.

[32] Le Roux WG, Meyer JP, Bello-Ochende T. Experimental testing of a tubular cavity receiver for a small-scale solar thermal Brayton cycle. In: Proceedings of the Southern African solar energy conference, SASEC, May, Skukuza, Kruger National Park, South Africa; 2015. p. 295—300.

[33] McDonald CG. Heat loss from an open cavity. Albuquerque (New Mexico): Sandia National Laboratories; 1995. SAND95—2939.

[34] Ho CK. Software and codes for analysis of concentrating solar power technologies. Sandia Laboratory Report. 2008. SAND2008—8053.

[35] Bode SJ, Gauché P. Review of optical software for use in concentrated solar power systems. In: Proceedings of the Southern African solar energy conference, SASEC, May, Stellenbosch, South Africa; 2012.

[36] Yilmaz M, Sara ON, Karsli S. Performance evaluation criteria for heat exchangers based on second law analysis, Exergy. An International Journal 2001;1(4):278—94.

[37] Oğulata RT, Doba F, Yilmaz T. Irreversibility analysis of cross flow heat exchangers. Energy Conversion and Management 2000;41(15):1585—99.

[38] McDonald CF. Low-cost compact primary surface recuperator concept for microturbines. Applied Thermal Engineering 2000;20:471—97.

[39] McDonald CF. Recuperator considerations for future higher efficiency microturbines. Applied Thermal Engineering 2003;23:1463—87.

[40] Kreith F, Kreider JF. Principles of solar engineering, Hemisphere, Colorado. 1978.

[41] Hesselgreaves JE. Rationalisation of second law analysis of heat exchangers. International Journal of Heat and Mass Transfer 2000;43(22):4189—204.

[42] Nellis GF, Pfotenhauer JM. Effectiveness-NTU relationship for a counterflow heat exchanger subjected to an external heat transfer. Journal of Heat Transfer 2005;127:1071—3.

[43] Zhuge W, Zhang Y, Zheng X, Yang M, He Y. Development of an advanced turbocharger simulation method for cycle simulation of turbocharged internal combustion engines. Proceedings of the Institution of Mechanical Engineers, Part D: Journal of Automobile Engineering 2009;223:661. http://dx.doi.org/10.1243/09544070JAUTO975.

[44] Westin F. Simulation of turbocharged SI-engines—with focus on the turbine [Doctoral thesis. KTH School of Industrial Engineering and Management, TRITA—MMK 2005:05]. Stockholm: Royal Institute of Technology; 2005.

[45] Wahlström J, Eriksson L. Modeling diesel engines with a variable-geometry turbocharger and exhaust gas recirculation by optimization of model parameters for capturing non-linear system dynamics. Proceedings of the Institution of Mechanical Engineers, Part D: Journal of Automobile Engineering July 7, 2011;225.

[46] Batteh JJ, Newman CE. Detailed simulation of turbocharged engines with Modelica. The Modelica Association; March 3 and 4, 2008.

[47] Guzzella L, Onder CH. Introduction to modeling and control of internal combustion engine systems. 2nd ed. Berlin (Germany): Springer-Verlag; 2010. http://dx.doi.org/10.1007/978-3-642-10775-7.

[48] Snyman JA. Practical mathematical optimization. Pretoria: University of Pretoria; 2009.

[49] Wilson DG, Korakianitis T. The design of high-efficiency turbomachinery and gas turbines. 2nd ed. New Jersey: Prentice Hall; 1998.

[50] Lidsky LM, Lanning DD, Staudt JE, Yan XL, Kaburaki H, Mori M, et al. A direct-cycle gas turbine power plant for near-term application: MGR-GT. Energy 1991;16:177—86.

[51] MATLAB. Version 7.6.0.324 Natick. Massachusetts: The MathWorks Inc.; 2008.

[52] Colebrook CF. Turbulent flow in pipes, with particular reference to the transition between the smooth and rough pipe laws. Journal of the Institute of Civil Engineers London 1939;11:133—56.

[53] Weston KC. Energy Conversion. [e-book]. Tulsa: Brooks/Cole; 2000. Available at: http://www.personal.utulsa.edu/~kenneth-weston.

CHAPTER SEVEN

Heat-Driven Cooling Technologies

R. Narayanan
Central Queensland University, Bundaberg, QLD, Australia

7.1 INTRODUCTION

Air conditioning accounts for a major part, up to 40%, of the energy use in buildings [1]. At the same time, the demand for air conditioning is steadily increasing: world sales in 2011 were up 13% from 2010 and that growth is expected to accelerate in coming decades as ambient temperature levels rise throughout the world and personal income levels are increasing with smaller families living in larger houses in hotter places. Widespread air-conditioning system installation not only increases the total energy consumption, but also raises the peak load demand. The peak demand in many countries and many Australian cities is growing at about 50% more than the base demand growth and, in some areas, up to four times the average, which in turn requires higher generation capacity [2]. The generation of power accounts for the majority of greenhouse gas emissions and associated climate change, which is the most serious challenge currently faced by mankind. These challenges emphasize the necessity of developing a green and sustainable system that minimizes the negative human impacts on the natural surroundings, materials, resources, and processes that prevail in nature [3]. Heat-driven air-conditioning systems are systems that use low-grade thermal energy like solar energy and waste heat rather than electricity. These systems can produce significant energy savings, have low global warming and ozone depletion potential, and also ensure higher indoor air quality. Studies have demonstrated that such cooling equipment is applicable to wide ranges of climatic conditions and offers a feasible alternative to conventional air-conditioning systems [4–10].

The purpose of air conditioning is to control the temperature, humidity, filtration, and air movement of the indoor environment. The atmospheric air always contains moisture in the form of water vapor, but the maximum amount of water vapor that may be contained in the air depends on its temperature and so the higher the temperature, the more water vapor that can be contained in it. At high temperatures and moisture content levels, extreme discomfort is experienced as the evaporation of moisture from the body into the atmosphere by the process of perspiration becomes difficult. The term "humidity ratio" indicates the mass of water vapor present per kilogram of dry air, whereas the term "relative humidity" is the ratio between the actual moisture content of the air

Clean Energy for Sustainable Development
ISBN 978-0-12-805423-9, http://dx.doi.org/10.1016/B978-0-12-805423-9.00007-7

and the moisture content of the air required for saturation at the same temperature. The purpose of comfort conditioning is solely to provide a comfortable environment for the majority of occupants. Humans are reasonably tolerant to humidity and may be comfortable from a range of between 55% and 20% relative humidity at normal comfort temperatures [11].

In air conditioning, the humidification of air to increase the moisture content is achieved by the use of a humidifier, and in reverse, the moisture content of the air is reduced by a dehumidification process. The independent control of the air temperature by heating and cooling processes and of relative humidity by the humidifying and dehumidifying processes constitutes full air conditioning, but this control is not always exercised. The more often used vapor compression refrigeration equipment is capable of cooling and then dehumidifying.

Another method of cooling air is by an environmentally friendly evaporative cooler that uses the latent heat of evaporation of water to cool the air. In direct evaporative cooling, the moisture content of the air is increased. However, adding moisture to the air is avoided through indirect evaporative cooling [12]. The performance of evaporative systems deteriorates as the inlet air humidity increases, so evaporative cooling systems are not suitable for humid climates [13]. To enable the use of evaporative cooling, the air has to be dehumidified.

The cooling load is the amount of heat energy to be removed from the air confined in the building by the air-conditioning equipment to maintain the indoor air at the desired temperature. The total cooling load has two components, which are sensible and latent cooling loads. The sensible cooling load refers to reducing the dry bulb temperature of the air and the latent cooling load refers to reducing the moisture content. Even though the water vapor content of atmospheric air is small compared with the mass of dry air, due to its very high heat of evaporation, the latent heat content in air conditioning is generally of the same order as the sensible load [14]. Therefore dehumidification considerably reduces the total cooling load and enables the use of air cooling with the less energy-intensive option of evaporative cooling.

The air can be dehumidified by two methods. The first one is by cooling air below dew point temperature to condense water vapor in the air using a cooling coil using a vapor compression system. Besides being energy intensive, these systems are not effective in dealing with higher latent loads, which require a lower dew point temperature because they are not suitable when the required dew point temperature is lower than $0°C$, the freezing point [15]. Condensation of water during this process causes mildew. This process also may necessitate heating after reducing the moisture content. Another method for dehumidification is by sorption of water vapor in the air with a sorbent or a desiccant. This method has no limitation of dew point temperature as there is no condensation of water and has greater energy saving potential.

Based on the type of input energy used, air-conditioning systems can be classified generally into vapor compression systems, which use electricity and thereby mechanical work to achieve cooling, and heat-driven cooling systems, which use low-grade thermal energy.

7.2 HEAT-DRIVEN AIR CONDITIONING

The simplest heat-driven system transfers heat at three temperature levels and the basic thermodynamic scheme is as shown in Fig. 7.1.

The coefficient of performance (COP) of this system is different from the vapor compression system and is defined as the ratio of cooling effect produced to the heat input provided to produce the cooling. So it can be expressed as:

$$COP = \frac{Q_{cold}}{Q_{heat}} \tag{7.1}$$

By applying the first and second laws of thermodynamics, the above-mentioned system will give the ideal COP as:

$$COP_{Ideal} = \frac{T_c}{T_h} \frac{(T_h - T_m)}{(T_m - T_c)} \tag{7.2}$$

where T_c, T_h, and T_m are the temperatures of the cold source, driving heat source, and intermediate temperature level where heat is rejected [16].

Thermally driven chillers can be classified according to the cycle of operation into two categories, namely, closed cycle and open cycle systems. Absorption and adsorption

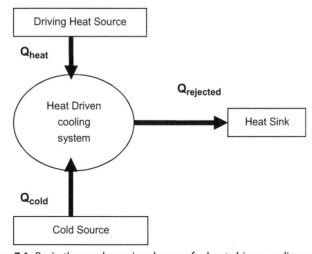

Figure 7.1 Basic thermodynamic scheme of a heat-driven cooling system.

chillers come under the first category, whereas desiccant evaporative cooling (DEC) systems are open cycle systems. When these air-conditioning systems use solar thermal energy to provide the heat required to drive the system, they are called solar air-conditioning systems.

7.2.1 Closed Cycle Systems

Closed cycle thermally driven chillers produce chilled water, which is used in the air handling units for cooling and dehumidification. These chillers can be categorized into absorption chillers and adsorption chillers based on the process used for producing chilled water. Most of the present closed cycle systems use absorption technology, and only very few use adsorption technology [17].

7.2.2 Absorption Chillers

The basic operation of a vapor absorption cycle is shown in Fig. 7.2. Absorption chillers differ from the compression chillers in that the cooling system is driven by heat energy, rather than mechanical energy. The compressor is replaced by an absorber and a generator. The evaporator allows the refrigerant to evaporate and to be absorbed by an absorbent fluid. This process produces the cooling effect. The combined fluids then go to the generator where it is heated, thus driving the refrigerant out of the absorbent. The refrigerant vapor then goes to a condenser and gets cooled down to liquid phase, whereas

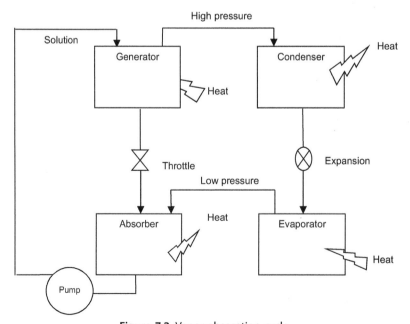

Figure 7.2 Vapor absorption cycle.

the absorbent is pumped back to the absorber. The cold liquid refrigerant is released through an expansion valve into the evaporator, and the cycle continues. The absorption chillers use either a water—lithium bromide pair (H_2O/LiBr) or an ammonia—water pair (NH_3—H_2O). The H_2O/LiBr system uses water as refrigerant and lithium bromide as absorbent. The ammonia—water system uses water as absorbent and ammonia as refrigerant. For air-conditioning applications, the water—lithium bromide pair is mainly applied [4].

In direct-fired absorption units, the heat source is gas or some other fuel that is burned in the unit. Indirect-fired units use steam, hot water, or some other transfer fluid that brings in heat from a separate source, such as a boiler or waste heat recovered from an industrial process or solar energy. Absorption chillers are categorized into single, double, or triple effect according to the number of stages of absorption. Theoretical and experimental studies show that single-effect absorption chillers have a maximum COP of 0.85 [18]. The low COP of single-effect absorption systems has made them uncompetitive and they are used only if a cheap heat source such as waste heat, district heat, or heat from a cogeneration plant is available. Under normal conditions, such machines typically need driving temperatures of 80—100°C. The desire for higher efficiencies in absorption chillers led to the development of double-effect systems. The double-effect chiller differs from the single-effect one in that there are two condensers and two generators to allow for more refrigerant boil-off from the absorbent solution. The double-effect absorption chillers have a COP of 1.1—1.3 [17]. Even though a double-effect chiller has higher COP, the driving temperature required is in the range of 140—160°C. The three- or four-effect system will have higher COPs from 1.7 to 2.2, but requires a higher driving temperature [4].

The major limitations for these chillers are that most of the systems available in the market are of large capacities (30 kW—5 MW) and have a high cost. In contrast to large-capacity applications, the small-capacity systems have difficulty entering the market due to relatively low efficiencies, control issues, and high initial cost [19,20].

The working of an adsorption chiller is shown in Fig. 7.3. In an adsorption chiller, the refrigerant is adsorbed onto the internal surfaces of a highly porous solid adsorbent. The most widely used adsorbate—adsorbent pair is water/silica but other pairs like water/zeolite and ammonia/activated carbon are also in use. The solid sorbent will be cooled and heated to be able to adsorb and desorb the refrigerant. So the operation is cyclic and not continuous.

At first, the refrigerant adsorbed in the right compartment is driven off by the hot water circulated. This refrigerant vapor enters the condenser and gets condensed by the cooling water. The condensate is sprayed into the evaporator and it evaporates under low pressure to produce the cooling effect. The refrigerant vapor produced is adsorbed onto the left compartment adsorber. Once the left compartment is fully charged and the right compartment is fully regenerated, their functions are interchanged.

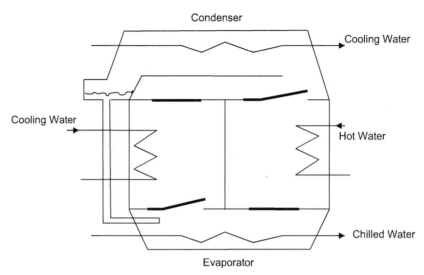

Figure 7.3 Adsorption chiller.

There are a few manufacturers of adsorption chillers most of which are of large cooling capacity, around 70 kW. The driving temperature is in the range of 55–95°C. The COP is around 0.4–0.7 [17]. The main limitation of an adsorption chiller is the limited heat transfer between the granular sorption material and the water that flows through the heat exchanger in the compartments containing the heat exchange [15]. The development of small-scale systems is discussed in Section 1.4.3.

7.2.3 Open Cycle Systems

DEC systems are open heat-driven cycles consisting of a combination of a dehumidifier, a sensible heat exchanger, and evaporative coolers. In this environmentally friendly system, dehumidification of air is achieved first to low humidity levels so that evaporative cooling can be employed effectively to reduce the air temperature. The DEC systems can be classified into two categories according to the type of desiccant material used. They are solid and liquid desiccant systems. Both these systems have evaporative cooling stages. So it is important to look at the theory of evaporative cooling system.

7.2.3.1 Evaporative Cooling

An evaporative cooling system works by employing water's large enthalpy of vaporization. The temperature of dry air can be reduced significantly through the phase transition of liquid water to water vapor, which can cool air using much less energy than refrigeration. In extremely dry climates, evaporative cooling of air has the added benefit of conditioning the air with more moisture for the comfort of building occupants.

Evaporative coolers can be categorized as direct, indirect, and two- and multistage. Direct evaporative air conditioners are the most popular in the market. In this system, the outside air is drawn through wetted filter pads, where the hot dry air is cooled and humidified through water evaporation. The evaporation of water takes some heat away from the air making it cooler and more humid. The dry-bulb temperature of the air leaving the wetted pads approaches the wet-bulb temperature of the ambient air. Direct evaporative air conditioners are more effective in dry climates. As they produce warmer, more humid air in comparison with refrigerated air conditioners, considerably more air volumes are required to produce the same cooling effect. The cool/humid air is used once and cannot be reused. Evaporation (saturation) effectiveness is the key factor in determining the performance of evaporative air conditioners. This property determines how close the air being conditioned is to the state of saturation. Usually, the effectiveness is 85−95% [21].

A direct evaporative air conditioning is ideal for arid climates where water is available. The direct evaporative air conditioners currently produced have, by and large, overcome the drawbacks associated with older systems. In addition to more efficient fan and duct designs and control systems, the use of plastics for the bodywork and cellulose and other synthetic materials for the pads together with automatic water bleeding or flushing has resulted in more reliable operation with little maintenance. Many of today's evaporative air conditioners have quite sophisticated control systems with variable air speeds and pad wetting rates. The one remaining drawback associated with direct cooling is the water saturation limit inherent in the process. Even with saturation efficiency over 80%, which is common for many modern systems, the air supplied may not provide cooling comfort if the outside air temperature is high and/or its moisture content is high and close to saturation with water vapor. The lowest possible temperature limit attained by direct evaporative cooling is the wet-bulb temperature at which the delivered air is fully saturated with moisture [22].

The saturation effectiveness also has an impact on water consumption. Increased saturation effectiveness is associated with higher water consumption. However, as higher saturation effectiveness produces conditioned air at lower temperatures, the overall impact of having higher saturation effectiveness is usually an improved energy and water consumption per unit cooling output.

The principle of operation of indirect evaporative cooling is the use of cool air produced by direct evaporative cooling to cool the air stream that is used for space cooling by the use of a heat exchanger. As cooling of the primary air stream takes place by heat transfer across the heat exchanger walls without the mixing of the two air streams, the primary air stream becomes cooler without an increase in its humidity. Indirect evaporative air conditioners are effective in regions with moderate/high humidity. According to manufacturers' rating, this effectiveness ranges from 40% to 80% [21].

The two-stage evaporative air conditioners combine both direct and indirect evaporative principles. In two-stage evaporative air conditioners, the first stage (indirect) sensibly cools the primary air (without increasing its moisture content) and the air is evaporatively cooled further in the second stage (direct). The dry-bulb temperature of the supplied primary air can be reduced to 6 K or more below the secondary air wet-bulb temperature without adding too much moisture. As two-stage evaporative air conditioners produce lower temperatures, they consequently require less air delivery in comparison with the direct systems. Experimental studies show that the saturation effectiveness of the indirect/direct evaporative air conditioner varies in a range of $108 \sim 111\%$ [22]. Also, over 60% energy can be saved using this system compared with a vapor compression system. However, it consumes 55% more water in comparison with direct evaporative cooling system for the same air delivery rate. Monitoring the electricity consumption of evaporative and conventional refrigerated cooling systems in a small commercial building has demonstrated considerable energy savings and improved thermal comfort with evaporative cooling. Indirect evaporative cooling can also be used as a component of multistage air-conditioning systems, which also include refrigerated cooling stages. In such cases, the indirect evaporative cooling may be sufficient for the provision of typical summer cooling requirements. The refrigerated stage operation is limited to peak demand days.

Types of evaporative air conditioners range from portable units, window/wall units, and ducted units for residential and commercial use. Portable units cool one room at a time. They are fitted with legs and wheels and can be moved easily from room to room. A small pump is used to keep the cooling pads wet and water is needed to be periodically filled manually in the internal water storage tank. Window/wall evaporative air conditioners are mounted through exterior windows or walls and they can cool larger areas than portable units. Ducted evaporative air conditioners are usually mounted on the roof and the cooled air is delivered through ducts to each room in the building. Both window/wall and ducted units have water bleeding systems to control the water salinity under a certain level [22].

7.2.4 Liquid Desiccant Systems

A liquid desiccant air-conditioning system removes moisture and latent and sensible heats from process air through a liquid desiccant material. In the basic configuration, concentrated and cooled liquid desiccant flows into the absorber and down through a packed bed of granular particles enhanced mass transfer surface or packing. Return air passes up through the bed, transferring both moisture and heat to the counter-flowing liquid desiccant. The liquid desiccant leaves the bottom of the packed bed diluted by the water absorbed from the air, and flows into the regenerator. A heat source such as gas- or oil-fired, waste heat, solar heat in the regenerator heats the weak liquid desiccant solution, which is then sprayed on another packed bed. The heated solution transfers the absorbed moisture to a counter-flowing scavenger airstream to regenerate a concentrated liquid desiccant solution. After the return feed from the regenerator passes through a cooling tower or chiller, the cooled liquid desiccant solution returns to the absorber to complete the cycle. Designs often include a counterblow heat exchanger between the absorber and the regenerator to reduce the amount of external heating and cooling required. Industrial units for deep drying and applications requiring precise humidity control account for most of liquid desiccant air-conditioning market. Commercial air-conditioning units are becoming available but currently have a very small market share [23]. The operation of a liquid desiccant system is shown in Fig. 7.4. The system is an open cycle absorption system in which water serves as a refrigerant. Due to the direct contact with air, the absorbent has to be nontoxic and environmentally friendly and commonly used liquid desiccants are lithium chloride and calcium chloride. The process air is passed through a desiccant spray in the conditioner module to get dehumidified. The outside air is passed through a warm desiccant spray in the regenerator module to regenerate the desiccant.

The main advantages of the system are the possibility of removal of the heat of evaporation and mixing simultaneously with the dehumidification process, ability to store cooling capacity by means of regenerated desiccant, modular design, easy monitoring of desiccant quality, and reduced regeneration air equipment. The studies show that these systems have potential use in hot humid climates [24,25]. But the liquid desiccants are less effective than solid desiccants in dehumidifying the air and have the

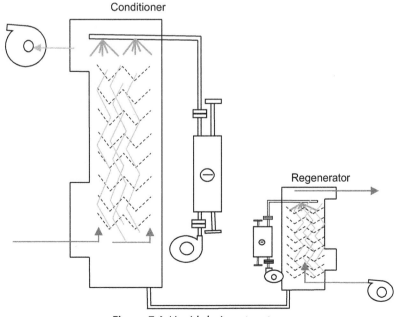

Figure 7.4 Liquid desiccant system.

possibility of liquid desiccant carryover by the air to the conditioned space, which can be hazardous to human health [26].

7.2.5 Solid Desiccant Systems

In a solid desiccant system, air is passed through the surface of a solid hygroscopic material like silica gel or zeolite for dehumidification. The different types of solid desiccant are solid packed tower system, multiple vertical bed system, and rotating desiccant wheel system.

7.2.5.1 Solid Packed Tower System

The schematic diagram of this system is shown in Fig. 7.5. In a solid packed tower system, there are two towers filled with solid desiccant with one performing dehumidification and the other for regeneration. The position will be alternated between process air path and regeneration air path. The main advantage of this system is that it can achieve very low dew points. The system's disadvantage is that the output conditions will vary with the level of moisture trapped and consequently the air velocity is critical to achieve optimum performance.

7.2.5.2 Multiple Vertical Bed System

The multiple vertical bed system is a combination of packed tower and rotating beds design through the use of a rotating carousel of many towers as shown in Fig. 7.6. The

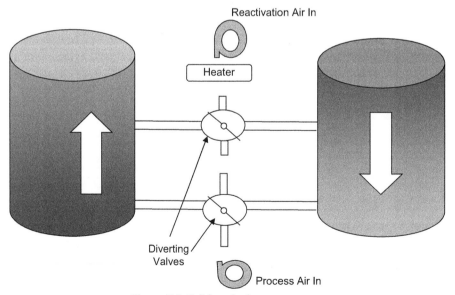

Figure 7.5 Solid packed tower system.

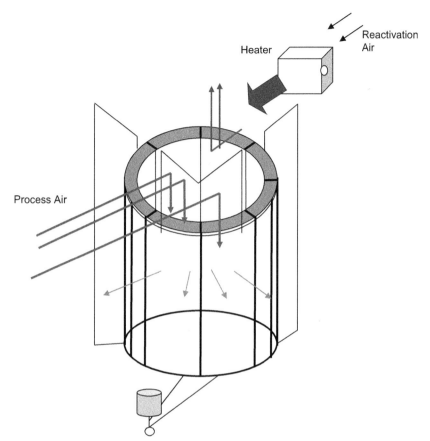

Figure 7.6 Multiple vertical bed system.

process air is dehumidified when it passes through desiccant held by the stacked perforated trays of the rotating carousel in the dehumidification side. Similarly, the heated air is passed through the stacked perforated trays of the rotating carousel in the reactivation side to regenerate the desiccant.

Even though this system has the advantages of constant outlet moisture level, high performance, and low dew point, it is a complex mechanical system that incurs increased maintenance and higher initial cost [27].

7.2.6 Rotating Desiccant Wheel System

The operation of a DEC system with a rotating desiccant wheel is shown in Fig. 7.7. The basic cycle is called a Pennington cycle or ventilation cycle. This was introduced by N.A. Pennington in 1955 and since then there have been many modifications to this cycle to suit the climatic conditions prevailing at different places. The ambient air is blown

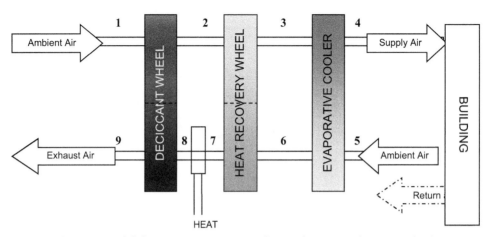

Figure 7.7 Solid desiccant evaporative cooling with rotating desiccant wheel.

through the dehumidification part of the desiccant wheel and is dehumidified and heated. This dry and hot air is passed through the heat recovery wheel where the air is partially cooled. It is followed by evaporative cooling.

The warm and humid air enters the slowly rotating desiccant wheel through the dehumidification section and is dehumidified by the desiccant material in process 1−2 shown in Fig. 7.7. During this process, air is also heated up. Air is then passed through a heat recovery wheel during process 2−3, resulting in significant precooling of the supply air stream. Subsequently, the air is humidified and thus further cooled by an evaporative cooler during process 3−4. The exhaust air from the building is cooled and humidified in process 5−6, close to saturation point to have the full cooling potential and to allow an effective heat recovery process 6−7 in the heat recovery wheel. The hot air coming out of the heat recovery wheel is heated to the regeneration temperature in a heater in process 7−8. Finally, in process 8−9, the hot air at regeneration temperature is passed through the regeneration section of the desiccant wheel, driving away the humidity from the wheel and reactivating the solid desiccant. The process is called regeneration. The rotation of the wheel allows continuous operation of the dehumidification system. These processes are shown in the psychrometric chart given in Fig. 7.8.

Compared with other desiccant systems, this system is much superior because of the low pressure drop across the wheel, low dew point, high capacity, continuous operation, light weight, and simplicity of design, which make it easy to maintain. Most of the systems available in the market are large-scale systems. A typical system will have an air flow of 4500 m^3 per hour and a cooling power of 22−60 kW.

These large desiccant systems have become more widely applied in spaces where humidity levels are high such as supermarkets and theatres in the United States and Europe [28−30]. Desiccant systems can be driven by conventional fuel, waste heat, and

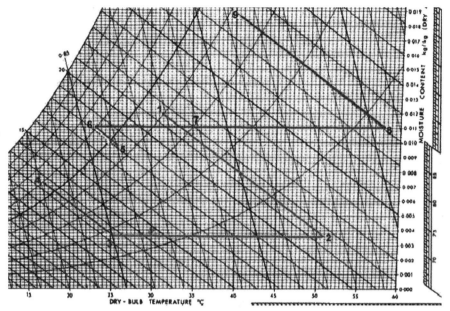

Figure 7.8 Desiccant evaporative cooling in psychrometric chart.

solar heat. A typical example of solar installation in Europe is a commercial room in Fribourg in Germany where an autonomous solar desiccant cooling system provides cooling for two meeting rooms of 65 and 148 m^2 containing 120 persons. It has 100 m^2 of solar collectors and 60 kW of cooling capacity. The reduction in primary energy consumption is estimated at about 30,000 kWh and the CO_2 emission is reduced by about 8800 kg/year [31].

7.3 DESICCANT WHEEL

A desiccant wheel is a common type of sorption dehumidifier using a solid desiccant. It is also known as a rotary dehumidifier. The desiccant material is coated, impregnated, or formed in place on the supporting rotor structure. The desiccant wheel is called a passive desiccant wheel or enthalpy wheel or rotary energy wheel, when there is no regeneration air heater. It is called an active desiccant wheel when it is provided with an air heater and the regeneration and process air side are separated by clapboard. The wheel is installed with thermal insulation and air-proof material, so no mass and energy exchange takes place with the surroundings.

The operation of the desiccant wheel is shown in Fig. 7.9. The various components of the wheel are:

• wheel matrix consisting of supporting materials and desiccant material;
• clapboard for the division of regeneration and process air side;

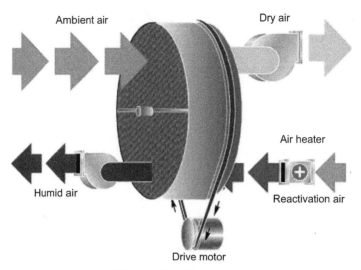

Figure 7.9 Operation of a desiccant wheel.

- wheel case;
- air heater;
- driving motor.

The matrix of the desiccant wheel has many channels parallel to the axis of rotation and has a honeycomb structure. The flow channels of the desiccant wheels available are of different shapes, namely, triangular, sinusoidal, and hexagonal. A desiccant wheel functions as a heat and mass exchanger between the process and return air streams. The desiccant wheels are not only used for air dehumidification, but also for enthalpy recovery. When it is used for enthalpy recovery, there is no heater and it rotates between process air and exhaust air to facilitate the transfer of heat and humidity between the streams. So the enthalpy wheel is used in winter to recover heat and moisture from the exhaust air, whereas in summer it is used to cool and dehumidify the process air. The rotational speed of the wheel depends on the chosen operating mode: it varies commonly within the range of 6–12 rotations per hour in the dehumidifier mode and 8–14 rotations per minute for the enthalpy recovery mode [4]. The storage of water vapor and energy in the air and axial conduction through the matrix are more important in the case of rotary heat exchangers, whereas for dehumidification purposes with high temperature desorption, it will have different sorption characteristics to energy wheels as it operates over a small temperature range.

The physical process causing dehumidification of air in a desiccant wheel is adsorption, which is a process where molecules from a gas phase or from a solution bind in a condensed layer on the surface of a solid or a liquid [32]. The molecule that binds to the surface is called adsorbate, whereas the substance that holds the adsorbate is called adsorbent. The substance that adsorbs water vapor forms a separate category of materials called desiccant and these substances can induce or sustain a state of dryness desiccation in its local vicinity through adsorption of water. The process of adsorption of water is a result of the tendency of the water vapor pressure in the adsorptive to reach equilibrium with the water vapor pressure on the surface of the adsorbent. If the vapor pressure on the surface of the adsorbent is lower than the partial pressure of water vapor in the air, adsorption occurs and water is extracted from the air. Adsorption progresses until the water vapor pressures in the air and at the desiccant surface reach equilibrium. Desorption takes place if the water vapor pressure on the surface of the desiccant exceeds the water vapor pressure of the air. In this case, water is released from the desiccant surface to the surrounding air.

A desiccant is a substance that has a high affinity for water and is a dehumidifying agent. The common desiccant materials in use are silica gel, zeolite molecular sieve, lithium chloride, and activated alumina. The desiccant's ability to attract moisture depends on the amount of water on the desiccant surface in comparison with the amount of water in the gas. The difference between these amounts of water is described by the vapor pressure in the air and at the surface of the desiccant [7].

7.3.1 Issues in Desiccant Evaporative Cooling Using Desiccant Wheel

There are many issues in a DEC system that make the performance lower and act as barriers for the emergence of this system as an alternative to the vapor compression chillers. These issues are related to the adsorption ability of the desiccant wheel, heating of the supply air, and pressure drop.

Commonly used materials for desiccant rotors are silica gel, zeolite molecular sieve, and lithium chloride (LiCl). The isotherm curve that shows the mass of water adsorbed in grams per 100 g of desiccant at various relative humidities for these materials is shown in Fig. 7.10. It is clear from the figure that silica gel and lithium chloride have high capacities of adsorption throughout the range of relative humidities.

A large specific surface area is preferable for providing large adsorption capacity, so the presence of large numbers of microspores and pore size distribution are important properties deciding the ability of adsorbents. Most of the adsorbents used in modern processes are microspore adsorbents. Pores can be classified into three categories, namely, micropore (<2 nm), mesopore (2–50 nm), and macropore (>5 nm).

The most suitable desiccant material for DEC systems is silica gel [33]. Silica gel is an amorphous form of silicon dioxide, synthetically produced in the form of hard irregular granules and is a hard glassy substance that is a milky white color. A microporous structure of interlocking cavities gives a very high surface area of $750-800 \text{ m}^2/\text{g}$. It is this structure that makes silica gel a high-capacity desiccant. Water molecules adhere to its surface because it exhibits a lower vapor pressure than the surrounding air. Adsorption of water vapor to its surface occurs until equilibrium of equal pressure is reached and then no more adsorption occurs. Thus the higher the relative humidity of the surrounding air,

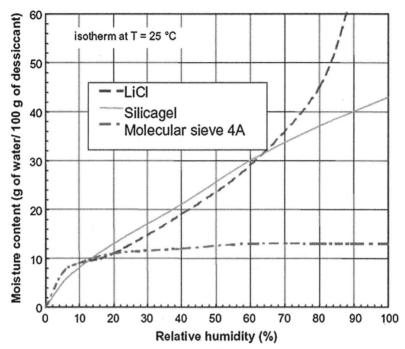

Figure 7.10 Isotherms of different materials [32].

the greater the amount of water that is adsorbed before equilibrium is reached. With silica gel, only the physical adsorption of water vapor occurs and there is no chemical reaction, by-products, or side effects. Even when saturated with water vapor, silica gel still has the appearance of a dry product, its shape unchanged. The process of adsorption and desorption in silica gel follows almost an isotherm curve [34]. These advantages make it a superior adsorbent.

There are two types of silica gels, namely, regular density (RD) silica gel and intermediate density silica gel. RD silica gel has a higher adsorption capacity due to its smaller prediameter and larger internal surface area and it has the most favorable isotherm for air-conditioning applications [35]. Typical properties of both types of silica gels are given in Table 7.1.

7.3.1.1 Adsorption Capacity

Evaporative cooling is an ecofriendly cooling system, but this system will not be effective in humid climates. But in a DEC system, the desiccant wheel lowers the humidity of air before doing the evaporative cooling. So the success of the DEC system largely depends on the ability of the desiccant wheel to dehumidify the air to such a low humidity level that evaporative cooling can be employed effectively. It has the advantage that sensible

Table 7.1 Properties of Silica Gel [36]

Property	RD Silica Gel	ID Silica Gel
Pore volume (m³/kg)	0.43×10^{-3}	1.15×10^{-3}
Surface area (m²/g)	750–800	340
Pore radius (Å)	11	68
Specific heat (J/kg/K)	921	921
Thermal conductivity (W/mK)	0.144	0.144
Bulk density (kg/m³)	721.1[a]	400.6

ID, intermediate density; *RD*, regular density.
[a]Grade 1 RD silica gel.

and latent heat portions can be addressed separately. Past researchers have shown that the desiccant system is quite efficient in dealing with the latent heat load, but considerably less efficient in the case of the sensible heat load [37]. So the performance of the desiccant wheel must be improved such that it can do the dehumidification more effectively and produce sufficiently dry air so that sensible heat can be effectively removed by evaporative cooling. The sorption capacity of the desiccant wheel is a critical parameter in the performance of the DEC system. The desiccant material's properties that most influence the sorption capacity are given in the following sections.

7.3.1.2 Isotherm Shape

The adsorption isotherm is a plot of the amount of adsorbent that adsorbs as a function of pressure of the gas. The shapes of the equilibrium isotherm, moisture, and heat diffusion rates of the desiccant are important factors in determining the sorption performance of the desiccant wheel [33].

The optimal desiccant for space comfort conditioning will be the one that contains the moisture wave front within a practical depth of the desiccant wheel. The wave fronts of materials like silica gel cannot be contained regardless of operational condition. Further studies have identified a type I material with a separation factor between 0.07 and 0.1 as the optimal for air-conditioning application using high temperature regeneration [34]. A desiccant with such an isotherm is called a Type 1M (moderate) isotherm. Type 1M isotherm has been identified as the preferred shape of isotherm for solid desiccants in cooling applications [37].

7.3.1.3 Maximum Uptake

Maximum uptake is the property of the desiccant that is defined by the fraction of the maximum amount of water vapor adsorbed to the mass of the desiccant material. This property is considered one of the most important; however, with increased heat capacity, it results in poor performance. The heat capacity of the supporting materials adds to the heat capacity of the material. So material with uptake and low heat capacity deposited on the low heat capacity substrate will result in better dehumidification performance.

7.3.1.4 Temperature Rise of Supply Air Due to Heat of Adsorption

During the dehumidification process in a desiccant wheel, there will be a rise in the air temperature. This heating is caused by the heat of adsorption and the carryover heat from the regeneration air.

7.3.1.5 Heat of Adsorption

The process of adsorption releases heat, which is called the heat of adsorption. It is the sum of the latent heat of the vaporization of water and the heat of wetting, which is the heat that is released upon contact of the water vapor with the desiccant surface. The released heat is transferred to the desiccant material and the air flow that causes an increase in the temperature of the air stream.

7.3.1.6 Carryover Heat

During the operation of the system, the wheel matrix passes first through the supply air stream and then it passes through the hot regeneration air. This process reactivates the desiccant, at the same time it heats up the matrix. After this process, the matrix enters into the supply air stream. During this dehumidification, the matrix dissipates a part of the sensible heat to the supply air stream and increases the air temperature. This heating increases the cooling requirement in the heat recovery wheel of the evaporative cooler and lowers the performance of the system.

7.3.1.7 Pressure Drop

When the air passes through the small channels of the porous wheel, a significant amount of the pressure drop will be incurred. Above this, the further stages of cooling have also significant pressure drops. This increases the fan pumping power and increases the overall power consumption of the system, thereby lowering its advantage over the vapor compression system. The pressure drop largely depends on the depth of the channel and so the wheel thickness is to be carefully selected to ensure sufficient dehumidification without producing an excessive pressure drop.

7.3.1.8 Performance Optimization

The COP of the open cycle DEC system with desiccant wheel is in the range of 0.5−0.6 and the driving temperature is around 60°C [17]. The performance of the desiccant wheel depends on many factors, which can be classified into design and operational factors. The major design factors are the diameter and thickness of the wheel, flow configurations, supply to regeneration area ratio, shape and dimensions of the channels, desiccant loading, and desiccant layer thickness. Operational factors include wheel rotational speed, humidity ratio and temperature of supply air and regeneration air, and the velocity of flow for the two streams. The modeling of heat and mass transfer behavior is extremely important for the optimization of these parameters [38]. There are several difficulties in numerical modeling of the desiccant wheel such as lack of knowledge of the properties of a porous medium, numerical difficulties in handling the coupling between heat and mass transfers, and time required for computations [39].

7.3.2 Disadvantages of Desiccant Systems

There are many successful installations around the globe. In Europe, of the 450 solar cooling systems installed, there are 16 DEC systems using sorption wheels [40]. But the application of the desiccant evaporative system is limited to temperate climates, since the dehumidification is insufficient to enable evaporative cooling of the supply air at conditions with higher values of humidity of ambient air. The conventional systems installed in these locations show that there are many disadvantages in using this technology to develop a small-scale system. The practical drawbacks are:

1. Air leakage between the supply and the return air, leading to reduction in the performance of small-scale systems.
2. The heat of adsorption and the heat carryover from the regeneration section increase the temperature of the process air leaving the desiccant wheel. This creates the need for extra cooling efforts.
3. The process of sorption is adiabatic and increased temperature of the desiccant leads to reduced dehumidification potential.

To overcome the disadvantages, new heat exchanger concepts with higher dehumidification potential are being investigated as part of Task 38 of the International Energy Agency's solar heating and cooling program. One of this kind of heat exchanger is called evaporative cooled sportive heat exchanger. It is a counter flow air-to-air heat exchanger. The process is based on simultaneous sportive dehumidification and indirect evaporative cooling of the process air. The heat exchanger has a separate sportive channel with desiccant material on it and a cooling channel and both these channels can exchange heat. During dehumidification phase, the heater will be switched off and process air is passed through sportive channel to get dehumidified. In the cooling channel continuous humidification of cooling stream takes place with heat received from the air in the sportive channel. To facilitate quick evaporation, the air passed through the cooling channel must be close to saturation. After the process in the sportive section, the process air is cooled in a direct evaporative cooler. During regeneration, the evaporative cooler will be inactive and the heater will be switched on. The heated return air is passed through the sportive channel to regenerate the desiccant. The regeneration phase is followed by a precooling process to remove the heat carryover in the sportive section. This ensured proper dehumidification during

Table 7.2 Comparison of Small-Scale Sorption Chillers Available in the Market [31]

Company	EAW	Sonnenklima	Rotortica	Climatewell	SolarNext	SorTech	SJTU
Product name	Wegracal SE 15	Suniverse	Solar 045	Climatewell 10	Chilli PSC10	Chilli STC8	SWAC-10
Technology	Absorption	Absorption	Absorption	Absorption	Adsorption	Adsorption	Adsorption
Working pair	$H_2O/LiBr$	$H_2O/LiBr$	$H_2O/LiBr$	$H_2O/LiCl$	$H_2O/Silica$ gel	$H_2O/Silica$ gel	$H_2O/Silica$ gel
Cooling capacity (kW)	15	10	4.5	10	10	7.5	10
Heating temperature (°C)	90/80	75/65	90/85	83/—	85/78	75/67	85/79
Recooling temperature (°C)	30/35	27—35	30—35	30/—	24/29	27/32	30/36
Cold water temperature (°C)	17/11	18/15	13/10	—/15	12/6	18/15	15/10
COP	0.71	0.77	0.67	0.68	0.63	0.53	0.39

next phase. But to have continuous operation two heat exchangers have to be operated periodically.

7.3.3 Recent Development of Small-Scale Heat-Driven Chillers

There are many manufacturers of large heat-driven air-conditioning systems with cooling capacity above 50 kW and available in the market. Considering the huge market potential for small-scale solar cooling units, several companies started development of small-scale systems with cooling capacity below 50 kW and down to 5 kW and a few products are available now as shown in Table 7.2. Further development of small-scale systems still remains of high interest [40].

7.4 CONCLUSIONS

There are many alternative heat-driven cooling technologies and these systems are available in the market. Among these absorption technology with lithium bromide—water solution and water—ammonia as working pair are well developed in the large-capacity range and widely used. Small-capacity systems have been developed recently and the first prototypes or small series products are being tested and demonstrated. An alternative is adsorption technology. This technology is in the early stage of development and is considered to have a high technical potential. Large amount of research and development activities are being carried out in this technology. Open cycles, best known through the DEC system, not only produce cold but can also dehumidify the air and are thus especially interesting in cases where a high amount of treated air is necessary and dehumidification is mandatory. Different cycles that are adapted to a wide variety of climates make these systems suitable for air conditioning of buildings.

REFERENCES

[1] Chua KJ, Chou SK, Yang WM, Yan J. Achieving better energy-efficient air conditioning — a review of technologies and strategies. Applied Energy 2013;104(0):87—104.

[2] Saman WY, Halawa E. NATHERS—peak load performance module research. Institute of Sustainable Systems and Technologies, USA; 2009.

[3] Green guide — design, construction and operation of sustainable buildings. ASHRAE; 2010.

[4] Henning H-M. Solar assisted air conditioning of buildings: an overview. Applied Thermal Engineering 2007;27(10):1734—49.

[5] Daou K, Wang RZ, Xia ZZ. Desiccant cooling air conditioning: a review. Renewable and Sustainable Energy Reviews 2006;10(2):55—77.

[6] Balaras CA, Grossman G, Henning H-M, Infante Ferreira CA, Podesser E, Wang L, et al. Solar air conditioning in Europe—an overview. Renewable and Sustainable Energy Reviews 2007;11(2):299—314.

[7] Narayanan R. Development of a thermally-driven solid desiccant system for dehumidification/cooling [Ph.D. thesis]. Univeristy of South Australia; 2012.

[8] Narayanan R, Saman WY, White SD, Goldsworthy M. Comparative study of different desiccant wheel designs. Applied Thermal Engineering 2011;31(10):1613—20.

[9] Narayanan R, Saman WY, White SD, Goldsworthy M. A non-adiabatic desiccant wheel: modeling and experimental validation. Applied Thermal Engineering 2013;61(2):178—85.

[10] Narayanan R. Theoretical modelling of silica gel desiccant wheels. Applied Mechanics and Materials 2015;787:311—7.

[11] McQuiston FC, Parker JD, Spitler JD. Heating, ventilating and air-conditioning systems. 6th ed. 2005.

[12] Delfani S, Esmaeelian J, Pasdarshahri H, Karami M. Energy saving potential of an indirect evaporative cooler as a pre-cooling unit for mechanical cooling systems in Iran. Energy and Buildings 2010;42(11):2169—76.

[13] Jain S, Dhar PL, Kaushik SC. Evaluation of solid-desiccant-based evaporative cooling cycles for typical hot and humid climates. International Journal of Refrigeration 1995;18(5):287—96.

[14] McDowall R. Fundamentals of HVAC systems. Elsevier Publications; 2006.

[15] Liu W, Lian Z, Radermacher R, Yao Y. Energy consumption analysis on a dedicated outdoor air system with rotary desiccant wheel. Energy 2007;32(9):1749—60.

[16] Henning HM, Morgenstern A, Nunez T. Thermodynamic analysis of solar thermally driven cold production. In: 2nd international conference on solar airconditioning; 2007 [Costa Dorada, Spain].

[17] Task 38-solar airconditioning and refrigeration, IEA solar heating and cooling program. International Energy Agency; 2009.

[18] Xie G, Sheng G, Bansal PK, Li G. Absorber performance of a water/lithium-bromide absorption chiller. Applied Thermal Engineering 2008;28(13):1557—62.

[19] Moser H, Rieberer R. Small capacity ammonia-water absorption heat pump for heating and cooling-used for solar cooling applications. In: 2nd international conference on solar air-conditioning; 2007 [Tarragona, Costa Dorada, Spain].

[20] Heidarinejad G, Pasdarshahri H. Potential of a desiccant-evaporative cooling system performance in a multi-climate country. International Journal of Refrigeration 2011;34(5):1251—61.

[21] ASHRAE Handbook. HVAC applications. Atlanta: American Society of Heating, Refrigerating and Air-Conditioning Engineers, Inc.; 2007.

[22] Saman W, Bruno F, Liu M. Technical background research on evaporative air conditioners and feasibility of rating their water consumption. Institute for Sustainable Energy Systems and Technologies; 2009.

[23] Dieckmann J, Roth KW, Brodrick J. Emerging technologies. ASHRAE; 2004.

[24] Saman WY, Alizadeh S. Modelling and performance analysis of a cross-flow type plate heat exchanger for dehumidification/cooling. Solar Energy 2001;70(4):361—72.

[25] Saman WY, Alizadeh S. An experimental study of a cross-flow type plate heat exchanger for dehumidification/cooling. Solar Energy 2002;73(1):59—71.

[26] Alizadeh S. Development of a dehumidification/indirect evaporative cooling system using liquid desiccant. University of South Australia; 2002.

[27] Torrey M, Westerman J. Desiccant cooling technology resource guide. 2000.

[28] Saman WY. Solar energy efficient air-conditioning system. University of Sharjag Journal of Pure and Applied Scineces; 2004. p. 17—33.

[29] Narayanan R, Saman WY, White SD, Goldsworthy M. Modeling and parametric analysis of silica-gel desiccant wheels. In: Proceedings of international institute of refrigeration Gustav Lorentzen conference, April 2010, Sydney; 2010.

[30] Narayanan R, Saman WY, White SD, Goldsworthy M. Adsorption analysis and modeling of silica-gel desiccant wheels. In: Proceedings of EUROSUN 2010, September 2010, Graz, Austria; 2010.

[31] Jacob U. New concepts and promising technologies. In: Sustainable cooling system conference; 2008 [Vienna].

[32] Stabat P, Marchio D. Heat-and-mass transfers modelled for rotary desiccant dehumidifiers. Applied Energy 2008;85(2—3):128—42.

[33] Chung JD, Lee D-Y. Effect of desiccant isotherm on the performance of desiccant wheel. International Journal of Refrigeration 2009;32(4):720—6.

[34] Payne FW. User's guide to natural gas technologies. Faremont Press International; 1999.

[35] Pesaran AA. Moisture transport in silica gel particle beds. Los Angles: University of California; 1983.

[36] Do DD. Adsorption analysis: equilibria and kinetics, vol. 2. London: Imperial College Press; 2008.

[37] Jia CX, Dai YJ, Wu JY, Wang RZ. Use of compound desiccant to develop high performance desiccant cooling system. International Journal of Refrigeration 2007;30(2):345—53.

[38] Ruivo CR, Costa JJ, Figueiredo AR. Numerical study of cyclic behaviour of a desiccant layer of a hygroscopic rotor. Numerical Heat Transfer 2008;48(6):1037—53.

[39] Ruivo CR, Costa JJ, Figueiredo AR. Analysis for simplifying assumptions for numerical modelling of heat and mass transfer in porous medium. Numerical Heat Transfer 2006;49:851—72.

[40] Jacob U, Kolenbach P. Recent developments of sorption chillers in Europe. In: IIR Gustav Lorentzen conference; 2010 [Sydney].

CHAPTER EIGHT

Solar Pyrolysis: Converting Waste Into Asset Using Solar Energy

M.U.H. Joardder[1,2], P.K. Halder[3], M.A. Rahim[1] and M.H. Masud[1]
[1]Queensland University of Technology, Brisbane, QLD, Australia
[2]Rajshahi University of Engineering and Technology, Rajshahi, Bangladesh
[3]Jessore University of Science and Technology, Jessore, Bangladesh

8.1 INTRODUCTION

Biomass produces a huge amount of renewable energy that reduces the dependency on fossil fuels and on advantage it does not affect the environment with any new carbon dioxide [1]. For this production of renewable energy (e.g., gaseous and liquid fuels) pyrolysis is one of the thermochemical conversion processes [2].

Pyrolysis can be defined as a process in which carbon-based matter is decomposed in the absence of oxygen and at high temperature into its constituent elements, such as bio-oil, syngas, and bio-char, as shown in Fig. 8.1. Generally, the higher heating value of bio-oil ranges between 15 and 38 MJ/kg while this value for solid char is about 17−36 MJ/kg [3]. On the contrary, the higher heating value of pyrolysis gas fraction is approximately 6.4−9.8 MJ/kg [4]. The energy of the char and gas products can be recovered and utilized for further heating the pyrolysis reactor.

Typically, pyrolysis processes can be classified as slow pyrolysis and fast pyrolysis depending on the time taken for the completion of thermal decomposition of the feed materials. On the other hand, a wide range of reactors are being used for the production of liquid bio-oil from different waste biomass [5]. The production rate and properties of this product will depend on the design of pyrolysis reactor, reaction temperature, heating rate, residence time, pressure, catalyst, type of biomass, and particle size, shape, and structure [6]. As, for example, a low temperature with slow rate pyrolysis normally produces higher amounts of bio-char, and the opposite condition produces higher amounts of bio-oil.

High sensible heat is required to increase the biomass temperature to thermal decomposition temperature to enhance the pyrolysis reaction. In the current pyrolysis system, the reactor is heated using conventional energy sources which associated energy expense and environmental pollution. This high expense and environmental threat are the major challenges to refrain pyrolysis process as a member of industrial level energy production option. Although, for maintaining the high-temperature condition for

Clean Energy for Sustainable Development
ISBN 978-0-12-805423-9, http://dx.doi.org/10.1016/B978-0-12-805423-9.00008-9

213

Figure 8.1 Schematic diagram of biomass pyrolysis.

producing more bio-oil, the syngas that is produced in pyrolysis can be burned within the system, it is a challenge to design such kind of complex pyrolysis reactor. Therefore, a novel concept of solar heating is assimilated to the biomass heating system to mitigate the problems.

In this chapter, the overview of solar pyrolysis including history, design, challenges, and its solutions is presented. The chapter is organized as follows. Section 8.2 discusses the historical background and current state of biomass pyrolysis. Section 8.3 presents the major challenges of conventional pyrolysis technology highlighting the issues of reactor heating. Section 8.4 illustrates the possible means of heating the reactor for thermal decomposition of biomass. Section 8.5 depicts the history of solar heating approach and its classification. Additionally, Section 8.6 describes the incorporation of solar heating into pyrolysis technology including different heating system and updated design concepts. Moreover, present applications and feasibility of solar-assisted pyrolysis technology are presented in Sections 8.7 and 8.8, respectively. Finally, the possible challenges and future development scope of solar-integrated pyrolysis technology are demonstrated in Sections 8.9 and 8.10, respectively.

8.2 HISTORY OF PYROLYSIS

Pyrolysis has an unknown and untold long history. It is hard to trace back when exactly human being started using this technique to retrieve energy to liquid phase from solid phase. An attempt of representation of the history of pyrolysis is shown in Fig. 8.2.

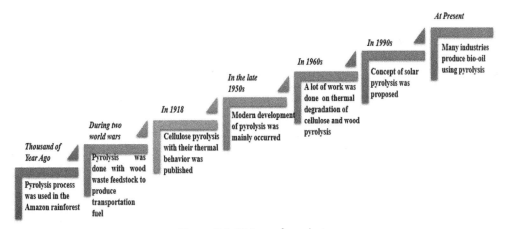

Figure 8.2 History of pyrolysis.

By having the aim to produce bio-char, which is a charcoal-like product and used to both enrich and stabilize the nutrient-poor rainforest soils, the pyrolysis process was used near about thousands of years ago in the Amazon rainforest. At that time, people were starting fires by covering the fuel with earth surface for making a low amount of oxygen environment so that in the absence of oxygen bio-char was produced rather than ash by the breaking down of fuel continuously.

During the two world wars, pyrolysis was done with wood waste in order to reduce transportation fuel when fossil fuels were limited. In the late 1950s, the modern development of pyrolysis mainly started its journey. Along with liquid yield expectation, pyrolytic gasification was first introduced in the United States at Bell Laboratories in 1958 [7].

Although pyrolysis gasification is one of the common thermochemical processes, further research is still required to improve the quality of intermediate and final products. As an example, while producing synthetic hydrocarbons by using pyrolytic gasification process, highly contaminant content gas is produced [8]. In addition, the oxygen content of the obtained product from pyrolytic gasification is too high [9]. In recent time, different laboratories has started working on it to improve the pyrolysis gasification quality.

The Bell Laboratories in the United States with some renowned universities and organizations of the world in 1958 had carried out research and development programs to determine the usefulness of pyrolysis. Production of gas by using waste materials is the purpose of the system. Then in the 1970s a pyrolysis plant producing 200 ton/day RDF (refuse-derived fuel) was built by the Occidental Research Corporation in San Diego, United States [10]. During late 1970s to early 1980s a flow pyrolysis process was developed by the Georgia Tech [11].

A more efficient batch systems gave way to continuous feed systems with a cone design that made the evacuation of the gases, which was developed during late 1970s to early 1980s. An indirect pyrolytic gasification system was designed by the Balboa Pacific Corporation in the United States in the mid-1980s. This indirect pyrolytic gasification system was patented in 1988. The work of Scotland with his group developed a flow pyrolysis system at the University of Waterloo, Canada, in the Georgia Tech, which was the next significant development in the field of pyrolysis. A great number of inspirable quasi-commercial scale pyrolysis systems were developed in the 1980s and 1990s [10]. Then during 1993–95, a two-year period, 50 ton/day pilot plant was refined and constructed, which was partially funded by Southern California Gas Company and private investors. For a period of 18 months, the plant operated in California, during which time, the system's ability to process, on a continuous feed basis, a wide variety of toxic and nontoxic, liquid or solid organic waste streams was exhaustively studied by different nation's environmental engineers such as Dames & Moore.

Besides all of this, pyrolysis was occurring with a lot of things, such as sugarcane, tire, plastic, wood, cellulose, as well as with different types of waste in different places of the world. For example, pyrolysis with sugarcane was published in 1980s. For a very long time truly from the ancient time wood pyrolysis was used by Mediterranean and northern civilizations after separating the charcoal from the furnace. During 18th and 19th centuries, a great number of experiments on wood pyrolysis show that the liquids obtained by it include several phases, such as water, wood spirit, light oils, pyrolytic acid, tars, and so on. [12].

A significant work on cellulose pyrolysis with their thermal behavior was published during 19th and 20th centuries, and sometime in 1918 this was reported by Pictet and Sarasin. They also discussed the thermal behavior of Amidon and ultimately points out on the industrial preparation of glucose and alcohol by using cellulose pyrolysis [13]. In 1920, Clément and Rivière recapitulated this issue in their book named "La cellulose." Three articles during mid-1960s and 1970s report about the cellulose pyrolysis evidencing the phase changing phenomena in this pyrolysis process [12]. In the articles of Hawley and Stamm, in 1956, several references are found on thermal degradation of cellulose and wood pyrolysis [14].

M. J. Antal has been producing a lot of fluid fuels by using concentrated radiant energy [15]. In Copper Mountain's meeting, M. J. Antal described the solar furnace and bench-scale facilities which he performed at the Princeton University, in French. He demonstrated that the product of the concentrated radiant energy pyrolysis also known as flash pyrolysis can be either liquid syrups or permanent gases depending upon the condition of the reactor, which was made more clear by Dejenga and Antal in 1982 who showed that the majority of the gaseous product is LVG [16]. Then in 1983 to decipher the result acquired in a continuous solar reactor, Antal et al. used the BS model [15]. In 1984, Hopkins et al. interpreted about the cellulose flash pyrolysis which is carried out in a spouted bed settled at the focus of an arc image furnace (xenon lamp) [17].

In the same year, Mok and Antal calculated the requirement of heat in cellulose flash pyrolysis. After calculating the required heat, Mok and Antal proposed a new detailed model that can control the pressure and flow rate of the working fluid as per demand [18]. Few years later, in 1993, cellulose flash pyrolysis, which was proposed by Mok and Antal was scrutinized by Varhegyi et al. [19]. In 1993 inside a transport flow reactor under a maximum temperature of 1473K, Vladars-Usas studied the fast thermal decomposition of Avi-cel cellulose [20]. Then, in 2002 and 2004 fast pyrolysis in the heated reactor was carried out by Boutin et al. [21] and Luo et al. [22], respectively. The melting phenomena of cellulose were discussed by Schoeter and Felix in 2005 [23].

The pyrolysis technology with different method using different sources of energy was investigated by different people in several places of the world [24]. However, the development of advanced technologies for upgradation of bio-oil and to increase the percentage of bio-oil production are the next issues for the researchers, yet to be solved. Therefore, researchers have devoted themselves to investigate different technologies for upgrading the bio-oil [25]. Currently, the researchers have given emphasis to blend bio-oil with other oils to enhance their fuel properties [26]. Additionally, attention has been given to catalytic fast pyrolysis for production of hydrocarbons from biomass [27].

Since mid-2010s, mathematical modeling studies have been conducted to identify the nature of the chemical reaction during thermal decomposition and to understand the kinetics of biomass reactor. The analysis of hydrodynamic modeling and simulation of particles in pyrolysis, their interaction, and the effect on the performance of the reactors revealed that the characteristics of feedstock, physical and chemical properties of feedstock, and residence times are major factors for the desired performance of the pyrolysis process [28−31].

8.3 CHALLENGES OF EXISTING PYROLYSIS SYSTEMS

From the early time of pyrolysis, it pesented different challenges to be a user-friendly application. As pyrolysis, the process can provide solid, liquid, and gas phases, the optimization of the process depends on the yield product expected. A sound determination ofspecific yield product is required to allow any of these products to be maximized, and it is the designers' challenge to optimize the process. There are still many challenges associated with the existing pyrolysis systems; some of the prime challenges are discussed in the following sections:

8.3.1 Reactor Heating System

A wide number of factors make the design of the heating system of pyrolysis including heating rate, temperature control, and cooling rate [32]. Thermal degradation of biomass feed materials in the absence of oxygen demanded high heat transfer rate and controlled high reactor temperature, usually around $500°C$, which is a critical task. In addition, the reactor must be designed in such a way that the yield vapor can leave the reactor within 2−5 s to avoid further cracking reactions. As biomass possesses very low thermal conductivity, feed material size needs to be significantly small to achieve high heat transfer rate between carrier gas—solid interfaces. Similar to the heating system, cooling system, condenser, needs to have rapid cooling capacity in order to transfer the vapor phase to liquid oil.

Apart from considering gas flow path, efficient char removal is a crucial factor in avoiding a secondary cracking reaction. As hot char is highly catalytically active, rapid removal of char ensures minimal contact with produced volatile materials and further cracking reaction. Energy source and mode of heat transfer affect the design and yields of the pyrolysis process. Depending on the heat transfer mode, different types of paralytic reactors are available. Table 8.1 demonstrates the generic mode of heat transfer applied in the different reactors and their corresponding characteristics.

8.3.2 Waste Disposal: Environmental Effect

The degree of environmental effects depends on the type of biomass heating. In both external and internal heating, pyrolysis system uses conventional fuels and electricity that results in emission of harmful gases. The reactor that is heated by biomass burning is a remarkable source of CO_2, CH_4, and N_2O. On the other hand, the permanent gases produced from pyrolysis (from feed materials) depend on heating temperature and feed materials. For example, when the temperature exceeds 800°C, CO_2, CH_4, and H_2O production decreases and essentially becomes zero whereas H_2 and CO production increases significantly. The overall conclusion is that bio-oil, as compared with other conventional fuels, offers insignificant environmental, health, and safety risks [33,34].

Table 8.1 Nature of Heat Transfer and General Characteristics of Different Pyrolytic Reactor

Reactor Type	Suggested Mode of Heat Transfer	Advantages/Disadvantages/Features
Ablative	95% conduction, 4% convection, and 1% radiation	Accepts large-sized feedstock, although very high mechanical char abrasion from biomass. Compact design, however, heat supply problematic.
Circulating fluid bed	80% conduction, 19% convection, and 1% radiation	High heat transfer rates, however, high char abrasion increased complexity of system.
Fluid bed	90% conduction, 9% convection, and 1% radiation	Heat supply to fluidizing gas directly causes high heat transfer rates; it limits char abrasion; the reactor configuration is very simple.
Entrained flow	4% conduction, 95% convection, and 1% radiation	Low heat transfer rates.

8.3.3 Yield Product Application

Application of pyrolyzed oil is still challenging due to its high acidity (pH ~ 2) and gel formation characteristics. Although extensive research has been carried out, due to these properties, bio-oil does not appear to be a good replacement for No.2 fuel oil in the home-heating application. However, bio-oil can be utilized in industrial boilers that are highly corrosion resistant and equipped with a preheating system [35].

In order to overcome the negative effect of certain properties of the pyrolytic oil, catalytic upgrading is a promising method for converting liquid yield into higher quality fuel [36,37]. Catalytic hydro-treatment and catalytic cracking are two widely practiced upgrading methods [38]. In brief, the challenges mainly depend on pyrolysis reactor design and heating system. Heat transfer modes directly influence the overall design and performance of the pyrolysis process. Solar pyrolysis is a potential candidate to minimize environmental pollution and heating cost, and for design simplicity.

As reactor heating is the crucial part of pyrolysis, incorporation of the solar heating system mainly deals with this core part. Prior to discussing solar heating, a brief discussion of conventional heating is discussed in the following section.

8.4 HEATING OF PYROLYSIS REACTOR

Extensive heat flux is necessary to heat up the pyrolysis reactor to an elevated temperature for converting the feed materials into bio-oil. Conventional external heating sources (gas heater or electric heater) have been used for the last two decades for pyrolysis of biomass [39,40]. Generally, there are three methods for providing heat energy to the reactor, namely auto-thermal heating system, indirect heating system, and direct heating system. The characteristics of some heating approaches are presented in Table 8.2.

Table 8.2 Some Common Heating Sources and Their Characteristics

Heating Methods	Remarks
Biomass/NG	This indirect heating process is exothermic in nature, responsible for environmental pollution and produces soot and coke. Difficulties in rapid change in temperature for thermal runaway prevention.
Electric heater	The external electric heating furnace is used to heat the feed materials within the pyrolysis reactor and the reactor itself.
MW heating	The MW-absorbing material (dielectric material) along with the feed materials inside the pyrolysis heating chamber directly absorbs electromagnetic wave in the form of thermal energy and transfer throughout materials through molecular interaction with the electromagnetic field.
Induction heating	The feed substances within the heating chamber absorb heat from the magnetic field produced by passing alternating current through the induction coil inside the chamber.

In the auto-thermal heating method, the heat flux required for the pyrolysis is supplied by the partial combustion of feed materials and yield bio-char. This method reduces the amount of char production as the heat indispensable for pyrolysis is produced from combustion of biochar itself thereby decreasing the fuel cost. On the other hand, in the case of the direct heating method, the hot carrier gas or solid heat carrier is introduced into the reactor and the feed materials absorb the heat for decomposition. The hot carrier gas produces CO_2 and H_2O during the combustion reaction which results in the reduction in the concentration of gas and heating value of oil. Besides, the use of solid heat carrier increases the oil yield, however, degrades the quality of oil due to the intake of a large amount of dust [41]. Contrarily from the direct heating method, in indirect heating, the reactor is heated from an external energy source where the heat is generated from the combustion of biomass or natural gas (NG) burning heater and electric heater. The heat is transferred through the reactor wall to raw materials by conduction and convection heat transfer principle. The oil yield and the efficiency of this heating system are affected by the thermal conductivity of reactor material, thermal diffusivity, and the temperature gradient between the reactor surface and center of the material inside the heating chamber. These constraints restrict the rapid heating of biomass and results in substantial input energy loss.

Therefore, the indirect heating system is less efficient; however, it produces less diluted gas compared to direct method. This reduces oil yield as well as increases char production that causes coking and fouling in the system [42]. Furthermore, the uneven nature of temperature profile causes an undesired secondary reaction in different regions of the pyrolysis reactor which produces toxic compounds [43,44]. On the contrary, this heating method has the advantages of producing retorting gas of high quality from a wide variety of particle size simultaneously. Since early 2000s, microwave (MW) heating pyrolysis is drawing the attention of researchers to overcome the limitations of conventional external heating of pyrolysis reactor. In 2001, this method was first implemented in pyrolysis of waste plastic [45]. The application of MW heating in pyrolysis reduces significant amount of energy input and time required for pyrolysis process due to direct and rapid heating. It also offers the advantages of uniform heating and less equipment dimension. In addition to this, the method improves the efficiency of the system as well as the quality of pyrolysis products. The MW heating enhances the high-grade product yield in reduced reaction time [46,47]. However, the foremost problems associated with this method are: different design approach considering the dielectric properties of biomass and the complication in heat and transfer mechanism and also in chemical reactions inside the different sections of heating chamber. Other electromagnetic heating, induction heating, is becoming popular in pyrolysis reactor heating due to its rapid heating rate and high efficiency, although the method is less flexible and requires high energy input and cost [48].

From the previous discussion, it is clear that a constant energy source is required to run the pyrolysis system, which involves consumption of conventional fuel. This means, a high-quality fuel is consumed to produce relatively less-quality fuel. Reactor heating with conventional fuel is not the economic approach of production of bio-oil. Therefore, reactor heating with another renewable source such as solar heating would be a great choice.

8.5 SOLAR HEATING APPROACH

8.5.1 History of Solar Heating

Although the direct heating approaches have some advantages over the indirect and autothermal heating approaches, all these have several shortcomings. These heating methods consume a huge amount of nonrenewable energy, such as biomass, or electricity for heating the reactor to produce bio-oil and also have some environmental hazards. Hence, the concept of incorporation of solar energy to heat the reactor is a possible solution to resolve these problems. The idea of solar energy utilization started at 212 BC by the Greek scientist Archimedes [49]. During the 18th century, the solar furnace was used for melting different metals, while during the 19th century solar energy was used for steam generation for a steam engine [50].

The concepts of a solar tower, parabolic dish, and parabolic trough were explored for solar heating in 1878, 1882, and 1883, respectively. Consequently, in 1913 the first concentrated power plant was established in Cairo, Egypt. Different technologies were developed during the mid-19th century for harnessing solar energy while in 1958 the first solar panel was used to launch a space satellite. Currently, solar energy is utilized in industrial process heating, space heating, as well as producing electricity by various technologies. However, the assimilation of concentrated solar energy in biomass gasification started during the end of the 19th century [51]. A great number of researchers recommended utilizing solar energy for renewable energy production through biomass gasification and pyrolysis during the last century [17,21,52−55]. On the contrary, solar thermochemical processes were investigated in vertical axis solar furnace at PSA, Spain. An artificial concentrated solar energy from the high flux solar simulator is used to provide heat in a horizontal off-axis solar furnace for the thermochemical process at German Aerospace Center (DLR), Germany, as presented in Fig. 8.3 [56].

Utilization of concentrated solar energy in biomass pyrolysis offers several advantages which are summarized as follows:

- High heat flux for heating the pyrolysis reactor rapidly to elevated temperature.
- Reduction of secondary reactions in different regions of the heating chamber due to comparatively small focal area.
- Renewable heat source lessens the heating cost of the pyrolysis reactor and also protects the reserve of nonrenewable energy sources.
- No burning of fossil fuels, hence the system produces no emission and is considered environmentally sustainable.

Figure 8.3 High-flux solar simulator at German Aerospace Center (DLR), Germany [56].

- No contamination of pyrolysis gas with combustion products, thus improves the quality of the yield.
- Relatively less-complicated heating system.
- As the reactor and its surrounding gas do not require to be maintained at a high temperature, the effective thermal mass of the pyrolysis system is minimal [15].
- Focused solar radiation can exactly be placed to the biomass sample where it is required while rest of the part can be kept in cold [21].

8.5.2 Classification of Solar Heating Reactor

In broad scenes, the solar heating system can be classified as a directly heated system and indirectly heated system. In the directly heated system the pyrolysis reactor is directly exposed to the sun shine and heated by solar radiation as illustrated in Fig. 8.4A [56]. On the contrary, in the indirectly heated system, as shown in Fig. 8.4B, an opaque surface is exposed to the incident sun rays which transfers the heat to the reactor surface [56].

Besides these, solar-assisted pyrolysis are further classified as partial heating and continuous heating as presented in Fig. 8.5. In partial heating, the reactor is heated partially to a certain temperature by solar concentrator and then by another heating system.

On the other hand, in continuous heating, the reactor is heated continuously by solar concentrator with another heating system throughout the pyrolysis cycle.

8.6 INTEGRATION OF SOLAR ENERGY WITH PYROLYSIS

8.6.1 Solar Concentrating System

Despite solar energy being the prominent clean and low-cost renewable energy on the earth, high-temperature process heat application is a great challenge. Solar collectors are

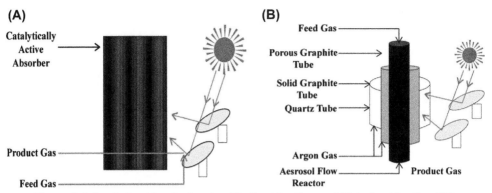

Figure 8.4 Pyrolysis with solar heating (A) direct heating and (B) indirect heating [56].

promising technologies for harnessing solar energy to process heating as well as electricity production. However, stationary flat plate collectors are applicable only for low-temperature application. Therefore, solar concentrating systems have been investigated for several decades for utilization of concentrated solar energy for high-temperature application. Solar concentrators are the devices that collect solar radiation and

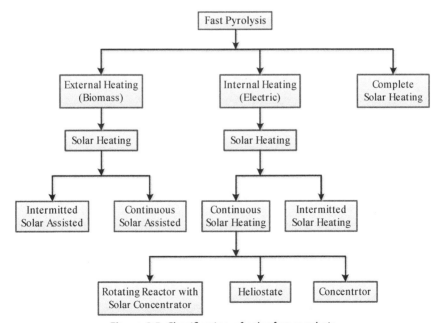

Figure 8.5 Classification of solar fast pyrolysis.

concentrate at a single focal point. The devices are mainly comprised of series of lens or mirror assembly, heat receiver, and the tracking system. The performance of the concentrator depends on the intensity of solar radiation, the incident angle of radiation, and its relative position to the sun and the reactor to be heated.

Solar concentrators are classified into four categories based on their optical characteristics, such as the concentration ratio, focal shape, and optical standard. These concentrators can be single-axis tracking or two-axis tracking to the sun as follows:
1. parabolic-trough concentrator (PTC);
2. parabolic-dish concentrator (PDC);
3. heliostat-field concentrator (HFC);
4. double-concentration concentrator (DCC).

Different types of concentrators provide different temperature ranges. In addition to this, a variation of concentration ratio is a distinctive feature of the concentrators. Table 8.3 presents some basic characteristics of solar concentrators.

8.6.1.1 Parabolic-Trough Solar Concentrator

Parabolic-trough solar concentrating systems are parabolic-shaped collectors made of reflecting materials. The collectors reflect the incident solar radiation onto its focal line toward a receiver that absorbs the concentrated solar energy to raise the temperature of the fluid inside it as shown in Fig. 8.6 [57]. Because of its single-axis tracking, all the solar radiations fall parallel to its axis.

The operating temperature of the system is in the range of 500−700K and the geometric concentration ratio of the parabolic-trough system is in the range of 30−100 (Table 8.3) [57]. Parabolic-trough collector is a viable technology for harnessing solar energy for industrial process heat in Cyprus [58]. Parabolic-trough solar concentrating systems are an advanced and matured technology. PTCs are suitable for heating between 100 and 250°C temperature as well as for concentrated solar power plants; Southern California power plants are the largest application of this system [59].

Table 8.3 Specifications of Common Solar Concentrators

Tracking Mechanism	Concentrator Type	Focal Type	Concentration Ratio	Operating Temperature (K)
Single-axis	Parabolic trough	Focal line	30−100	500−700
	Double concentration	Focal line	5000−10000	Over 1500
Two-axis	Parabolic dish	Focal point	1000−5000	Over 1800
	Heliostat field	Focal point	300−1500	Up to 1600

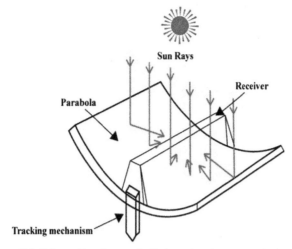

Figure 8.6 Schematic of a parabolic-trough solar concentrator [57].

8.6.1.2 Parabolic-Dish Solar Concentrator

Parabolic-dish solar concentrators are two-axis solar tracking systems that concentrate the solar radiations toward the thermal receiver located on the focal point of the dish collector as demonstrated in Fig. 8.7 [57].

These collectors consist of a set of parabolic dish—shaped mirrors. The operating temperature of the systems is over 1800K while the concentration ratio typically is in the

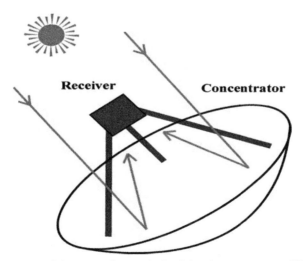

Figure 8.7 Schematic of a parabolic-dish solar concentrator [57].

range between 1000 and 5000K. Solar parabolic-dish concentrating systems are most appropriate for concentrated photovoltaic application due to its high concentration ratio and operating temperature. These systems are very bulky and high-cost device. Kussul et al. (2008) proposed a prototype of flat facet PDC with 24 mirrors to address these problems [60]. The large energy project operated with PDC was constructed in Shenandoah, Georgia between 1982 and 1989.

8.6.1.3 Heliostat Fields Solar Concentrator

Heliostat-field solar concentrating systems are focal point collectors that track the sun in two axes. HFCs consist of a large array of mirrors called heliostats which are distributed around a receiver avoiding shading. These heliostats reflect the directly incident solar rays toward receiver located on top of a tower as shown in Fig. 8.8 [57]. These systems usually achieve temperatures up to 1600K and concentration ratios of 300–1500. Although the system is very expensive, the large commercial power plant was demonstrated in California [61].

8.6.1.4 Double Concentration Solar Concentrator

Double concentration solar concentrators are focal line and single-axis solar collectors with heliostat, reflective tower, and ground receiver as presented in Fig. 8.9 [57]. The heliostats redirect the directly incident solar radiation toward a hyperboloidal reflector that reflects the solar beams downward to the receiver. On the ground, the compound

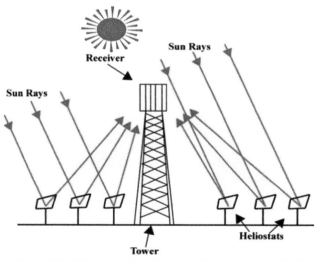

Figure 8.8 Schematic of a heliostat solar concentrator [57].

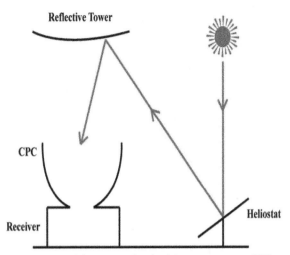

Figure 8.9 Schematic of a double concentrator [57].

parabolic concentrator acts as a secondary concentrator to enhance the further concentration of the reflected solar beams. These systems can operate in excess of 1500K and has the concentration ratio between 5000 and 10000.

In the seabelt, the solar concentrating technologies were found to be suitable for effective utilization of high-temperature heat for conversion into other probable energy applications [62,63]. Therefore, the concentrated solar energy can be effectually used as heat source for thermochemical processes such as pyrolysis. Although solar thermal electricity is generating from concentrated solar energy, the solar thermochemical process is still in developing stage.

8.6.2 Conceptual Design and Updated Concepts of Solar Pyrolysis

Although many attempts of incorporating solar energy in pyrolysis heating system have been made for a long time, there is no industrial implementation still that is successful. The main constraint in this regard is the design of solar concentrator and pyrolysis reactor. The reactor is considered as the heart of the pyrolysis system. Hence, the appropriate design is important to transfer the heat throughout the biomass during pyrolysis. The sole solar heating or combined solar with other heating sources can be very effective for pyrolysis. This heating method is energy efficient and attractive considering its minimized heating cost and renewable energy production consuming heat from a renewable source. In this process, the reactor along with the inside feed materials are heated to an elevated temperature by the incident solar energy to a concentrator that reflects the rays to the focal point on the reactor. A conceptual design of solar pyrolysis system is depicted in Fig. 8.10 [52].

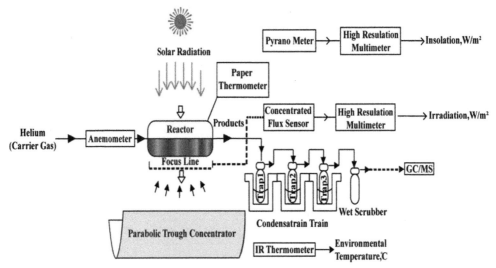

Figure 8.10 Conceptual design of solar pyrolysis system [52].

Although the solar heating of pyrolysis offers some advantages over the other heating methods, there still exists a difficulty in obtaining uniform heating throughout the reactor surface. The weather condition and the position of the sun are responsible for the fluctuation of solar intensity. On the contrary, the stationary solar concentrating system provides single-point heating to the reactor surface. Therefore, extensive researches are required to resolve this problem and to heat the reactor surface evenly for proper decomposition of feed materials. Several reactor geometries [64], internal surface coatings [65], and concentration methods [66] have been recommended for maintaining the isothermal internal condition of pyrolysis reactor. An updated conceptual design of solar heating system where a rotating reactor is heated continuously by the incident solar radiation to a sliding solar concentrator is revealed in Fig. 8.11A [67].

In addition, an electric motor—driven octagonal mirror field was designed to keep the focal point stationary throughout the day [15]. Moreover, an automated sun tracking system of two degrees-of-freedom was developed where the Fresnel lens of 91 × 69 cm always faces the radiation for ensuring maximum solar radiation concentration for pyrolysis of scrap rubber shown in Fig. 8.11B [68].

The heating performance of the solar concentrating system in pyrolysis can be improved by using mirrors and lens together at a time. Fig. 8.12 shows the front view of the proposed model of solar concentrated pyrolysis reactor.

The system consists of a cylindrical shaped, tilting type colorful glass-made reactor surrounded by spherical shaped mirrors bottom the reactor and convex lenses top the reactor. The top lenses will concentrate the huge amount of solar rays into the reactor,

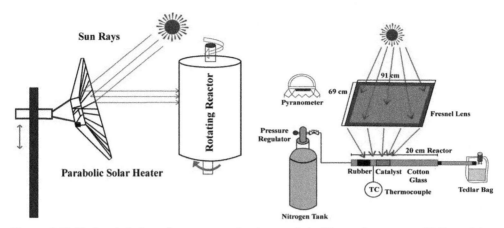

Figure 8.11 Updated design of concentrated solar pyrolysis (A) rotating reactor (B) Fresnel lens [67,68].

and the bottom mirrors from both sides of the reactor will also concentrate the incident solar rays and redirect it to the center of the reactor.

8.7 CURRENT RESEARCH AND APPLICATION OF SOLAR PYROLYSIS

Although the incorporation of concentrated solar energy began over almost 37 years ago, still the system is in the embryonic stage. Only laboratory-scale projects have been experimentally carried out, that is, at a fundamental level, and no pilot

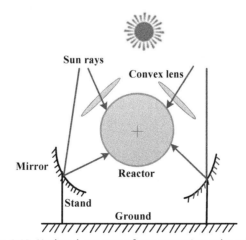

Figure 8.12 Updated concept of concentrating solar collector.

industrial project has been implemented yet. However, in spite of the suggestions and great attention of using solar heating approach for biomass pyrolysis, laboratory-based research studies are available for upgrading the system which indicates solar energy as an effective rapid heating source for biomass pyrolysis [15]. The incorporation of solar heating in coal gasification was investigated in the Lawrence Livermore National Laboratory, whereas coal pyrolysis with solar furnace was studied in the Los Mamos National Scientific Laboratory [69,70]. In addition, solar or other lighting source—driven image furnace was used for biomass pyrolysis to investigate the physical and chemical process and their modeling [17,71−73]. Concentrated solar heating in pyrolysis enhances the calorific value yield and decreases pollution [74] and is found to be significantly different from conventional reactors [15]. Some studies have reported the characteristics of product yield from solar pyrolysis of biomass residues [75,76].

On the other hand, Morales et al. [52] examined the use of laboratory-scale real solar furnace heated with the parabolic-trough solar concentrator. Besides, Zeng et al. [77] designed a 1.5 kW vertical-axis solar furnace and investigated the heating effects for bio-char and bio-oil production. Haueter et al. [78] developed an improved design of a direct heating solar thermal chemical reactor (called ROCA reactor) for chemical decomposition of ZnO. The reactor has low thermal inertia and outstanding thermal shock resistance capability. However, the most promising design of rotating cavity solar reactor, namely ZIRRUS reactor, was invented by Müller et al. [79] to address the problem of inability to recuperate a large percentage of products from the reactor associated with the ROCA reactor.

8.8 CONSIDERATIONS FOR FEASIBILITY OF SOLAR PYROLYSIS

Although the model of integrating the solar energy into pyrolysis is a sustainable solution for thermochemical decomposition of biomass, it is still in the developing stage. Therefore, it is necessary to analyze the feasibility of solar energy incorporation for heating the reactor. The requirement of energy input for proper decomposition of feed materials, environmental sustainability, cost and complexity of reactor design, and cost of heating the feed materials are the major parameters for investigating the feasibility of solar-assisted pyrolysis implementation.

Generally, a substantial amount of heat energy from biomass or electric heating sources is used for thermal decomposition of feed materials inside the reactor in pyrolysis. Therefore, complete or partial inclusion of solar energy in pyrolysis instead of these nonrenewable heating sources may reduce the amount of energy input as well as fuel cost significantly and thus increase the effectiveness of pyrolysis. In addition to these, the combustion of biomass is responsible for environmental pollution as it emits considerable amount of CO_2, CH_4, N_2O, NO_x, CO, and volatile carbon compounds, which is the major concern in the 21st century. Therefore, the heating of pyrolysis reactor with green and renewable solar energy will be an effective option to reduce environmental pollution.

8.9 CHALLENGES IN SOLAR PYROLYSIS

Radiant energy from the sun hits the earth in wavelengths in orders ranging from 0.1 to 10 mm. The primary challenge in the industrial application of solar energy is to match a location where high solar radiation is available [53]. The other challenges associated with solar pyrolysis are as follows:

- The design of right solar collector for the right reactor is a crucial factor. This needs a considerable attention of researchers.
- Uneven heating throughout the day due to variable weather condition is another great challenge for continuous heating system such as pyrolysis process. This uneven heating may cause incomplete pyrolysis of the feed materials.
- Even modeling of solar pyrolysis is not widely investigated. It would be obviously a complicated task when solar radiation is introduced in the regular reactor. Therefore, a numerical model considering heat and mass transfer, chemical kinetics, and solar radiation needs to be investigated to check the feasibility of the solar fast pyrolysis.
- Sufficient basic data concerning the techno-economic feasibility of solar pyrolysis at pilot level are not available in the literature. This lack of experimental investigation at pilot plants level constrains the implementation of the concept of solar pyrolysis at industrial level.

8.10 FUTURE SCOPE OF SOLAR PYROLYSIS

In order to overcome current challenges and to implement solar pyrolysis at industry level, the following R&D can be recommended:

- As there is uncertainty of solar intensity over the period of pyrolysis, there is the possibility of an incomplete process of feed materials. An automatically controlled feed supply system that maintains feed flow with solar intensity can be a great addition in current solar pyrolysis system.
- Distributed solar energy using movable solar concentrator and pyrolytic heater can be implemented in order to overcome uneven heating by the solar concentrator.
- Complete solar pyrolysis is still under investigation level. Suitable backup energy source can be linked in order to minimize uneven heating from the solar concentrator.
- Solar pyrolysis numerical modeling is still at the preliminary level as many data and parameters are unavailable in the literature, such as kinetics and thermodynamics and kinetics of chemical pathways; optical properties (absorptivity, emissivity, and

reflectivity) of all the associated species; and physicochemical properties. More complication arises when the model considers real-time value of these properties. Therefore, extensive modeling investigation is needed to ensure feasibility of solar pyrolysis.

8.11 CONCLUDING REMARKS

A great number of studies concerning solar pyrolysis for bio-oil production have been reported since mid-1980s; however, they are only at the fundamental and laboratory levels. Therefore, investigation of the feasibility test at pilot-plant level needs special attention of the researchers. The successful implementation of solar pyrolysis would help to meet our growing need of renewable energy sources. Incorporation of solar radiation instead of conventional heating system in pyrolysis reactor would minimize harmful emissions and assist to reduce greenhouse gasses.

REFERENCES

[1] McKendry P. Energy production from biomass (part 1): overview of biomass. Bioresource Technology 2002;83:37—46.
[2] White JE, Catallo WJ, Legendre BL. Biomass pyrolysis kinetics: a comparative critical review with relevant agricultural residue case studies. Journal of Analytical and Applied Pyrolysis 2011;91:1—33.
[3] Asadullah M, Rahman MA, Ali MM, Rahman MS, Motin MA, Sultan MB, et al. Production of bio-oil from fixed bed pyrolysis of bagasse. Fuel 2007;86:2514—20.
[4] Jung S-H, Kang B-S, Kim J-S. Production of bio-oil from rice straw and bamboo sawdust under various reaction conditions in a fast pyrolysis plant equipped with a fluidized bed and a char separation system. Journal of Analytical and Applied Pyrolysis 2008;82:240—7.
[5] Kumaravel ST, Murugesan A, Kumaravel A. Tyre pyrolysis oil as an alternative fuel for diesel engines — a review. Renewable and Sustainable Energy Reviews 2016;60:1678—85.
[6] Stedile T, Ender L, Meier HF, Simionatto EL, Wiggers V. Comparison between physical properties and chemical composition of bio-oils derived from lignocellulose and triglyceride sources. Renewable and Sustainable Energy Reviews 2015;50:92—108.
[7] Bridgwater A. Renewable fuels and chemicals by thermal processing of biomass. Chemical Engineering Journal 2003;91:87—102.
[8] Anis S, Zainal ZA. Tar reduction in biomass producer gas via mechanical, catalytic and thermal methods: a review. Renewable and Sustainable Energy Reviews 2011;15:2355—77.
[9] Goyal HB, Seal D, Saxena RC. Bio-fuels from thermochemical conversion of renewable resources: a review. Renewable and Sustainable Energy Reviews 2008;12:504—17.
[10] Radlein D, Quignard A. A short historical review of fast pyrolysis of biomass. Oil & Gas Science and Technology — Revue d'IFP Energies Nouvelles 2013;68:765—83.
[11] Research on the pyrolysis of hardwood in an entrained bed process development unit. Washington, DC: United States. Dept. of Energy ; 1991.
[12] Lédé J. Cellulose pyrolysis kinetics: an historical review on the existence and role of intermediate active cellulose. Journal of Analytical and Applied Pyrolysis 2012;94:17—32.
[13] Pictet A, Sarasin J. Sur la distillation de la cellulose et de l'amidon sous pression re duite. Helvetica Chimica Acta 1918;1:87—96.
[14] Stamm AJ. Thermal degradation of wood and cellulose. Industrial & Engineering Chemistry 1956;48:413—7.
[15] Antal MJ, Hofmann L, Moreira J, Brown CT, Steenblik R. Design and operation of a solar fired biomass flash pyrolysis reactor. Solar Energy 1983;30:299—312.

[16] De Jenga CI, Antal MJ, Jones M. Yields and composition of sirups resulting from the flash pyrolysis of cellulosic materials using radiant energy. Journal of Applied Polymer Science 1982;27:4313–22.

[17] Hopkins MW, DeJenga C, Antal MJ. The flash pyrolysis of cellulosic materials using concentrated visible light. Solar Energy 1984;32:547–51.

[18] Mok WS-L, Antal MJ. Effects of pressure on biomass pyrolysis. II. Heats of reaction of cellulose pyrolysis. Thermochimica Acta 1983;68:165–86.

[19] Mok WSL, Antal MJ, Szabo P, Varhegyi G, Zelei B. Formation of charcoal from biomass in a sealed reactor. Industrial & Engineering Chemistry Research 1992;31:1162–6.

[20] Shen DK, Gu S. The mechanism for thermal decomposition of cellulose and its main products. Bioresource Technology 2009;100:6496–504.

[21] Boutin O, Ferrer M, Lédé J. Flash pyrolysis of cellulose pellets submitted to a concentrated radiation: experiments and modelling. Chemical Engineering Science 2002;57:15–25.

[22] Luo W, Liao C. Mechanism study of cellulose rapid pyrolysis. Industrial & Engineering Chemistry Research 2004;43:5605–10.

[23] Schroeter J, Felix F. Melting cellulose. Cellulose 2005;12:159–65.

[24] Rasul MG, Jahirul MI, Science W. Recent developments in biomass pyrolysis for bio-fuel production: its potential for commercial applications pyrolysis process description pyrolysis classification. Recent Researches in Environmental and Geological Sciences Recent 2012:256–65.

[25] Gollakota ARK, Reddy M, Subramanyam MD, Kishore N. A review on the upgradation techniques of pyrolysis oil. Renewable and Sustainable Energy Reviews 2016;58:1543–68.

[26] Krutof A, Hawboldt K. Blends of pyrolysis oil, petroleum, and other bio-based fuels: a review. Renewable and Sustainable Energy Reviews 2016;59:406–19.

[27] Resende FLP. Recent advances on fast hydropyrolysis of biomass. Catalysis Today 2016;269:148–55.

[28] Sharma A, Pareek V, Zhang D. Biomass pyrolysis—a review of modelling, process parameters and catalytic studies. Renewable and Sustainable Energy Reviews 2015;50:1081–96.

[29] Papari S, Hawboldt K. A review on the pyrolysis of woody biomass to bio-oil: focus on kinetic models. Renewable and Sustainable Energy Reviews 2015;52:1580–95.

[30] Leal LAB, Janna FC. Modeling and simulation of biomass fast pyrolysis in a fluidized bed reactor. Journal of Analytical and Applied Pyrolysis 2016;118:105–14.

[31] Ranganathan P, Gu S. Computational fluid dynamics modelling of biomass fast pyrolysis in fluidised bed reactors, focusing different kinetic schemes. Bioresource Technology 2016;213:333–41.

[32] Bridgwater AV, Meier D, Radlein D. An overview of fast pyrolysis of biomass. Organic Geochemistry 1999;30:1479–93.

[33] Bridgwater AV. Review of fast pyrolysis of biomass and product upgrading. Biomass and Bioenergy 2012;38:68–94.

[34] Kong S-H, Loh S-K, Bachmann RT, Rahim SA, Salimon J. Biochar from oil palm biomass: a review of its potential and challenges. Renewable and Sustainable Energy Reviews 2014;39:729–39.

[35] Laird DA, Brown RC, Amonette JE, Lehmann J. Review of the pyrolysis platform for coproducing bio-oil and biochar. Biofuels, Bioproducts and Biorefining 2009;3:547–62.

[36] Adjaye JD, Bakhshi NN. Production of hydrocarbons by catalytic upgrading of a fast pyrolysis bio-oil. Part I: conversion over various catalysts. Fuel Processing Technology 1995;45:161–83.

[37] Vitolo S, Seggiani M, Frediani P, Ambrosini G, Politi L. Catalytic upgrading of pyrolytic oils to fuel over different zeolites. Fuel 1999;78:1147–59.

[38] Elliott DC, Beckman D, Bridgwater AV, Diebold JP, Gevert SB, Solantausta Y. Developments in direct thermochemical liquefaction of biomass: 1983–1990. Energy & Fuels 1991;5:399–410.

[39] Balat M, Demirbas MF. Pyrolysis of waste engine oil in the presence of wood ash. Energy Sources, Part A: Recovery, Utilization, and Environmental Effects 2009;31:1494–9.

[40] Kim YS, Jeong SU, Yoon WL, Yoon HK, Kim SH. Tar-formation kinetics and adsorption characteristics of pyrolyzed waste lubricating oil. Journal of Analytical and Applied Pyrolysis 2003;70:19–33.

[41] Lin L, Lai D, Guo E, Zhang C, Xu G. Oil shale pyrolysis in indirectly heated fixed bed with metallic plates of heating enhancement. Fuel 2016;163:48–55.

[42] Ramasamy K, Traissi A. Hydrogen production from used lubricating oils. Catalysis Today 2007;129:365—71.

[43] Miura M, Kaga H, Sakurai A, Kakuchi T, Takahashi K. Rapid pyrolysis of wood block by microwave heating. Journal of Analytical and Applied Pyrolysis 2004;71:187—99.

[44] Domeño C, Nerín C. Fate of polyaromatic hydrocarbons in the pyrolysis of industrial waste oils. Journal of Analytical and Applied Pyrolysis 2003;67:237—46.

[45] Ludlow-Palafox C, Chase HA. Microwave-induced pyrolysis of plastic wastes. Industrial & Engineering Chemistry Research 2001;40:4749—56.

[46] Budarin VL, Clark JH, Lanigan BA, Shuttleworth P, Breeden SW, Wilson AJ, et al. The preparation of high-grade bio-oils through the controlled, low temperature microwave activation of wheat straw. Bioresource Technology 2009;100:6064—8.

[47] Zhang Z, Zhao ZK. Microwave-assisted conversion of lignocellulosic biomass into furans in ionic liquid. Bioresource Technology 2010;101:1111—4.

[48] Henkel C, Muley PD, Abdollahi KK, Marculescu C, Boldor D. Pyrolysis of energy cane bagasse and invasive Chinese tallow tree (*Triadica sebifera* L.) biomass in an inductively heated reactor. Energy Conversion and Management 2016;109:175—83.

[49] Anderson BN. Solar energy: fundamentals in building design. 1977. New York.

[50] Kalogirou SA. Solar thermal collectors and applications, vol. 30; 2004.

[51] Antal MJJ, Princeton UN, Royere C, Vialaron A. Biomass gasification at the focus of the Odeillo (France 1-MW (thermal) solar furnace. ACS Symposium Series; (United States) 1980:130.

[52] Morales S, Miranda R, Bustos D, Cazares T, Tran H. Solar biomass pyrolysis for the production of bio-fuels and chemical commodities. Journal of Analytical and Applied Pyrolysis 2014;109:65—78.

[53] Murray J. Reaction of steam with cellulose in a fluidized bed using concentrated sunlight. Energy 1994;19:1083—98.

[54] Taylor RW, Berjoan R, Coutures JP. Solar gasification of carbonaceous materials. Solar Energy 1983;30:513—25.

[55] Zeng K, Gauthier D, Lu J, Flamant G. Parametric study and process optimization for solar pyrolysis of beech wood. Energy Conversion and Management 2015;106:987—98.

[56] Yadav D, Banerjee R. A review of solar thermochemical processes. Renewable and Sustainable Energy Reviews 2016;54:497—532.

[57] Kodama T. High-temperature solar chemistry for converting solar heat to chemical fuels. Progress in Energy and Combustion Science 2003;29.

[58] Kalogirou SA. Parabolic trough collectors for industrial process heat in Cyprus. Energy 2002;27:813—30.

[59] Fernández-García A, Zarza E, Valenzuela L, Pérez M. Parabolic-trough solar collectors and their applications. Renewable and Sustainable Energy Reviews 2010;14:1695—721.

[60] Kussul E, Baidyk T, Makeyev O, Lara-Rosano F, Saniger JM, Bruce N. Flat facet parabolic solar concentrator with support cell for one and more mirrors. WSEAS Transactions on Power Systems 2008;3:577—86.

[61] Barlev D, Vidu R, Stroeve P. Innovation in concentrated solar power. Solar Energy Materials and Solar Cells 2011;95:2703—25.

[62] Lewis NS. Solar energy use. 2007. p. 798—802.

[63] Steinfeld A, Schubnell M. Optimum aperture size and operating temperature of a solar cavity-receiver. Solar Energy 1993;50:19—25.

[64] Shuai Y, Xia X-L, Tan H-P. Radiation performance of dish solar concentrator/cavity receiver systems. Solar Energy 2008;82:13—21.

[65] Steinfeld AA. Solar receiver-reactor with specularly reflecting walls for high-temperature thermo-electrochemical and thermochemical processes. Energy 1988;13:301—11.

[66] Solar heat aperture control apparatus. 1980.

[67] Joardder MUH, Halder PK, Rahim A, Paul N. Solar assisted fast pyrolysis: a novel approach of renewable energy production. Journal of Engineering 2014;2014. ID252848.

[68] Zeaiter J, Ahmad MN, Rooney D, Samneh B, Shammas E. Design of an automated solar concentrator for the pyrolysis of scrap rubber. Energy Conversion and Management 2015;101:118—25.

[69] Beattie WH. Laser simulation of solar pyrolysis and gasification using static coal samples. 1981.

[70] Gregg DW, Taylor RW, Campbell JH, Taylor JR, Cotton A. Solar gasification of coal, activated carbon, coke and coal and biomass mixtures. Solar Energy 1980;25:353—64.

[71] Chan W-CR, Kelbon M, Krieger BB. Modelling and experimental verification of physical and chemical processes during pyrolysis of a large biomass particle. Fuel 1985;64:1505—13.

[72] Lédé J. Radiant flash pyrolysis of cellulose pellets: products and mechanisms involved in transient and steady state conditions. Fuel 2002;81:1269—79.

[73] Pozzobon V, Salvador S, Bézian JJ, El-Hafi M, Le Maoult Y, Flamant G. Radiative pyrolysis of wet wood under intermediate heat flux: experiments and modelling. Fuel Processing Technology 2014;128:319—30.

[74] Nzihou A, Flamant G, Stanmore B. Synthetic fuels from biomass using concentrated solar energy — a review. Energy 2012;42:121—31.

[75] Li R, Zeng K, Soria J, Mazza G, Gauthier D, Rodriguez R, et al. Product distribution from solar pyrolysis of agricultural and forestry biomass residues. Renewable Energy 2016;89:27—35.

[76] Zeng K, Gauthier D, Li R, Flamant G. Solar pyrolysis of beech wood: effects of pyrolysis parameters on the product distribution and gas product composition. Energy 2015;93:1648—57.

[77] Zeng K, Minh DP, Gauthier D, Weiss-Hortala E, Nzihou A, Flamant G. The effect of temperature and heating rate on char properties obtained from solar pyrolysis of beech wood. Bioresource Technology 2015;182:114—9.

[78] Haueter P, Moeller S, Palumbo R, Steinfeld A. The production of zinc by thermal dissociation of zinc oxide - solar chemical reactor design. Solar Energy 1999;67:161—7.

[79] Müller R, Haeberling P, Palumbo RD. Further advances toward the development of a direct heating solar thermal chemical reactor for the thermal dissociation of ZnO(s). Solar Energy 2006;80:500—11.

Wind Energy Systems

CHAPTER NINE

Grid Integration of Wind Energy Systems: Control Design, Stability, and Power Quality Issues

H.M. Boulouiha[1], A. Allali[2] and M. Denai[3]
[1]University of Relizane, Relizane, Algeria
[2]University of Science and Technology of Oran, Oran, Algeria
[3]University of Hertfordshire, Hatfield, United Kingdom

9.1 INTRODUCTION

Wind energy is among the most viable renewable energy resources in the world and its installed capacity is projected to grow substantially in the years ahead. Although wind energy technology has already reached a mature stage of development, there are, however, key challenges associated with the control complexity of wind energy—conversion systems (WECSs) for a successful integration into the electric power grid. The world's global wind power cumulative capacity has expanded from 3 GW to 370 GW at the end of 2014 (see Fig. 9.1). In Europe, 130 GW of both onshore and offshore wind installations are connected to the grid, and six countries (Denmark, Portugal, Ireland, Spain, Romania, and Germany) generate between 10% and 40% of their electricity from wind [1]. The global wind power capacity has increased by 50 GW from 2013 to 2014 reaching 365.4 GW.

Onshore wind energy is currently an established technology that is still undergoing extensive improvements. Presently, research and development is first and foremost focused on maximizing the assessment of wind energy and includes offshore technology, where public opinion surveys show strong support of new wind farm installations. While its share of the total wind capacity residue is small, the offshore wind production has experienced a net increase in year 2013, with 1.6 GW of new capacity linked to the grid. Offshore wind power installations accounted for nearly 14% of the total EU (European Union) wind power installations in 2013, that is, increased by 4% from 2012 [1].

The year 2014 brought a novel record in wind power installations; around 50 GW of capacity were added, bringing the total wind power capacity close to 370 GW. The market volume for new wind capacity was 40% bigger than in 2013, and significantly bigger than in the previous record year 2012, when 44.6 GW were installed [3].

Fig. 9.2 shows the top 10 countries with a total generation of 44.8 GW from new wind power plants, half of them setting new national records [3]. China added 23.3 GW,

Clean Energy for Sustainable Development
ISBN 978-0-12-805423-9, http://dx.doi.org/10.1016/B978-0-12-805423-9.00009-0

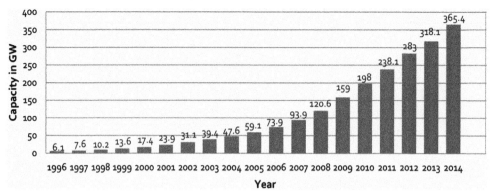

Figure 9.1 Globally installed wind power cumulative capacity from 1996 to 2014 [1,2].

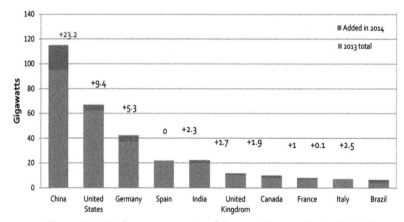

Figure 9.2 Wind power capacity of the top 10 countries in 2014 [1].

the largest capacity a country has ever produced within 1 year, reaching a total capacity of 115 GW. Germany has become the second largest market for new wind turbines, with a combined total of onshore and offshore wind generation of 5.8 GW. The US market recovered from its previous decline and reached 4.9 GW. In 2014, Brazil produced an additional capacity of 2.8 GW and became the first Latin American country that reached such a figure. New installation records were also achieved in Canada (1.9 GW) and Sweden (1 GW). Denmark has set a new world record by reaching a wind power share of 39% used for domestic power supply. Among the top 12 countries, Spain, Denmark, and Italy saw a stagnation in terms of new wind power installations.

The Global Wind Energy Council (GWEC) [4] produced the data of Fig. 9.3 on wind energy annual market and its forecast for the years ahead by continent.

The aim of this chapter is to present the latest developments of wind energy technology and provide an in-depth overview of the principles, modeling, and control

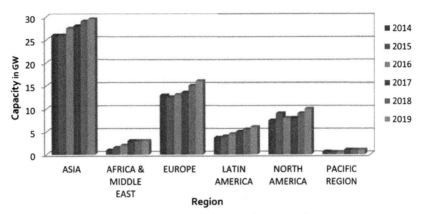

Figure 9.3 Annual market forecast by region for 2014–19.

strategies of variable speed WECS. The stability and power quality issues of grid-connected WECSs are discussed in detail, and control solutions are proposed to improve their performances.

The chapter provides an extensive coverage of the fundamental principles on the design of WECSs. A simplified methodology for identifying the parameters of the main component of a WECS including the turbine pale radius, DC bus voltage, and the impedance characteristics of the transmission line are presented.

The proposed WECS consists of a three-bladed wind turbine connected to the variable-speed squirrel-cage induction generator (SCIG). The mathematical models of the turbine, generator, and converter are derived. Due to the inherently intermittent nature of wind energy and continuously fluctuating wind velocities, maximum power point tracking (MPPT) strategies are employed to capture the maximum power from the wind turbine. The proposed MPPT method searches for a pseudo maximum power point based on the knowledge of the characteristic curve of the wind turbine to be driven.

The wind turbine-generator set is connected to the supply through two converters AC/DC/AC. In the generator mode, the first converter is used as a pulse width modulation (PWM) rectifier that ensures current flow from the induction generator AC side to the DC side. The grid-side converter is used for controlling the DC-side voltage magnitude, the active and reactive powers by adjusting the modulation index of the inverter, and the phase shift between the voltage and current components of the grid. For the generator-side converter two-level and three-level topologies are proposed to control the torque and flux of the SCIG driven by a variable speed wind turbine. Two control strategies, namely vector control or field-oriented control (FOC) and direct torque control (DTC), are designed, evaluated, and compared. A comparative study between the conventional DTC and the DTC-SVPWM (space vector-PWM) for two- and three-level inverter topologies to improve the energy efficiency and performance characteristics of the variable speed WECS is presented. A simple pole placement control technique is used to design the controller for the grid-side converter.

The PWM technique requires a programmed control law for the grid-side converter and the DC-link voltage based on the modulation index in amplitude (*MI*) and phase angle. The programmed PWM technique, developed and implemented as a lookup table, is used to control the six inverter switches. In the case of the three-level converter the SVPWM is used.

The main objective of this study is to address the stability and the quality of energy of the grid-integrated WECS. The chapter includes several simulation results to demonstrate the impact of the proposed control strategies and converter topologies on the performance of the system in terms of harmonics distortion and generator torque ripples. Simulation scenarios replicating different fault conditions including symmetrical and asymmetrical faults on the network side are examined and discussed in terms of the dynamic performance, robustness, and stability of the overall power system.

This chapter is organized as follows. Section 9.2 shows the major components of a WECS and the different wind turbine configurations. In addition, the modeling and MPPT control of the wind turbine is developed and the different methods for the aerodynamic protection of the turbine are presented. Section 9.3 overviews the most recent results related to the most popular wind energy systems currently available. The dynamic models of the synchronous and asynchronous generator are developed in this section.

The topologies of converters and their modulation techniques used to control the switching of the back-to-back AC/DC/AC converter are detailed in Section 9.4. Also included in this section are the various frequently used PWM techniques and the modeling of the two- and three-level converters. The proposed space vector PWM (SVPWM) method for the three-level converter is performed essentially to balance the DC-side voltage and minimize its fluctuations in order to improve the quality of the energy.

The design methodology of indirect field—oriented control (IFOC) and the classical DTC, DTC-SVPWM for the two- and three-level topologies are developed in Section 9.5. This section also presents the modeling and control design of the DC voltage and currents of the grid side. The design of proportional—integral (PI) feedback controllers for the generator and grid systems are presented in this section. To minimize the effects of ripples due to symmetrical and asymmetrical faults of grid voltages or wind speed variations, control mechanisms are necessary to diminish the maximum harmonics of the back-to-back AC/DC/AC converters. Solutions to enhance the power quality and improve the overall stability of the grid-integrated wind energy system is clearly explained in Section 9.6. Protection systems against grid faults are presented in this section. Finally, a conclusion with a little synthesis of the work performed in this chapter is presented in Section 9.7.

9.2 WIND TURBINE TECHNOLOGIES

Wind turbines convert the kinetic energy of the wind into mechanical energy which is then converted to electrical energy. The main components of a modern horizontal–axis wind turbine (HAWTs) are shown in Fig. 9.4. The wind measurement system (anemometer) monitors the wind speed and transfers the measurements to the control panel. The angle of the blade is controlled to maximize the aerodynamic efficiency of the wind turbine and limit the load on mechanical transmission in strong wind conditions.

Wind turbine technology has evolved rapidly during mid-1990s; however, the basic principle is almost unchanged. The simplest and most commonly used system consists of mechanically coupling the generator rotor to the wind turbine driveshaft via a multiplier (or gearbox). However, in some cases, gearboxes are undesirable because they are expensive, bulky, and heavy. Multipolar generators based on gearless technology are an alternative in this case.

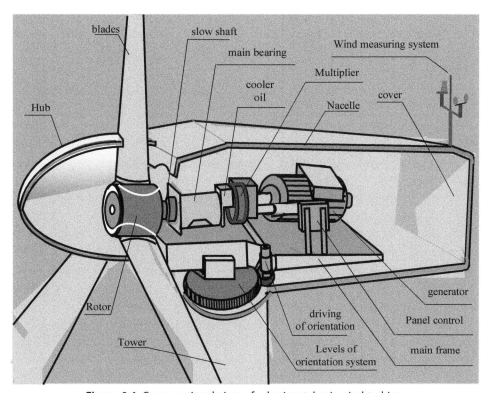

Figure 9.4 Cross-sectional view of a horizontal-axis wind turbine.

Wind turbines are equipped with a robust security system including an aerodynamic braking system.

The blades are among the major components of a wind turbine. They evolved considerably in aerodynamic design and materials. The first windmill blades were made of wood and cloth, while modern blades are usually made of aluminum, fiberglass, or carbon-fiber providing high strength—weight ratio, leading to improved fatigue life and rigidity while minimizing the weight [5]. The three-bladed rotor is considered as the industry standard for large wind turbines. Single and two-bladed wind turbines are used for higher rotational speeds and low torques applications. The key advantage of these blade configurations is the speed reduction ratio compared to three-bladed turbines, and consequently a lower cost and reduced size of the gearbox. However, there is a rise in the acoustic noise which varies in proportion to the blade tip speed ratio. Rotors with more than three blades are less common because they are more expensive and the air turbulence in the interblade space tends to slow down the blade. Therefore, the three-bladed rotor is the best compromise between the mechanical force, acoustic noise, cost, and speed of rotation of large wind turbines.

The power of a mass air flow at a wind speed v through a surface S can be determined by:

$$P_v = \frac{1}{2} \cdot \rho \cdot S \cdot v^3 \tag{9.1}$$

where ρ represents the air density (kg/m^3) which depends on the atmospheric pressure and air temperature (e.g., $\rho = 1.225$ kg/m^3 at a temperature of 15°C and a pressure of 1013 mbar), S is the area swept by the rotor (m^2), v is the wind speed (m/s).

The power delivered by the wind turbine rotor is:

$$P_m = \frac{1}{2} \rho S v^3 C_p \tag{9.2}$$

where C_p is the power coefficient of the wind turbine. This factor has a maximum theoretical value of 0.59 according to Betz's limit. With today's technology, the power coefficient of modern turbines is usually in the range of 0.2—0.5 and depends on the rotation speed and the number of blades.

Eq. (9.2) shows that there are three possible ways to increase the power P_m captured by a wind turbine: the wind speed v, the power coefficient C_p, and the area S swept by the rotor blades. The wind speed cannot be controlled; however, wind turbines are usually installed in areas with higher and more regular average wind velocities such as offshore. The wind turbine can be designed with a larger swept area using rotor blades with a larger radius to capture more wind power. Since the swept area is $S = \pi R^2$, the blade radius R has a quadratic effect on S and hence on the power captured. This explains a tendency in recent years to increase the diameter of the rotor blades. Finally, the power captured can be increased by improving the power coefficient through a better aerodynamic design.

Large blade manufacturing, transportation, and installation are complex and challenging process requiring careful planning. Fig. 9.5 illustrates the gradual development of wind turbine blades [6,7]. The blade element momentum (BEM) theory [8,9] is generally used to design and optimize blade shapes.

9.2.1 Wind Turbine Configurations

Wind turbines are broadly categorized into two types, based on the orientation of the shaft and rotational axis, namely the vertical-axis and horizontal-axis wind turbines (VAWTs and HAWTs).

9.2.1.1 Vertical-Axis Wind Turbines

VAWTs have their shafts perpendicular to the ground and are suitable for low power applications with efficiencies limited to 25% [7]. VAWTs have a simple construction, a high starting torque, and operate at low speeds (Savonius model in Fig. 9.6). However, VAWTs have several disadvantages including the mechanical guide, in particular the bottom bearing, which must support the total weight of the turbine. Furthermore, some designs (Darrieus model in Fig. 9.6) are not self-starting and require an external power source. Another type of VAWT is known as Musgrove blades with H-shaped rotor. For high wind speeds the rotor blades are rotated about a horizontal point caused by the centrifugal force. This eliminates the risk of higher aerodynamic forces on the blades.

9.2.1.2 Horizontal-Axis Wind Turbines

HAWTs have their shafts mounted horizontally parallel to the ground. Similar to VAWTs, the HAWTs can be built with two or three blades. The largely dominant

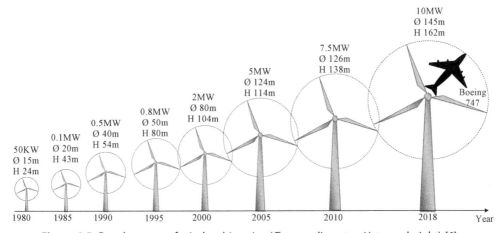

Figure 9.5 Development of wind turbine size (Ø, rotor diameter; H, tower height) [6].

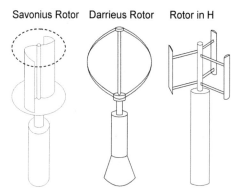

Figure 9.6 VAWT technologies.

technology today is the three-bladed HAWT although the two-bladed rotor and the rotor facing the wind models have also been popular. The turbine can be at the front of the nacelle (upwind) or at the back (downwind) (Fig. 9.7). Downwind devices automatically face the wind direction and therefore do not require mechanical orientation system. The major disadvantage is increased fatigue due to frequent oscillations caused by wind fluctuations. The downwind or wind downstream model are less used than the upwind or wind upstream [10].

The reduced number of blades theoretically reduces the cost but leads to irregular torque. The power coefficient C_p is also considerably lower, around 5% difference between the three-blade and two-blade configurations.

Turbines equipped with a large number of blades operate at low speeds. Their power coefficient quickly reaches its maximum value initially when the speed increases but decreases rapidly thereafter. While turbines operating at high speeds have a lesser number of blades, the power coefficient assumes large values and slowly decreases as the speed

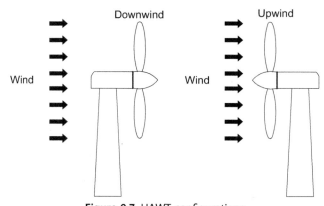

Figure 9.7 HAWT configurations.

increases. Fig. 9.8 gives a comparison of different wind turbine configurations with respect to the power coefficient versus the tip speed ratio.

From Fig. 9.8, it can be noticed that the curves $C_p(\lambda)$ clearly show the advantage of HAWTs with regard to the aerodynamic performance and power output. $C_p(\lambda)$ curves are flatter for HAWTs with a small number of blades (3, 2, and 1) as compared to the VAWTs or multiblades. They are less sensitive to the variations of λ around its optimal value λ_{opt}.

Wind turbines of American type have a large number of blades because they operate at low speeds. They develop a large aerodynamic torque in order to produce mechanical energy and they have been more popular for pumping applications. Finally, one can observe the influence of the number of blades on aerodynamic efficiency.

9.2.2 Mathematical Modeling of Wind Turbine Characteristics

The relationship between wind speed and aerodynamic mechanical power extracted from the wind can be described as follows [11,12]:

$$P_m = \frac{1}{2}\pi\rho C_p(\lambda, \beta)R^2 v^3 \tag{9.3}$$

where P_m (W) is the mechanical power of the wind turbine and β (degrees) represents the pitch angle of the blades. The power coefficient C_p defines the aerodynamic efficiency of

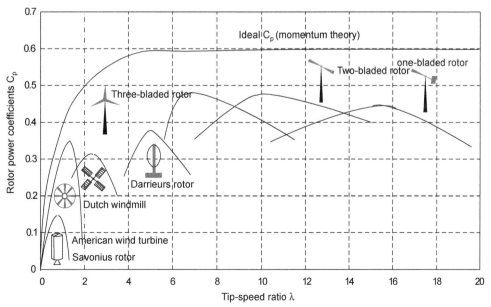

Figure 9.8 Power coefficient and torque based on the standard speed λ for different types of turbines.

the wind turbine. One commonly used equation to model C_p, which is a function of speed ratio λ and the pitch angle β of the blade, can be expressed as follows:

$$C_p(\lambda, \beta) = 0.5 \left[\frac{33}{\lambda_i} - 0.2\beta - 0.4 \right] \exp\left(-\frac{12.7}{\lambda_i} \right) \tag{9.4}$$

$$\frac{1}{\lambda_i} = \frac{1}{\lambda + 0.08\beta} - \frac{0.035}{\beta^3 + 1} \tag{9.5}$$

$$\lambda = \frac{\Omega_m R}{v} \tag{9.6}$$

where Ω_m is the mechanical speed of the turbine (rad/s).

A simplified dynamic model of the wind turbine is shown in Fig. 9.9.

In the model of Fig. 9.9, wind speed dynamics can be recorded from real measurements or reproduced using analytical models. A commonly used model of wind speed is based on the spectral characteristic of Van Der Hoven [13]. In this model, the turbulence part is considered as a stationary random process and is independent of the fluctuations of the mean of the wind speed. This model is defined as follows:

$$v(t) = v_l(t) + v_t(t) = \frac{2}{\pi} \sum_{k=0}^{N_i} A_k \cos(w_k t + \theta_k) + \frac{2}{\pi} \sum_{N_i}^{N} A_k \cos(w_k t + \theta_k) \tag{9.7}$$

where A_k is the magnitude of each spectral component, w_k represents the pulse (rad/s), and θ_k is the phase (rad). The floating wind speed fluctuates between 2 and 16 m/s. The resulting wind speed waveform is shown in Fig. 9.10.

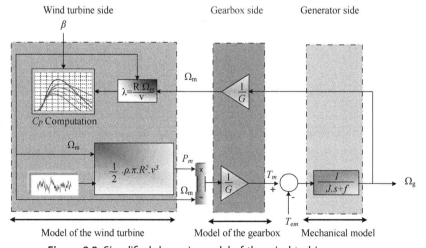

Figure 9.9 Simplified dynamic model of the wind turbine.

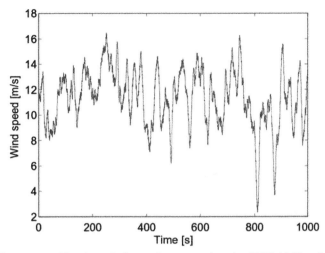

Figure 9.10 Floating wind speed generated under MATLAB/Simulink.

9.2.3 Maximum Power Point Tracking Strategies

Depending on the wind aerodynamic conditions, there exists an optimal operating point for which maximum power can be extracted from the turbine. The power captured by the wind turbine Eq. (9.3) can be substantially maximized by adjusting the coefficient C_p which represents the aerodynamic efficiency of the wind turbine and is dependent on the speed of the generator (or the speed ratio λ). It is necessary to design control strategies to maximize the power generated (thus the torque) by adjusting the speed of the turbine to a reference value regardless of the disturbances acting on the wind speed. Several methods exist for adjusting the wind turbine at partial load following the trajectory of maximum power point. This can be achieved either through the control of the rotational speed or the power of the turbine. In this work, two different controllers have been considered: The indirect speed controller (ISC) and direct speed controller (DSC).

There are several MPPT methods in the literature: those that are not based on the knowledge of the wind turbine characteristics (direct methods) and those that use the wind turbine characteristics (indirect methods). Direct methods usually lead to a complex control structure depending on the approach used to search for the MPPT. The indirect method, such as the one used in this work, searches for a pseudo-maximum power point from the knowledge of the characteristic curve of the wind turbine to be driven. These methods move rapidly toward the optimum without the need to capture the wind speed, in other words, without the need to capture the wind speed.

The optimum angular speed $\Omega_{m,opt}$ for the mechanical transmission of the maximum wind turbine is given by:

$$\Omega_{m,opt} = \frac{\lambda_{opt} v}{R} \qquad (9.8)$$

The following relation can be deduced:

$$P_{m,\max} = K_{p,opt}\Omega_{m,opt}^3 \qquad (9.9)$$

where

$$K_{p,opt} = \frac{1}{2}\rho C_{p,\max}\frac{R^2}{\lambda_{opt}^3} \qquad (9.10)$$

Thus, the corresponding optimum torque is:

$$T_{m,\max} = \frac{P_{m,\max}}{\Omega_{m,opt}^2} \qquad (9.11)$$

The control objectives are to ensure security of the wind turbine at high wind speeds and limit the power to its nominal value P_n developed at a nominal speed v_n. When the wind speed exceeds v_n, the turbine settings should be adjusted to prevent the mechanical destruction.

Four main operation regions can be distinguished for the wind turbine as shown in Fig. 9.11:

where v_d denotes the initial speed at which the wind turbine begins to supply power, v_m is the maximum wind speed beyond which the wind turbine will have to stop generating for safety reasons.

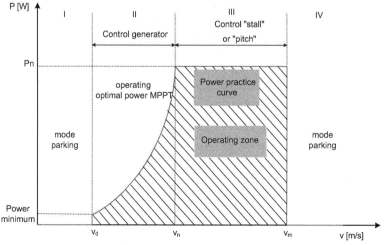

Figure 9.11 Operating regions of the wind turbine: power versus wind speed characteristics.

With reference to Fig. 9.11, Zone I corresponds to $P = 0$ (the turbine does not work). In Zone II, the power supplied by the turbine depends on the wind speed v. When the generator speed reaches a threshold value, a control algorithm for extracting the maximum power from the wind is applied. To extract maximum power, the angle of the blade is kept constant at its minimum, usually the pitch angle is fixed at $\beta 4 = 0$ degrees. In Zone III, the rotational speed is kept constant and the power supplied is maintained equal to P_n. Zone IV corresponds to excessive wind speed values for which the wind turbine is shut down.

9.2.4 Aerodynamic Protection Methods

Most large wind turbines use two aerodynamic control principles to limit the extracted power to the rated power of the generator. (1) *Variable pitch* or simply *pitch control* which adjusts the lift blades to the wind speed to maintain a substantially constant power in zone III speed. (2) *Aerodynamic stall* or simply *stall control* which is more robust as it is the blade shape which leads to a loss of lift beyond a certain wind speed. However, the power curve falls faster in this case.

Fig. 9.12 shows the maximum power characteristics of the turbine obtained using Eqs. (9.9) and (9.10).

9.2.4.1 Aerodynamic Stall Control Method

Most wind turbines connected to the electrical grid need a fixed rotational speed for reasons of frequency coherence with the network. The load stall control is a passive control

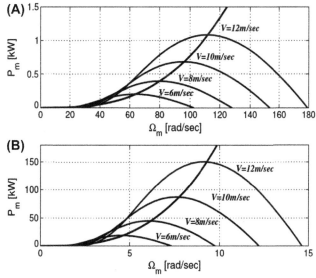

Figure 9.12 Optimum operating characteristics of the turbine. Nominal power of (A) 2 kW and (B) 149.2 kW.

system that reacts beyond a certain wind speed [14]. The rotor blades are locked and cannot rotate around their longitudinal axis. The pitch angle is chosen to allow the stall of air flow on the blade surface for wind speeds larger than the rated speed. This reduces the lift and increases the drag force. Aerodynamic stall turbines have the following advantages: No pitch angle control system, easier and less expensive construction of the rotor, easier maintenance, and better reliability (smaller number of moving mechanical parts).

There are two types of stall control strategies: *passive-stall control* and *active-stall control*. For passive (passive-stall) control the blade is fixed on the rotor hub at an optimal angle of attack. When the wind speed is below its nominal value, the turbine blades with the optimal angle of attack can capture the maximum power available depending on the wind speed. If the wind speed exceeds the nominal value, the strong wind can cause turbulence on the surface of the blade. The operating principle of passive control is shown in Fig. 9.13, where the lift force generated by the nominal wind speed $F_{w,n}$ is smaller than the force $F_{w,stall}$ produced by large wind speeds. This will ensure a braking of the turbine for wind speeds above rated wind speed v_n.

As shown in Fig. 9.14, passive- and active-stall, and the pitch control cannot maintain the power P_m constant and its rated value may be exceeded for some wind speed values which is not desirable.

At higher wind speeds, when the turbine speed is adjusted to start slowly, this is called the *active-stall* control. In this case, the rotor blades are oriented in the direction of the stall (negative angle) and not in the opposite direction (positive angle), as with variable-pitch wind angles.

The principle of passive-stall control is illustrated in Fig. 9.15. If strong wind speeds, the blade loses interaction with the wind and causes the turbine rotor to stop. This action can be used above the nominal speed to protect the wind turbine.

9.2.4.2 Pitch Control Method
Normally, for the pitch of the blades to vary as a function of the wind speed, the changes in air characteristics and surface condition of the blades, which influence the behavior

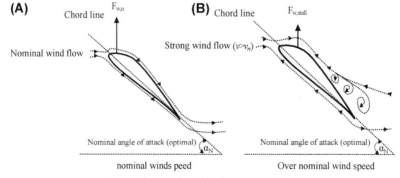

Figure 9.13 Principle of passive-stall control [6].

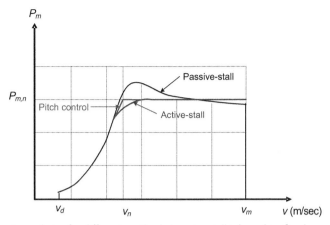

Figure 9.14 Power boundaries for different methods (passive stall is based on fixed speed operation) [15].

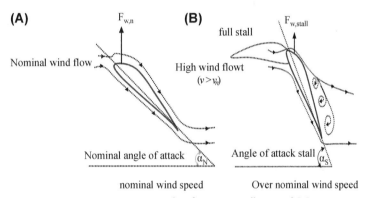

Figure 9.15 Principle of passive-stall control [6].

aerodynamic, should be taken into account. The blades face the wind at low speed (from the minimum speed to the base speed) and then incline to reach the "flag" position at maximum speed. This control system offers several advantages including (1) the possibility to perform active power control under all wind conditions (out of the safe limited speed), including a partial power, (2) it can provide the rated power even when air density is low (sites with hills, high temperatures), (3) greater energy production of stall wind turbines under the same conditions, (4) quick start by changing the pitch angle, (5) there is no need for powerful brakes for emergency stops, (6) physical constraints on the blades are lower when the powers are greater than the nominal value, and (7) the reduced mass of the rotor blades leads to reduced mass of the wind turbine. The operation principle of the pitch control is shown in Fig. 9.16. When the wind is below the rated speed (nominal), the angle of blade attack is held at its optimal value α_P. For wind speeds

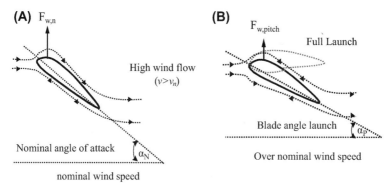

Figure 9.16 Principle of aerodynamic control of the pitch angle [6].

higher than the rated wind speed, the angle of attack of the blade is lowered to reduce the lift force F_w. When the blade is fully launched, the angle of attack of the blade is aligned with the wind direction, as shown by Fig. 9.16A, and no lifting force will be produced. The turbine stops turning and will be locked by the mechanical brake to protect the turbine and the blades. The execution of the *launch control* is represented by Fig. 9.16B where the mechanical transmission of the turbine operation above rated wind speed can be well controlled.

The pitch control reacts faster than stall control and provides better controllability. It is now widely adopted in large wind energy systems.

In general, changing the pitch angle of the blade has four distinct objectives: (1) start at a reduced wind speed v_d; (2) speed control Ω, for $v > v_n$; (3) optimization of the energy conversion regime when the wind speed varies between the limits $[v_d, v_n]$; and (4) protection against large wind speeds ($v > v_m$) by orienting the blades in the "flag" position.

This control strategy was studied in Refs. [16–19]. In Ref. [16], the authors refer to the advantages of variable speed wind turbine with pitch angle control. For low wind speeds, the generator and the power converter control the wind turbine to maximize the energy captured by maintaining the rotor speed at a predetermined optimum value. At high wind speeds, the wind turbine is controlled to maintain the aerodynamic power developed by the turbine by controlling the pitch angle. Refs. [17] and [18] propose a detailed model of the wind turbine operating at variable speed and controlled by an appropriate pitch angle to study transient stability of wind turbines with a power of few megawatts. The simulated model is based on a set of nonlinear curves relating the tip speed ratio of the blade, the power coefficient, and the pitch angle of the wind turbine. The references also present a detailed model of the wind dynamics. In Ref. [17] the model of the electric generator driven by the wind turbine is represented by transfer functions, while in Ref. [18] a $d-q$ mode of the synchronous generator is used. A PI controller is used to adjust the pitch angle to limit the power of the turbine under high wind speeds conditions. Ref. [18] presents a general model for all types of variable speed wind turbines. The proposed wind turbine model maintains the control of the pitch angle which reduces the performance of the turbine rotor at high wind velocities, as described in Refs. [17] and [19]. The wind turbine dynamics are represented using nonlinear curves, which are numerical approximations to estimate the turbine power coefficient for a given tip speed ratio and pitch angle data values. The authors present a comparison between the power curves in per-unit of two commercial wind turbines and those obtained theoretically using numerical approximations. The results indicate that numerical approximations can be used to simulate different types of wind turbines.

The timing angle control, illustrated in Fig. 9.17 is aimed to limit the power taken by adjusting the pitch angle β of the blades. The positioning mechanism is to guide the blades toward a reference angle (β_{ref}) via a hydraulic or electric system. The choice of this angle is usually achieved via an external loop to regulate either the speed of the turbine or the mechanical power generated. In our model, the latter method is used to generate the reference of the pitch angle (β_{ref}).

It can be noted from Fig. 9.18 that the power is maintained at 5 MW, for the nominal wind speed or for wind speed above the nominal value using the control of the pitch angle. Similar remark applies to the rotor speed of the generator.

9.3 GENERATOR TYPES IN THE WIND ENERGY CONVERSION SYSTEMS

The different types of electric generators used in WECSs are presented in Fig. 9.19.

A WECS can be operated at fixed speed or variable speed depending on the configuration of the wind turbine. The main benefits of both configurations are summarized as folows:

- *Fixed speed operation*: (1) simple electric system, (2) greater reliability, (3) low probability of the excitation of resonant frequencies of wind turbine elements, (4) no need for power electronics system, and (5) low cost.

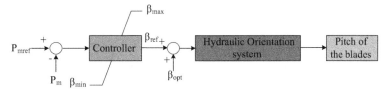

Figure 9.17 Control of the mechanical power P_m.

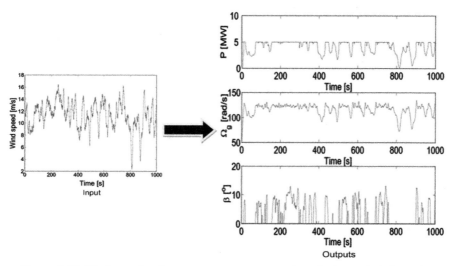

Figure 9.18 Wind speed, active power, generator angular velocity, and pitch angle.

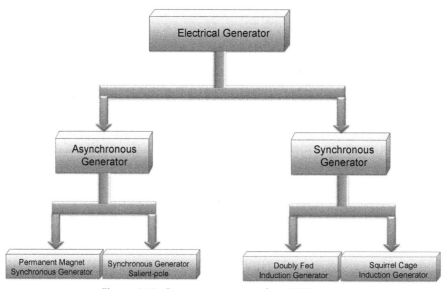

Figure 9.19 Generator types used in WECSs.

- *Variable speed operation*: (1) increased energy efficiency, (2) reduced torque oscillations in the power train, (3) reduction of forces applied to the power train, and (4) generation of electric power of better quality.

The most popular electric machines in the wind industry are the synchronous and asynchronous machines and their variants [20,21].

9.3.1 Synchronous Generators

Synchronous generators are particularly employed in direct drives (i.e., without mechanical multiplier). Synchronous generators are very advantageous when they have a large number of poles, however, in this case the frequency becomes incompatible with the network's frequency, hence an inverter is required. Therefore, direct drive machines are all of variable speed. Fig. 9.20 shows the basic structure of a WECS based on permanent magnet synchronous generator (PMSG).

Direct drive synchronous generators have a wound inductor (rotor) and require brush rings to provide DC supply. PMSGs are becoming more and more popular for variable speed applications and are expected to become increasingly important in the future.

The aerodynamic-axis of the wind turbine rotor and the generator can be connected directly (i.e., without gearbox). In this case, the generator is a multipolar synchronous generator designed for low speed. Alternatively, they may be coupled through a gearbox which allows the use of a generator with a larger number of poles. For variable speed operation, the synchronous generator is connected to the grid via two power converters to adjust the frequency which completely decouples the speed of the generator and the grid frequency. Consequently, the generator frequency will vary with the wind speed, whereas the grid frequency will remain constant.

The power converter system has two converters, grid-side and generator-side, connected back-to-back by a DC link.

The major drawback of this technique is the size of the bidirectional converter which should match the power of the alternator. Also, one must remove the distortion caused by the harmonics due to bidirectional converter using a filter system. Another drawback is that a multipolar machine requires a large number of poles, which increases the size of the machine as compared to the generators with transmission coupling.

Active and reactive power control for a PMSG was studied in Refs. [22–28]. In Ref. [22], the author proposed a method to control the wind power system which is connected to the PMSG under grid fault conditions. The authors proposed the use of a capacitor in the DC-side for short-term energy storage to compensate for the oscillations of the torque and

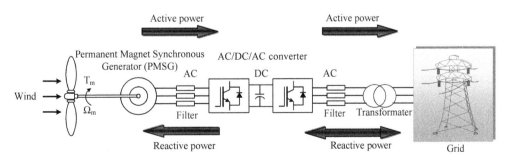

Figure 9.20 Synchronous generator (wound rotor) and frequency inverter.

speed, and to ensure stable operation of the wind turbine under the grid faults. The author in Ref. [23] proposed a current control strategy to limit the grid current to the inverter and reduce the power output of the machine during grid faults.

The inverter control strategy for PMSG-based wind power system under unbalanced three-phase voltage was studied in Ref. [24]. The negative sequence fault current is decomposed and added to the current calculated by the phase-locked loop (PLL). This control method ensures a three-phase sinusoidal, balanced current for the grid side, however, the control of the DC-link voltage is not addressed. The model proposed in Ref. [25–27] neglects power exchange with the inductors. Thus, for a highly unbalanced case or for a system with a high inductance value, this method is not effective. In Ref. [28], the author proposed a control strategy with two setting modes to separately control the positive and negative sequence fault current. The first mode achieves balanced currents on the mains side while the second mode reduces the ripples in the DC-link voltage under unbalanced grid conditions.

Using Park transformation, the actual stator voltage and current are converted to their d–q counterparts as illustrated by Fig. 9.21.

The stator quantities are expressed in Park reference frame linked to the rotor:

$$\begin{cases} v_{sd} = R_s i_{sd} + \dfrac{d\varphi_{sd}}{dt} - \omega_g \varphi_{sq} \\[2mm] v_{sq} = R_s i_{sq} + \dfrac{d\varphi_{sq}}{dt} - \omega_g \varphi_{sd} \end{cases} \tag{9.12}$$

Similarly, the stator fluxes are:

$$\begin{cases} \varphi_{sd} = L_d i_{sd} + \varphi_f \\[2mm] \varphi_{sq} = L_q i_{sq} \end{cases} \tag{9.13}$$

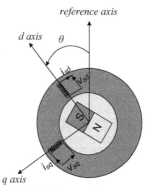

Figure 9.21 Park model of the synchronous machine.

L_d and L_q are the components of the inductance on the direct and quadrature axis. The machine is assumed to have smooth poles, hence $L_d = L_q$, and φ_f represents the mutual flux.

Substituting Eq. (9.12) into Eq. (9.13) gives:

$$\begin{cases} v_{sd} = R_s i_{sd} + L_d \dfrac{di_{sd}}{dt} - \omega_g L_q i_{sq} \\[4mm] v_{sq} = R_s i_{sq} + L_q \dfrac{di_{sq}}{dt} + \omega_g (L_d i_{sd} + \varphi_{sd}) \end{cases} \tag{9.14}$$

The electromagnetic torque produced is:

$$T_{em} = \frac{3P}{2} \left((L_d - L_q) i_{sd} i_{sq} + \varphi_f i_{sq} \right) \tag{9.15}$$

The final forms of the PMSG equations in the $d-q$ reference frame are:

$$\begin{cases} \dfrac{di_{sd}}{dt} = -\dfrac{R_s}{L_d} i_{sd} + \dfrac{L_q}{L_d} \omega_g i_{sq} + \dfrac{1}{L_d} v_{sd} \\[4mm] \dfrac{di_{sq}}{dt} = +\dfrac{L_d}{L_q} \omega_g i_{sd} - \dfrac{R_s}{L_q} i_{sq} - \omega_g \varphi_f + \dfrac{1}{L_q} v_{sq} \\[4mm] T_{em} = \dfrac{3}{2} P \left((L_d - L_q) i_{sd} i_{sq} + \varphi_f i_{sq} \right) \\[4mm] T_{em} - T_m - f\Omega_g = J \dfrac{d\Omega_g}{dt} \end{cases} \tag{9.16}$$

9.3.2 Self-Excited Induction Generator

SEIG are robust and reliable machines for wind energy conversion applications, particularly in isolated areas, because they do not require an external supply to produce the excitation magnetic field. Self-excitation is achieved by capacitors directly connected to the load or using a converter to provide the required reactive power. Self-excitation is however difficult to control because the load variations directly influence the voltage and frequency. In addition, a device for the orientation of the wind turbine blades is required to maintain the rotational speed, and hence a constant frequency. To determine the performance and the operating limits of this device, a dynamic model of the SEIG in a two-phase stationary reference frame linked to a reference is developed. The load model and the capacitor values for the self-excitation are also determined independently of the SEIG model. For self-excitation to occur, the rotor should have sufficient residual magnetization and the value of three capacitors should be sufficient [29].

The structure of the SEIG-based WECS is shown in Fig. 9.22. For direct connection, the induction generator must be synchronized to the grid frequency. Therefore, a gearbox is required to increase the speed of the rotor blades to the frequency of the grid.

There is a vast literature related to SEIGs [16,30,31]. The authors in Ref. [30] highlight an early work on self-excited induction machines. In this paper, the self-excited induction machine is driven by a motor and is self-excited by a capacitor bank connected via its stator terminals to provide the required reactive power. The authors also discuss the importance of the saturation curve and the reactance of the excitation capacity and conclude that the induction generator works with any type of load, provided that the loads are compensated to ensure a unity power factor for the generator. Following this work [31], the author proposed a method for predetermining the characteristic of the induction machine operating as a SEIG and it is emphasized that the value of the terminal voltage of the SEIG increases due to the fact that self-excitation depends on the rotor speed, the capacitor value, and the load. In a SEIG, the terminal voltage and frequency are unknown and must be calculated from the rotor speed, capacity, and load values. The analysis of SEIGs is complex due to magnetic saturation of the machine. In Ref. [16], the authors proposed an analytical technique based on Newton—Raphson method to identify the parameters in saturation mode, and the generator frequency for given operating conditions (speed value, load, and capacity). This method uses an equivalent circuit of the induction machine to estimate the values obtained from the analytical analysis, however, the method is applicable only to balanced systems.

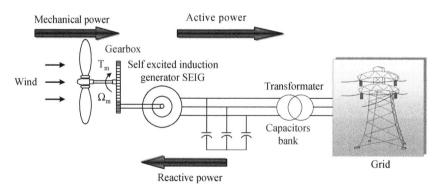

Figure 9.22 SEIG directly connected to the grid.

Two-phase $d-q$ models give a simplified representation of the three-phase induction machine. Fig. 9.23 shows the equivalent circuit of a SEIG in the $d-q$ reference Park connected with an inductive load.

The general equations of the induction machine are [5,32,33]:

$$\begin{cases} v_{sd} = R_s i_{sd} - \omega_s \varphi_{sq} + \dfrac{d\varphi_{sd}}{dt} \\[2mm] v_{sq} = R_s i_{sq} + \omega_s \varphi_{sd} + \dfrac{d\varphi_{sq}}{dt} \\[2mm] v_{rd} = 0 = R_r i_{rd} - (\omega_s - \omega_g)\varphi_{rq} + \dfrac{d\varphi_{rd}}{dt} \\[2mm] v_{rq} = 0 = R_r i_{rq} + (\omega_s - \omega_g)\varphi_{rd} + \dfrac{d\varphi_{rq}}{dt} \end{cases} \qquad (9.17)$$

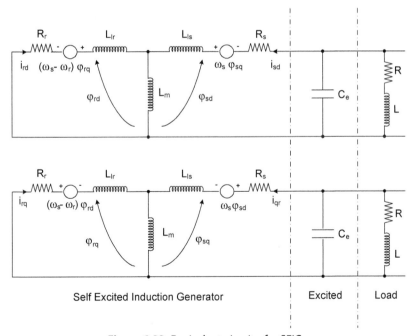

Figure 9.23 Equivalent circuit of a SEIG.

The rotor speed is:

$$\omega_r = \omega_s - \omega_g \qquad (9.18)$$

where ω_g is the SEIG generator's speed and ω_s is the stator speed.

Different models for the induction machines are obtained depending on the orientation of the $d-q$ reference frame which can be linked to the stator (*stator reference frame*), rotor (*rotor reference frame*), or rotating at the synchronous speed (*synchronously rotating reference frame*). In each of these references, the model of the machine becomes simpler than in the case of a reference frame rotating at an arbitrary speed.

The *stator reference frame* is used to study the starting and braking of the machine in other words, this reference is better suited for representing instantaneous quantities. It is characterized by $\omega = \omega_s = 0$ and consequently $\omega_r = -\omega_g$ (where ω is the speed of the arbitrary rotating frame). The *rotor reference frame* is used to simulate the machine's transient regime with a speed assumed to be constant. For this reference frame ($\omega = \omega_s = \omega_g$ and $\omega_r = 0$).

The *synchronously rotating reference frame* is used for vector control implementation and is characterized by $\omega = \omega_s$ implying that the control variables are continuous. The advantage of using this reference is that all quantities are constant in steady state which makes them easier to control. In this chapter, the stationary reference linked to the stator ($\omega = \omega_s$) is employed to simulate the model of the SEIG.

The stator and rotor voltages of the SEIG in the stator reference frame are obtained by replacing $\omega_s = 0$ in Eq. (9.17):

$$\begin{cases} v_{sd} = R_s i_{sd} + \dfrac{d\varphi_{sd}}{dt} \\[2mm] v_{sq} = R_s i_{sq} + \dfrac{d\varphi_{sq}}{dt} \\[2mm] v_{rd} = 0 = R_r i_{rd} - \omega_r \varphi_{rq} + \dfrac{d\varphi_{rd}}{dt} \\[2mm] v_{rq} = 0 = R_r i_{rq} + \omega_r \varphi_{rd} + \dfrac{d\varphi_{rq}}{dt} \end{cases} \qquad (9.19)$$

R_s and R_r are the resistances of the stator and rotor, respectively.
Writing these equations in state space form gives:

$$\frac{d}{dt}\begin{bmatrix} i_{sd} \\ i_{sq} \\ i_{rd} \\ i_{rq} \end{bmatrix} = K_L \left\{ \begin{bmatrix} R_s L_r & -\omega_r L_m^2 & -R_r L_m & -\omega_r L_m L_r \\ \omega_r L_m^2 & R_s L_r & \omega_r L_m L_r & -R_r L_m \\ -R_s L_m & \omega_r L_m L_s & R_r L_r & \omega_r L_s L_r \\ -\omega_r L_m L_s & -R_s L_m & -\omega_r L_s L_r & R_r L_r \end{bmatrix} \begin{bmatrix} i_{sd} \\ i_{sq} \\ i_{rd} \\ i_{rq} \end{bmatrix} + \begin{bmatrix} -L_r & 0 \\ 0 & -L_r \\ L_m & 0 \\ 0 & L_m \end{bmatrix} \begin{bmatrix} v_{sd} \\ v_{sq} \end{bmatrix} \right\}$$

(9.20)

The gain K_L is defined by:

$$K_L = \frac{1}{L_m^2 - L_s L_r}$$

where the inductances L_s of the stator and L_r of the rotor are expressed by the leakage inductance of the stator and rotor windings L_{ls} and L_{lr}, respectively, and the magnetizing inductance L_m by:

$$\begin{cases} L_s = L_{ls} + L_m \\ L_r = L_{lr} + L_m \end{cases}$$

(9.21)

Stator and rotor fluxes are written as:

$$\begin{cases} \varphi_{sd} = L_{ls}i_{sd} + L_m(i_{sd} + i_{rd}) \\ \varphi_{sq} = L_{ls}i_{sq} + L_m(i_{sq} + i_{rq}) \\ \varphi_{rd} = L_{lr}i_{rd} + L_m(i_{rd} + i_{rd}) \\ \varphi_{rq} = L_{lr}i_{rq} + L_m(i_{rq} + i_{sq}) \end{cases}$$

(9.22)

v_{sd}, v_{sq}, v_{rd}, v_{rq}, i_{sd}, i_{sq}, i_{rd}, i_{rq}, φ_{sd}, φ_{sq}, φ_{rd}, and φ_{rq} are direct and quadratic components of voltages and currents, and the stator and rotor fluxes, respectively.

The d–q components of the magnetizing branch fluxes are the product of the magnetizing inductance L_m by the sum of the stator and rotor currents of the same axis in the equivalent circuit of the SEIG shown in Fig. 9.23.

$$\begin{cases} \varphi_{md} = L_m(i_{sd} + i_{rd}) \\ \varphi_{mq} = L_m(i_{sq} + i_{rq}) \end{cases} \tag{9.23}$$

The SEIG currents in the stator and rotor are deduced from Eq. (9.20) as:

$$\begin{cases} i_{sd} = \dfrac{\varphi_{sd}}{L_{ls}} - \dfrac{\varphi_{md}}{L_{ls}} \\[2mm] i_{sq} = \dfrac{\varphi_{sq}}{L_{ls}} - \dfrac{\varphi_{mq}}{L_{ls}} \\[2mm] i_{rd} = \dfrac{\varphi_{rd}}{L_{lr}} - \dfrac{\varphi_{md}}{L_{lr}} \\[2mm] i_{rq} = \dfrac{\varphi_{rq}}{L_{lr}} - \dfrac{\varphi_{mq}}{L_{lr}} \end{cases} \tag{9.24}$$

Substituting Eq. (9.24) into Eq. (9.22), the magnetizing branch fluxes along d and q axes are obtained as:

$$\begin{cases} \varphi_{md} = L_{mm}\left(\dfrac{\varphi_{sd}}{L_{ls}} + \dfrac{\varphi_{rd}}{L_{lr}}\right) \\[3mm] \varphi_{mq} = L_{mm}\left(\dfrac{\varphi_{sq}}{L_{ls}} + \dfrac{\varphi_{rq}}{L_{lr}}\right) \end{cases} \tag{9.25}$$

where

$$\frac{1}{L_{mm}} = \frac{1}{L_m} + \frac{1}{L_{ls}} + \frac{1}{L_{lr}}$$

The equation for the electromagnetic torque (positive sign for the motor operation and negative sign for generator operation) is [34]:

$$T_{em} = \left(\frac{3}{2}\right)\left(\frac{p}{2}\right) L_m(i_{sq}i_{rd} - i_{sd}i_{rq})\,(\text{motor}) \tag{9.26}$$

$$T_{em} = -\left(\frac{3}{2}\right)\left(\frac{p}{2}\right)L_m(i_{sq}i_{rd} - i_{sd}i_{rq})(\text{generator}) \tag{9.27}$$

The speed ω_g of the SEIG is obtained from the dynamics of the mechanical system as follows:

$$\frac{d\omega_g}{dt} = \frac{p}{2J}\left(T_m - T_{em} - \frac{2}{p}F\omega_g\right) \tag{9.28}$$

where T_{em} is the electromagnetic torque (Nm), T_m is the mechanical torque produced by turbine (Nm), J is the inertia of the machine (kg m^2), and F is coefficient of viscous friction (Nm s/rad).

In most situations, the linear model of the induction machine is sufficient to analyze the dynamics of the machine. This model considers the magnetizing inductance constant which is not exact, since the magnetic material used is not perfectly linear. However, in certain cases, the effect of saturation of the magnetic circuit and consequently the change in magnetizing inductance cannot be neglected. The model considered here takes into account the magnetic saturation that allows us to choose an optimal path for the flux of the machine. Since the saturation is better represented with continuous quantities, the $d-q$ model is preferred to the abc model.

It is assumed that all system parameters are constant and independent of the magnetic saturation, except the magnetizing inductance L_m. The following model is used to represent the nonlinear relationship between the magnetizing inductance L_m and the magnetizing current I_m [36]:

$$L_m = 1.3e^{-0.05545\ I_m} - 0.95e^{-(I_m+1)} \tag{9.29}$$

where the magnetizing current is defined by:

$$I_m = \sqrt{i_{md}^2 + i_{mq}^2} = \sqrt{(i_{ds} + i_{rd})^2 + (i_{sq} + i_{rq})^2} \tag{9.30}$$

Fig. 9.24 shows a plot of L_m vs. I_m relationship given by Eq. (9.29).

When capacitors are connected across the terminals of the stator of an induction machine driven by an external device, a voltage will be induced across the terminals. The presence of a residual field in the magnetic circuit of the machine develops an electromagnetic torque producing an electromotive force (EMF) on the stator windings. The capacitors connected across the stator create a reactive current which increases the magnetic field of the machine and consequently the EMFs. It is this cyclic reaction that allows the machine to reach a steady state located in the saturated zone. Thus, in an autonomous mode of operation, the induction generator should be operated in the saturation region. This ensures that only one intersection point exists between the magnetization curve and the reactance line of the capacitor, and

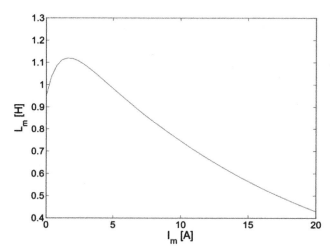

Figure 9.24 Plot of the magnetizing inductance L_m versus the magnetizing current I_m.

guarantees the stability of the output voltage under the applied load as shown in Fig. 9.25. The voltage V_1 depends on the magnetizing current I_m, rising linearly until the point of operation.

Without load, the capacitor current is $I_c = V_1/X_c$. This current must equal the magnetizing current $I_m = V_1/X_m$. The SEIG output frequency is $f = 1/(2\pi C_e X_m)$ and $\omega_s = 2\pi f$, where C_e is the self-exciting capability.

When the operating point is reached, the machine develops a stator voltage with constant RMS value as shown in Fig. 9.26. If the magnetizing inductance is considered constant and equal to its nonsaturated value, then the magnetization characteristic has no saturation bend, and there is no intersection with the external characteristic of the capacitor. Self-priming is possible but the stator voltage then increases to an infinite value as shown in Fig. 9.26 [32,35–37].

Self-priming of the asynchronous generator is ensured by the presence of the capacitors and saturation. The capacitors deliver the necessary reactive power, however,

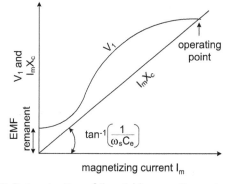

Figure 9.25 Determination of the stable operating point of the SEIG.

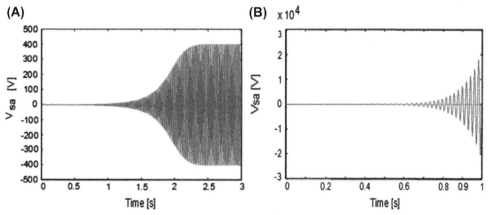

Figure 9.26 Self-priming of the SEIG (A) with saturation and (B) without saturation.

the generator voltage cannot increase indefinitely because it will be limited by the saturation in the machine. This process is governed by the following relationship, known as the priming the generator equation:

$$
\begin{cases}
v_{Cd} = v_{sd} = \dfrac{1}{C_e} \displaystyle\int_0^t i_{Cd}\,dt + v_{sd0} \\[4mm]
v_{Cq} = v_{sq} = \dfrac{1}{C_e} \displaystyle\int_0^t i_{Cq}\,dt + v_{sq0}
\end{cases}
\tag{9.31}
$$

with v_{sd0} and v_{sq0} representing the nonzero initial voltages of the induction machine (residual flux), i_{Cd} and i_{Cq} are the capacitors' currents.

The capacitors' currents are given by:

$$
\begin{cases}
i_{Cd} = i_{sd} - i_{Ld} \\[2mm]
i_{Cq} = i_{sq} - i_{Lq}
\end{cases}
\tag{9.32}
$$

The capacitor and load voltages are equal to the stator voltages and can be expressed as follows:

$$
\begin{cases}
v_{Ld} = v_{sd} = R i_{Ld} + L\dfrac{d i_{Ld}}{dt} \\[4mm]
v_{Lq} = v_{sq} = R i_{Lq} + L\dfrac{d i_{Lq}}{dt}
\end{cases}
\tag{9.33}
$$

R and L are, respectively, the resistance and inductance of the inductive load connected to the SEIG.

From Eq. (9.33), the load currents i_{Ld} and i_{Lq} are obtained as:

$$\begin{cases} \dfrac{di_{Ld}}{dt} = \dfrac{1}{L}v_{Ld} - \dfrac{R}{L}i_{Ld} \\[2ex] \dfrac{di_{Lq}}{dt} = \dfrac{1}{L}v_{Lq} - \dfrac{R}{L}i_{Lq} \end{cases} \tag{9.34}$$

The approximate value of the minimum capacitance required for the self-excitation at no load condition can be calculated as follows [38]:

$$C_{\text{emin}} \approx \frac{1}{\omega_g^2 L_{mn}} \tag{9.35}$$

where L_{mn} is the unsaturated value of the magnetizing inductance.

However, it is not recommended to use the minimum capacitance value, because any variation of the load or the rotor speed can result in the loss of self-excitation.

When the generator is excited by a capacitance value $C = 200\ \mu\text{F}$ and the rotor speed is increased from zero to 295 rad/s at $t = 0.75\ s$, the voltages produced by the self-excited generator without load reach 400 V at time $t = 2.3\ s$ as shown in Fig. 9.27. After introduction of the inductive load 1 ($R = 1\ \text{k}\Omega$ and $L = 10\ \text{mH}$) at $t = 4\ s$, the voltage drops to

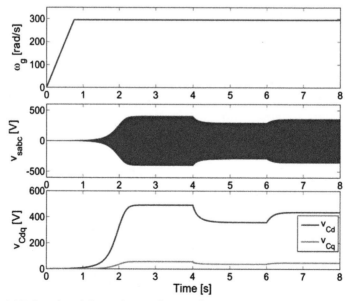

Figure 9.27 Speed and three-phase voltages of the SEIG and the capacitor voltages.

298 V due the presence of the inductance. A small increase of 97 V in the voltage amplitude is observed when load 2 ($R = 200\ \Omega, L = 100$ mH) is applied. Fig. 9.28 shows the response of the electromagnetic torque T_{em}, the current I_m, the magnetizing inductance L_m, and the three-phase stator currents i_{sabc} of the self-excited generator. This figure also shows the torque produced by the SEIG where the steady-state value reached almost -15 Nm without load. For both inductive loads 1 and 2 the electromagnetic torque of the SEIG takes values -11.91 and -8.35 Nm, respectively. The three-phase balanced currents, respectively, reach 34 A (without load), 24.6 A (load 1), and 30 A (load 2). The load variations are directly affected by the amplitude of the stator currents of the generator. These changes can also be visualized on the circular plot of the direct-axis current i_α versus the quadratic-axis current i_β as shown in Fig. 9.29.

9.3.3 Squirrel-Cage Induction Generator

The low cost and standardization of asynchronous machines has led to a very large domination of induction generators with squirrel cage for powers above the megawatt. SCIGs only require a fairly basic installation. They are often associated with a capacitor bank for the compensation of reactive power or with a soft-starter to limit transients during the connection to the grid (Fig. 9.30).

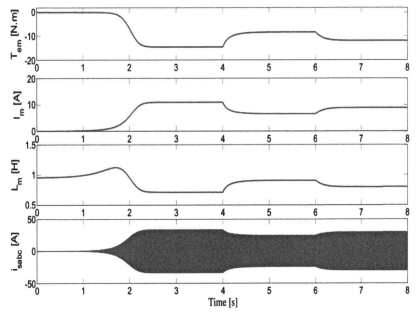

Figure 9.28 Responses of the electromagnetic torque, magnetizing current, magnetizing inductance, and three-phase stator currents of SEIG.

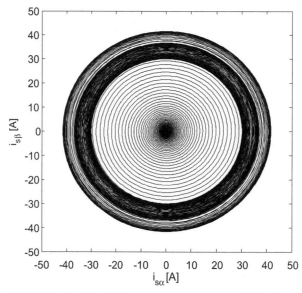

Figure 9.29 Circular plot of the stator current vector of the SEIG.

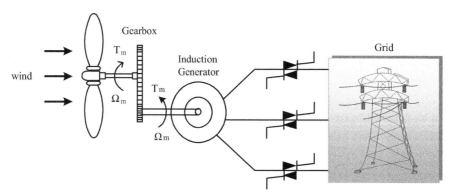

Figure 9.30 Soft-starter of the SCIG.

A wind turbine with fixed-speed SCIG is directly connected to the grid through a transformer. Since the SCIG requires reactive power from the grid, it will create unacceptable conditions, such as voltage drop and low power factor.

In a variable speed SCIG, the power converter must supply reactive power to the generator. The structure of the SCIG-based WECS is shown in Fig. 9.31. This topology has the following advantages: (1) the SCIG is directly connected to the grid through a transformer and (2) the SCIG requires minimal maintenance. However, the disadvantages are: (1) the SCIG requires a speed limiting control to avoid dangerous speeds; (2) the SCIG pulls the reactive power of the grid which is compensated by the capacitor

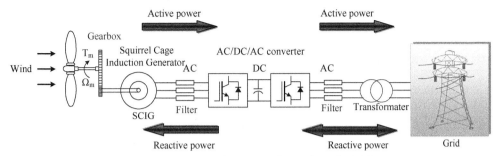

Figure 9.31 Grid-connected SCIG-based WECS.

banks (in the absence of the capacitor bank voltage, fluctuations, and inevitable electric line losses). Hence, this topology requires reactive power excitation, and (3) fixed speed operation.

The stator winding is connected to the grid via the power converters AC/DC/AC. The control system of the generator-side converter controls the electromagnetic torque and provides reactive power to maintain magnetization of the machine. The grid-side converter controls the active and reactive powers and regulates the level of the DC-link voltage. This topology has some advantages [39]: (1) the rectifier can produce soft excitation for the generator, (2) it has fast transient response, and (3) the inverter can operate as a reactive power compensator (harmonic reduction) when capacity is available. The disadvantages of this type are: (1) a complex control (field-oriented control) requiring the knowledge of the generator parameters which vary with temperature and frequency [40]. In this case, DTC (direct torque control) is an alternative approach to overcome the dependency on the machine parameters, and (2) the generator-side converter must be oversized for 30–50% of nominal power, to provide the required magnetization for the machine [41].

The control of variable speed SCIG-based WECS has been extensively studied [41–43]. Fuzzy control of the turbine to maximize the power of a variable speed SCIG has been used in Ref. [41]. The generator-side and grid-side converters are controlled by conventional PI controllers. The author demonstrated the superiority of using the fuzzy logic controller against a PI controller. The control of the SCIG coupled to a variable speed wind turbine based on DTC combined with an MPPT strategy was studied in Ref. [42]. In Ref. [43], the interaction of two wind turbines coupled to SCIGs and connected to the grid via static converters has been studied. The author proposed a control strategy for the two WECSs using a single grid-side converter. For the regulation of the DC bus voltage and the exchange of active and reactive powers, the author used a simple PI controller.

The equivalent circuit of the SCIG is shown in Fig. 9.32.

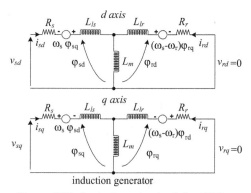

Figure 9.32 Equivalent circuit of the SCIG.

The equations of the SCIG of Fig. 9.32 written in the Park system are:

$$\begin{cases} v_{sd} = R_s i_{sd} + \dfrac{d\varphi_{sd}}{dt} - \omega_s \varphi_{sq} \\[3mm] v_{sq} = R_s i_{sq} + \dfrac{d\varphi_{sq}}{dt} + \omega_s \varphi_{sd} \end{cases} \tag{9.36}$$

where v_{sd}, v_{sq} are the stator voltage components in the Park system, ω_s is the synchronous speed of the generator, and φ_{sd}, φ_{sq} and i_{sd}, i_{sq} are, respectively, the flux and stator current components in d-axis and q-axis of the Park frame.

Similar equations can be written for the rotor:

$$\begin{cases} v_{rd} = 0 = R_r i_{rd} + \dfrac{d\varphi_{rd}}{dt} - \omega_r \varphi_{rq} \\[3mm] v_{rq} = 0 = R_r i_{rq} + \dfrac{d\varphi_{rq}}{dt} + \omega_r \varphi_{rd} \end{cases} \tag{9.37}$$

The rotor voltages v_{rd} and v_{rq} are set to zero since the rotor is short circuited, ω_r is the rotor speed of the generator, and φ_{rd}, φ_{rq} and i_{rd}, i_{rq} are, respectively, the flux and rotor current components in d-axis and q-axis of the Park frame.

The stator and rotor fluxes of the machine in the $d-q$ system are:

$$\begin{cases} \varphi_{sd} = L_s i_{sd} + L_m i_{rd} \\[3mm] \varphi_{sq} = L_s i_{sq} + L_m i_{rq} \end{cases} \tag{9.38}$$

$$\begin{cases} \varphi_{rd} = L_m i_{sd} + L_r i_{rd} \\ \varphi_{rq} = L_m i_{sq} + L_r i_{rq} \end{cases} \tag{9.39}$$

and the mechanical equation is:

$$T_{em} - T_m = f\Omega_g + J\frac{d\Omega_g}{dt} \tag{9.40}$$

where T_{em} is the electromagnetic torque (Nm) and T_m denotes the mechanical torque produced by the turbine (Nm).

The electromagnetic torque is expressed as:

$$T_{em} = p\frac{3}{2}\frac{L_m}{L_r}\left(\varphi_{rd}i_{sq} - \varphi_{rq}i_{sd}\right) \tag{9.41}$$

The choice of the state variables, inputs, and outputs of the system depends on the control objectives. For control applications with a reference linked to the rotating field the appropriate choice for the state vector is:

$$x = \begin{bmatrix} i_{sd} & i_{sq} & \varphi_{rd} & \varphi_{rq} \end{bmatrix}^T \tag{9.42}$$

Substituting the fluxes in the voltage equations gives the complete model of the induction machine:

$$\begin{cases} \dfrac{di_{sd}}{dt} = -\left(\dfrac{1}{T_s\sigma} + \dfrac{(1-\sigma)}{T_r\sigma}\right)i_{sd} + \omega_s i_{sq} + \dfrac{(1-\sigma)}{T_r L_m\sigma}\varphi_{rd} + \dfrac{(1-\sigma)}{L_m\sigma}\omega_m\varphi_{rq} + \dfrac{1}{L_s\sigma}v_{sd} \\[3mm] \dfrac{di_{sq}}{dt} = -\omega_s i_{sd} - \left(\dfrac{1}{T_s\sigma} + \dfrac{(1-\sigma)}{T_r\sigma}\right)i_{sq} - \dfrac{(1-\sigma)}{L_m\sigma}\omega_m\varphi_{rd} + \dfrac{(1-\sigma)}{T_r L_m\sigma}\varphi_{rq} + \dfrac{1}{L_s\sigma}v_{sq} \\[3mm] \dfrac{d\phi_{rd}}{dt} = \dfrac{L_m}{T_r}i_{sd} - \dfrac{1}{T_r}\varphi_{rd} + \omega_r\varphi_{rq} \\[3mm] \dfrac{d\phi_{rq}}{dt} = \dfrac{L_m}{T_r}i_{sq} - \omega_r\varphi_{rd} - \dfrac{1}{T_r}\varphi_{rq} \\[3mm] T_{em} - T_r = J\dfrac{d\Omega_m}{dt} + f\Omega_m \end{cases}$$

$$\tag{9.43}$$

where $\omega_m = p\,\Omega_m$; $\omega_r = [\omega_s - \omega_m]$; $\sigma = 1 - \frac{L_m^2}{L_s L_r}$; $T_r = \frac{L_r}{R_r}$; $T_s = \frac{L_s}{R_s}$.

This model structure reduces the number of variables required to simulate the machine. Indeed, only the instantaneous values of the stator voltages and load torque need to be defined. Therefore, the knowledge of the stator pulsation or slip is not required as in the case of the model equations written in the synchronously rotating frame.

9.3.4 Doubly Fed Induction Generator

The doubly fed induction generator (DFIG) has been used for many years in variable speed drives' applications. The stator is connected directly to the source and the rotor is supplied from a bidirectional converter which is also connected to the source, as shown in Fig. 9.33.

Using vector control techniques, the bidirectional converter will generate the required energy at the nominal frequency and voltage of the source, independent of the rotor speed [44].

The main feature of the DFIG-based system is that it is the only scheme where the generated power can exceed the nominal power of the machine used. The rotor frequency can vary from a value lower than the grid frequency (if the wind turbine is connected), it is called *hypo-synchronous* operation. It can also exceed the grid frequency (up to twice the grid frequency when the *slip* = −1), which corresponds to the *hyper-synchronous* regime.

The power electronic converter which controls power flows from the rotor is designed for a power rating much lower (around 20%) than that of the stator. The cost of the system is therefore reduced.

The main characteristics can be summarized as follows: (1) limited speed operation range (−30% to +20%); (2) the power electronic converter can be sized to suit the rating of the system, reduced power losses, and cost; (3) active and reactive powers exchanged with the source; and (iv) slip rings are required.

Two control approaches can be distinguished in Refs. [45–48]. The first approach in Ref. [45] is based on the control of active power via the blade angle of attack, the speed of rotation is controlled by the quadratic component of the rotor current, and the control of

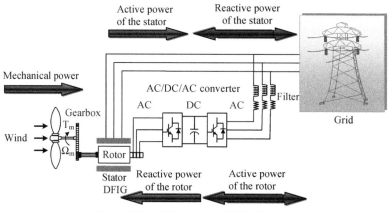

Figure 9.33 Grid-connected DFIG-based WECS.

the reactive power is through the direct component of the rotor current. The second approach, used in Refs. [46–48], is based on the control of active power by the quadratic component of the rotor voltage, the speed of rotation is controlled by the blade angle of attack, and the reactive power is controlled by the direct voltage component of the rotor.

The control of the DFIG under unbalanced conditions has been briefly studied in Ref. [49], while the DFIG protection under asymmetrical grid perturbations has been studied in Ref. [50]. The operation and control of the DFIG under unbalanced networks were studied in Refs. [51–56]. In Refs. [51,52] the work was limited to the control of the grid-side converter to achieve a functionality similar to that of a STATCOM (Static Compensator) [53]. The control of rotor-side converter for compensating the torque ripple in the DFIG during unbalanced voltage condition has been studied in Refs. [54,55]. In Ref. [56] the authors provided a detailed analysis of the impact of unbalanced stator voltages on the ripples of the DFIG's rotor current, torque, and active and reactive stator powers. Different control designs have been proposed to minimize the imbalance in the stator/rotor currents, the active/reactive powers, and torque oscillations.

The equivalent circuit of the DFIG is shown in Fig. 9.34.

The conventional electric equations of the DFIG in Park reference frame are written as follows:

$$
\left\{
\begin{aligned}
v_{sd} &= R_s i_{sd} - \omega_s \varphi_{sq} + \frac{d\varphi_{sd}}{dt} \\[2mm]
v_{sq} &= R_s i_{sq} + \omega_s \varphi_{sd} + \frac{d\varphi_{sq}}{dt} \\[2mm]
v_{rd} &= R_r i_{rd} - \omega_r \varphi_{rq} + \frac{d\varphi_{rd}}{dt} \\[2mm]
v_{rq} &= R_r i_{rq} + \omega_r \varphi_{rd} + \frac{d\varphi_{rq}}{dt}
\end{aligned}
\right.
\tag{9.44}
$$

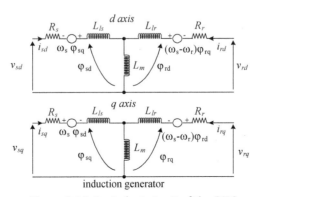

Figure 9.34 Equivalent circuit of the DFIG.

The stator and rotor fluxes are expressed by:

$$\begin{cases} \varphi_{sd} = L_s i_{sd} + L_m i_{rd} \\ \varphi_{sq} = L_s i_{sq} + L_m i_{rq} \end{cases} \tag{9.45}$$

$$\begin{cases} \varphi_{rd} = L_r i_{rd} + L_m i_{sd} \\ \varphi_{rq} = L_r i_{rq} + L_m i_{sq} \end{cases} \tag{9.46}$$

The final model of the stator flux and the rotor currents is given by the following system of equations:

$$\begin{cases} \dfrac{d\varphi_{sd}}{dt} = -\dfrac{1}{T_s}\varphi_{sd} + \omega_s \varphi_{sq} + \dfrac{L_m}{T_s} i_{rd} + v_{sd} \\[2mm] \dfrac{d\varphi_{sd}}{dt} = -\omega_s \varphi_{sd} - \dfrac{1}{T_s}\varphi_{sq} + \dfrac{L_m}{T_s} i_{rq} + v_{sq} \\[2mm] \dfrac{di_{rd}}{dt} = \dfrac{(1-\sigma)}{L_m T_s \sigma}\varphi_{sd} - \dfrac{(1-\sigma)}{L_m \sigma}\omega_g \varphi_{sq} - \left(\dfrac{1}{T_r \sigma} + \dfrac{(1-\sigma)}{T_s \sigma}\right) i_{rd} + \omega_r i_{rq} - \dfrac{(1-\sigma)}{L_m} v_{sd} + \dfrac{1}{L_r \sigma} v_{rd} \\[2mm] \dfrac{di_{rd}}{dt} = \dfrac{(1-\sigma)}{L_m \sigma}\omega_g \varphi_{sd} + \dfrac{(1-\sigma)}{L_m T_s \sigma}\varphi_{sq} - \omega_r i_{rd} - \left(\dfrac{1}{T_r \sigma} + \dfrac{(1-\sigma)}{T_s \sigma}\right) i_{rq} - \dfrac{(1-\sigma)}{L_m} v_{sq} + \dfrac{1}{L_r \sigma} v_{rq} \end{cases} \tag{9.47}$$

The active and reactive powers of the stator and rotor are defined as follows:

$$\begin{cases} P_s = \dfrac{3}{2}\left(v_{sd}\, i_{sd} + v_{sq}\, i_{sq}\right) \\[3mm] Q_s = \dfrac{3}{2}\left(v_{sq}\, i_{sd} - v_{sd}\, i_{sq}\right) \end{cases} \tag{9.48}$$

$$\begin{cases} P_r = \dfrac{3}{2}\left(v_{rd}\, i_{rd} + v_{rq}\, i_{rq}\right) \\[3mm] Q_r = \dfrac{3}{2}\left(v_{rq}\, i_{rd} - v_{rd}\, i_{rq}\right) \end{cases} \tag{9.49}$$

The mechanical and electromagnetic torque relationship is given by:

$$T_{em} - T_m = f\Omega_g + J\dfrac{d\Omega_g}{dt} \tag{9.50}$$

with

$$T_{em} = -p\frac{L_m}{L_s}\left(\varphi_{sq}\,i_{rd} + \varphi_{sd}\,i_{rq}\right) \tag{9.51}$$

Fig. 9.35 shows the responses of direct (i_{rd}) and quadratic (i_{rq}) rotor currents, and the stator active (P_s) and reactive (Q_s) powers of the DFIG. The direct component i_{rd} controls the active power P_s. The reactive power Q_s is controlled via the quadratic component of the current i_{rq}. The reference reactive power is kept zero for this simulation.

Initially, the stator power before the turbine started is zero. Once the turbine is started, the DFIG supplies -1.2 and -1.5 MW of active power for wind speeds of 10 and 11 m/s, respectively. For the rated power $P_s = 1.5$ MW the direct current reaches a value of 1.5 kA.

The waveforms of the voltage v_{dc}, the modulation index (MI), and the three-phase rotor current i_{rabc} are illustrated in Fig. 9.36. One can notice that the DC-link voltage follows the reference 2 kV after a short transient.

Fig. 9.37 shows the electromagnetic torque T_{em}, the mechanical torque T_m, the angular velocity ω_r of the rotor, the tip speed ratio λ, and the power coefficient C_p. It is noted that the power coefficient C_p reaches its optimal value of 0.48 after a short transient. Similarly, the tip speed ratio λ reaches a maximum value of 8.1.

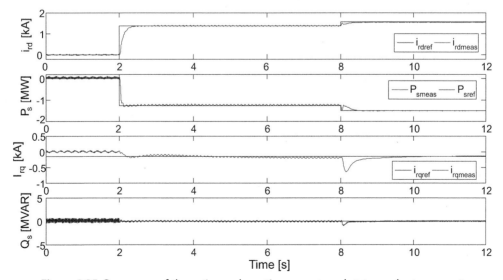

Figure 9.35 Responses of the active and reactive powers, and stator and rotor currents.

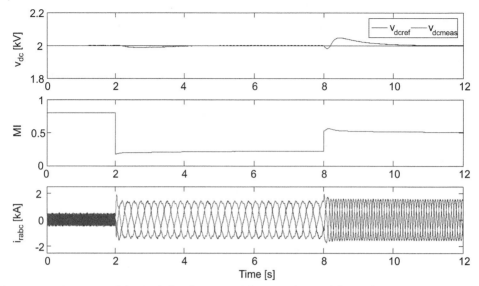

Figure 9.36 Response of the DC-link voltage, modulation index, and three-phase rotor currents.

9.4 CONVERTER TOPOLOGIES AND MODULATION TECHNIQUES

The back-to-back (AC/DC/AC) converters form the grid-side converter (inverter) and the generator-side converter (rectifier). The converter transforms the DC input voltage into a symmetrical AC output voltage of desired magnitude and frequency. The output voltage can be constant or variable at a fixed or variable frequency. A variable output voltage can be obtained by varying the input DC voltage and maintaining the gain of converters constant. The output voltage waveforms of ideal converters should be sinusoidal. But, the waveforms of realistic converters are nonsinusoidal and contain certain harmonics [57,58].

9.4.1 Two-Level Topology

The circuit diagram of a three-phase inverter is shown in Fig. 9.38. It consists mainly of six power switches. V_{dc} is divided into two source voltages $V_{dc}/2$ creating a fictitious neutral point O.

Several modulation techniques exist for the control of this converter and the most popular is the control by PWM. The PWM techniques used in this chapter to control the switches of such converter are: (1) sine PWM (SPWM), (2) PWM selective harmonic elimination programmed (SHE-PWM), and (3) space vector (SVPWM).

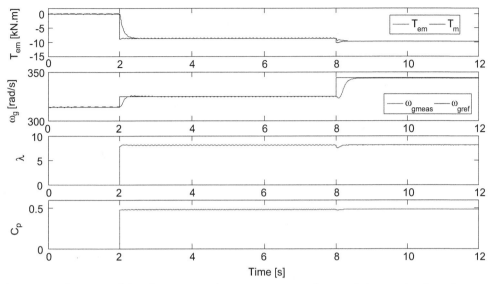

Figure 9.37 Response of the electromagnetic torque, rotor angular speed, tip speed ratio, and power coefficient.

The inverter line-to-line voltages are:

$$\begin{cases} V_{ab} = V_{aO} - V_{bO} = V_{dc}(S_a - S_b) \\ V_{bc} = V_{bO} - V_{cO} = V_{dc}(S_b - S_c) \\ V_{ca} = V_{cO} - V_{aO} = V_{dc}(S_c - S_a) \end{cases} \qquad (9.52)$$

V_{ao}, V_{bo}, and V_{co} being the DC input voltages of the inverter. They are referenced to a midpoint "O" of a shadow input divider. To evaluate the inverter phase voltage (V_{an}, V_{bn}, V_{cn}), the voltage V_{no} is calculated as:

$$V_{nO} = \frac{1}{3}(V_{aO} + V_{bO} + V_{cO}) \qquad (9.53)$$

The phase voltages are obtained as:

$$\begin{cases} v_{an} = V_{aO} - V_{nO} \\ v_{bn} = V_{bO} - V_{nO} \\ v_{cn} = V_{cO} - V_{nO} \end{cases} \qquad (9.54)$$

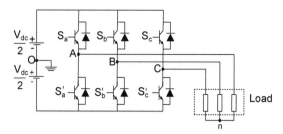

Figure 9.38 Topology of a two-level inverter.

Finally, the model of the inverter is:

$$
\begin{bmatrix} v_{an} \\ v_{bn} \\ v_{cn} \end{bmatrix} = \frac{1}{3} \begin{bmatrix} 2 & -1 & -1 \\ -1 & 2 & -1 \\ -1 & -1 & 2 \end{bmatrix} \begin{bmatrix} V_{aO} \\ V_{bO} \\ V_{cO} \end{bmatrix} = \frac{V_{dc}}{3} \begin{bmatrix} 2 & -1 & -1 \\ -1 & 2 & -1 \\ -1 & -1 & 2 \end{bmatrix} \begin{bmatrix} S_a \\ S_b \\ S_c \end{bmatrix} \qquad (9.55)
$$

9.4.1.1 Sine PWM

In this case, the reference signal is sinusoidal and the voltage output waveform of the inverter is a variable pulse width train as shown in Fig. 9.39. The switching instants are determined by the intersection between the carrier and the modulating signals. The switching frequency is dictated by the carrier. This PWM method is the most widely used as it has proved very effective for the elimination of harmonics.

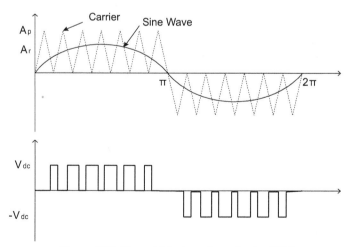

Figure 9.39 Sinusoidal-triangular unipolar PWM.

The essential parameters of the PWM are the modulation index: $m_f = \frac{f_p}{f_r}$, where f_p is the modulation frequency (carrier), f_r is the fundamental frequency of the inverter output voltage, and the control factor or amplitude modulation index: $MI = \frac{A_r}{A_p}$, where A_r is the peak value of the fundamental of the desired load voltage and A_p is the amplitude of the output voltage.

Increasing m_f rejects the first nonzero harmonics to higher frequencies which makes filtering easier. However, m_f cannot be increased above certain values due to the constraints on the converter switches commutations and the minimum pulse width.

Fig. 9.40 shows the state S_a of the switch of phase A and the output voltage for $MI = 0.8$ and $m_f = 40$ when the three-phase inputs voltage is sinusoidal with a frequency of 50 Hz and the DC voltage is equal to 500 V. This simulation shows that MI rejects the harmonics of the inverter output voltage to the high frequencies. This reduces the effect of these harmonics and improves the quality of current waveform and makes their filtering easier. The maximum line voltage of the fundamental is $V_{1\max} = (V_{dc}/2)MI\sqrt{3} = (500/2)0.8\sqrt{3} = 346.4\ V$. This is very close to the voltage obtained by Fourier decomposition which is 345.6 V.

9.4.1.2 Programmed or Selective Harmonic Elimination PWM

The SHE-PWM method is a programmed modulation technique based on precalculated switching instants defined by an objective function. The objective function generates a control signal based on the prescribed switching pattern to eliminate a set of selected harmonics according to the requirements of the application. It is unusual to

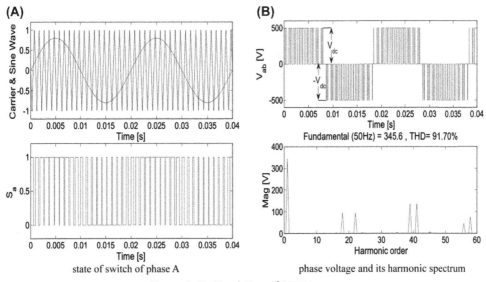

(A)

state of switch of phase A

(B)

Fundamental (50Hz) = 345.6 , THD= 91.70%

phase voltage and its harmonic spectrum

Figure 9.40 Simulation of SPWM.

attempt to minimize the harmonic components present in the spectrum of output voltage of the inverter. The advantages of the programmed PWM are [59]: (1) about 50% reduction of the switching frequency compared to conventional PWM sinusoidal; (2) over modulation is possible; (3) the DC-side current fluctuations are reduced due to the high quality of voltage and output current of the inverter; and (4) reducing the PWM frequency reduces the inverter switching losses and allows the use for high-power converters and also the elimination of lower-order harmonics which eliminates resonance with external passive components.

Generally, only one cycle is used which has a symmetry relative to a quarter of a period and then the other angles are deduced by symmetry. The decomposition in Fourier series of a bipolar PWM signal which is symmetrical relative to quarter period and antisymmetrical to the half period shows the existence of harmonics of odd order. Adjusting the switching angles $\alpha_1, \alpha_2, \alpha_3, \alpha_4,\ldots,\alpha_N$ allows the elimination of harmonics of order $(N-1)$ and the control of the fundamental.

The Fourier transform of a periodic alternating signal is given by:

$$u(t) = \frac{a_0}{2} + \sum_{n=1}^{\infty} a_n \sin(n\omega t) + \sum_{n=1}^{\infty} b_n \cos(n\omega t) \tag{9.56}$$

The Fourier coefficients are given by:

$$a_0 = 1/T \int_0^T u(t)\, dt, \quad a_n = \frac{2}{T} \int_0^T u(t)\cos(n\omega t)dt, \quad b_n = \frac{2}{T} \int_0^T u(t)\sin(n\omega t)\, dt$$

For a periodic signal with a symmetry on the quarter period and an antisymmetry at the half period:

$$a_n = 0 \text{ and } b_n = \frac{4}{\pi} \int_0^{\frac{\pi}{2}} V_s \sin(n\omega t) \tag{9.57}$$

Eq. (9.57) contains K equations with N unknowns. The fundamental can be controlled and $(K-1)$ harmonics can be eliminated.

Therefore,

$$u(t) = \sum_{n=1}^{\infty} \left[\frac{4}{n\pi} \left(1 + 2 \sum_{i=1}^{K} (-1)^i \cos(n\omega t)\sin(n\omega t) \right) \right] \tag{9.58}$$

and the RMS values are:

$$U_1 = \frac{b_1}{\sqrt{2}}, \quad U_3 = \frac{b_3}{\sqrt{2}}, \ldots, U_n = \frac{b_n}{\sqrt{2}} \tag{9.59}$$

Based on the previous equations, the following system of equations is obtained:

$$\begin{cases} U_1 = \dfrac{2\sqrt{2}}{\pi}\left[1 - 2\cos(\alpha_1) + 2\cos(\alpha_2) - 2\cos(\alpha_3) + 2\cos(\alpha_4) - \ldots(-1)^K 2\cos(\alpha_K)\right] \\[2ex] U_3 = \dfrac{2\sqrt{2}}{3\pi}\left[1 - 2\cos(3\alpha_1) + 2\cos(3\alpha_2) - 2\cos(3\alpha_3) + 2\cos(3\alpha_4) - \ldots(-1)^K 2\cos(3\alpha_K)\right] \\[2ex] \qquad\qquad\qquad\vdots \\[2ex] U_n = \dfrac{2\sqrt{2}}{n\pi}\left[1 - 2\cos(n\alpha_1) + 2\cos(n\alpha_2) - 2\cos(n\alpha_3) + 2\cos(n\alpha_4) - \ldots(-1)^K 2\cos(n\alpha_K)\right] \end{cases}$$

$$(9.60)$$

For full-wave bridge circuit the RMS of the fundamental component is:

$$U_{10} = \frac{2\sqrt{2}}{\pi}V_s \tag{9.61}$$

Combining Eqs. (9.59), (9.60), and (9.61) gives the RMS of the output voltages. The solution sought is one that must verify the following condition: $0 < \alpha_1 < \alpha_2 < \ldots < \alpha_n < \pi/2$.

For the elimination of $(K - 1)$ harmonics one has to solve the following equations iteratively using some numerical methods, such as Newton–Raphson:

$$\frac{U_1}{2U_{10}} = \frac{\pi}{4}MI = \left[\frac{1}{2} - \cos(\alpha_1) + \cos(\alpha_2) - \cos(\alpha_3) + \cos(\alpha_4) - \ldots \pm \cos(\alpha_K)\right]$$

$$0 = \left[\frac{1}{2} - \cos(3\alpha_1) + \cos(3\alpha_2) - \cos(3\alpha_3) + \cos(3\alpha_4) - \ldots \pm \cos(3\alpha_K)\right]$$

$$\vdots$$

$$0 = \left[\frac{1}{2} - \cos(n\alpha_1) + \cos(n\alpha_2) - \cos(n\alpha_3) + \cos(n\alpha_4) - \ldots \pm \cos(n\alpha_K)\right]$$

$$(9.62)$$

The programmed PWM lookup table was organized for 4 kbytes ($4096 = 4 \times 1024$) points per cycle per MI. The MI was varied from 0 to 1.15 in steps of 0.01. Each MI value is represented by a table of $\omega_e t$ for a period of 0.02 s with a sampling period of 0.02/ 4096. The solutions obtained by Newton–Raphson method to eliminate 10 harmonics for each MI are stored in an array of size 4 kbytes, that is, 4096 points per cycle.

Two counters were then designed; a horizontal one to detect the *MI* and a vertical counter to detect the instant of time $(\omega_e t + \theta)$ (θ is the phase shift between the fundamental component of the side grid converter voltage and the current of the grid), this will also correspond to the start for a counting period of 0.02 s in steps of 0.02/4096. Once the *MI* is found, the pointer of the second counter is fixed on the table corresponding to the *MI* obtained by the first counter from the instant $(\omega_e t + \theta)$. Fig. 9.41 illustrates the PWM generation principle for phase A; the other phases B and C are generated in the same fashion and are shifted by ±120 degrees, respectively [60,61].

In the simulation, a two-level inverter supplied with 500 V DC is used to generate a line phase voltage output of 439 V at 50 Hz. Fig. 9.42 (A) shows the 14 firing angles of the three-phase two-level inverter. The spectral decomposition is shown in Fig. 9.42 (B) where the result were obtained with $MI = 1$.

The results clearly show the behavior of a programmed PWM: (1) there is a fundamental modulation frequency, with an amplitude directly proportional to the modulation index *MI*; (2) the low harmonics are almost nonexistent; and (3) the 14 firing angles (stored in lookup table) are adjusted to eliminate the first 13 harmonics with the fundamental control angle of the voltage.

9.4.1.3 *Space Vector PWM*

For a two-level voltage source inverter (VSI), there are eight possible voltage vectors that can be represented in space vector as shown in Fig. 9.43. The $\alpha-\beta$ plane is divided into six sectors of 60 degrees angle each and are defined in Table 9.1.

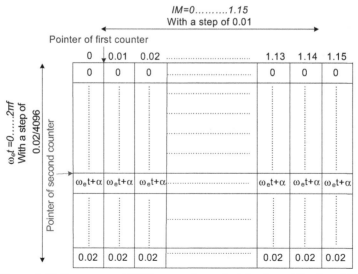

Figure 9.41 Programmed PWM generation principles for a $MI = 0,\dots,1.15$.

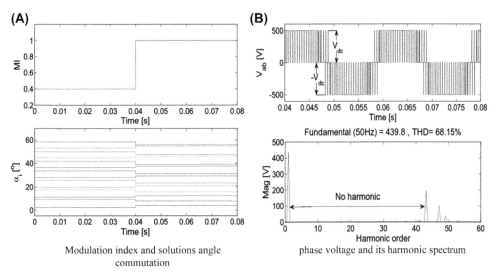

Modulation index and solutions angle commutation

phase voltage and its harmonic spectrum

Figure 9.42 Voltage output waveform based on SHE-PWM.

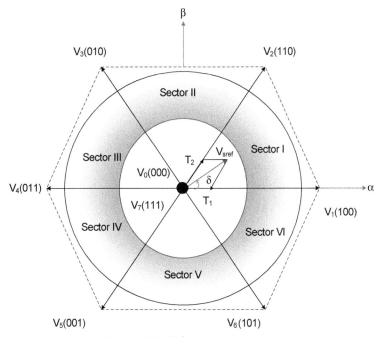

Figure 9.43 Voltage space vectors.

Where V_1 to V_6 represent the active voltage vectors and V_0 and V_7 represent the zero vectors.

Table 9.1 Space Vector Sectors

If Angle δ is	Then V_{sref} Is in Sector:
0 degree $\leq \delta < 60$ degrees	I
60 degrees $\leq \delta < 120$ degrees	II
120 degrees $\leq \delta < 180$ degrees	III
-180 degrees $\leq \delta < -120$ degrees	IV
-120 degrees $\leq \delta < -60$ degrees	V
-60 degrees $\leq \delta < 0$ degrees	VI

SVPWM seeks to average the adjacent vectors for each sector to produce a vector that transitions smoothly between sectors and thus provide sinusoidal line-to-line voltages. Assuming balanced three-phase inverter voltages:

$$v_a(t) + v_b(t) + v_c(t) = 0 \tag{9.63}$$

where $v_a(t)$, $v_b(t)$, and $v_c(t)$ are the instantaneous voltages of three-phase load.

The *abc* to $\alpha\beta$ transformation is defined by:

$$\begin{bmatrix} v_\alpha(t) \\ v_\beta(t) \end{bmatrix} = \frac{2}{3} \begin{bmatrix} 1 & \frac{-1}{2} & \frac{-1}{2} \\ 0 & \frac{\sqrt{3}}{2} & \frac{-\sqrt{3}}{2} \end{bmatrix} \begin{bmatrix} v_a(t) \\ v_b(t) \\ v_c(t) \end{bmatrix} \tag{9.64}$$

A space vector is expressed as:

$$\overrightarrow{v}(t) = v_\alpha(t) + jv_\beta(t) \tag{9.65}$$

Substituting Eq. (9.63) into Eq. (9.64) gives:

$$\overrightarrow{v}(t) = \frac{2}{3}[v_a(t)e^{j0} + v_b(t)e^{j2\pi/3} + v_c(t)e^{j4\pi/3}] \tag{9.66}$$

The six active vectors can be derived as [26]:

$$\overrightarrow{V}_k = \frac{2}{3}V_{dc}e^{\frac{j(k-1)\pi}{3}}k = 1 \div 6 \tag{9.67}$$

Any given reference voltage V_{ref} can be synthesized by selecting two adjacent voltage vectors and zero voltage vectors. Within a sampling period T_s, the reference voltage V_{ref} is generated by combining vector V_1 (to the right), vector V_2 (to the left), and a vector zero (either V_0 or V_7). The voltage V_{ref} can be approximated by applying V_1 for a period

T_1, V_2 for a period T_2, and (V_0 or V_7) for the remaining period T_0 of the sampling period [27,28]:

$$T_1 = \frac{2\sqrt{3}}{\pi} MI \cdot \sin(60° - \delta) \cdot T_s \tag{9.68}$$

$$T_2 = \frac{2\sqrt{3}}{\pi} MI \cdot \sin(\delta) \cdot T_s \tag{9.69}$$

$$T_0 = T_1 + T_2 + T_0 \tag{9.70}$$

where MI is the modulation index and defined by $MI = \pi V_{ref}/2V_{dc}$.

Sectors are determined using the algorithm of Fig. 9.44.

Fig. 9.43 shows the waveforms of the phase angle, the sector related to the displacement of the voltage vector in the space, and the voltages of both phases A and B.

The reference voltage vector moves in the space by an angle δ which varies from 0 to 360 degrees (or from 180 to −180 degrees as shown in Fig. 9.45). The determination of

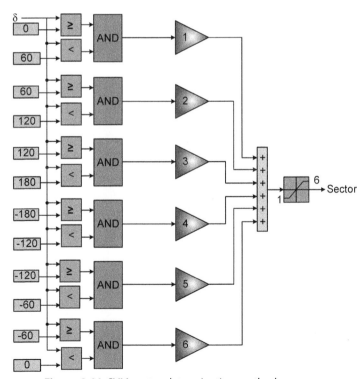

Figure 9.44 SVM sector determination method.

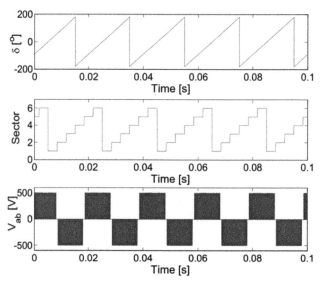

Figure 9.45 Phase shift, sector (N), and output voltage of the converter.

the sector is based on the knowledge of angle δ and the use of the algorithm of Fig. 9.44. The converter output voltage is obtained by the switching sequences shown in Fig. 9.43.

The maximum reference voltage $V_{PWM,\,S}$ (voltage of the sinusoidal PWM) can be defined by $v_{dc}/2$. However, the maximum of reference voltage $V_{PWM,\,V}$ (voltage of the space vector PWM) corresponding to the angle 30 degrees is given by $v_{dc}/\sqrt{3}$. This means that for a given DC-link voltage v_{dc}, vector PWM can produce a voltage reference $2/\sqrt{3}$. Details of the calculation of these values are given in Appendix B.

It can be shown that M_c, the index that measures the ability of the inverter and MI the amplitude modulation index are given by:

$$M_c = \frac{V_{PWM}}{V_\pi} MI = \frac{V_m}{V_p} \tag{9.71}$$

where V_{PWM} is the PWM voltage, V_π is the full-wave voltage, V_m is the modulating voltage, and V_p is the carrier voltage.

For a sinusoidal PWM, the maximum capacity index $M_{c,max}$ corresponding to $MI = 1$ is obtained by:

$$M_{c,\max} = \frac{V_{PWM,S}}{V_\pi} = \frac{v_{dc}/2}{2v_{dc}/\pi} = \frac{\pi}{4} = 0.7854 \tag{9.72}$$

The fundamental of the sinusoidal PWM output voltage can therefore reach 78.54% of the driving voltage ($v_{dc}/2$).

However, the maximum capacity index $M_{c,\max}$ for vector PWM is defined by:

$$M_{c,\max} = \frac{V_{PWM,V}}{V_\pi} = \frac{v_{dc}/\sqrt{3}}{2v_{dc}/\pi} = \frac{\pi}{2\sqrt{3}} = 0.907 \qquad (9.73)$$

Fig. 9.46 gives a plot of the capacity index of the inverter as a function of the amplitude modulation index for areas of modulation and overmodulation.

Four regions are considered:

- **Zone I**: it is a linearity zone with a sinusoidal reference voltage ($MI_{\max} = 1$, $M_{c,\max} = 0.785$). In this area, the fundamental of the inverter phase voltage V_1 increases linearly with the magnitude of the reference voltage (modulating signal $V_{1,ref}$). If the peak of $V_{1,ref}$ exceeds $v_{dc}/2$ ($MI > 1$), the system operation is no longer linear. $MI > 1$ corresponds to 78% of maximum voltage with a two-level voltage inverter.

- **Zone II**: this is an extension of the linear zone I obtained by adding harmonic of order 3 and its multiples (zero sequence—zero voltage) to the modulating sinusoid. This leads to a reference voltage fundamental with $MI_{\max} = 1.1547$. The voltage vector magnitude increases linearly from $v_{dc}/2$ to $v_{dc}/\sqrt{3}$. The capacity index reaches $M_{c,\max} = 0.907$. This gives an increase of 15.47% in the gain.

- **Zone III**: this is the overmodulation region where the current and voltage waveforms are deteriorated. For this reason a linear region is preferred to an overmodulation region.

- **Zone IV**: this is the full-wave operating area (firing angle of 180degrees or "six-step").

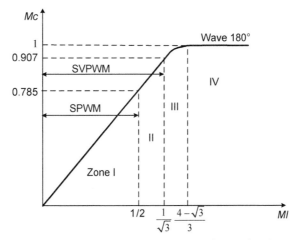

Figure 9.46 Modulation and overmodulation zones of a two-level inverter voltage.

The advantages of SVPWM compared to SPWM are [61—63]: (1) SVPWM minimizes voltage harmonics and increases the inverter efficiency by 15.47% compared to SPWM, (2) the magnitude of the line voltage can reach the value v_{dc}, hence it is possible to use the maximum of the input voltage in the linear operating region, (3) it is possible with the control of a single reference voltage vector to create three sine waves, (4) the real–time implementation of SVPWM is easier, (5) flexibility of selecting inactive states (state 0) and their distribution in sample period leads to two degrees of freedom, and (6) since the reference voltage vector is a two-dimensional quantity (v_α and v_β), it is possible to implement the SVPWM with the existing advanced vector control techniques for AC machines.

9.4.2 Three-Level Topology

In the structure of the NPC three-level inverter, the DC bus was split into two parts using a capacitor divider bridge as shown in Fig. 9.47. When the bridge is composed of two capacitors, the voltages across the capacitors are naturally balanced for a modulation with a modulating signal with zero mean value.

In Fig. 9.47 the positive state is denoted by P (defined by 1), the negative state is denoted by N (defined by −1), and the zero state is denoted by O (set 0).

PWM methods used in multilevel inverters are classified into two large families: fundamental switching frequency and high switching frequency PWM [64]. These methods are presented in Fig. 9.48 [65,66].

The classical PWM technique, sinusoidal carrier-based PWM or sine PWM (SPWM), is the most widely used modulation method in multilevel inverters for reducing harmonics in the load voltage [67—70]. Another interesting alternative is the space vector modulation (SVM) approach, which has been used in three-level inverters [48].

The key criteria to consider when choosing a modulation method are: (1) reduction of load current harmonics and switching frequency, (2) regular switching frequency for the entire switching strategy, and (3) DC link—capacitor voltage balancing.

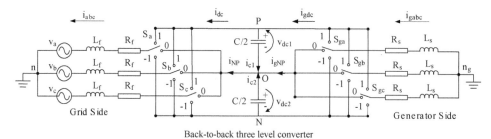

Back-to-back three level converter

Figure 9.47 Three-phase three-level NPC back-to-back converter topology.

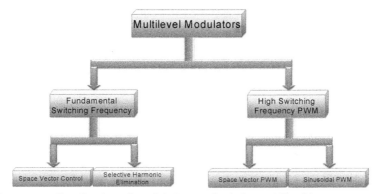

Figure 9.48 Classification of modulation methods for multilevel inverters.

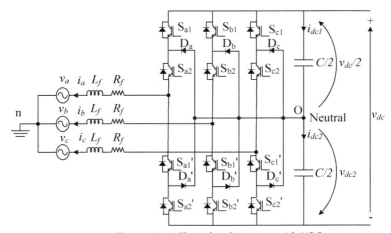

Figure 9.49 Three-level inverter with NPC.

The basic circuit diagram of a three-level NPC inverter topology is shown in Fig. 9.49.

The voltages at nodes A, B, and C of the three-phase converter, based on the middle point O are given by:

$$
\begin{cases}
V_{ao} = S_{a1} S_{a2} v_{dc1} - S'_{a1} S'_{a2} v_{dc2} \\
V_{bo} = S_{b1} S_{b2} v_{dc1} - S'_{b1} S'_{b2} v_{dc2} \\
V_{co} = S_{c1} S_{c2} v_{dc1} - S'_{c1} S'_{c2} v_{dc2}
\end{cases}
\tag{9.74}
$$

From Eq. (9.74), it can be noticed that the converter corresponds to the series connection of two two-level converters, one consisting of the upper half-arm and supplied by v_{dc1} and the other formed of the lower half-arm and supplied by $-v_{dc2}$.

The different line-to-line voltages of three-phase converter are expressed as follows:

$$
\begin{cases}
V_{ab} = V_{aO} - V_{bO} = (S_{a1}S_{a2} - S_{b1}S_{b2})v_{dc1} - (S'_{a1}S'_{a2} - S'_{b1}S'_{b2})v_{dc2} \\
V_{bc} = V_{bO} - V_{cO} = (S_{b1}S_{b2} - S_{c1}S_{c2})v_{dc1} - (S'_{b1}S'_{b2} - S'_{c1}S'_{c2})v_{dc2} \\
V_{ca} = V_{cO} - V_{aO} = (S_{c1}S_{c2} - S_{a1}S_{a2})v_{dc1} - (S'_{c1}S'_{c2} - S'_{a1}S'_{a2})v_{dc2}
\end{cases}
\tag{9.75}
$$

With $v_{dc1} = v_{dc2} = v_c$, Eq. (9.75) becomes:

$$
\begin{cases}
V_{ab} = V_{aO} - V_{bO} = [(S_{a1}S_{a2} - S_{b1}S_{b2}) - (S'_{a1}S'_{a2} - S'_{b1}S'_{b2})]v_c \\
V_{bc} = V_{bO} - V_{cO} = [(S_{b1}S_{b2} - S_{c1}S_{c2}) - (S'_{b1}S'_{b2} - S'_{c1}S'_{c2})]v_c \\
V_{ca} = V_{cO} - V_{aO} = [(S_{c1}S_{c2} - S_{a1}S_{a2}) - (S'_{c1}S'_{c2} - S'_{a1}S'_{a2})]v_c
\end{cases}
\tag{9.76}
$$

The phase voltages are related to the line voltages by the following relationships:

$$
\begin{cases}
v_{an} = \dfrac{1}{3}(V_{ab} - V_{ac}) \\[2mm]
v_{bn} = \dfrac{1}{3}(V_{bc} - V_{ab}) \\[2mm]
v_{cn} = \dfrac{1}{3}(V_{ca} - V_{bc})
\end{cases}
\tag{9.77}
$$

The equations relating the DC-side currents i_{dc1} and i_{dc2} of the inverter and the current i_a, i_b, and i_c of the AC-side are:

$$
\begin{cases}
i_{dc1} = S_{a1}S_{a2}i_a + S_{b1}S_{b2}i_b + S_{c1}S_{c2}i_c \\
i_{dc1} = S'_{a1}S'_{a2}i_a + S'_{b1}S'_{b2}i_b + S'_{c1}S'_{c2}i_c
\end{cases}
\tag{9.78}
$$

The final model is given by:

$$
\begin{bmatrix} v_{an} \\ v_{bn} \\ v_{cn} \end{bmatrix} = \frac{1}{3} \begin{bmatrix} 2 & -1 & -1 \\ -1 & 2 & -1 \\ -1 & -1 & 2 \end{bmatrix} \left\{ \begin{bmatrix} S_{a1}S_{a2} \\ S_{b1}S_{b2} \\ S_{c1}S_{c2} \end{bmatrix} v_{dc1} - \begin{bmatrix} S'_{a1}S'_{a2} \\ S'_{b1}S'_{b2} \\ S'_{c1}S'_{c2} \end{bmatrix} v_{dc2} \right\}
\tag{9.79}
$$

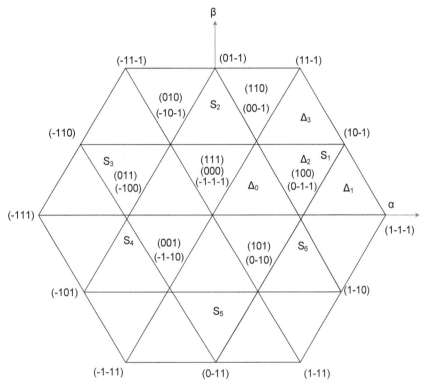

Figure 9.50 Space vector diagram of three-level inverter.

9.4.2.1 DC Link Capacitor Voltage Balancing Method

The commutation of the inverter in normal operation produces a current flow through the neutral n, for all different operating conditions. This current flow causes an unbalance (charge and discharge of the DC-side of capacitors) in the DC voltage across each capacitor branch $C/2$. Therefore, the converter requires a special modulation algorithm to balance the DC-side voltage [71].

One way to achieve this is to use the SVM [72]. The imbalance in the DC link caused by internal vectors can be canceled out using the same vectors and this can be done by the symmetrical SVM technique. The method is advantageous since it can be applied to various structures with only minor adjustments in the control strategy. Balancing is achieved by using different redundant states for the same point in the $\alpha-\beta$ plane.

Fig. 9.50 shows the space vector configuration for three-level inverter. The space vector diagram for a three-level inverter is divided into six sectors (S_1, S_2, S_3, S_4, S_5, and S_6) and each sector is divided to four regions ($\Delta_0, \Delta_1, \Delta_2, \Delta_3$) [73]. There is a total of $3^3 = 27$ possible states for the three-level inverter switches.

Each sector of the external hexagon of the three-level SVM diagram is divided into four smaller triangles, arranged as shown in Fig. 9.51. All voltage vectors can be divided

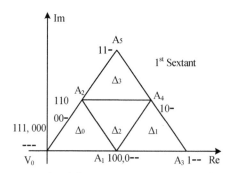

Figure 9.51 Normalized first sextant and voltage reference vector.

into four groups: the *zero voltage vectors* (000,111,-1-1-1), the *small voltage vectors* (110,00-1,…,0-1-1), the *medium voltage vectors* (10-1,01-1…,1-10), and *large voltage vectors* (1-1-1,11-1,…,1-11). All vectors are zero at the center of the hexagon, and all the other voltage vectors (24 active voltage vectors) are found at the corners of each triangle of the hexagon. All possible states are shown in Fig. 9.51.

Each of these triangles is then considered as a sector of a hexagon with two levels, with the same original redundancy. Recall that, the principle of SVPWM is to approximate the voltage control vector using three adjacent vectors. The following two equations should be satisfied for SVPWM for conventional two-level inverters [74–77]:

$$T_1 V_1 + T_2 V_2 + T_3 V_3 = T_s V_{ref}$$

$$T_1 + T_2 + T_3 = T_s$$

(9.80)

where V_1, V_2, and V_3 are vectors that define the area of triangle where the reference voltage V_{ref} is located. T_1, T_2, and T_3 are the respective time courses of these vectors. The three-level inverter is treated like a two-level inverter, each vector space diagram is divided into six sectors. The generation of the switching pattern is presented only for sector I as the calculations are similar for other sectors. Sector I is divided into four regions as shown in Fig. 9.51 where all possible switching states for each region are given as well.

To illustrate the voltage balancing strategy in the capacitors by SVM, the first sextant is taken as example. There are [78,79]:

- Large vectors: V_1 (1-1-1) or (PNN) and V_2 (11-1) or (PPN) are vectors with the highest amplitude. They do not connect to AC–side at the neutral point, so they do not contribute to the charge or discharge of the DC-side capacitors.
- Medium vectors: The amplitude of V_{12} (10-1) or (PON) defines the maximum attainable voltage output vector. It is adjusted to the AC-side at the neutral point, so it contributes to the charge and discharge of the DC bus capacitors.
- Small vectors: V_{10} (100) or (POO), V_{10} (0-1-1) or (PNN), V_{20} (110) or (PPO), and V_{20} (00-1) or (OON) have half the amplitude of large vectors. They charge and

discharge the capacitors. They appear in pairs, hence two different switching combinations generate the same voltage vector in the inverter output. The main advantage of using these vectors is to ensure a balancing of the voltages of DC-side capacitors.

- Zero vectors: V_0 (000) or (OOO), V_0 (111) or (PPP), and V_0 (-1-1-1) or (NNN) are used to produce a voltage to abort the current in the neutral point n.

For proper operation of DC-link capacitors, the two capacitor voltages should be equal (Fig. 9.52). Any deviation in one of these voltages will affect the currents and voltages in the same way the charge and discharge of the DC-link capacitors do.

9.4.2.2 SVPWM-Based Three-level Converter

The topology of a three-level neutral point clamped (NPC) voltage source converters (VSC) for both generator-side and grid-side AC/DC/AC converters is shown in Fig. 9.53.

The control blocks (either generator- or grid-side) outputs are obtained via Park voltages $v_{d,q}$ in the $d-q$ rotating frame. The control algorithm of the PWM vector technique is given by Park transformation of the voltages $v_{d,q}$ to a fixed frame $\alpha-\beta$ using the following matrix relationship:

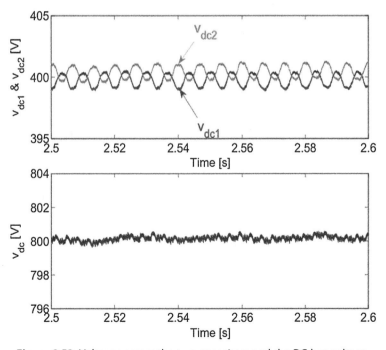

Figure 9.52 Voltages across the two capacitors and the DC bus voltage.

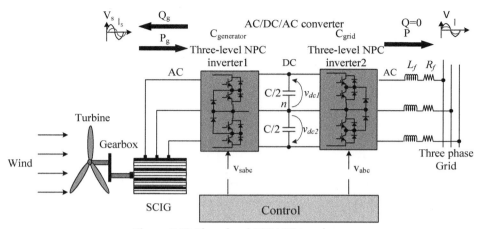

Figure 9.53 Three-level NPC VSC topology.

$$\begin{bmatrix} v_\alpha \\ v_\beta \end{bmatrix} = \begin{bmatrix} \cos(\delta_e) & -\sin(\delta_e) \\ \sin(\delta_e) & \cos(\delta_e) \end{bmatrix} \begin{bmatrix} v_d \\ v_q \end{bmatrix} \tag{9.81}$$

where δ_e is the electrical angle or the phase shift between the reference voltage V_{ref} and the α-axis voltage. δ is the angle of the grid-side voltage and δ_g is the angle of the generator-side voltage. These angles are given by:

$$\begin{cases} \delta = \tan^{-1}\left(\dfrac{v_\beta}{v_\alpha}\right) \\[2mm] \delta_g = \tan^{-1}\left(\dfrac{v_{s\beta}}{v_{s\alpha}}\right) \end{cases} \tag{9.82}$$

SVPWM for three-level inverters can be applied using the following steps: (1) determination of the sector, (2) determination of the region in the sector, (3) calculation of the switching times T_a, T_b, and T_c, and (4) determination of the switching states.

The first step is to determine the sector from δ or δ_g using the same SVPWM algorithm for the two-level inverter. The second step is to determine the region in sector I. The duty cycles or the projections of the normalized reference space vectors can be calculated as follows [54]:

$$a = m_2 = \frac{b}{\sin(\pi/3)} = \frac{2}{\sqrt{3}}b = \frac{2}{\sqrt{3}}m_n\sin\delta_n \tag{9.83}$$

$$m_1 = m_n \cos(\delta_n) - \left[\frac{2}{\sqrt{3}}m_n \sin(\delta_n)\right]\cos\left(\frac{\pi}{3}\right) \tag{9.84}$$

$$m_1 = m_n \left[\cos(\delta_n) - \frac{\sin(\delta_n)}{\sqrt{3}}\right] \tag{9.85}$$

where δ_n is the equivalent angle of the voltage space vector in the first sextant (Table 9.3), m_n is the normalized voltage amplitude of the reference space vector, and in the case of three-level inverter it is defined by:

$$m_n = \begin{cases} \dfrac{|\overrightarrow{v}_{ref}|^*}{v_{dc}/3} & forgridside \\[4mm] \dfrac{|\overrightarrow{v}_{refs}|^*}{v_{dc}/3} & forgeneratorside \end{cases} \tag{9.86}$$

where $|\overrightarrow{v}_{ref}|^*$ and $|\overrightarrow{v}_{refs}|^*$ are, respectively, the reference vector commands of the generator side and grid side.

Overmodulation is obtained if the normalized reference vector assumes lengths longer than $\sqrt{3}$ for some positions of this vector; however it can never be outside of the hexagon.

When:

$$\begin{cases} |\overrightarrow{v}_{ref}|^* = V_{ref}e^{j\delta} \\ |\overrightarrow{v}_{refs}|^* = V_{refs}e^{j\delta_s} \end{cases} \tag{9.87}$$

V_{refs} and V_{ref} represent, respectively, the magnitude of the reference voltage of the generator side and grid side and are calculated using:

$$\begin{cases} V_{ref} = \sqrt{v_\alpha^2 + v_\beta^2} \\ V_{refs} = \sqrt{v_{\alpha s}^2 + v_{\beta s}^2} \end{cases} \tag{9.88}$$

As the references voltages (V_{ref}, V_{refs}) rotate, $0 \le MI \le 0.907$ (where MI is defined by $MI = \pi V_{ref}/2V_{dc}$) remains in the hexagon corresponding to the region of linear operation. Table 9.2 summarizes the information needed to identify the region where the reference vector lies in the first sextant.

Thus, considering the hexagon symmetry, the times are calculated using vector space equivalent (V_{ref}) in the first sextant of the hexagon. Fig. 9.54 and Table 9.3 present the angle of the equivalent vector v_{eq} in sextant I. The vectors in the other sextants (II, III, IV,

Table 9.2 Determination of the Regions of the
Reference Vector

Case	Region
$(m_1 + m_2) \leq 0.5$	Δ_0
$m_1 > 0.5$	Δ_1
$m_2 > 0.5$	Δ_2
$(m_1 + m_2) > 0.5$	Δ_3

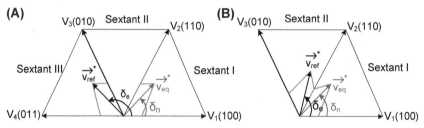

Figure 9.54 Equivalent space vector in sextant I. (A) When the space vector is in sextant II and (B) when the space vector is in sextant III.

Table 9.3 Equivalent Angle δ_n in the First Sextant $\delta_n \rightarrow (0, \pi/3)$

	Sextant					
Sector	1	2	3	4	5	6
δ_e	$(0, \pi/3)$	$(\pi/3, 2\pi/3)$	$(2\pi/3, \pi)$	$(\pi, 4\pi/3)$	$(4\pi/3, 5\pi/3)$	$(5\pi/3, 2\pi)$
δ_n	δ_e	$\delta_e - \pi/3$	$\delta_e - 2\pi/3$	$\delta_e + \pi$	$\delta_e + 2\pi/3$	$\delta_e + \pi/3$

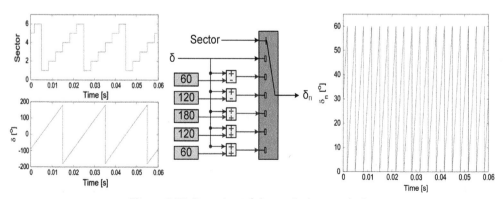

Figure 9.55 Detection of the equivalent angle δ_n.

Table 9.4 Switching Times for Sector I

Region	T_a	T_b	T_c
Δ_0	$1.1MIT_s\sin[(\pi/3)-\delta]$	$T_s/2[1-(2.2\sin(\pi/3))])$	$1.1\,T_s\sin\delta$
Δ_1	$T_s[1-1.1MI\sin(\delta+\pi/3)]$	$1.1\,T_sMI\sin\delta$	$T_s/2[(2.2MI\sin(\pi/3-\delta)])$
Δ_2	$T_s/2[1-2.2MI\sin\delta]$	$T_s/2[2.2MI\sin(\pi/3+\delta)]$	$T_s/2[1+2.2MI\sin(\delta-\pi/3)]$
Δ_3	$T_s/2[2.2MI\sin(\delta)-1]$	$1.1MIT_s\sin[(\pi/3)-\delta]$	$T_s[[1-1.1MI\sin(\delta+\pi/3)]$

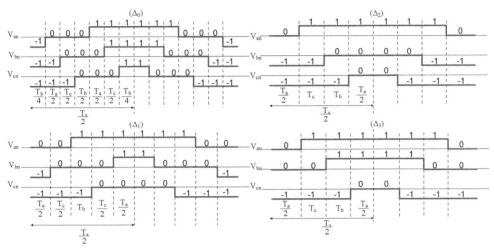

Figure 9.56 Symmetrical SVPWM sequence for selecting voltage vectors for the four regions in sextant I.

V, and VI) are reduced to sextant I to minimize the size of the algorithm which determines the regions (Δ_0, Δ_1, Δ_2, Δ_3).

The calculation of the equivalent angle with Table 9.3 is illustrated in Fig. 9.55.

The switching commands given below are obtained for each region in sector I when all switching states in each region are employed. The switching times T_a, T_b, and T_c for the four regions of sector I are then calculated as shown in Table 9.4.

The modulation index MI varies between 0 and 1 in the linear region. The reference voltage V_{ref} for $MI \leq 0.907$ remains in the hexagon corresponding to the region of linear operation.

The switching commands given in Fig. 9.56 are obtained for each region located in sector 1.

Fig. 9.57 illustrates how the algorithm for generating the pulses for the three-level inverter has been implemented.

In Fig. 9.57 S_{a1}, S_{a2}, S_{b1}, S_{b2}, S_{c1}, and S_{c2} are, respectively, the switches of the upper arms of the three-phase inverter for phases A, B, and C.

Fig. 9.58 shows the waveforms of the sector, angle, equivalent angle, and output line-to-line voltage.

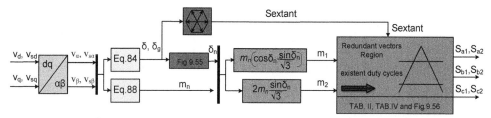

Figure 9.57 Simplified block diagram of the SVPWM technique for the three-level NPC generator-side or grid-side converters.

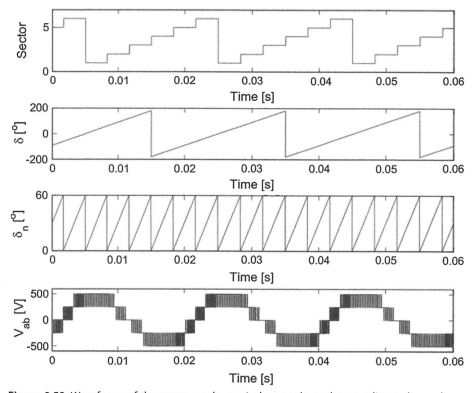

Figure 9.58 Waveforms of the sector, angle, equivalent angle, and output line-to-line voltage.

9.5 CONTROL DESIGN

9.5.1 Proportional–Integral Control Design

In the case of the stator flux orientation in the d–q reference frame, the d-axis component of the stator current i_{sd} is directly proportional to the magnitude of the stator flux φ_s. The d-axis component of the stator current i_{sd} is then directly proportional to the magnitude stator flux. By regulating and keeping constant i_{sd}, the decoupling of the torque control and the flux of the induction machine is achieved. Hence, if the stator flux is oriented along the d-axis, then $\varphi_s = \varphi_{sd}$ and $\varphi_{sq} = 0$.

The transfer functions obtained from this calculation of the flux and torque are then used to design the corresponding PI controllers as depicted in the generic bloc diagram of Fig. 9.59. All these results are summarized in Table 9.5.

In Table 9.5 the stator and rotor time constants are given by: $T_s = \frac{L_s}{R_s}$, $T_r = \frac{L_r}{R_r}$, $\sigma = 1 - \frac{L_m^2}{L_s L_r}$ is the dispersion coefficient, and $\omega_{slip} = \omega_s - \omega_g$ with s being the Laplace operator.

9.5.2 Grid-Side Converter Control

The grid-side converter is used to adjust the voltage of the capacitor and the d-axis and q-axis currents. The control structure of the grid-side converter shown in Fig. 9.60 includes an outer loop to adjust the DC-link voltage. The quadratic component i_q of the grid current is used to modulate the flow of reactive power. Here, the reactive power reference is fixed at zero to achieve a unity power factor. The reference voltage for the DC-side v_{dcref} is generally set larger than the voltage v_d of the grid. For instance, for a grid voltage of $v_s = 380$ or 400V $\left(\text{that is to say } v_d = \sqrt{2/3}v_s\right)$, the reference voltage for the DC-link voltage is set to $v_{dcref} = 800$ V.

A Park transformation is applied to the grid side with a reference linked to the rotating field of the source. With this approach, the d-axis is aligned with the source

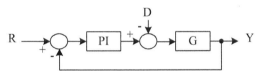

Figure 9.59 PI-based feedback control loop.

Table 9.5 Transfer Functions for the Design of PI Controllers

Regulator of	Reference R	Disturbance D	Transfer Function G	Output Y
Speed	Ω_{ref}	T_m	$\frac{1}{Js+f}$	Ω_g
Torque	T_{em}	ω_g where $\omega_g = p\Omega_g$	$\frac{3p}{2} \times \frac{(1-\sigma)T_r/L_s\varphi_s^2}{1+2\sigma T_r s}$	$\hat{T}_{em}$
Flux	φ_{sref}	$-\frac{\sigma R_s T_r i_{qs}\omega_{slip}}{1+\sigma T_r s}$	$\frac{T_s(1+\sigma T_r s)}{1+(T_s+T_r)s+\sigma T_s T_r s^2}$	$\hat{\varphi}_s$
DC voltage	v_{dcref}	0	$\frac{1}{Cs}$	v_{dc}
Currents Grid side	d-axis i_{dref}	0	$\frac{1}{L_f s}$	i_d
	q-axis i_{qref}	0		i_q

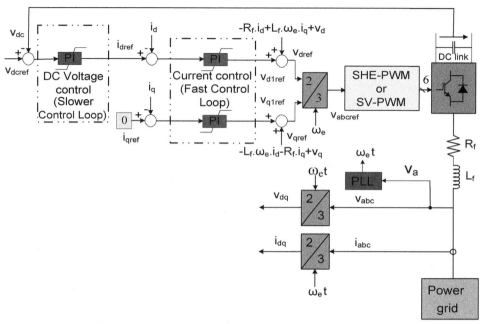

Figure 9.60 Controller system for the DC link and grid side.

voltage vector, and the voltage vector of the q-axis is zero ($v_q = 0$). Under these conditions, the active power will be controlled from the direct current component of the converter side i_d, whereas the reactive power is controlled from the quadratic component of converter current i_q. The difference between the reference and measured voltages of the DC-side ($v_{dcref} - v_{dc}$) provides the input to the PI controller of the DC voltage. The regulator output produces the reference for the direct component of the source current i_{dref}. Choosing $i_{qref} = 0$, the currents and voltages of the source are in phase and the proposed system produces only the real power to the grid.

Two decoupling terms are introduced at the output of the current regulators in order to improve the dynamic performance of the system.

9.5.3 Generator-Side Converter Control

The generator-side converter regulates the speed and torque of the generator and provides reactive power support when required as in the case of SCIG. However, its control is more difficult than other electric machines.

The most popular control methods for the generator-side converter are illustrated in Fig. 9.61.

Scalar control is the simplest and most widely used method in most induction motor drive applications. It gives a reasonably good steady-state speed and torque control. The

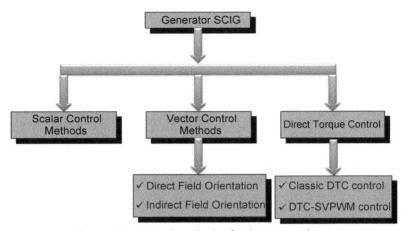

Figure 9.61 Control methods of induction machines.

stator flux and torque are not directly controlled and the parameters of the machine must be known. The speed accuracy is low and the dynamic response is slow. Vector control by rotor flux orientation or DTC are considered as high–performance control methods. The vector control concept proposed by Hasse in 1969 [80] and Blaschke in 1972 [81] produces a decoupled control of torque and flux in the induction machine. However, vector control requires the rotor position angle to maintain the decoupling and also depends on some of the machine parameters. DTC was proposed by Depenbrock and Takahashi [82]. It overcomes the problems associated with vector control since it does not need a position sensor, and the stator resistance of the machine is the only parameter needed to estimate the flux and torque. The objective of DTC is to optimize the control of the inverter switches to ensure the decoupling between the torque and stator flux even under large variations. The weak point is the fluctuations in the torque and flux due to the variation of the switching frequency. Further development of the basic DTC has resulted in various new control strategies. In this work, two DTC control approaches are proposed: an improved version of the classical DTC and DTC-SVPWM.

9.5.3.1 Indirect Field–Oriented Control

Vector control or FOC consists of controlling independently the flux and torque by the d- and q-components of the current, respectively. If the $d-q$ reference frame is linked to the rotating field with an orientation of the rotor flux (d-axis aligned with the direction of the rotor flux), then:

$$\varphi_{rd} = \varphi_r, \ \varphi_{rq} = 0 \tag{9.89}$$

In this approach, the reference axis of the rotating armature is aligned with rotor flux vector, $\varphi_r(\varphi_{rq} = 0)$. Under these conditions, the q-axis component of the stator current

i_{sq} controls the electrical torque of the machine component and the d-axis component of the stator current i_{sd} controls the speed of the generator.

If the flux is kept constant, the developed torque given in Eq. (9.46) becomes:

$$T_{em} = p\frac{3}{2}\frac{L_m}{L_r}\varphi_r i_{sq} \tag{9.90}$$

The speed controller is used to determine the electromagnetic torque reference. The speed may be controlled by a PI controller which will be derived subsequently.

With indirect vector control, the induction machine voltages v_{sd} and v_{sq} in steady state are:

$$\begin{cases} v_{sd} = R_s i_{sd} - \omega_s L_s \sigma i_{sq} \\ v_{sq} = \omega_s L_s i_{sd} + R_s i_{sq} \end{cases} \tag{9.91}$$

with σ being the dispersion coefficient defined as:

$$\sigma = 1 - \frac{L_m^2}{L_s L_r} \tag{9.92}$$

The steady-state value of the flux becomes:

$$\varphi_r = L_m i_{sd} \Rightarrow i_{sd} = \frac{\varphi_r}{L_m} \tag{9.93}$$

Fig. 9.62 shows the block diagram of the rotor IFOC strategy of the variable speed induction generator.

The strategy of indirect vector control of the SCIG is based on the classical PI control design using pole placement technique. The stator flux is maintained at its rated value (φ_{sn}) for the system to operate in the range of the base speed. For speeds higher than the base speed, the flux cannot be maintained constant; it must be reduced to limit the voltage at the machine terminals. Under these conditions, the reference flux is defined by:

$$\varphi_{sref} = \begin{cases} \varphi_n & if \quad |\Omega_g| \leq \Omega_n \\ \dfrac{\Omega_n}{|\Omega_n|}\varphi_n & if \quad |\Omega_g| \geq \Omega_n \end{cases} \tag{9.94}$$

Fig. 9.63 shows the active and reactive powers and phase voltages of the generator. Initially, before the wind turbine is started, the active power reaches a low value representing the iron and copper losses of the generator. Once the wind turbine is started at $t = 0.3\ s$, the generator delivers an active power equal to $-2\ kW$ at a wind velocity of $v = 12\ m/s$. At time $= 0.8\ s$, the wind speed is changed to $v = 10\ m/s$, this results in a fall

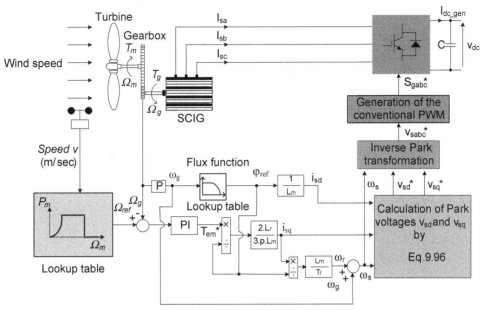

Figure 9.62 Indirect vector control of the induction machine.

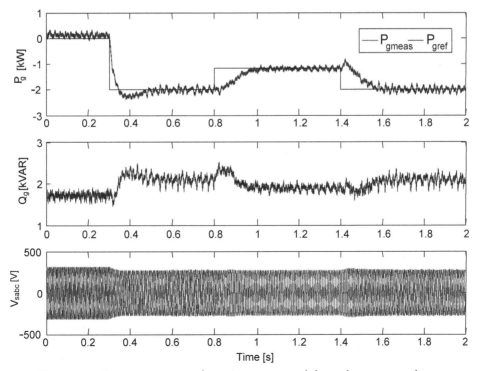

Figure 9.63 Generator active and reactive powers and three-phase stator voltage.

of almost half the active power (i.e., $P_g \approx -1$ kW). The active power returns to its rated value of -2 kW when the wind velocity returns to $v = 12$ m/s at $t = 1.4$ s.

The reference power P_{ref} is obtained via the MPPT strategy according to the wind speed available. Note that, based on the characteristic of the wind turbine, each wind speed corresponds to a given reference power. For example, for a speed of 12 m/s, the reference power is -2 kW.

Fig. 9.64 shows the rotor speed, the electromagnetic torque, and the three-phase stator currents of the generator for the same changes in wind characteristics. In this result, the turbine was subjected to a disturbance simulated as an increase or decrease of the wind speed (see Fig. 9.63).

The voltage waveform and power of the DC side are shown in Fig. 9.65. It can be noted that the DC voltage reference is maintained at 800 V after a short transient despite changes in wind speed from 12 m/s to 10 m/s. The power and current of the DC-side is almost zero in steady state. The responses of the grid active and reactive powers and waveforms of grid phase currents and load are illustrated in Fig. 9.66.

Before the start of the wind turbine, the source must supply an active power to the load and compensate for the losses of the generator. Once the turbine is started for $v = 12$ m/s, the power of the grid is expected to be close to zero and the generator will provide the necessary power (-2 kW, in this case) to the load. After a drop in the wind speed, the generator is able to supply only half of that power thus forcing its source to provide the other half of the power required by the load.

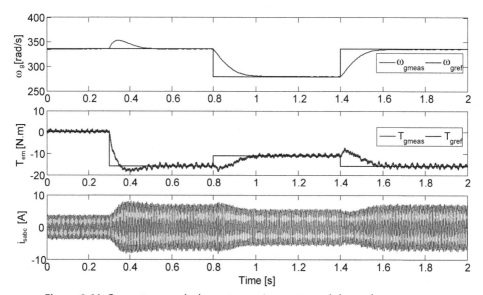

Figure 9.64 Generator speed, electromagnetic torque, and three-phase currents.

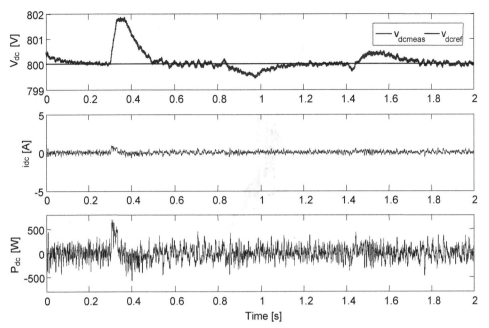

Figure 9.65 DC-link voltage, current, and power waveforms.

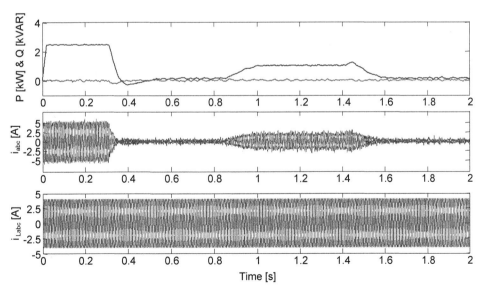

Figure 9.66 Active and reactive powers of the grid, and grid and load phase currents.

Fig. 9.67 shows the Park current with three–phase voltages of the grid for the same change of wind speed considered in the previous simulation. The current i_d follows its reference obtained by the voltage regulator. The quadratic current i_q follows its reference which is 0. The power coefficient C_p and the tip speed ratio λ are plotted in Fig. 9.68. As expected, the power coefficient C_p is maintained at an optimal value of 0.411 after a short transient while the speed λ settles at a maximum value of 8 [61].

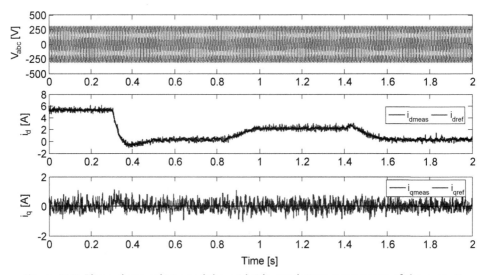

Figure 9.67 Three-phase voltage and the grid-side quadrature component of the current.

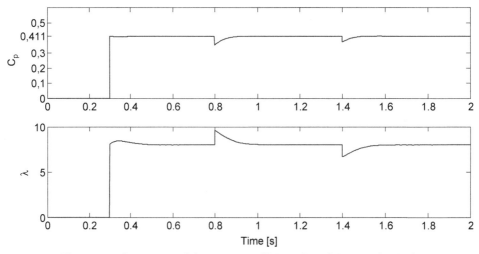

Figure 9.68 Responses of the power coefficient C_p and tip speed ratio λ.

Fig. 9.69 shows the responses of voltage, phase angle, and amplitude modulation index of the grid side. The phase shift θ controls the power flow between the side grid converter and the grid. Since the reference for reactive current component is set to zero, the θ phase angle is almost zero. On the other hand, the control of the grid active power and the voltage level of DC link are performed via the amplitude modulation index MI [61]. It can be noted that MI changes when the wind speed is stepped from 12 to 10 m/s and back to 12 m/s.

9.5.3.2 Model of the Induction Generator in the α–β Reference Frame

The dynamic model of an induction generator can be derived from its basic electrical and mechanical equations. In the stationary reference frame, the voltages are expressed as follows:

$$\begin{cases} v_{s\alpha} = R_s i_{s\alpha} + \dfrac{d}{dt}\varphi_{s\alpha} \\[2mm] v_{s\beta} = R_s i_{s\beta} + \dfrac{d}{dt}\varphi_{s\beta} \\[2mm] v_{r\alpha} = 0 = R_r i_{r\alpha} + \dfrac{d}{dt}\varphi_{r\alpha} + \omega_r \varphi_{r\beta} \\[2mm] v_{r\beta} = 0 = R_r i_{r\beta} + \dfrac{d}{dt}\varphi_{r\beta} - \omega_r \varphi_{r\alpha} \end{cases} \tag{9.95}$$

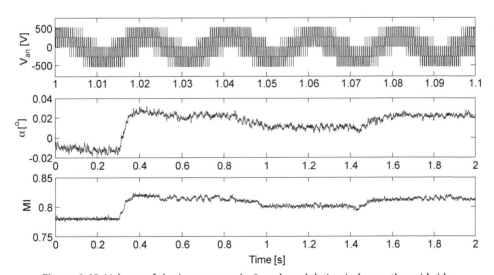

Figure 9.69 Voltage of the inverter, angle θ, and modulation index on the grid side.

The equations of stator flux and rotor are:

$$\varphi_{s\alpha} = L_s i_{s\alpha} + L_m i_{r\alpha}$$

$$\varphi_{s\beta} = L_s i_{s\beta} + L_m i_{r\beta}$$

$$\varphi_{r\alpha} = L_r i_{r\alpha} + L_m i_{s\alpha} \tag{9.96}$$

$$\varphi_{r\beta} = L_r i_{r\beta} + L_m i_{s\beta}$$

where R_s, R_r, L_s, and L_r are the stator and rotor resistances and inductance, respectively, L_m is the mutual inductance, $\omega_r = P\Omega_r$ is the rotor speed, P being the number of pole pairs, ω_s is the synchronous angular speed, and $v_{s\alpha}$, $v_{s\beta}$, $v_{r\alpha}$, $v_{r\beta}$, $i_{s\alpha}$, $i_{s\beta}$, $i_{r\alpha}$, $i_{r\beta}$, $\varphi_{s\alpha}$, $\varphi_{s\beta}$, $\varphi_{r\alpha}$ and $\varphi_{r\beta}$ are respectively the direct and quadratic components of the voltages, currents, and flux of the stator and rotor.

And the mechanical equation is:

$$T_{em} - T_m = f\Omega_g + J\frac{d\Omega_g}{dt} \tag{9.97}$$

where T_{em} is the electromagnetic torque (N·m) and T_m denotes the mechanical torque produced by the turbine (N·m).

The electromagnetic torque is:

$$T_{em} = \frac{3}{2}p\left(\varphi_{s\alpha}i_{s\beta} - \varphi_{s\beta}i_{s\alpha}\right) \tag{9.98}$$

The stator flux estimate is obtained from the voltage vectors and the stator current as:

$$\varphi_{s\alpha} = \int \left(v_{s\alpha} - R_s i_{s\alpha}\right)dt$$

$$\varphi_{s\beta} = \int \left(v_{s\beta} - R_s i_{s\beta}\right)dt \tag{9.99}$$

The magnitude and phase of the stator flux are:

$$|\varphi_s| = \hat{\varphi}_s = \sqrt{\varphi_{s\alpha}^2 + \varphi_{s\beta}^2} \quad \delta = \tan^{-1}\left(\frac{\varphi_{s\beta}}{\varphi_{s\alpha}}\right) \tag{9.100}$$

9.5.3.3 Classical Direct Torque Control

In the classical DTC, the reference values for the magnitudes of the stator flux and torque are compared to the estimated values to produce the error magnitude of the stator flux and electromagnetic torque to be fed to the respective hysteresis controllers [83]. The aim is to achieve the control of the system both in steady-state and during transient conditions by combining different switching strategies. With the SVM strategy, at each sample time, the voltage vector is selected to maintain the torque and flux within the two hysteresis bands. The selection of the voltage vector is made on the basis of the instantaneous flux error φ_s and the electromagnetic torque T_{em}. More than one voltage vectors can be selected for a given combination of flux and torque.

To fix the amplitude of the stator flux vector, the extremity of the flux vector must draw a circular path. Therefore, the voltage vector applied must always remain perpendicular to the flux vector. Thus selecting a suitable vector, the end of the flux vector can be controlled and displaced so that the amplitude of the flux vector remains within a certain range. The selection of V_s depends on the desired variation of the magnitude of the flux, but also the desired changes of the rotation speed and consequently the torque. Generally, the space of evolution of φ_s in the fixed reference (stator) frame can be delimited by dividing the space into six symmetrical zones with respect to the directions of the nonzero voltage. When the flux vector is in area i, the two vectors V_i and V_{i+3} have the largest flux component. In addition, their effect on the torque depends on the position of the flux vector in this zone. Flux and torque control is achieved by choosing one of the four nonzero or one of two zero vectors. The role of the voltage vector selected is described in Fig. 9.70.

The choice of the vector V_s depends on (1) the position of φ_s in the reference system(s), (2) the desired variation in the magnitude of φ_s, the desired variation in the torque, and (3) the direction of rotation of φ_s.

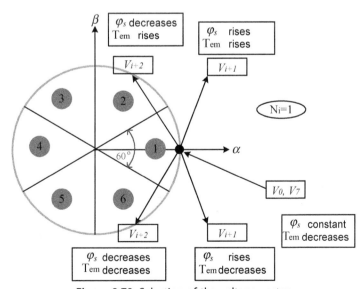

Figure 9.70 Selection of the voltage vector.

When φ_s flow is in zone i, then flux and torque control can be achieved by selecting one of the following eight voltage vectors:

- if V_{i+1} is selected, then both φ_s and T_{em} will increase;
- if V_{i-1} is selected, then φ_s will increase and T_{em} will decrease;
- decrease of T_{em} while φ_s remains unchanged.

The switching table of the DTC is given in Table 9.6.

By selecting one of the zero vectors, the rotation of stator flux is stopped and thus causes a decrease in the torque. V_0 or V_7 are selected to minimize the number of switching times of the same inverter switch.

The structure of the DTC is illustrated in Fig. 9.71.

Fig. 9.72 shows the wind speed profile and the generator speed. The wind speed varies from 12 to 10 m/s at time $t = 2.5$ s resulting in a variation of the generator speed from 1550 to 1300 rpm.

Table 9.6 Switching Table of DTC

	N_i	1	2	3	4	5	6
	$\Delta T_{em} = 1$	V_2	V_3	V_4	V_5	V_6	V_1
$\Delta \varphi_s = 1$	$\Delta T_{em} = 0$	V_7	V_0	V_7	V_0	V_7	V_0
	$\Delta T_{em} = -1$	V_6	V_1	V_2	V_3	V_4	V_5
	$\Delta T_{em} = 1$	V_3	V_4	V_5	V_6	V_1	V_2
$\Delta \varphi_s = -1$	$\Delta T_{em} = 0$	V_0	V_7	V_0	V_7	V_0	V_7
	$\Delta T_{em} = -1$	V_5	V_6	V_1	V_2	V_3	V_4

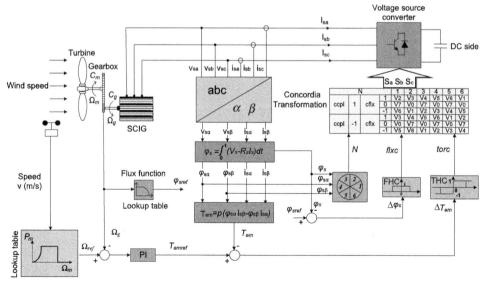

Figure 9.71 General structure of DTC strategy.

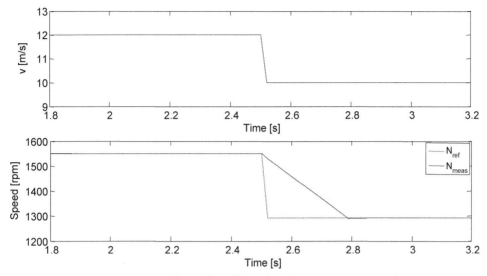

Figure 9.72 Wind speed profile and resulting generator speed.

Fig. 9.73 shows the DC voltage, the current of phase *A*, and the electromagnetic torque of the generator. These results clearly reflect the effects of the variable switching frequency on the quality of the waveform.

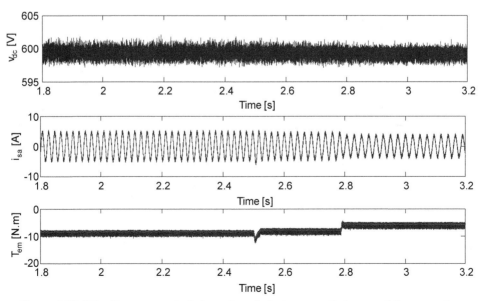

Figure 9.73 DC voltage, current of phase A, and electromagnetic torque of the generator.

9.5.3.4 Direct Torque Control Based on SVPWM

To preserve the basic idea of DTC in the proposed method, stator flux orientation technique is used. Thus, the control voltages may be initially generated by a simple PI controller and imposed by the SVPWM technique. The combination of a PI controller with the vector modulation technique is used to obtain a fixed switching frequency and reduce ripples in the torque and flux responses. This control structure retains the advantages of both vector control and DTC concepts and therefore overcomes the problems associated with the conventional DTC. The overall control structure is shown in Fig. 9.74.

The speed reference Ω_{ref} is obtained from the MPPT strategy. The speed regulator generates the torque reference $T_{em,\,ref}$ which is then compared with the estimated torque T_{em}, obtained from Eq. (9.98). The output of the torque controller produces the q-component of the stator voltage of the induction generator v_{qs}. Stator flux estimation is based on Eq. (9.99). A lookup table giving the flux reference as a function of the speed is developed based on Eq. (9.94). The flux regulator output gives the d-component of the stator voltage v_{ds}. A transformation from $d-q$ rotating frame to a fixed reference $\alpha-\beta$ is then applied. The algorithm requires the angle δ which is obtained from the estimation block of the flux and torque.

In this arrangement, two PI controllers are used to control the magnitude of the torque and stator flux instead of the original hysteresis band controllers. As compared to FOC, DTC does not require the use of a decoupling mechanism, and stator flux and torque values can be regulated independently by the PI controllers. The technique is

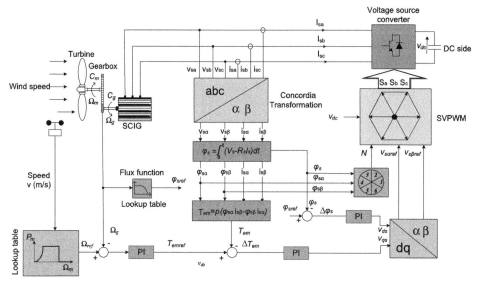

Figure 9.74 Torque and flux control based on DTC-SVPWM.

based on the calculation of the angle δ and the reference voltage V_{ref}. The SVPWM technique applied to a three-level inverter has already been detailed previously.

Fig. 9.75 shows the wind speed profile and the rotation speed of the generator. The wind speed varies from 12 to 10 m/s at time $t = 2.5$ s resulting in a variation of the generator speed from 1550 rpm to 1300 rpm.

Fig. 9.76 shows the DC voltage, the current of phase A, and the electromagnetic torque of the generator. These results demonstrate a considerable reduction in the current and torque ripples and hence an improved quality of the energy delivered by the SCIG.

9.6 STABILITY AND POWER QUALITY STUDIES

9.6.1 Floating Wind Speed Conditions

To demonstrate the performance of the proposed method under disturbance conditions, the system was simulated with a floating wind speed and the results are presented for the DTC-SVPWM with two and three levels.

For a floating wind speed, the turbulence component is considered as a stationary random process and so does not dependent on the mean value of the wind speed. The floating wind speed fluctuates between 10 and 12 m/s. The simulation results are shown in Fig. 9.77. The results show a clear improvement in the response of the DC-link voltage for the SVPWM with three-level topology and only few ripples can be seen in the transient responses of active and reactive powers of the grid side [84].

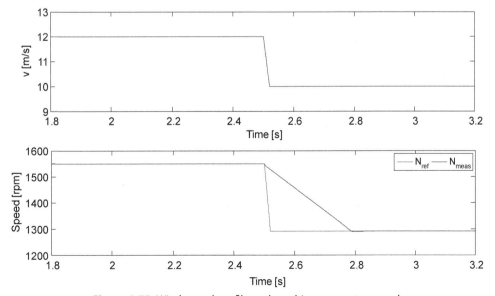

Figure 9.75 Wind speed profile and resulting generator speed.

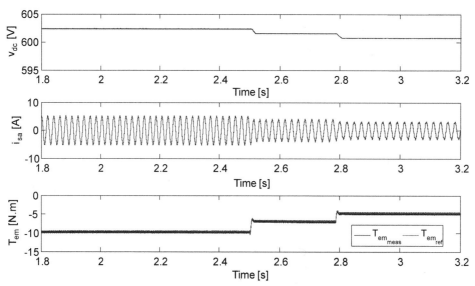

Figure 9.76 DC voltage, current of phase A, and electromagnetic torque of the generator.

9.6.2 Electric Protection System

This block provides the protection of the whole system based on some prescribed criteria including:

- the line voltage negative and positive sequences;
- DC-link voltage;

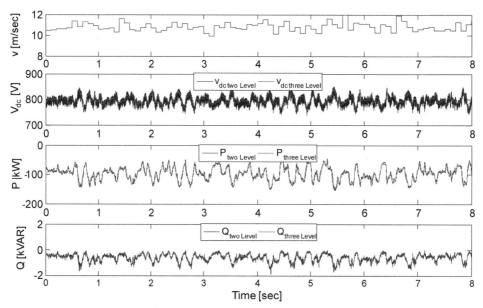

Figure 9.77 Wind speed profile, DC-link voltage, and active and reactive powers of the grid.

- the line current negative and positive sequences;
- the speed of the generator negative and positive sequences.

The protection system (PS) block of Fig. 9.78 includes the following subblocks (see Fig. 9.79): (1) two blocks to compute magnitudes of the voltage and current; (2) comparators to confront the positive and negative sequences of voltage magnitudes, current; (3) rotor speed and the DC voltage side (positive sequence); (4) an OR logic gate; (5) a timer; and (6) discrete bistable flip—flop with set or reset priority inputs.

This block is responsible for the protection of the overall system based on the following criteria: The line voltage

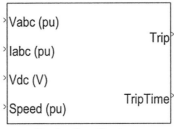

Protection System

Figure 9.78 Protection system block in Simulink.

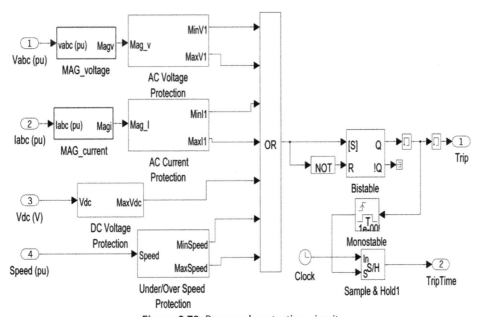

Figure 9.79 Proposed protection circuit.

- sequences negative and positive [0.5 1.5 pu] with the delay of 0.3 s;
- DC-link voltage $v_{dc\,max} = 1300$ V with the delay of 0.01 s;
- the line current sequence negative and positive [0.5 1.5 pu] with the delay of 5 s;
- the speed of the generator sequences negative and positive [0.3 1.5 pu] with the delay of 5 s.

9.6.3 Symmetrical and Asymmetrical Fault Analysis

The fault created on the grid side generates fluctuations in the DC-link voltage which will affect the torque response and consequently induce ripples on the rotational speed of the generator's rotor. However, these ripples can be reduced by means of: (1) a good selection of the DC-side capacitor, (2) the control algorithm used to minimize ripples mainly for the three-level inverter, and (3) a proper choice of the reference voltage for V_{dc} to ensure the proper functioning of the system in case of faults operating in the linear region ($0 \leq MI \leq 0.907$ in practice and $0 \leq MI < 1$ in simulations).

9.6.3.1 Symmetrical Faults

The proposed control systems were tested in the presence of a symmetrical fault at the source voltage to show their effects on the grid active and reactive powers' responses. A symmetrical fault was created by changing the source voltage by $\pm50\%$ of its maximum value for a nominal wind speed of 12 m/s and the results are shown in Fig. 9.80. The following conclusions can be drawn:

- The intervals $1.8\,s \leq t \leq 2$ s and $2.4\,s \leq t \leq 2.6$ s correspond to regular system operation, where the real power produced by the generator SCIG matches its nominal value.
- In the interval $2\,s \leq t \leq 2.2$ s, the source voltage is increased by 50%. This causes a rise in the real power for both inverter topologies and a short transient response in the DC-side voltage.
- In the interval 2.2 s $\leq t \leq 2.4$ s, the source voltage is decreased by 50%. This leads to a decrease in the real power for both inverter topologies and a short transient response in the DC-side voltage.

9.6.3.2 Asymmetrical Faults

A single-phase fault simulated as a voltage drop of phase A voltage ($\Delta V = 20\,\%$ of the maximum value). Fig. 9.81 shows waveforms of phase A grid voltage, DC-link voltage, and the active and reactive powers of the grid following the fault applied at $t = 0.9$ s and which lasted till $t = 1.2$ s. This results in large ripples in the DC-link voltage v_{dc} and in the active and reactive powers of the grid in the case of the classical DTC. The DTC-SVPWM with a two-level inverter resulted in a significant reduction of these ripples. The DTC-SVPWM with a three level, on the other hand, resulted in much better quality of the signal waveforms.

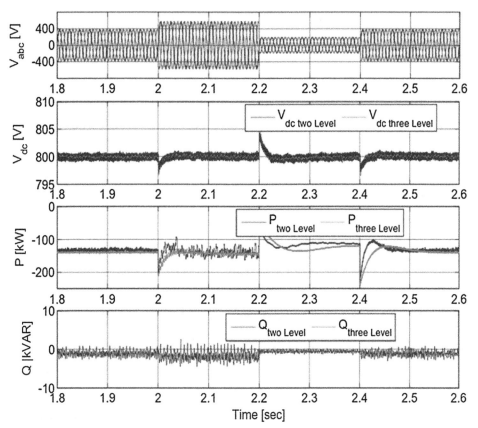

Figure 9.80 Response of the grid phase voltages, DC-link voltage, and active and reactive powers of the grid with three-phase fault.

In order to increase the severity of the disturbance, the asymmetrical fault was applied to the system under the same fluctuating wind conditions of Fig. 9.77. In this simulation, the phase unbalance of the grid voltage of 20% was applied during the time interval $3\,\text{s} < t < 4\,\text{s}$. As shown in Fig. 9.82, both methods perform well when the source is unbalanced, but the real and reactive powers from the source in the case of two-level SVPWM become highly unbalanced and large oscillations can be noticed. A huge reduction in the fluctuations of the response has been achieved in the case of the three-level SVPWM [84].

9.6.3.3 Simulation With the Electrical System Protection

A fault was simulated at $t = 2\,\text{s}$ as shown. The delay was set to 17 times the period of the grid voltage, that is, $D_f = [2 - 2 + 17/50]$ as shown in Fig. 9.83, where 50 is the grid frequency and D_f is the fault period. It is interesting to note that the protection

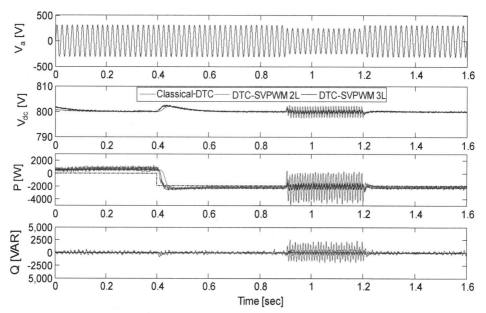

Figure 9.81 Response of the grid phase A voltage, DC-link voltage, and the active and reactive powers of the grid with single-phase fault.

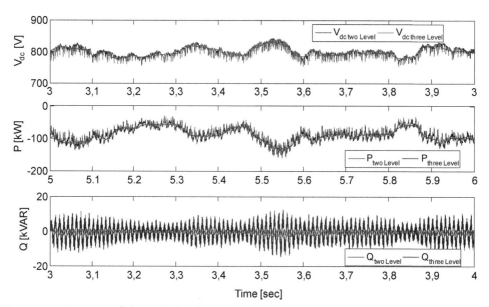

Figure 9.82 Response of the DC-link voltage and the active and reactive powers of the grid under single-phase fault.

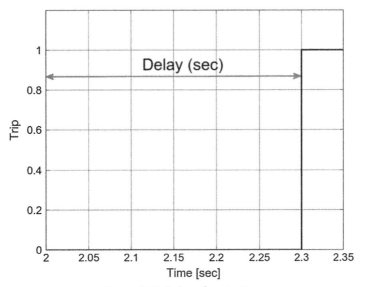

Figure 9.83 Delay of protection.

circuit of the system is local (i.e., PS essentially protects the turbine-generator system). The output of the PS circuit is a logic state (either "0" or "1"). If PS is ON, i.e., in logic state "1", the circuit is continually activated despite the elimination of the fault. The switching on of the generator is possible only when the protection system is switched off (i.e., PS set at logic "0").

Fig. 9.84 shows the three-phase voltages of the grid and the active power of the grid. Despite the fact that the fault has been eliminated, the active power transferred from the generator to the grid goes to zero after activating the protection circuit. Fig. 9.85 shows the rotor speed, the torque, and the current of the phase of the generator. After activating the protection circuit, those three variables tend to zero.

9.6.4 Power Quality Analysis

9.6.4.1 Harmonic Analysis

This section analyses the performance of the inverter and the quality of the output signal, in terms of the total harmonic distortion (THD) defined as:

$$THD = \sqrt{\frac{V_{rms}^2 - V_1^2}{V_1^2}} \tag{9.101}$$

where V_1 is the RMS value of the fundamental of the output voltage and V_{rms} is the RMS value of the output voltage.

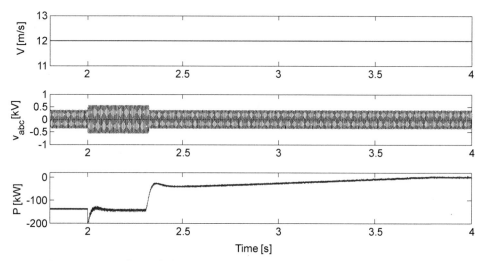

Figure 9.84 Wind speed, three-phase voltage, and active power on the grid side.

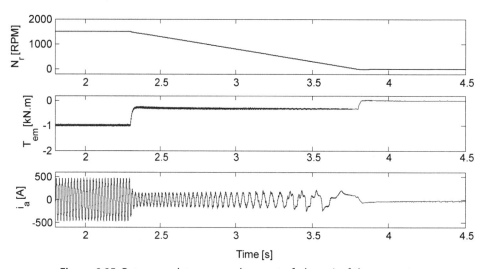

Figure 9.85 Rotor speed, torque, and current of phase A of the generator.

Fig. 9.86 shows the THD factor based on the amplitude modulation index for the two-level and three-level inverters.

These modulation strategies are developed for high-power inverters operating with a switching frequency below 1 kHz to reduce switching losses [6]. It may be noted that the three-level inverter gives a better performance as compared to the two-level inverter as shown in Fig. 9.86.

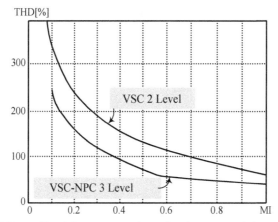

Figure 9.86 THD profile of the output voltage for the two- and three-level inverters.

The principal sources of harmonics are devices that contain switching elements (static converters) and devices that feature nonlinear voltage—current characteristics (rotating machines).

These electric equipment are considered as nonlinear loads that generate harmonic currents with frequencies integer multiples of the fundamental frequency, or sometimes even arbitrary frequencies. The passage of these harmonic currents through the electric network impedances can result in harmonic voltages at the connection points and then pollute consumer appliances connected to the same network. These harmonics cause overload (the neutral conductor, sources, and so on), unwanted tripping, accelerated aging, and performance degradation of the system. Consequently, it is necessary to reduce dominant harmonics below 5% as specified in the IEEE harmonic standards [85].

The harmonic spectrum of the current of phase A of the grid are shown in Fig. 9.87. From this figure, the THD is 15.41% for the DTC-SVM, 6.57% for the DTC-SVPWM two-level inverter, and 1.51% for three-level inverter. The lower THD is obtained with DTC-SVPWM for three-level inverter. In this case, the group of harmonics can be displaced further away from the fundamental frequency [84]. This leads to a good quality of the currents at the output of the inverter.

9.6.4.2 Power Quality Analysis

Power quality in electric power systems refers to maintaining a sinusoidal voltage with nominal amplitude and frequency. The concept of power quality has become very important, because of large-scale use of systems based on power electronics by both users and suppliers. Electric power must be supplied as a balanced three-phase system characterized by a balanced three voltages with the same amplitude and frequency, and having a sinusoidal waveform.

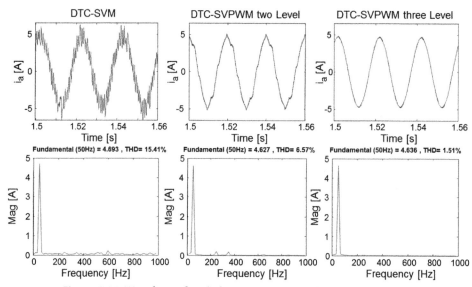

Figure 9.87 Waveform of grid phase A current and harmonic spectrum.

Power quality is a broad concept that covers various aspects of the electric supply. Improved power quality on electrical distribution networks is today a major issue for both network managers and operators of electrical energy. In normal operation, the power quality reduces mainly the quality of the voltage waveform supplied. The main disturbances which can affect power quality are dips voltage, short or long periods, the voltage fluctuation (or flicker), overvoltage, imbalance, or harmonics.

The main parameters that characterize a three-phase voltage is the frequency, amplitude, sinusoidal waveform, and the symmetry of the system characterized by equal amplitude of the three voltages and their relative phase shifts. Any physical phenomena affecting one or all of these settings is considered as a perturbation. The quality of the power delivered to users depends on these four parameters.

Fig. 9.88 shows the DC link—voltage, the grid d—q current components, and the waveform of grid phase A current with classical DTC and DTC-SVPWM for the two- and three-level inverters. The DC-link voltage v_{dc} shows good tracking of the 800 V reference voltage and the real and reactive currents follow their references (recall that reference i_d is provided by the voltage regulator while i_q is set to zero). With the classical DTC we can notice ripples in the voltage response and the i_q current components of the Park portion of the grid. The DTC-SVPWM for both two- or three-level inverter, on the other hand, produced significant reduction in these ripples. The short delay appearing after $= 1$ s, in the responses of the power, current, and voltage on the DC side is caused by the following constraints imposed on the torque: To avoid large excursions in the torque during the speed transient responses, the torque is maintained constant to

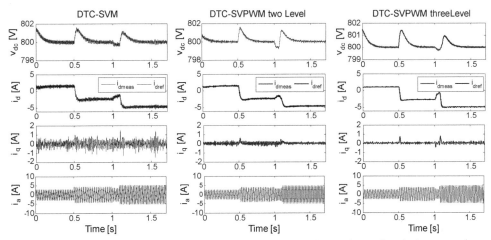

Figure 9.88 DC-link voltage, grid $d-q$ current components, and waveform of grid phase A current.

its current value. Once the speed reaches its steady state, this action on the torque is removed and the torque is allowed to vary normally.

9.7 DISCUSSIONS AND CONCLUSIONS

This chapter presented a comprehensive overview of WECSs focusing on control design methods and modulation strategies, stability, and power quality. The SCIG-based variable speed WECS was analyzed in the simulation studies. A new MPPT control strategy based on the knowledge of the wind turbine characteristics has been designed and implemented to maximize wind power extraction from the turbine. The PI controllers for the DC-side voltage as well as the direct and quadrature current components in the grid side have been designed using a simple pole placement technique. The grid quadratic current component is used to modulate the flow of reactive power. In our case, the reactive power reference is set to zero, to obtain a unity power factor. The switching signals for the grid-side three-phase inverter, which controls the DC voltage level, the direct current component (for controlling the active power of the grid), and the quadratic current component (for the reactive power flow) are obtained via the selective harmonic PWM (SHE-PWM) and space vector PWM (SVPWM) for two- and three-level inverter topologies, this has reduced the level of harmonics in the voltage output waveforms. The control system was able to maximize the energy extracted from the wind as reflected from the power coefficients obtained during the simulation scenarios considered.

Classical DTC is simple to implement. Moreover, this control strategy does not need the position sensor, and the stator resistance of the machine is the only parameter needed to estimate the flux and torque. However, two major disadvantages exist. First, the determination of the switching states is based on information from the trends in the flux and torque which characterize a hysteresis type of nonlinearity. Second, as the duration of switching is variable this leads to oscillations torque and flux. To overcome these problems associated with classic DTC-Control, a new control strategy, based on a fixed frequency, termed DTC-SVPWM (or DTFC, direct torque and flux control) is applied. This control offers several advantages:

1. the flux and torque are very well controlled to their respective reference values;
2. the flux trajectory is perfectly circular without any ripples at the start;
3. the stator current is sinusoidal and the switching frequency is constant but a little bit higher. Another approach for improving DTC control consists of using multilevel inverter topologies. This type of inverter has many advantages over standard two-level VSIs including reduced dv/dt, the output voltage waveform of the inverter is nearly sinusoidal, less distortions in the current waveform, and reduced switching frequency. The problem of multilevel NPC (Neutral Point Clamped) structure is the presence of antiparallel diodes in the inverters' topologies. Furthermore, this topology naturally limits the over voltage imposed on the power components (low value of dv/dt at the components' terminals) and introduces additional switching states that can be used to help maintain load balancing in the capacitors.

This chapter presented a comparison between classical DTC-SVM and direct torque and flux DTC-SVPWM with two- and three-level inverters in terms of the quality of the power delivered by the variable speed SCIG to the grid. The wind turbine SCIG (WT-SCIG) system is controlled via an MPPT strategy and operates at maximum power.

Compared to the DFIG which works with a range of rotor speed variations of ±30% about the synchronous speed (this is the range for generating electric power), the use of SCIG with this topology has the advantage of working with a larger variation range for the rotor speed. Since the total power delivered by the generator is sent to the grid via the AC/DC/AC converter, the latter should be oversized to 100% of the rated power. For high-power applications the DTC-SVPWM control structure based on multilevel inverters is well adapted, as compared to the conventional two-level structure, since the output voltages and current have a much lower THD. For example, the voltage across each switch of a three-level inverter is halved and the chopping frequency is reduced. The simulation results obtained for the three control methods show a significant reduction in the torque and flux ripples using DTC-SVPWM with two- and three-level inverter and therefore a net improvement in the quality of energy of the system. The simulation results with different symmetrical and asymmetrical faults on the network side show a significant improvement in dynamic performance, robustness, and stability with the use of DTC-SVPWM based on three levels as compared with conventional DTC or DTC-SVPWM with two levels.

APPENDIX

A. Parameters of the Simulation Models

Table 9.A.1 Parameters of the Simplified Model (Turbine and Generator)

Parameter	Value
Density area, ρ (kg/m)	1.225
Nominal mechanical power, $P_{m,n}$ (MW)	5
Radius of the turbine, R (m)	63
Nominal wind speed, v_n (m/s)	12
Gain of the multiplier, G	97
Inertia of the turbine, J_t (kg/m)	35.44×10^6
Inertia of the generator, J_g (kg/m)	534.116
Shaft spring constant, K (N/m/rad)	867.64×10^6
Shaft mutual damping, D (N/m/s/rad)	6.217×10^6
Hydraulic time constant, τ_H (s)	0.05

Table 9.A.2 Parameters of the Detailed Model

Parameters	SCIG 2 kW	SCIG 149.2 kW	DFIG 1.5 MW
Grid			
Effective voltage, V_s (V)	380	460	610
Frequency, f_s (Hz)	50	50	50
Transformer			
Leakage resistance, R_f (Ω)	3.4	0.1	3
The leakage inductance, L_f (mH)	3.3	2	2.5
Turbine			
Density area, ρ (kg/m²)	1.225	1.225	1.225
Nominal mechanical power, $P_{m,n}$ (kW)	2.68	149.2	1.5
Radius of the turbine, R (m)	1.4	10.5	35
Nominal wind speed, v_n (m/s)	12	12	11
Gain of the multiplier, G	2.445312	17.1806	75
SCIG and DFIG			
Nominal frequency, $f_{g,n}$ (Hz)	50	50	50
Stator resistance, R_s (mΩ)	4850	14.85	12
Stator leakage inductance, L_{ls} (mH)	16	0.3027	13.7
Rotor resistance, R_r (mΩ)	3805	9.295	21
Rotor leakage inductance, L_{lr} (mH)	16	0.3027	13.7
Cyclic mutual inductance, L_m (mH)	258	10.46	13.5
Inertia, J (kg/m²)	0.031	3.1	1000
Viscous friction coefficient, f (Nm s/rad)	0.00114	0.08	0.0024
Number of pole pairs, p	2	2	2

(Continued)

Table 9.A.2 Parameters of the Detailed Model—cont'd

Parameters	SCIG 2 kW	SCIG 149.2 kW	DFIG 1.5 MW
DTC-SVM			
Flux hysteresis band, $\Delta\varphi$	±0.01 Wb		
Torque hysteresis band, ΔT_{em}	±0.5 Nm		
SVM switching frequency (Hz)	2000	2000	

Table 9.A.3 Parameters of the PI Controllers

Parameters	SCIG 2 kW	SCIG 149.2 kW
DTC-SVPWM Control Block		
Proportional gain of speed controller, K_p	1	30
Integral gain of the speed controller, K_i	15.872	200
Proportional gain of torque controller, K_{pt}	2	1.5
Integral gain of torque controller, K_{it}	150	100
Proportional gain of flux controller, K_{pf}	200	250
Integral gain of flux controller, K_{if}	1200	4000
DC-Side Control Block		
Proportional gain of DC voltage controller, K_{pdc}	2	2
Integral gain of DC voltage controller, K_{pdc}	25	33.33
Source-Side Control Block		
Proportional gain of current controller, K_{pc}	6	6
Integral gain of current controller, K_{ic}	4500	4.50

Table 9.A.4 Parameters of the PI Controllers of the IFOC Blocks
IFOC Control Block

Regulator Parameters	SCIG 2 kW
Proportional gain of current controller, K_{pc}	1
Integral gain of current controller, K_{ic}	15.872

B. Influence of Order 3 Harmonic on the Behavior of the SVPWM

In three phase, harmonics can be reduced without reducing the amplitude of the output voltage because the harmonic of order 3 or a multiple of 3 are eliminated from output voltages. A harmonic of order 3 can be added to a sinusoid of frequency f to form the reference waveform. This harmonic appears in the three fictitious voltages V_{ao}, V_{b0}, and V_{c0} with respect to the fictitious midpoint 0, but does not appear in the phase output voltages V_{an}, V_{bn}, and V_{cn} and line-to-line output voltages V_{ab}, V_{bc}, and V_{ca}.

The addition of harmonic of order 3 increases the maximum amplitude of the fundamental in the output voltages.

With reference to Fig. 9.B.1, the reference voltage is composed of two sinusoids: one for the fundamental and the other for the harmonic of order 3.

The new reference signal for the PWM is:

$$(V_a - V_o)_w = \frac{U}{2}(MI\sin(\omega t) + k\sin(3\omega t)) \tag{9.B.1}$$

This is called suboptimal control.

The maximum value of MI occurs for $k = 1/3\sqrt{3}$ and is found as:

$$MI_{max} = \frac{2}{\sqrt{3}} \tag{9.B.2}$$

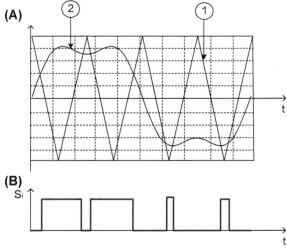

Figure 9.B.1 Suboptimal SVPWM. (A) Triangular carrier $V_t(t)$ and reference $(V_a - V_o)_w$. (B) Control pulse for phase i (where $i = a, b, c$).

Therefore, with suboptimal control, the maximum amplitude of the fundamental output voltage V_1 corresponding to MI_{max} is:

$$V_{1,\, MI_{max}} = 2\Big/\sqrt{3}\ V_{dc} \qquad\qquad (9.B.3)$$

NOMENCLATURE

PI	Proportional integral
BEM	Blade element momentum
MPPT	Maximum power point tracking
IGBT	Insulated gate bipolar transistor
AC	Alternative current
DC	Direct current
PMSG	Permanent magnet synchronous generator
SEIG	Self-excited induction generator
SCIG	Squirrel-cage induction generator
DFIG	Doubly-fed induction generator
GSC	Grid-side converter
RSC	Rotor-side converter
ISC	Indirect speed controller
DSC	Direct speed controller
STATCOM	Static synchronous condenser
FOC	Field-oriented control
IFOC	Indirect field—oriented control
DTC	Direct torque control
DTFC	Direct torque and flux control
PLL	Phase-locked loop
PWM	Pulse width modulation
FHC	Flux hysteresis controller
THC	Torque hysteresis controller
Flxc	Flux controller
Torc	Torque controller
NPC	Neutral point clamped
DTC-SVPWM	Direct torque control space vector pulse width modulation
d, q	Quantities in d-axis and q-axis
α, β	Quantities in a-axis and b-axis
s, r	Stator and rotor
C	DC-link capacitance (F)
MI	Modulation index
P_m	Mechanical power of the wind turbine (W)
P_s	Stator active power (W)
Q_s	Stator reactive power (VAR)
P_r	Rotor active power (W)
Q_r	Rotor reactive power (VAR)
P_n	Nominal power
v_n	Nominal wind speed
β	Pitch angle (°)
C_p	Aerodynamic efficiency of the wind turbine

P	Density area (kg/m^3)
S	Swept area of the wind turbine blades (m^2)
R	Radius of the turbine (m)
V	Wind speed (m/s)
Ω_m	Mechanical speed of the turbine (rad/s)
$\Omega_{m,opt}$	Optimum angular speed (rad)
Λ	Tip speed ratio
V_{PWM}	Voltage of the PWM
V_π	Full-wave voltage
V_m	Modulating voltage
V_p	Carrier voltage
V_1	RMS value of the fundamental voltage
V_{rms}	RMS value of the output voltage
M_c	Inverter capacity index
$M_{c,max}$	Maximum capacity index
MI_{max}	Maximum amplitude index
A_k	Magnitude of each spectral component
w_k	Angular speed (rad/s)
θ_k	Phase angle (rad)
P_n	Nominal power (W)
β_{ref}	Reference of pitch angle
v_{dc}	DC voltage
f_P	Carrier frequency
f_r	Fundamental frequency of the inverter output variables
A_r	Fundamental peak value of the desired load voltage
A_p	Amplitude of the output voltage
m_f	Modulation index
v_{sd}, v_{sq}	Stator voltage components in the Park system
ω_s	Synchronous speed
$\varphi_{sd}, \varphi_{sq}$	Stator flux components in d-axis and q-axis of the Park frame
i_{sd}, i_{sq}	Stator current components in d-axis and q-axis of the Park frame
v_{rd}, v_{rq}	Rotor voltage components in the Park system
ω_r	Rotor speed
$\varphi_{rd}, \varphi_{rq}$	Rotor flux components in d-axis and q-axis of the Park frame
i_{rd}, i_{rq}	Rotor current components in d-axis and q-axis of the Park frame
T_{em}	Electromagnetic torque (N/m)
T_m	Mechanical torque (N/m)
J, f_c	Inertia and viscous friction coefficient, respectively
G	Gain of the multiplier
σ	Coefficient of Blondel
R, L	Resistance and inductance of load, respectively
L_{ls}, L_{lr}	Leakage inductances of stator and rotor windings, respectively
L_m	Mutual inductance
ϕ_{sn}	Nominal stator flux
δ	Angle of stator flux
$|\varphi_s|$	Module of stator flux
V_{ref}	Reference voltage
V_s	RMS voltage
S	Laplace operator

REFERENCES

[1] Lacal Arántegui R, Serrano González J. Technology, market and economic aspects of wind energy in Europe, JRC wind status report. 2015. http://www.evwind.es/2015/06/26/the-technology-market-and-economic-aspects-of-wind-energy/52993.

[2] Robertscribbler. Denmark kicking fossil fuels addiction with record 39 percent (and Growing) wind generation (http://robertscribbler.com/tag/wind-energy/).

[3] New record in worldwide wind installations, February 5, 2015 (http://www.wwindea.org/new-record-in-worldwide-wind-installations/).

[4] Hill JS. A record year for wind, emerging markets driving global wind energy growth, CleanTechnica report, April, 2015. 2014. http://cleantechnica.com/2015/04/02/emerging-markets-driving-global-wind-energy-growth/.

[5] Burton T, Sharpe D, Jenkins N, Bossanyi E. Wind energy handbook. John Wiley & Sons; 2001.

[6] Wu B, Lang Y, Zargari N, Kouro S. Power conversion and control of wind energy systems. Wiley-IEEE Press; 2011.

[7] Blaabjerg F, Ma K. Future on power electronics for wind turbine systems. IEEE Journal of Emerging and Selected Topics in Power Electronics September 2013;1(3).

[8] Edon M. 38 meter wind turbine blade design, Internship report. Denmark: Folkecenter for Renewable Energy; 2007.

[9] Jureczko M, Pawlak M, Mezyk A. Optimization of wind turbine blades. Journal of Materials Processing Technology 2005;167:463−71.

[10] Xudong W, Zhong Shen W, Zhu WJ, Nørkær Sørensen JN. Blade optimizations for wind turbines. Wind Energy 2009;12:781−803.

[11] Satean T, Boonruang W, Chakrapong C, Suksri T. Grid connected based six-pulse converter applied a self-excited induction generator for wind turbine applications. Energy Procedia 2011;9:128−39.

[12] Deraz SA, Abdel Kader FE. A new control strategy for a stand-alone self-excited induction generator driven by a variable speed wind turbine. Renewable Energy 2013;51:263−73.

[13] Multon B, Gergaud O, Ben Ahmed H, Roboam X, Astier S, Dakyo B, et al. Etat de l'art dans les aérogénérateurs électriques, l'electronique de puissance vecteur d'optimisation pour les energies renouvelables. 2002. p. 97−154.

[14] Abdullah MA, Yatim AHM, Tan CW, Saidur R. A review of maximum power tracking algorithms for wind energy systems,. Renewable Sustainable Energy Reviews 2012;16(5):2355−3558.

[15] Blaabjerg F, Chen Z, Kjaer SB. Power electronics as efficient interface in dispersed power generation systems. IEEE Transactions on Power Electronics 2004;19(4):1184−94.

[16] Lei Y, Mullane A, Lightbody G, Yacamini R. Modeling of the wind turbine with a doubly fed induction generator for grid integration studies. IEEE Transactions on Energy Conversion 2006;21(1):257−64.

[17] Xiu-xing Y, Yong-gang L, Wei L, Ya-jing G, Xiao-jun W, Peng-fei L. Design, modeling and implementation of a novel pitch angle control system for wind turbine. Renewable Energy 2015;81:599−608.

[18] Girsang IP, Dhupia JS. Pitch controller for wind turbine load mitigation through consideration of yaw misalignment. Mechatronics 2015;32:44−58.

[19] Wei X, Pan Z, Liping L. Wind tunnel experiments for innovative pitch regulated blade of horizontal axis wind turbine. Energy 2015;91:1070−80.

[20] Anderson PM, Bose A. Stability simulation of wind turbine systems. IEEE Transactions on Power Apparatus and Systems 1983;PAS-102(12):3791−5.

[21] Wasynczuk O, Man DT, Sullivan JP. Dynamic behavior of a class of wind turbine generators during random wind fluctuations. IEEE Transactions on Power Apparatus and Systems 1981;PAS-100(6):2837−45.

[22] Slootweg JG, De Haan SWH, Polinder H, Kling WL. General model for representing variable speed wind turbines in power system dynamics simulations. IEEE Transactions on Power Systems 2003;18(1):144−51.

[23] Ackermann T, Söder L. An Overview of wind energy-status 2002. Renewable and Sustainable Energy Reviews 2002;6:67−127.

[24] Heier S. Grid integration of wind energy conversion systems. Wiley; 2006.

[25] Chinchilla M, Arnaltes S, Burgos JC. Control of permanent-magnet generator applied to variable-speed wind-energy systems connected to the grid. IEEE Transactions on Energy Conversion 2006;21(1):130—5.

[26] Hansen AD, Michalke G. Multi-pole permanent magnet synchronous generator wind turbines' grid support capability in uninterrupted operation during grid faults. IET Renewable Power Generation 2009;3(3):333—48.

[27] Abedini A, Nasiri A. PMSG wind turbine performance analysis during short circuit faults. In: Proceedings of 2007 IEEE Canada electrical power Conference, 160—65, Canada, 25—26; October 2007.

[28] Zhang Y, Gong J, Xie DJ. Inverter control strategy for direct-drive permanent magnet wind generator under unbalance of three-phase source voltage. In: Proceedings of 11th International Conference on electrical machines and systems, 2497—501, Wuhan, China; 2008.

[29] Ng CH, Li R, Bumby J. Unbalanced-grid-fault ride-through control for a wind turbine inverter. IEEE Transactions on Industrial Electronics 2008;44(3):845—56.

[30] Song HS, Nam V. Dual current control scheme for PWM converter under unbalanced input voltage conditions. IEEE Transactions on Industrial Electron 1999;46(5):953—9.

[31] Yazdani A, Iravani R. A unified dynamic model and control for the voltage-source converter under unbalanced grid conditions. IEEE Transactions on Power Delivery 2006;21(3):1620—9.

[32] Bhattacharya L, Woodward JL. Excitation Balancing of a self-excited induction generator for maximum power output. IEE Proceedings 1998;135. Part C.

[33] Manwell JF, McGowan JG, Rogers AL. Wind energy explained — theory, design and application. John Wiley & Sons; 2002.

[34] Vongmanee V, Monyakul V. A modeling of self-excited induction generators driven by compressed air energy based on field oriented control principle. In: 2nd IEEE International Conference on power and energy, Johor Baharu, Malaysia, December 1—3; 2008.

[35] Ong CM. Dynamic simulation of electric machinery. Prentice Hall; 1997.

[36] Novotny DW, Lipo TA. Vector control and dynamics of AC drives. Clarendon Press Oxford; 1997.

[37] Krause PC, Wasynczuk O, Sudhoff SD. Analysis of electric machinery. IEEE Press; 1994.

[38] Murray A, Mirzaeva G. Renewable energy source emulator, Final Report. 2010.

[39] Mekhtoub S, Khaldi T, Ivanes M. Amplitude des courants et du couple de reconnexion d'une machine asynchrone auto-amorcée. Revue Internationale de Génie Electrique (RIGE) 2001;4(1—2):149—72.

[40] Al-Bahrani AH. Analysis of Self-excited induction generators under unbalanced conditions. Electric Machines and Power Systems 1996;24:117—29.

[41] Seyoum D. The dynamic analysis and control of a self-excited induction generator driven by a wind turbine [Ph.D. thesis]. Sydney, Australia: School of Electrical Engineering and Telecommunications, UNSW; 2003.

[42] Simões MG, Bose BK, Spiegel RJ. Fuzzy logic based intelligent control of a variable speed cage machine wind generation system. IEEE Transactions on Power Electronics 1997;12(1):87—95.

[43] Martins CA, Carvalho AS. Technological trends in induction motor electrical drives. Power Tech Proceedings, IEEE Porto 2001;2.

[44] Adzic E, Ivanovic Z, Adzic M, Katic V. Maximum power search in wind turbine based on fuzzy logic control. Acta Polytechnica Hungarica 2009;6(1).

[45] Yidong C, Yulin Y, Liqiao W, Zhiyun J, Weiyang W. Grid-connected and control of MPPT for wind power generation systems based on the SCIG. In: 2nd International Asia Conference on informatics in control. Hong Kong: Automation and Robotics; March 2010. p. 6—7.

[46] Krichen L, François B, Ouali A. Modélisation, commande et interaction de deux éoliennes à vitesse variable. Revue des Energies Renouvelables 2007;10(2):225—30.

[47] Abad G, Rodríguez MA, Poza J. Three level NPC converter based predictive direct power control of the doubly fed induction machine at low constant switching frequency. IEEE Transactions on Industrial Electronics 2008;55(12):4417—29.

[48] Nicolás CV, Lafoz M, Iglesias J. Guidelines for the design and control of electrical generator systems for new grid connected wind turbine generator. In: Proceedings of 28th IECON Conference, vol. 4; November 2002. p. 2898—902.

[49] Rodriguez-Amenedo JL, Arnalte S, Burgos JC. Automatic generation control of a wind farm with variable speed wind turbines. IEEE Transactions on Energy Conversion 2002;17(2):279−84.

[50] Hansen AD, Sørensen P, Iov F, Blaabjeg F. Centralised power control of wind farm with doubly fed induction generators. Renewable Energy 2006;31(7):935−51.

[51] Sørensen P, Hansen AD, Iov F, Blaabjerg F, Donovan MH. Wind farm models and control strategies. report Risø-R-1464(EN). Roskilde, Denmark: Risø National Laboratory; August 2005.

[52] De Almeida RG, Castronuovo ED, Lopes JAP. Optimum generation control in wind parks when carrying out system operator requests. IEEE Transactions on Power Systems 2006;21(2):718−25.

[53] Pena R, Cardenas R, Escobar E, Clare J, Wheeler P. Control system for unbalanced operation of stand-alone doubly fed induction generators. IEEE Transactions on Energy Conversion 2007;22(2):544−5.

[54] Seman S, Niiranen J, Arkkio A. Ride-through analysis of doubly fed induction wind-power generator under unsymmetrical network disturbance. IEEE Transactions on Power Systems 2006;21(4):1782−9.

[55] Nass BI, Undeland TM, Gjengedal T. Methods for reduction of voltage unbalance in weak grids connected to wind plants. In: Proc. IEEE Workshop wind power impacts power systems, Oslo, Norway; 17−18 June 2002. p. 1−7.

[56] Rathi MR, Jose PP, Mohan N. A novel H∞ based controller for wind turbine applications operating under unbalanced voltage conditions. In: Proc. 13th International Conference in intelligent systems applications in power systems; 6-9 November 2005. p. 355−60.

[57] Brekken T, Mohan N. A novel doubly-fed induction wind generator control scheme for reactive power control and torque pulsation compensation under unbalanced grid voltage conditions. In: Proceedings of PESC, vol. 2; 2003. p. 760−4.

[58] Brekken T, Mohan N. Control of a doubly-fed induction wind generator under unbalanced grid voltage conditions. IEEE Transactions on Energy Conversion 2005;22(1):129−35.

[59] Xu L, Wang Y. Dynamic modeling and control of DFIG based wind turbines under unbalanced network conditions. IEEE Transactions on Power Systems 2007;22(1):314−23.

[60] Tahri A, Merabet Boulouiha H, Allali A, Tahri FA. Multi-variable LQG controller-based robust control strategy applied to an advanced static VAR compensator, Acta Polytechnica Hungarica. Journal of Applied Sciences 2013;10(4).

[61] Merabet Boulouiha H, Allali A, Tahri A, Draou A, Denai M. A simple MPPT based control strategy applied to a variable speed squirrel cage induction generator. Journal of Renewable and Sustainable Energy 2012;4.

[62] Rashid MH. Power electronics, circuits, devices and applications. 3rd ed. Prentice-Hall; 2007.

[63] Espinoza JR. Inverters. In: Rashid MH, editor. Power electronics handbook. Academic Press; 2001. p. 225−67.

[64] Patel HS, Hoft RG. Generalized of harmonic elimination and voltage control in thyristor inverters Part I-Harmonic elimination. IEEE Transactions on Industrial Applications 1973;IA-9:310−7.

[65] Rathnakumar D, Perumal JL, Srinivasan T. A new software implementation of space vector PWM. In: Proceedings of IEEE SouthEast Conference; 8-10 April, 2005.

[66] Khalfallah M, El-Afia A, Saad A, Ghouili J. Etude comparative des performances de la modulation sinusoïdale régulière et la modulation vectorielle d'un onduleur à MLI. In: Canadian conference on electrical and computer Engineering; 1-4 May, 2005.

[67] Parekh R. V/F control of 3-phase induction motor using space vector modulation. AN 955. Microchip Technology Inc.; 2005.

[68] Rodríguez J, Lai JS, Peng FZ. Multilevel inverters: a survey of topologies, controls, and applications. IEEE Transactions on Industrial Electronics 2002;49(4).

[69] Rodríguez J, Correa P, Morán L. A vector control technique for medium voltage multilevel inverters. IEEE Transactions on Industrial Electronics 2002;49(4):882−8.

[70] Celanovic N, Boroyevic D. A fast space vector modulation algorithm for multilevel three-phase converters. In: Conf. Rec. IEEE-IAS annual meeting, Phoenix, AZ, Oct. 1999; 1999. p. 1173−7.

[71] Hammond P. A new approach to enhance power quality for medium voltage ac drives. IEEE Transactions on Industrial Applications 1997;33:202−8.

[72] Tolbert L, Habetler TG. Novel multilevel inverter carrier-based PWM method. IEEE Transactions on Industrial Applications 1999;35:1098−107.

[73] Liang Y, Nwankpa CO. A new type of STATCOM based on cascading voltage-source inverters with phase-shifted unipolar SPWM. IEEE Transactions on Industrial Applications 1999;35:1118—23.

[74] Saeedifard M, Iravani R, Pou J. Analysis and control of DC capacitor-voltage-drift phenomenon of a passive front-end five level converter. IEEE Transactions on Industrial Electronics 2007;54(6):3255—66.

[75] Abad G, Lopez J, Rodriguez M, Marroyo L, Iwanski G. Doubly fed Induction machine modeling and control for wind energy generation. Wiley-IEEE Press; 2011.

[76] Guerrero JM, de Vicuna LG, Uceda J. Uninterruptible power supply systems provide protection. IEEE Industrial Electronics Magazine 2007;1(1):28—38.

[77] Kocalmis A, Sunter S. Modelling and simulation of a multilevel inverter using space vector modulation technique. In: Proceedings of the Int. Conf. on TPE, Ankara, Turkey, May 2006; 2006. p. 940—3.

[78] Chandra Sekhar O, Chandra Sekhar K. Simulation and comparison of 2-L & 3-L inverter fed induction motor DTC drives. International Journal of Computer and Electrical Engineering 2001;3(5):676—81.

[79] Habetler TG, et al. Direct torque control of induction machines using space vector modulation. IEEE Transactions on Industry Applications 1992;28(5):1045—53.

[80] Sapran S, Navani JP. Three level inverter using MATLAB. MIT International Journal of Electrical and Instrumentation Engineering 2011;1(1):20—4.

[81] Hasse K. About the dynamics of adjustable-speed drives with converter fed squirrel-cage induction motors (in German) [Ph.D. Dissertation]. Darmstadt Technische Hochschule; 1969.

[82] Blaschke. The principle of field orientation as applied to the new transvector closed loop control system for rotating field machines. Siemens Review 1972;34:217—20. Reprinted in: Bose BK. Adjustable Speed AC drive systems. IEEE Press 1980;99:162—5.

[83] Abdul Wahab HF, Sanusi H. Simulink model of direct torque control of induction machine. American Journal of Applied Sciences 2008;5(8):1083—90.

[84] Merabet Boulouiha H, Allali A, Laouer M, Tahri A, Denai M, Draou A. Direct torque control of multilevel SVPWM inverter in variable speed SCIG-based wind energy conversion system. Renewable Energy 2015;80:140—52.

[85] EEE Std 519-1992, IEEE recommended practices and requirements for harmonic control in electrical power systems.

CHAPTER TEN

The Hybrid Solar Power/Wind System for Energy Production, Observation, Application, and Simulation

M.S. Hossain[1], A.K. Pandey[1], M.A. Tunio[2], J. Selvaraj[1] and N.A. Rahim[1]
[1]UMPEDAC, University of Malaya, Kuala Lumpur, Malaysia
[2]Mehran University of Engineering Technology, Khairpur Mir's, Sindh, Pakistan

10.1 INTRODUCTION

The sunlight-based photovoltaic (PV) and wind turbine frameworks are being used in a few applications of solid renewable energy (RE) advancements that have been perceived by many nations for a long time [1]. These advancements have the most critical contributions in ensuring the ecological and environmental conservations of the earth. In other words, it is a safe and free process by collecting the free radiant energy from the sun and free kinetic energy from the wind. There are numerous applications from these sources of renewable energies being innovated by individuals as additional usages as and when required. In future, renewable energies are the only dependable energy sources that can satisfy the global ever rising energy need. The combination of PV and wind turbine framework stands out among the most efficient means to deliver energy from RE sources [2]. In this procedure, partial tests and repeat tests are being performed to determine the best method to fully exploit those sources.

Hybrid power systems are a combination of two or more energy conversion devices (e.g., electricity generators or storage devices), or two or more energizes for the same gadget that, when incorporated, will form the hybrid power system. Framework efficiencies are normally higher than those of the individual systems independently operated and a better dependability can be obtained with the energy stockpiling advances. Some crossbreed frameworks incorporate both the hybrid power systems and the energy stockpiling advances, which can gainfully enhance the quality and accessibility of the available power so obtained [3,4].

Samples of hybrid power systems include:

1. wind energy consolidated with diesel energy,
2. PV energy consolidated with battery stockpiling or diesel energy, and
3. energy components consolidated with smaller scale turbine energy.

Clean Energy for Sustainable Development
ISBN 978-0-12-805423-9, http://dx.doi.org/10.1016/B978-0-12-805423-9.00010-7

In the case of the wind power energy, variable wind velocities are ever present inconveniences as a result of which energy from the wind ventures cannot generate a constant flow of power for 24 h; subsequently they are viewed as "irregular" power sources [5]. In this case the preferred arrangement is to have the crossbreed framework, such as, the consolidated seaward wind and sun-oriented frameworks, otherwise called the half PV and half wind framework, which can produce constant sustainable energy when the wind and PV advancements have been properly established. Then, the sun-based energy may make up for the discontinuity of the seaward wind energy while the economies of scale created from the seaward wind energy and sun-based energy can both create a sizable cost reduction for the exploitation of available renewable power. Hybrid power plants that have been constructed economically are sunlight-based battery-diesel framework and wind-battery-diesel framework [6].

Al-Ashwal and Moghram displayed a strategy for appraisal on the premise of Loss of Load Probability for choice of an exploratory ideal extent of PV and wind generator limits in the hybrid PV/wind energy system as a result of which an optimal system blend was chosen on the premise of capital expense and yearly independence level [7]. Hennet and Samarakou examined ways to deal with the improvement on the half breed PV/wind/battery framework with the traditional power plant and computed ideal framework design on the premise of Life-Cycle Cost. The month-to-month blends of sun-oriented/wind assets lead to one-sided sunlight-based month; one-sided wind month, and months with even sun-based/wind asset. The absolute framework cost and the unit cost of power produced for a lifetime from the framework may be investigated on a yearly framework execution basis [8]. Celik presented technoeconomic analysis based on the solar/wind-based months for autonomous hybrid PV/wind energy system. The author has observed that an optimum combination of the hybrid PV/wind energy system provides a higher system performance than either of the single system individually operated, for the same system cost and battery storage capacity [9]. A decisive support model for a hybrid PV/wind energy system was discussed by Chedid et al. [10] based on the political, social, technical, and economic issues.

This study will discuss the existing model used in different countries on the solar/wind hybrid system and their applications on the power production. To identify PV/wind technology annual power production in Malaysia, the RETScreen software can be used. Finally, using LabView to simulate some PV panels maximum power point tracking (MPPT)/wind turbine for maximum power can also be performed through calculation.

10.2 HYBRID SOLAR/WIND ENERGY SYSTEMS

A hybrid renewable energy system (HRES) is a mix of low carbon energy advances that are consolidated to produce a higher energy. Hybrid energy systems are regularly utilized in smaller scale matrices, comprising renewables, for example, wind turbines, sun-based PV, and low carbon plant, such as, consolidated warmth and force frameworks to provide power. The vision behind the utilization of hybrid energy frameworks is the production of energy which is more than the aggregate of its parts [11]. The mix of distinctive power sources takes into consideration a more noteworthy efficiency of supply because of the way that they all create energy from diverse sources and this implies that the mix can invalidate any constraints brought about by a faulty gadget [12].

The HRES consists of three sections:

1. wind energy transformation framework with changeless magnet synchronous wind turbine generator, uncontrolled full extension rectifier, and proportional—integral (PI)-controlled buck converter,

2. solar energy conversion system containing solar panels, MPPT, and PI-controlled buck converter, and

3. interphase transformers (IPTs), sinusoidal pulse width modulation—controlled full bridge inverter, and three-phase transformer.

The output of those energy conversion systems are combined through the IPT to fixed DC link voltage for required input voltage of a full bridge inverter. The HRES is illustrated in Fig. 10.1 [13].

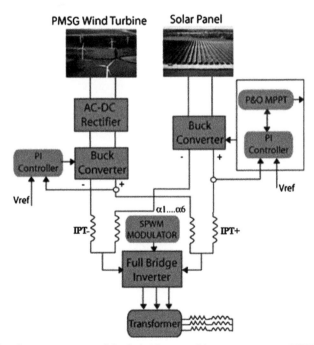

Figure 10.1 Graphical representation of the hybrid renewable energy system. *MPPT*, maximum power point tracking; *PI*, proportional—integral; *PMSG*, permanent-magnetic synchronous generator; *SPWM*, sinusoidal pulse width modulation. *Reprinted from Kabalci E. Design and analysis of a hybrid renewable energy plant with solar and wind power. Journal of Energy Conversion and Management 2013;72(0):51—9 with permission from Elsevier.*

10.2.1 Wind Energy Conversion System

A wind turbine framework produces energy by changing over the streaming wind speed into mechanical energy after which it is converted into power. The force contained in the wind dynamic energy is communicated by Refs. [13,14]:

$$p = \frac{1}{2} \times \rho \times A \times V^3 \tag{10.1}$$

where A is the area traversed by the wind (m^2), ρ is the air density (1.225 kg/m^3), and V the wind speed (m/s).

The electrical power is given by:

$$p_w = \frac{1}{2} \times \rho \times C_e \times A \times V^3 \times 10^{-3} \tag{10.2}$$

where C_e is the coefficient of the wind turbine's performance, according to Bertz $C_{e\text{-}Limit} = 0593$.

Therefore the energy produced by the wind turbine is given by:

$$E_w = P_w \times \Delta t \tag{10.3}$$

10.2.2 Photovoltaic System

The PV generator contains modules which are composed of many interconnected solar cells in series/parallel to form a solar array. The energy created by the PV generator is given by the mathematical statement [15]:

$$E_{PV} = A \times \eta_m \times P_f \times \eta_{PC} \times I \tag{10.4}$$

where A is the total area of the PV generator (m^2), η_m is the module efficiency (0.111), P_f is the packing factor (0.9), η_{PC} is the power conditioning efficiency (0.86), and I is the hourly irradiance (kWh/m^2).

10.2.3 Sizing of the Hybrid Power System

There are a few strategies for estimating the crossover PV/wind frameworks, for example, the yearly month-to-month normal system [16]; the most unfavorable month system [17], or the Loss of Power Supply Probability method [18]. Alternative techniques for the estimation can also use the programming bundles, for instance, the information of the most unfavorable months. From Table 10.1, the unfavorable light month and unfavorable wind speed month are December and January, respectively. Thus we can apply the most unfavorable strategy for our crossover framework measurement as shown by [17].

Table 10.1 Diesel Generator Consumption

Generator Loading (%)	Fuel Consumption (L/h)	Fuel Consumption (L/kWh)
75	1.98	0.26
50	1.56	0.31
30	1.04	0.52

Reprinted from Kanzumba K, Jacobus V. Hybrid renewable power systems for mobile telephony base stations in developing countries. Journal of Renewable Energy 2013;51:419–25 with permission from Elsevier.

The month-to-month energy delivered by the framework per unit territory is
$E_{PV,m}$ (kWh/m^2) for PV,
$E_{W,m}$ (kWh/m^2) for wind energy, and
$E_{H,m}$ (kWh/m^2) (where $m = 1, ..., 12$ refers to the month of the year).
The most noticeably bad month is an element of the month-to-month burden request, the RE assets and the framework parts' execution. The size (m^2) of the generator expected to guarantee full scope (100%) load (E_{Load}) amid a month is given by Ref. [19]:

$$A_i = \max \frac{E_{Load,m}}{E_{i,m}} \tag{10.5}$$

where A_i represents the size in square meters of the PV or wind component.

The total energy produced by the PV or wind generators and supplied to the load is expressed as [19]:

$$\sum E_i A_i = E_{Load} \tag{10.6}$$

with:

$$E_i \cdot A_i = f_i \cdot E_{Load} \tag{10.7}$$

where f_i is the part of the stock pile supplied by the PV or wind sources.

At each minute, the whole of the divisions of energy commitment from every segment supplied to the stock pile must be equivalent to 1.

$$\sum f_i = 1 \tag{10.8}$$

with $0 \leq f_i \leq 1$.

The genuine size is figured by the surface unit of the part ($S_{PV,u} = 0.3$ m^2 and $S_{W,u} = 0.65$ m^2). The span of the stockpiling battery is resolved from the month-to-month most extreme burden request $E_{L,max}$, and it can be communicated by Ref. [20] as:

$$C_{bat} = \frac{E_{L,max} \times 1000 \times \Delta t}{V_{sys} \times N_m} \tag{10.9}$$

where V_{sys} is the voltage of the framework; N_m the quantity of days in the most exceedingly bad month; Δt the time in days of independence required, which is characterized by the fashioner (for basic loads, for example, information transfers, this quality may be set to 5 days).

With respect to the PV and wind parts, the genuine size of the battery is resolved from the limit of a battery unit $C_{Bat,u}$.

From the site details, the capacity of the accessible RE assets and the stock pile estimation can be computed to measure the mixed power framework. It is made of two 7.5-kW wind generators, 8-kW PV model, 7.5-kW inverter (48 V DC data, 220 V air conditioning yield), and 114 batteries (6 V, 360 Ah) for a 48-V framework voltage [19].

10.2.4 Microgrid System

A microgrid is a system where energy is produced locally to supply local demands that are connected to the microgrid [21], which is often connected to the low carbon and RE systems that generate energy to meet the local energy demands [22]. A microgrid can be applied to a small level of energy demand, a single building, for example, providing energy up to large communities, such as, towns, villages, and islands [23]. As microgrids produce energy on-site, it is possible for heating demands to be met through the by-product of electricity generation when using cogeneration plant, such as combined heat and power [24].

Microgrids are decentralized from the traditionally centralized grid and can be considered as a single entity apart. However, microgrids can be connected to the centralized grid to allow for import and export of energy [25], which allows the microgrids to be considered as a source of highly reliable power as they can produce energy locally when it is possible while selling the surplus energy to the grid and draw from the grid when there is a deficit in the local supply. Furthermore, microgrids can take advantage of storage technologies, such as batteries and hot water storage systems to increase the efficiency of the energy produced [20]. It is for this reason that microgrids are being considered as one of the most important developments in future energy production.

10.2.4.1 Microgrid Feasibility Framework

The steps taken to locate a decent microgrid for a given plan are vital as they will choose whether or not a plan is going to address the issues of the considerable number of partners or just to figure out how to meet some of their requests. The necessities of partners in hybrid energy system microgrids fluctuate and differ with one another. The approach taken in carrying out a feasibility study of potential microgrids cannot just focus on meeting the demands alone, as expressed some time recently, because the plan must also be feasible monetarily with the goal that it can be viewed as a sound speculation and has more chances of being actualized in the pragmatic sense of economy. Fig. 10.2 shows a structural procedure in a small-scale framework [12].

Figure 10.2 Approach framework process in microgrid system. *Reprinted from Tait L. Decision making framework and financial tool for hybrid energy system micro-grids [A thesis submitted in partial fulfilment for the requirement of the degree Master of Science]. 2011:18.*

There is a need to establish the central acceptable parameters of the considerable number of requests of the partners, a signed up way to deal with the way that plans are envisioned from the initial plausibility for arrangement of onward actions. To this point, the accompanying signed up approach enumerates important steps to be attended to guarantee that a partner in a plan settles on an appropriately powerful choice. This procedure has been created to work in conjunction with the Legitimacy Energy Programming and the monetary related choice as a guiding apparatus to get the best choice [12].

There are a few models that are now available for the retrial, reconstruction, and examination of small-scale network frameworks that strictly consider variables, for example, magnitude of request, atmospheric conditions, site geographical conditions and accessibility, safety from seasonal floods, and availability of fund.

Some products from the above examples are [12]:

1. Energy generation device created by the College of Strathclyde, which permits the client to complete examination of RE plans through the utilization of interested profiles.

2. RETScreen created by the Canadian Government that permits clients to do examination on low carbon energy framework ventures. RETScreen permits the client to survey the possibility of a plan fiscally utilizing manufactured product as a part of monetary investigation.

3. Homer developed by National Renewable Energy Laboratory has been used to carry out examination on any framework to be mixed with low carbon energy frameworks. It permits the client to break down the electrical attributes of executing renewables into pragmatic structures and miniaturized scale lattices.

The subsequent stage in applying the system is to use a possibility instrument, for example, Legitimacy to locate various conceivable situations that could be connected to meet the association's requests. This examination ought to concentrate on the designing qualities that are required for the miniaturized scale lattice to be gainfully feasible, good competitive rate, low energy shortage, diminished CO_2, and so forth. There must be investigation of the monetary possibilities of a task to permit a financial specialist to judge the advantages and pitfalls of putting resources into such a plan. This is accomplished through the utilization of an Exceed expectations monetary choice guide instrument that has been created to permit the client to analyze the intricacy of the venture through the designing advantages in the Legitimacy concept [26].

10.3 HYBRID CONTROLLERS FOR SOLAR AND WIND ENERGY SYSTEMS

The controller is the main part of any hybrid RE system. The hybrid controller is an appropriate apparatus to control the proper production of energy from RE sources. The controller may be developed in many methods in tandem with the reason for using it. The advanced system of hybrid controller is developed for a comfortable interface between two or more systems at a time. The solar/wind hybrid controller main advantages are the increased power output; better interaction between wind turbine, solar PV, and battery or any necessary equipment monitoring the system; longer battery life; breaking routine, etc.

In China, the renewable solar/wind energy industry is developing rapidly but the problem is hard to sell the electricity to the power grid because of the instability and discontinuity in the electricity supply. To solve this problem, they used a hybrid power generator to break the development of bottleneck of the RE generation, especially, the unstable RE. For the hybrid control, the Ideal Matter-Element Extension method was used and it included the Fuzzy Comprehensive Judgment Method, the Analytic Hierarchy Process, Gray Clustering Method, and Geographic Information Systems. The hybrid control was applied on three different sites that were good for power generation and distribution [27]. Table 10.2 shows a comparative result of different sites and their power production capacities.

Applying different methods for improvising the energy level from the hybrid system can be quite costly with the resultant uncertainty of the energy availability. To solve this problem, it needs a proper guidance of power management, which includes a storage system or backup sources to provide the necessary power supply with the reliability of the RE system the total sum of which can reduce the cost of energy production by the hybrid RE system. Fig. 10.3 shows a management of renewable hybrid system strategy [28].

Table 10.2 Different Sites Power Production by Using Ideal Matter-Element Extension Method

Districts or Site	Solar Installed Capacity (MW)	Wind Installed Capacity (MW)	Extension Method	Ideal Extension Methods
Inner Mongolia (site 1)	230	1590	Great	1
Qinghai (site 2)	1003	770	Good	2
Tibet (site 3)	90	15	Good	3

Reprinted from Xiangjun L, Yong L, Xiaojuan H, Dong H. Application of fuzzy wavelet transform to smooth wind/PV hybrid power system output with battery energy storage system. Energy Procedia 2011;12:994−1001 with permission from Elsevier.

This is the basic configuration but different configurations can also be applied. It depends on the energy utilization, available RE sources, and the customer loads. The hybrid system can combine more than one energy source and the size optimization can be taken to reduce its capital cost [28].

10.3.1 Fuzzy Base Controller

The method of fuzzy controller is based on a logic control. It is applicable in the RE technology. Many traditional methods are already used in this RE field but fuzzy logic is a powerful controller to solve easily the energy management system problem. The researchers have developed many methods based on the artificial neural network for solving optimum operation, forecasting the meteorological data for energy performance and demand for the user of local energy generation. The development of unoptimized fuzzy logic controller (FLC) uses the optimal power flow and the optimal FLC is used for the Hopfield neural network control. From the researchers' report, the optimal FLC is better than the unoptimized FLC and the PI controller. Fig. 10.4 shows an example of using the FLC in hybrid solar/wind system [29].

As can be seen in Fig. 10.4, the system contains a PV array, wind turbine, an electrolyzer, a stack of batteries, a fuel cell, and a hydrogen storage tank that are connected to a DC bus and an inverter, where the power flow is regulated by using an optimized FLC. The wind turbine power production (kW) equation has been given by Ref. [30].

$$P_{wind}(t) = \begin{cases} P_{max,wind} \cdot \dfrac{V(t) - V_c}{V_R - V_c} & \text{if } V_c \leq V(t) \leq V_R \\[2mm] P_{max,wind} & \text{if } V_R \leq V(t) \leq V_F \\[2mm] 0 & \text{if } V(t) < V_C \cup V(t) > V_F \end{cases} \tag{10.10}$$

where $P_{max,wind}$ is a nameplate rating of wind turbine; $V(t)$ is wind speed defined at time t (m/s); V_c, V_R, and V_F are the characteristic parameters determined (m/s).

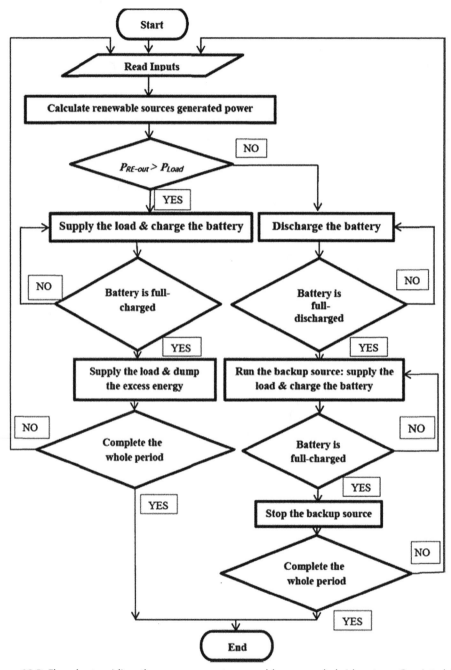

Figure 10.3 Flowchart guiding the management renewable energy hybrid system. *Reprinted from Ismail M, Moghavvemi M, Mahlia T, Muttaqi K, Moghavvemi S. Effective utilization of excess energy in standalone hybrid renewable energy systems for improving comfort ability and reducing cost of energy: a review and analysis. Journal of Renewable and Sustainable Energy Reviews 2015;42:726—34 with permission from Elsevier.*

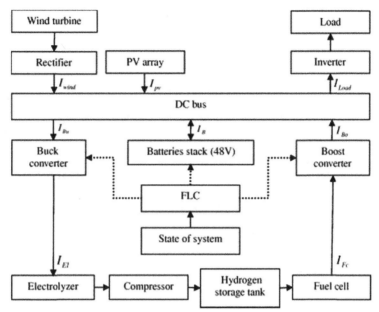

Figure 10.4 Block diagram of hybrid system using fuzzy logic controller (FLC). *Reprinted from Safari S, Ardehali M, Sirizi M. Particle swarm optimization based fuzzy logic controller for autonomous green power energy system with hydrogen storage. Journal of Energy Conversion and Management 2013;65:41−9 with permission from Elsevier.*

The solar PV power generated (kW) with N_S modules is produced in series, whereas the N_P modules are in parallel. The effects of temperature are accounted for from the given equation [18]:

$$P_{pv}(t) = N_P \cdot N_S \cdot \frac{\frac{V_{OC}}{\eta_{MPP}KT/q} - In\left(\frac{V_{OC}}{\eta_{MPP}KT/q} + 0.72\right)}{1 + \frac{V_{OC}}{\eta_{MPP}KT/q}}$$

$$\cdot \left(1 - \frac{R_s}{V_{OC}/I_{SC}}\right) \cdot I_{SCO}\left(\frac{G}{G_0}\right)^{\alpha} \cdot \frac{V_{OCO}}{1 + \beta In\frac{G_0}{G}} \cdot \left(\frac{T_0}{T}\right)^{\gamma} \cdot \eta_{MPPT}\eta_{oth}$$

(10.11)

where V_{OC} and I_{SC} are the open circuit voltage (V); η_{MPP} is the ideality factor ($1 < \eta_{MPP} < 2$); K is the Boltzmann constant (1.38×10^{-23} J/K); T, T_0 are the temperatures (K); q is the magnitude of the electron charge ($1.6 \times 10^{-19\circ}$C); R_s is the series resistance (Ω); α is the exponent nonlinear effects; β is a PV module coefficient and l is the exponent of nonliner temperature voltage effects; G_0, G are two different solar irradiance; η_{MPPT} is the MPPT; and η_{oth} is the factor reprinting other losses.

The electrical charge is stored by the battery stack and the battery current (A) is given by this equation [31]:

$$I_B(t) = I_{pv}(t) + I_{wind}(t) + I_{Bo}(t) - I_{Bu}(t) - I_{LOAD}(t) \qquad (10.12)$$

where I_{pv}, I_{wind}, I_{Bo}, I_{BU}, and I_{LOAD} are currents (A) of PV array, wind turbine boost converter, Buck converter, and load respectively.

The battery voltage (V) equation is as shown below:

$$U_B(t) = (1 + at)U_{B,0} + R_i I_B(t) + K_i Q_R(t) \qquad (10.13)$$

where a is the self-discharge rate (S^{-1}), $U_{B,0}$ is the open circuit voltage (V), R_i is the internal resistance (Ω), K_i is the polarization coefficient (Ω/h), and $Q_R(t)$ is the accumulated charge (Ah).

The battery energy storage is given by this equation (Ah):

$$E_B(t) = E_{B,0} + \frac{1}{3600} \int I_B(t) dt \qquad (10.14)$$

where $E_{B,0}$ is the battery initial stored energy (Ah) and the battery state of charge (SOC) (%) is defined in the following equation:

$$SOC(t) = 100 \times \frac{E_B(t)}{E_{B,\max}} \qquad (10.15)$$

where $E_{B,max}$ is the total capacity of the battery stack (Ah).

A practical process of fuzzy control system takes a given value after which the value is passed through a fuzzification process and it is then processed by an inference engine. Finally, that value is obtained through a defuzzification process [32]. Fig. 10.4 shows a fuzzy logic control process that can maintain the equations from (10.10) to (10.15). Here the input values of fuzzy logic control are the net power flow and the batteries, where the net power flow is the difference between the power productions and power consumed. The FLC input and output values are the optimized and the unoptimized results as shown in Fig. 10.5, where the (A) and (B) are the inputs and (C) is the output. As can be seen in Fig. 10.5, N is the Negative fuzzy set, Z is the Zero, P is the Positive, L is the Low, M is the Medium, and H is the High, respectively [31].

The fuzzy logic control base overall energy management system is as shown in Fig. 10.6. This management system is based on the initial control by using a battery or an electrolyzer, which is used to maintain the overall utilization costs. Fig. 10.6 shows on the right side the utilization costs by using FLC.

This controller is used for computing the net force (Pnet). To start with, first obtain the net force and then subtract the net force from the renewable wind/sunlight-based

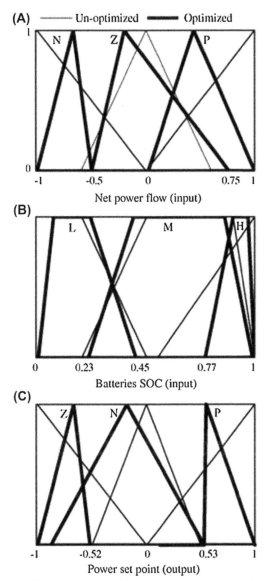

Figure 10.5 (A, B) Input and (C) output value of fuzzy logic controller. *Reprinted from Safari S, Ardehali M, Sirizi M. Particle swarm optimization based fuzzy logic controller for autonomous green power energy system with hydrogen storage. Journal of Energy Conversion and Management 2013;65:41—9 with permission from Elsevier.*

boards. If the net force is correct, it implies that the force requested by the stock pile is higher than the force generated by the renewable sources and the fuzzy control or battery will work. If the net force is negative it implies that the force generated by the renewable

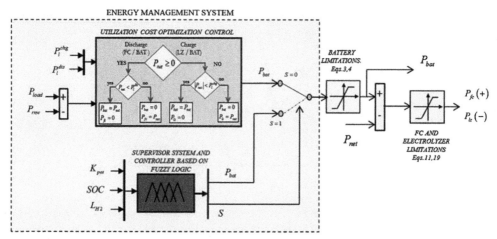

Figure 10.6 Overall energy management and utilization cost optimization system. *SOC*, state of charge. *Reprinted from García P, Torreglosa J, Fernández L, Jurado F. Optimal energy management system for stand-alone wind turbine/photovoltaic/hydrogen/battery hybrid system with supervisory control based on fuzzy logic. International Journal of Hydrogen Energy 2013;38(33):14146–58 with permission from Elsevier.*

wind/sunlight-based sources is higher than the force requested by the stock pile. This is the idea of the fuzzy logic base controller application [33].

10.3.2 Fuzzy-PID Base Controller

The fuzzy-PID (proportional–integral–derivative) hybrid controller in the hybrid solar radiation/wind system is a powerful controller in the electric power management system. The traditional control theories may not solve all the controlling problems. The combined fuzzy and PID control is presented to solve the problem of battery management models of wind turbine generator and PV array.

Most of the hybrid wind/solar radiation systems are using batteries for storing the electric charge. The DC rectified by three-phase form wind and PV charge the batteries via a controller which then controls the functions of electric charge storage and adjustment. When the sources (wind/solar radiation) supply electrical energy, the batteries store it and discharge it to the load side. The fuzzy intelligence controller is used to switch and regulate the working state of batteries or float charge after which the continuity of power generation and supply system are activated. Fig. 10.7 shows a combined system using the fuzzy and PID controller.

The controlling method of the combined system is using a double-closed loop, which is used for battery charge and discharge of renewable solar radiation/wind power plants. Fig. 10.8 shows a combined controller fuzzy and PID system.

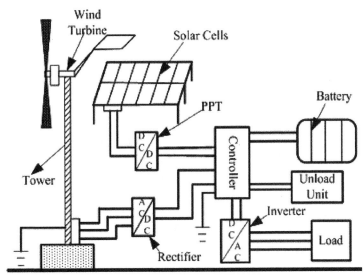

Figure 10.7 Wind—solar hybrid system by using combined fuzzy-PID controller. *PID*, proportional—integral—derivative; *PPT*, power point tracking. *Reprinted from Zhang F, Shi Q, Wang Y, Wang F. Simulation research on wind solar hybrid power system based on fuzzy-PID control Proceeding of international conference on electrical machines and systems 2007:338—42 (IEEE Xplore).*

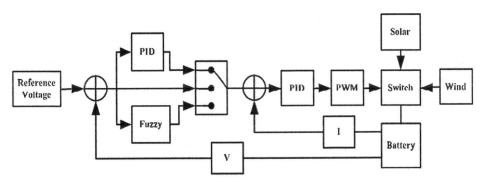

Figure 10.8 Diagram of fuzzy-PID combined control system. *PID*, proportional—integral—derivative; *PWM*, pulse width modulation. *Reprinted from Zhang F, Shi Q, Wang Y, Wang F. Simulation research on wind solar hybrid power system based on fuzzy-PID control Proceeding of international conference on electrical machines and systems 2007:338—42 (IEEE Xplore).*

The inner circle is controlled by the PI control for charged current circle and the outer circle is for the charged voltage circle. The MPPT is used to control the PID. The entire framework is controlled by the fluffy PID control individually by an exchanging structure. At the point when the battery needs a quick charge, the PID controls the huge current; however, when the charge is around 80—90% full, it will then switch to a charge mode control by using the fluffy control and the charge voltage becomes constant. These sorts of controllers have a great impact on the battery charging process and in the administration of renewable solar energy/wind hybrid system [34].

10.3.3 PID, Virtual Instrument and National Instrument Controller

Other consolidated controls are the PID and virtual instrument and national instrument (NI) controls. These blends are a development framework for controlling purposes in the sunlight-based wind half breed framework. The National Instrument is controlled by the microcontrollers that can be used for diverse examination reasons. The LabView is programmed for interfacing between the PC and the NI with different gadgets. It is used to decide on MPPT, I—V, P—V, most extreme current, voltage, force point, fill element (FF), and so on in the sunlight-based PV board and wind turbine for demonstration, force ascertainment and test by utilizing the LabView programming, and the NI gadget. In power reproduction cases, a few specialists have utilized the MATLAB. However, power reproduction utilizing the LabView in which the graphical programming has replaced the routine programming systems and viable client interface has become all the more simple to utilize. The sunlight-based PV cell circuit model and the mathematical statement relating to the current and voltage of the cell is as shown in Fig. 10.9.

The basic equation of the current from the circuit is defined by Kirehhoff's current law [35]:

$$I = I_{ph} - I_d - I_{sh} \tag{10.16}$$

The diode current and voltage are defined by this equation:

$$I_d = I_o[\exp(V_d q / nKT_c) - 1] \tag{10.17}$$

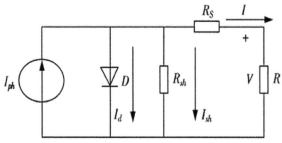

Figure 10.9 Single-diode equivalent circuit for photovoltaic cell, where I_{ph} is the solar cell photo-current, D is the diode, I_d is the diode current, R_{sh} and R_s are the series and shunt resistances, I_{sh} is the shunt current, and R is the load (V). *Reprinted from Chouder A, Silvestre S, Taghezouit B, Karatepe E. Monitoring, modelling and simulation of PV systems using LabVIEW. Journal of Solar Energy 2013;91:337—49 and Jaleel J, Nazar A, Omega A. Simulation on maximum power point tracking of the photovoltaic module using LabVIEW. International Journal of Advanced Research in Electrical, Electronics and Instrumentation Engineering 2012;1(3):190—9 (International conference).*

$$V_d = IR_s + V \tag{10.18}$$

where I_o is reverse saturation current at T_c (A) and V_d is diode voltage.

The single-diode equivalent circuit for the PV cell and I—V equation can be represented by:

$$I = I_{sc} - I_o \left[\exp\left\{ \frac{(V + IR_s)q}{nKT_c} \right\} - 1 \right] - \frac{V + IR_s}{R_{sh}} \tag{10.19}$$

when the PV cell is open circuited, the maximum voltage is produced. That is called V_{oc}.

$$V_{oc} = In\left(\frac{I_{sc}}{I_o} + 1 \right)\left(\frac{nKT_c}{q} \right) \tag{10.20}$$

The reverse saturation current is represented by this equation:

$$I_{or} = \frac{I_{scr}}{\left[Exp\left(\frac{qV_{ocr}}{nKT_r} \right) - 1 \right]} \tag{10.21}$$

where the fill variable of the PV cell is the proportion between the maximum force and the open circuit voltage and the short out current. The I—V bend of PV cell is made due to resistive losses. This is an important component to decide on the nature of the PV cell. The fill element is represented by this comparison:

$$FF = \frac{I_m \times V_m}{I_{sc} \times V_{oc}} \tag{10.22}$$

where V_m and I_m are defined as the voltage and current values at the maximum power point.

$$R' = \frac{V_m}{I_m} \tag{10.23}$$

where R' is the load resistance for operating the power point of the PV and the efficiency is given by this equation:

$$\eta = \frac{I_{sc} \times V_{oc} \times FF}{P_{in}} \tag{10.24}$$

The mathematical statements from (10.16) to (10.24) are expressions to the regular I—V and P—V portrayal utilizing the LabView programming, which are the same portrayal of the PV cell that appears in Fig. 10.10A. Fig. 10.10B illustrates diverse temperatures I—V and P—V focuses in which it can be seen that the most extreme force focuses are additionally distinctive. It is possible to estimate the MPPT utilizing the LabView and the National Instrument [36].

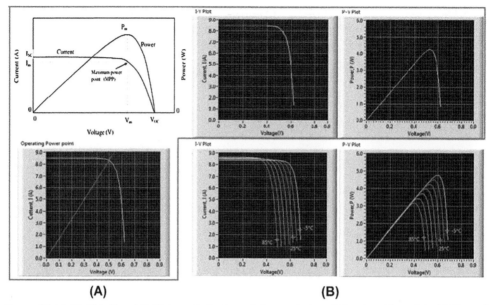

Figure 10.10 (A, B) I—V and P—V characterization of photovoltaic cell. *Reprinted from Jaleel J, Nazar A, Omega A. Simulation on maximum power point tracking of the photovoltaic module using LabVIEW. International Journal of Advanced Research in Electrical, Electronics and Instrumentation Engineering 2012;1(3):190—9 (International conference).*

The characteristic of the PV cell is most affected in an open circuit voltage (V_{oc}) because of its dependence on the saturation current (I_c) as given by the following equation:

$$I_o = I_{or} \left[\frac{T_c}{T_r}\right]^3 \exp\left[\frac{qE_g}{nK}\left\{\frac{1}{T_r} - \frac{1}{T_c}\right\}\right] \tag{10.25}$$

where I_{or} is the saturation current at T_r and E_g is the energy band gap.

The PV cell power output is dependent on the solar irradiation because the light generates the current I_{ph}, which is dependent on the solar irradiation and for this reason I_{sc} increases with the increase in the incident irradiation.

$$I_{sc} = \frac{G}{1000}\left[I_{scr} + K_i(T_c - T_r)\right] \tag{10.26}$$

where G is the incident solar irradiation in W/m^2, I_{scr} is the short circuit current, T_r is the reference current, K_i is the temperature coefficient, and T_c is the cell temperature.

The LabView for wind turbine control outline and reproduction model, including the mechanical drive train, generator, and power framework are highly sophisticated parts in the control framework that can interface with the principle turbine and that can then get the continuous information from the framework. The control outline and reproduction module to break down the communications between the hybrid mechanical—electrical systems also appears in Fig. 10.11A and B.

It is possible to enhance the existing models through innovative investigations and other controlling procedures for more sophisticated generators and similarly for more intricate drive train modules. The essential mathematical statement for the wind turbine generator model and power generated can be clearly understood by comparing Eqs.(10.1) and (10.2). The tip-speed proportion of wind turbine is given by this comparison [37]:

$$\lambda = \frac{R\omega}{v} \tag{10.27}$$

where ω is the wind turbine angular speed and the aerodynamic torque can be developed as:

$$T_a = \frac{1}{2}\rho\pi C_T(\lambda, \beta)v^3 R^3 \tag{10.28}$$

where $C_T(\lambda,\beta)$ is the torque coefficient, which can be expressed as:

$$C_T(\lambda, \beta) = \frac{1}{\lambda}C_P(\lambda, \beta) \tag{10.29}$$

The fitting functions of $C_P(\lambda,\beta)$ is written as:

$$C_P(\lambda, \beta) = C_1\left(\frac{C_2}{\lambda_i} - C_3\beta - C_4\right)e^{\frac{C_5}{\lambda_i}} + C_6\lambda \tag{10.30}$$

$$\frac{1}{\lambda_i} = \frac{1}{\lambda + 0.08\beta} - \frac{0.035}{\beta^3 + 1} \tag{10.31}$$

where $C_1, \ldots, C_6$ are the undetermined coefficients according to the characteristics of the wind turbine.

10.4 HYBRID SOLAR/WIND ENERGY APPLICATION

The technology of a hybrid system has been visualized concerning the distinctive national desire on getting information on the climatic conditions of the world due to the importance for diverse nations on the energy administration technique that uses a mix of wind turbine, PV, energy unit, or hitter. The blend of the two energies (wind turbine and PV) is also used to produce hydrogen. Presently hydrogen generation is viewed as a piece of RE. In Fig. 10.12, this is a case of a mix of PV and wind turbine framework to obtain hydrogen through electrolysis [38].

This is a 10-kW electrolyzer framework that can be used to get much hydrogen. The PV board slope is equivalent to the edge of the latitudinal position of Sahand (37.2°N).

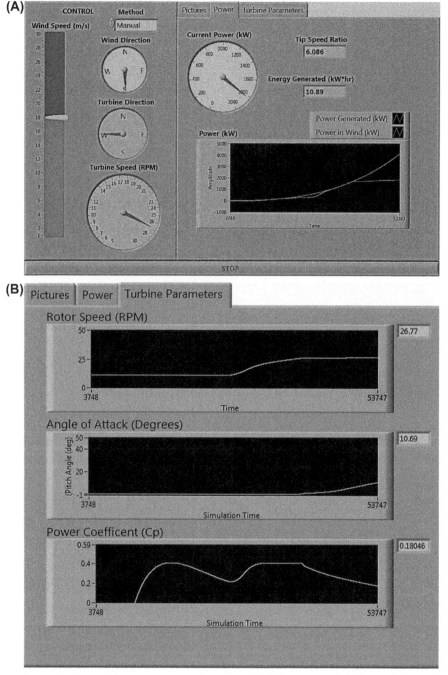

Figure 10.11 (A) Wind turbine power control system with LabVIEW Simulink. (B) Wind turbine parameters control system with LabVIEW Simulink. *Reprinted from LabVIEW. Simulating a wind turbine. National Instruments 2011 and Beula S, Arivuselvam B. Online conditional monitoring for solar and wind energy system. Proceeding of 5th national conference on VLSI 2014 (LabVIEW Website and International conference).*

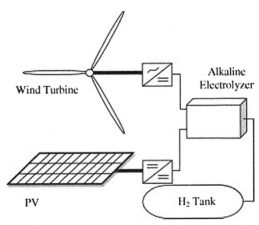

Figure 10.12 Combined hybrid wind and solar system. *PV, photovoltaic. Reprinted from Khalilnejad A, Riahy G. A hybrid wind-PV system performance investigation for the purpose of maximum hydrogen production and storage using advanced alkaline electrolyzer. Journal of Energy Conversion and Management 2014;80:398–406 with permission from Elsevier.*

The sun-powered radiation of this site is expected to be 1000 W/m². The attributes of PV modules are PV range 0.85 m², V_{oc} is 22.2 V, I_{sc} is 5.45 A, V_{mpp} is 17.2 V, and I_{mpp} is 4.9 A. The wind turbine qualities are as shown in Fig. 10.13.

Fig. 10.14A and B illustrates the force generation structure of hybrid sources and power utilization in the electrolyzer. The energy generation from RE sources can be found in Fig. 10.14A, where the energy generation is initiated from the stock pile before being supplied to the electrolyzer [38]. This unused energy is highly needed for

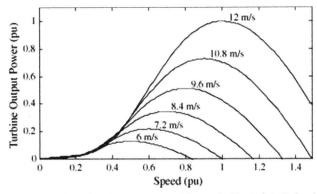

Figure 10.13 Characteristic of wind turbine. *Reprinted from Khalilnejad A, Riahy G. A hybrid wind-PV system performance investigation for the purpose of maximum hydrogen production and storage using advanced alkaline electrolyzer. Journal of Energy Conversion and Management 2014;80:398–406 with permission from Elsevier.*

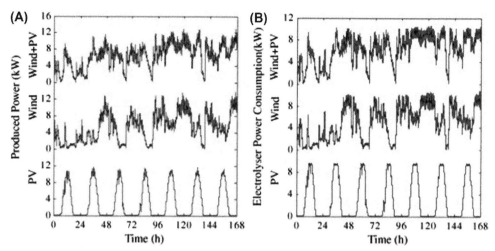

Figure 10.14 (A, B) Combined power production of renewable energy sources. *PV, photovoltaic. Reprinted from Khalilnejad A, Riahy G. A hybrid wind-PV system performance investigation for the purpose of maximum hydrogen production and storage using advanced alkaline electrolyzer. Journal of Energy Conversion and Management 2014;80:398—406 with permission from Elsevier.*

hydrogen generation. Fig. 10.14B illustrates the energy utilization by the electrolyzer where the energy generation and utilization do not have a large difference.

In the process, the normal rate of hydrogen generation from a hybrid system is 0.00173 mol/s and the energy generation from a wind turbine is (49.8%) higher than the PV in a mixed framework.

Another examination on utilizing the solar radiant energy and wind hybrid system for hydrogen generation has been done in the areas of Kopaonik National Park, Novi Sad, and Belgrade city. This undertaking used the PV module BP SX3200 (200 W). Fig. 10.15 shows a fundamental schematic outline of a mixed solar-powered wind framework that can replace any PV board module and it can also replace the wind turbine suitable for the climate condition and request.

The board is combined with a 50-cell module and the aggregate range is 1.406 m^2. This task used distinctive wind turbine limits and the wind turbine centers had three individual sizes of 10, 15, and 20 m, respectively. Table 10.3 illustrates the diverse WT limits and the electrical energy yield. The most extreme wind speeds for the three wind turbines were 3.85, 4.48, and 5.21 m/s, respectively, at 20 m height in Novi Tragic, Belgrade, and Kopaonik [39].

Table 10.3 clearly shows the 5-kW wind turbine that can generate more electrical energy. The investigation on the above set used the alkaline electrolyzer (HySTAT). The sun-oriented irradiance was found to be from 800 to 1000 W/m^2 and the cell temperature was 25°C. The photocopy of the solar-based electrical energy PV module is

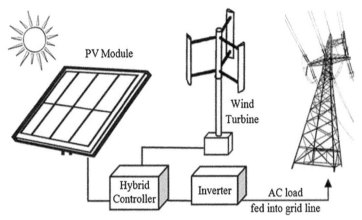

Figure 10.15 Basic schematic design of hybrid wind—solar system. *PV, photovoltaic. Reprinted from Chong W, Naghavi M, Poh S, Mahlia T, Pan K. Techno-economic analysis of a wind—solar hybrid renewable energy system with rainwater collection feature for urban high-rise application. Journal of Applied Energy 2011;88(11):4067—77 with permission from Elsevier.*

Table 10.3 Various Wind Turbine Capacities and Their Output Electrical Energy

	Wind Turbine Capacities (kW)	Novi Sad	Belgrade	Kopaonik
Bergey wind power	1	767.7	983.0	1550.2
West wind power	3	1472.1	1840.0	2890.6
West wind power	5	3086.2	3930.0	6179.2

Reprinted from Vukman V, Milada L, Marina P, Valentina M, Biljana S, Nikola S. Technical analysis of photovoltaic/wind systems with hydrogen storage. Journal of Thermal Science 2012;16:865—75.

as shown in Fig. 10.16. The electrical energy from the PV for the three spots were not all confirmed. The normal, most extreme, and least energy generation were recorded in July and December (40.6 and 10.4 kWh) at Belgrade city.

The aggregate electrical energy of crossover framework and hydrogen generation from diverse sources in Kopaonik park are shown in Table 10.4. It is additionally specified that the rate of renewable energy standard in three better places also vary between one and another.

The hydrogen generation and application use the crossbreed wind-close solar radiant energy system, which itself is a RE source. The blend of three sources, namely, wind, sun-oriented, and hydroelectric crossbreed framework, is material for open matrix or miniaturized scale network that is shown in Fig. 10.17 [40]. The use of half and half framework can permit the generation of energy from different sources while giving a nonstop supply. The solar radiation-based PV applications and uses are exceptionally acceptable innovations in the RE generation field. In any country, particularly which

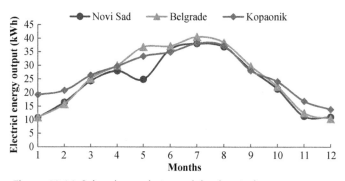

Figure 10.16 Solar photovoltaic module electrical energy output.

Table 10.4 Electrical Energy and Hydrogen Production

		PV With 1-kW Wind Turbine	PV With 3-kW Wind Turbine	PV With 5-kW Wind Turbine
Novi Sad city	Electrical energy output (kWh)	6728	7430	9050
	Hydrogen production (Nm³)	141	162	228
	Renewable energy annually (RES %)	64	71	86
Belgrade city	Electrical energy output (kWh)	7185	8040	10,130
	Hydrogen production (Nm³)	156	182	264
	Renewable energy annually (RES %)	68	76	96
Kopaonik park	Electrical energy output (kWh)	7950	9290	12,500
	Hydrogen production (Nm³)	172	219	361
	Renewable energy annually (RES %)	75	88	119

PV, photovoltaic, *RES,* renewable energy standard.
Reprinted from Vukman V, Milada L, Marina P, Valentina M, Biljana S, Nikola S. Technical analysis of photovoltaic/wind systems with hydrogen storage. Journal of Thermal Science 2012;16:865−75.

does not have any electrical power supply, it is prudent to mobilize the implementation of power source from half wind energy and half renewable solar energy (PV). It is to be noted that even now we have electric automobiles that use the hybrid system to charge their batteries.

10.4.1 Advanced Hybrid System

The development of half and half solar radiation-oriented energy and wind energy framework has supplied enormous amount of energy to the electric network through the use of half and half board sunlight-based PV, wind turbine, and electrolyzer. The controlling mechanism is in the middle of the boards using an electric matrix to apply some sort of control on the corresponding (PI) where the electrolyzer is joined with the DC side of the multilevel current source inverter, which supplies electric energy to the electrolyzer to produce hydrogen, which then goes to the energy component to convert

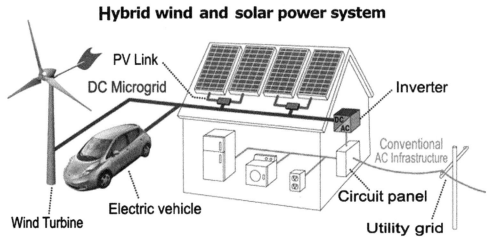

Figure 10.17 An application of hybrid solar and wind power system, vehicle charging. *Reprinted from Valigra L. Turbine maker links wind and solar with new technology. Pika Energy 2013; pika-energy.com (Internet).*

the electric energy for storage in the stock pile. The entire framework is as shown in Fig. 10.18 [41].

The first proposed ideas to utilize the wind energy in engineering structures were mooted in Germany in the 1930s and 1940s. The outline of this framework is created by

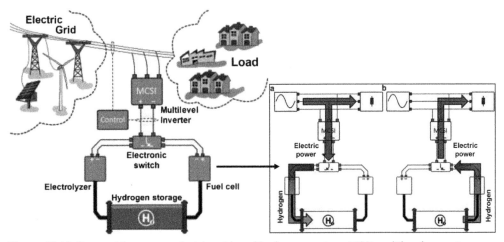

Figure 10.18 Renewable sources, electric grid, and hydrogen system. *MCSI*, multilevel current source inverter. *Reprinted from Aguirre M, Couto H, Valla M. Analysis and simulation of a hydrogen based electric system to improve power quality in distributed grids. International Journal of Hydrogen Energy 2012;37(19):14959—65 with permission from Elsevier.*

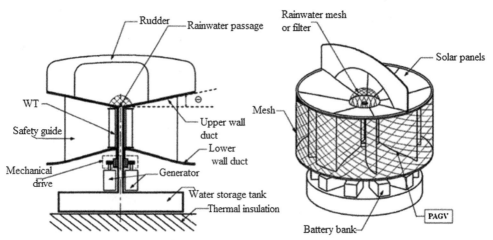

Figure 10.19 Basic arrangement of the hybrid solar–wind energy system with rain water collection. *PAGV*, power augmentation guide vane; *WT*, wind turbine. *Reprinted from Chong W, Naghavi M, Poh S, Mahlia T, Pan K. Techno-economic analysis of a wind–solar hybrid renewable energy system with rain-water collection feature for urban high-rise application. Journal of Applied Energy 2011;88(11):4067–77 and Zhou W, Lou C, Li Z, Lu L, Yang H. Current status of research on optimum sizing of stand-alone hybrid solar–wind power generation systems. Journal of Applied Energy 2010;87:380–9 with permission from Elsevier.*

Chong et al. to join a half and half solar radiant energy and wind energy framework at the highest point of an elevated structure. This framework can keep running with a low wind speed and can also take advantage of a rainy downpour to speed up the fan velocity as well as a water tank to store the rain water [42]. The graphical course of action is as shown in Fig. 10.19.

The above system is a cylindrical shape or different shape of design can be fixed up of the building roof design as well. A power augmentation guide vane (PAGV) is placed in the middle of the wind turbine together with the safety guide and a rudder, which is fixed to the PAGV, all of which can also be connected to an upper wall duct or a lower wall duct. The generator and a mechanical driver are then connected together to the lower wall duct below which is a water storage tank with a thermal insulation. This design is a combined system so that a few numbers of battery banks can be used to store the solar energy through the solar panel. A mesh or filter can help to filter the rain water. The simulation technology was applied in Kuala Lumpur, Malaysia. Fig. 10.20 shows a graphical 3D idea, the wind/solar power hybrid system with rain water storage tank on top of high-rise buildings (PAGV). The generator and a mechanical driver are placed together below the lower divider pipe under which is a water storage tank to set a warm protection. This outline is a consolidated framework in which a few numbers of battery banks are shown to store the sun-based energy system. A channel can discharge the

Figure 10.20 Graphical 3D design of the hybrid renewable system with storing rain water on top of high-rise buildings. *Reprinted from Chong W, Naghavi M, Poh S, Mahlia T, Pan K. Techno-economic analysis of a wind—solar hybrid renewable energy system with rainwater collection feature for urban high-rise application. Journal of Applied Energy 2011;88(11):4067—77 with permission from Elsevier.*

downpour water. The reconstructed version of the innovation was done in Kuala Lumpur, Malaysia. Fig. 10.20 demonstrates a graphical 3D concept, the wind/sunlight-based half breed framework with rain water discharge pipe on top of tall structures [43].

This sort of exploratory innovation is an ongoing process at present. All inquiries about the tests, where the tests are conducted for purposes of examination, controlling, and observing are appropriately recorded for reference. This is highly relevant and useful for remote correspondence to check or obtain information. We can call the half and half framework as a crossbreed unit framework that may generate energy from two or more sources of RE. Every unit exchanges information with the principle collector unit, whereas the collector unit can control the source unit and this simple procedure could be made possible by any technique. Fig. 10.21 illustrates a case that utilizes the LabView and the National Instrument programming and equipment.

The reasons of utilizing LabView and National Instrument programming and equipment are to examine the energy quality and to install the air conditioning and DC computerized board meters that can record and screen the energy values [44].

Solar power/wind hybrid systems have been seriously considered for further extensive and sophisticated developments as these two sources of RE supplement one another wherein if one is insufficient or inert the other framework becomes effective. Improvement of the framework is performed by examining the behavior of the sun-oriented/wind energy, which has been the fundamental focus of the specialists in this subject, and in view of this consideration further expansion on the sun—oriented energy needs the relevant information based on the multigenerational frameworks, thus, more researches are being directed on these frameworks consistently. Ultimately, this ought to

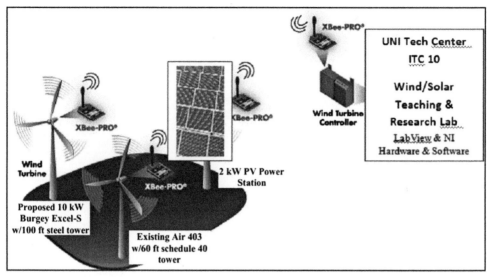

Figure 10.21 Hybrid wind–solar system data recovering, monitoring, and controlling using software and hardware. *Reprinted from Pecen R, Nayir A. Design and implementation of a 12 kW wind-solar distributed power and instrumentation system as an educational testbed for electrical engineering technology students. IEEE conference 2010:1—6 (IEEE conference).*

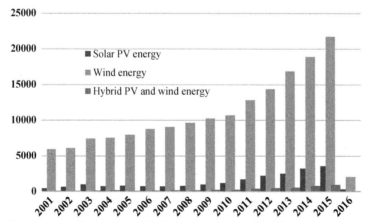

Figure 10.22 The number of articles published on solar and wind hybrid energies for the period from 2001 to 2016. *PV*, photovoltaic. *Science Direct, Solar PV energy, wind energy and Hybrid PV and wind energy.*

bring about finding new better systems, increment in the efficiencies, and a decline in working expenses.

10.5 CURRENT TRENDS

Fig. 10.22 shows the number of articles published between the years 2001 and 2016 on the application of hybrid solar power/wind system in the renewable energies sphere including all the articles published in solar PV, wind, and hybrid energy; from the year 2001 onward, the number of research articles continuously increase reflecting the interest of researchers around the world in this particular area, and especially, in recent years there is a significant growth in the number of research articles, which makes this topic the thrust area of research.

10.6 CONCLUSION

This hybrid technology will eventually lead to generation of energy for 24 h with resultant potential of less polluted environment. Hybrid systems are most suitable for small electric grids and isolated or stand-alone systems as hybrid power generation is, by definition, a solution for getting around problems where one energy source is not sufficient. The most suitable location for the hybrid solar power/wind system will have been identified first for the development of the most reliable and economic hybrid systems that can significantly contribute toward the future sustainable and better ecological and environmental conservations, improving healthy living and social services and reducing transmission costs.

ACKNOWLEDGMENTS

The authors are thankful to the University of Malaya, the Ministry of Higher Education of Malaysia (MOHE) (UM.C/HIR/MOHE/ENG/32), and UM Power Energy Dedicated Advanced Centre (UMPEDAC) for supporting this research project, which made the publication of this book chapter to be academically possible.

NOMENCLATURE

A	Area traversed by the wind, m^2
A	Area of the photovoltaic generator, m^2
A_i	Size of the PV or wind component, m^2
D	Diode
E_g	Energy band gap
$E_{B,0}$	Battery initial stored energy, Ah
$E_{B,max}$	Total capacity of the battery stack, Ah
G	Incident solar irradiation, W/m^2
G_0, G	Different solar irradiance, W/m^2

I	Hourly irradiance, kWh/m^2
I_m	Maximum current, A
I_{pv}, I_{wind}	Currents, A
I_{Bo}, I_{BU}, I_{LOAD}	Currents, A
I_d	Diode current, A
I_{sh}	Shunt current, A
I_{SC}	Short circuit current, A
I_o	Reverse saturation current, A
I_{or}	Saturation current, A
I_{scr}	Short circuit current, A
K	Boltzmann constant (1.38×10^{-23}), J/K
K_i	Polarization coefficient, Ω/h
MMPT	Maximum power point tracking
$Q_R(t)$	Accumulated charge, Ah
R	Load, V
R_i	Internal resistance, Ω
R_s	Series resistance, Ω
R_{sh} **and** R_s	Series and shunt resistances, Ω
R'	Load resistance for operating the power point of the PV, Ω
T, T_0	Temperatures, K
$U_{B,0}$	Open circuit voltage, V
V	Wind speed, m/s
V_d	Diode voltage, V
V_m	Maximum voltage, V
V_{sys}	Voltage of the framework, V
$V(t)$	Wind speed defined at the time t, m/s
V_c, V_R, V_F	Characteristic parameters determined, m/s
V_{OC}	Open circuit voltage, V

GREEK SYMBOLS

ρ	Air density (1.225), kg/m^3
Δt	Time in days of independence required
q	Magnitude of the electron charge (1.6×10^{-19}), °C
Ω	Wind turbine angular speed, S^{-1}

REFERENCES

[1] OzluI Dincer S. Development and analysis of a solar and wind energy based multigeneration system. Journal of Solar Energy 2015;122:1279–95.

[2] Ahmed N, Miyatake M, Al-Othman A. Power fluctuations suppression of stand-alone hybrid generation combining solar photovoltaic/wind turbine and fuel cell systems. Journal of Energy Conversion and Management 2008;49(10):2711–9.

[3] Dursun B. Determination of the optimum hybrid renewable power generating systems for Kavakli campus of Kirklareli University, Turkey. Journal of Renewable and Sustainable Energy Reviews 2012;16(8):6183–90.

[4] Jun Z, Junfeng L, JieH Ngan W. A multi-agent solution to energy management in hybrid renewable energy generation system. Journal of Renewable Energy 2011;36(5):1352—63.

[5] Lopez N, Espiritu J. An approach to hybrid power systems integration considering different renewable energy technologies. Journal of Procedia Computer Science 2011;6:463—8.

[6] Deshmukh M, Deshmukh S. Modeling of hybrid renewable energy systems. Journal of Renewable and Sustainable Energy Reviews 2008;12(1):235—49.

[7] Negi S, Mathew L. Hybrid renewable energy system: a review. International Journal of Electronic and Electrical Engineering 2014;7(5):535—42.

[8] Deshmukh M, Deshmukh S. Hennet and Samarakou (13) discussed approach to optimize hybrid PV/wind/battery system with conventional. Journal of Renewable and Sustainable Energy Reviews 2008;12:235—49.

[9] Celik A. Optimisation and techno-economic analysis of autonomous photovoltaic—wind hybrid energy systems in comparison to single photovoltaic and wind systems. Journal of Energy Conversion and Management 2002;43:2453—68.

[10] Chedid R, Akiki H, Rahman S. A decision support technique for the design of hybrid solar-wind power systems. Energy Conversion, IEEE Transactions 1998;13(1):76—83.

[11] Burch GD. Hybrid renewable energy systems presentation outline. Why hybrids ? Hybrid Study Conclusions. Power 2001.

[12] Tait L. Decision making framework and financial tool for hybrid energy system microgrids [A thesis submitted in partial fulfilment for the requirement of the degree Master of Science]. 2011. p. 18.

[13] Kabalci E. Design and analysis of a hybrid renewable energy plant with solar and wind power. Journal of Energy Conversion and Management 2013;72(0):51—9.

[14] Nrel. Wind turbine calculator. 2005. Available from: www.nrel.gov/analysis/power_databook/calc_wind.php.

[15] Razak JA, Sopian AY, Alghoul ME, Zaharim A, Ahmad IS. Optimization of PV-wind-hydrodiesel hybrid system by minimizing excess capacity. European Journal of Scientific Research 2009;25(4):663—71.

[16] Diaf S, Haddadi M, Belhamel M. Analyse technico économique d'un système hybride (photovoltaïque/éolien) autonome pour le site d'Adrar. Revue des Energies Renouvelables 2006;9(3):127—34.

[17] Khadimi AE, Bchir L, Zeroual A. Dimensionnement et Optimisation Technicoéconomique d'un système d'Energie Hybride photovoltaïque-Eolien avec Système de stockage. Energies Renouvelable 2004;7:73—83.

[18] Yang H, Zhou W, Lu L, Fang Z. Optimal sizing method for stand-alone hybrid solar wind system with LPSP technology by using genetic algorithm. Elsevier. Journal of Solar Energy 2008;82:354—67.

[19] Kanzumba K, Jacobus VH. Hybrid renewable power systems for mobile telephony base stations in developing countries. Journal of Renewable Energy 2013;51:419—25.

[20] Chris M, Giri V, Michael S, Afzal S, Ryan F, Bala C. Optimal technology selection and operation of micro-grids in commercial buildings. In: IEEE Power Engineering Society general meeting, vol. 282; 2007. p. 1—7. Available at: http://ieeexplore.ieee.org/lpdocs/epic03/wrapper.htm?arnumber=4275613.

[21] Epri DH. What is a micro-grid?. 2003. http://www.sandia.gov/ess/publications/ESHB%201001834%20reduced%20size.pdf.

[22] Sakis A. Consortium for electric reliability technology solutions white paper on integration of distributed energy resources the CERTS MicroGrid concept [Public interest]. 2002.

[23] Born FJ, Clarke JA, Johnstone CM, Smith NV. Merit — an evaluation tool for 100% renewable energy provision. Systems Research 2001:1—4.

[24] David JCM. Sustainable energy - without the hot air. Book; 2009.

[25] Farhangi H. The path of the smart grid. IEEE Power and Energy Magazine 2010;8(1):18—28. Available at: http://ieeexplore.ieee.org/lpdocs/epic03/wrapper.htm?arnumber=5357331.

[26] Gyuk I. EPRI-DOE handbook of energy storage for transmission & distribution applications. U.S. Department of Energy; 2003. http://www.sandia.gov/ess/publications/ESHB%201001834%20reduced%20size.pdf.

[27] Xiangjun L, Yong L, Xiaojuan H, Dong H. Application of fuzzy wavelet transform to smooth wind/PV hybrid power system output with battery energy storage system. Energy Procedia 2011;12:994−1001.

[28] Ismail M, Moghavvemi M, Mahlia T, Muttaqi K, Moghavvemi S. Effective utilization of excess energy in standalone hybrid renewable energy systems for improving comfort ability and reducing cost of energy: a review and analysis. Journal of Renewable and Sustainable Energy Reviews 2015;42:726−34.

[29] Safari S, Ardehali M, Sirizi M. Particle swarm optimization based fuzzy logic controller for autonomous green power energy system with hydrogen storage. Journal of Energy Conversion and Management 2013;65:41−9.

[30] Ren H, Gao W. A MILP model for integrated plan and evaluation of distributed energy systems. Journal of Applied Energy 2010;87:1001−14.

[31] Bilodeau A, Agbossou K. Control analysis of renewable energy system with hydrogen storage for residential applications. Journal of Power Sources 2006;162:757−64.

[32] Engelbercht A. Computational intelligence: an introduction. 2nd ed. John Wiley & Sons; 2007.

[33] García P, Torreglosa J, Fernández L, Jurado F. Optimal energy management system for stand-alone wind turbine/photovoltaic/hydrogen/battery hybrid system with supervisory control based on fuzzy logic. International Journal of Hydrogen Energy 2013;38(33):14146−58.

[34] Zhang F, Shi Q, Wang Y, Wang F. Simulation research on wind solar hybrid power system based on fuzzy-PID control. In: Proceeding of international conference on electrical machines and systems; 2007. p. 338−42.

[35] Gang M, Ming C. LabVIEW based simulation system for the output characteristics of PV cells and the influence of internal resistance on it. In: WASE international conference on information engineering, vol. 1; 2009. p. 391−4.

[36] Jaleel J, Nazar A, Omega A. Simulation on maximum power point tracking of the photovoltaic module using LabVIEW. International Journal of Advanced Research in Electrical, Electronics and Instrumentation Engineering 2012;1(3):190−9.

[37] LabVIEW. Wind turbine control methods. National Instruments; 2008. http://www.ni.com/white-paper/8189/en/.

[38] Khalilnejad A, Riahy G. A hybrid wind-PV system performance investigation for the purpose of maximum hydrogen production and storage using advanced alkaline electrolyzer. Journal of Energy Conversion and Management 2014;80:398−406.

[39] Vukman V, Milada L, Marina P, Valentina M, Biljana S, Nikola S. Technical analysis of photovoltaic/wind systems with hydrogen storage. Journal of Thermal Science 2012;16:865−75.

[40] Valigra L. Turbine maker links wind and solar with new technology. Pika Energy; 2013. pika-energy.com.

[41] Aguirre M, Couto H, Valla M. Analysis and simulation of a hydrogen based electric system to improve power quality in distributed grids. International Journal of Hydrogen Energy 2012;37(19):14959−65.

[42] Heymann M. Signs of Hubris: the shaping of wind technology styles in Germany, Denmark, and the United States, 1940−1990. Technology Culture 1998;39:641−70.

[43] Chong W, Naghavi M, Poh S, Mahlia T, Pan K. Techno-economic analysis of a wind−solar hybrid renewable energy system with rainwater collection feature for urban high-rise application. Journal of Applied Energy 2011;88(11):4067−77.

[44] Pecen R, Nayir A. Design and implementation of a 12 kW wind-solar distributed power and instrumentation system as an educational testbed for electrical engineering technology students. In: IEEE conference; 2010. p. 1−6.

[45] Chouder A, Silvestre S, Taghezouit B, Karatepe E. Monitoring, modelling and simulation of PV systems using LabVIEW. Journal of Solar Energy 2013;91:337−49.

[46] LabVIEW. Simulating a wind turbine. National Instruments; 2011.

[47] Beula S, Arivuselvam B. Online conditional monitoring for solar and wind energy system. In: Proceeding of 5th national conference on VLSI; 2014.

[48] Zhou W, Lou C, Li Z, Lu L, Yang H. Current status of research on optimum sizing of stand-alone hybrid solar−wind power generation systems. Journal of Applied Energy 2010;87:380−9.

CHAPTER ELEVEN

Study on Wind Energy Potential by Eight Numerical Methods of Weibull Distribution

A.K. Azad and M.G. Rasul
Central Queensland University, Rockhampton, QLD, Australia

11.1 INTRODUCTION

Wind energy is one of the pollution-free renewable energy sources in the world. Other renewable energies, such as solar [1], hydro [2], biomass [3], and biofuel [4,5], also contribute to the world total energy supply [6]. Wind, solar, and hydro energy are called zero emission energy [7]. The uses of these energies are growing day-by-day [8]. The production of wind power is growing rapidly due to the progression of wind power industries [9]. Therefore, accurate knowledge about the wind energy is needed for planning, design, and operation of wind turbines [9,10]. The main requirements are analysis of the wind speed data and proper assessment of wind energy potential for choosing suitable windy sites [11,12]. The wind characteristics can be analyzed using cumulative frequencies of the measured data and the frequency distribution calculated by Weibull distributions [13,14]. It is reported in the literature that there are some methods available to determine Weibull distribution [15]. Different research groups used different methods to analyze wind energy potential by Weibull distribution. For example, a study has been done for wind energy potential in Iran using two methods, namely standard deviation method and power density method [12]. In planning of offshore wind farms, Morgan et al. [16] analyzed the probability distributions for short-term wind speeds [16−18]. They have shown that the widely accepted Weibull distribution provides a poor fit to the distribution of wind speeds when compared with more complicated models [19,20]. Many researchers and engineers are investigating the potential of wind energy by developing new methods [21]. The most efficient method to asses wind power potential is using probability distribution functions. They are: Rayleigh distribution, Chi-squared distribution, Normal distribution, Binomial distribution, Poisson distribution, and Weibull distribution. Two-parameter Weibull distribution is commonly used in many fields including wind energy assessment, rainfall, water-level prediction, sky clearness index classification, and life length analysis of materials [22,23]. For more than half a century, the Weibull distribution has attracted the attention of statisticians working

Clean Energy for Sustainable Development
ISBN 978-0-12-805423-9, http://dx.doi.org/10.1016/B978-0-12-805423-9.00011-9

on theory and methods as well as in various fields of statistics [24,25]. Together with the normal exponential distributions, the Weibull distribution is without any doubt the most popular model in statistics [26,27]. All these distributions are used to determine the probability of occurrence. The nature of occurrence affects the shape of the probability curve and, in the case of a wind regime, the cumulative curve probability nature mostly fits Weibull functions. This function has much flexibility. Depending on the shape factor k value, it falls under an exponential distribution ($k = 1$), logarithmic normal distribution ($k = 1.5-3$), Rayleigh distribution ($k = 2.0$), logarithmic normal distribution ($k = 3.1-3.6$), and normal distribution ($k > 3.6$). Rocha et al. [28] and Chang [22] have used some methods for three and two statistical tests to determine the Weibull parameters in their research work, respectively. In this work, eight numerical methods and nine statistical test tools are used to rank the methods for determining the value of Weibull distribution precisely. The wind monitoring stations' information are given in Table 11.1. The next section offers a detailed outline of the methodology and associated theories for the statistical analysis. The results and discussions of this work is presented briefly in Section 11.3. Finally, concluding remarks are presented at the end of the chapter.

11.2 OUTLINE OF METHODOLOGY

11.2.1 Weibull Probability Density Function

The Weibull probability density function (WPDF) is a two-parameter function characterized by a dimensionless shape parameter (k) and scale parameter (c) (m/s). These two parameters determine the optimum performance of a wind energy conversion system [12,29]. The expression of WPDF is presented in Eq. (11.1), where v is the velocity of wind speed (m/s).

$$f(v) = \frac{dF(v)}{dv} = \frac{k}{c} \left(\frac{v}{c}\right)^{k-1} \times e^{\left[-\left(\frac{v}{c}\right)^k\right]} \tag{11.1}$$

Table 11.1 Site Name, Location, and Wind Speed Measuring Height

Wind Monitoring Station Number	Station Name	Latitude		Longitude		Wind Speed Measured Height (m)	
		Angle (degrees)	Direction	Angle (degrees)	Direction		
WM Station-I	Mongla	22°00.20′	N	89°02.36′	E	10	20
WM Station-II	Sandwip	22°00.29′	N	91°05.26′	E	10	20
WM Station-III	Khagrachari	23°00.45′	N	91°57.67′	E	10	20
WM Station-IV	Panchagarh	26°33.47′	N	88°00.50′	E	10	20

11.2.2 Cumulative Distribution Function or Weibull Function

The cumulative distribution function is the integration of WPDF. This is cumulative of the relative frequency of each velocity interval [30,31]. The expression of the Weibull function is presented in Eq. (11.2):

$$F(v) = 1 - e^{\left[-\left(\frac{v}{c}\right)^k\right]} \tag{11.2}$$

In this study, eight methods have been used to determine Weibull factors. They are:
1. Methods of moments (MOM)
2. Maximum likelihood method (MLM)
3. Modified maximum likelihood method (MMLM)
4. Least square method (LSM)
5. Power density method (PDM)
6. Equivalent energy method (EEM)
7. Graphical method (GM)
8. Empirical method (EM)

11.2.2.1 Method of Moments

The method of moments is one of the common techniques used in the field of parameter estimation. If $\bar{v}$ represents the mean wind speed data, then the values of k and c can be easily determined using Eqs. (11.3) and (11.4) [22]:

$$\bar{v} = c\Gamma\left(1 + \frac{1}{k}\right) \tag{11.3}$$

$$k = \left(\frac{0.9874}{\frac{\sigma}{\bar{v}}}\right)^{1.0983} \tag{11.4}$$

The following expressions show the coefficient of variation (COV) in Eq. (11.5) and the standard deviation in Eq. (11.6).

$$COV = \frac{\sigma}{\bar{v}} = \sqrt{\frac{\Gamma\left(1 + \frac{2}{k}\right)}{\left[\Gamma\left(1 + \frac{1}{k}\right)\right]^2} - 1} \tag{11.5}$$

$$\sigma = c\left[\Gamma\left(1 + \frac{2}{k}\right) - \Gamma^2\left(1 + \frac{1}{k}\right)\right]^{1/2} \tag{11.6}$$

An expression of the gamma function $\Gamma(x)$ is presented as: Eq. (11.7).

$$\Gamma(x) = \int_0^\alpha t^{x-1}e^{(-t)}dt \tag{11.7}$$

11.2.2.2 Maximum Likelihood Method

MLE has been the most widely used method due to its very desirable properties for estimating the parameters of the Weibull distribution [28]. Consider the WPDF given in Eq. (11.1), in MLM, the parameters k and c are determined using Eqs. (11.8) and (11.9), respectively [22]

$$k = \left[\frac{\sum_{i=1}^n v_i^k ln(v_i)}{\sum_{i=1}^n v_i^k} - \frac{\sum_{i=1}^n ln(v_i)}{n} \right]^{-1} \tag{11.8}$$

$$c = \left[\frac{1}{n} \sum_{i=1}^n (v_i)^k \right]^{1/k} \tag{11.9}$$

11.2.2.3 Modified Maximum Likelihood Method

The MMLM can be considered if the available data of wind speed are already in the shape of the Weibull distribution. The solution of the equations in the MLM requires some numerical iteration using the Newton–Raphson method [12,22,28]. The parameters can be estimated using Eqs. (11.10) and (11.11):

$$k = \left[\frac{\sum_{i=1}^n v_i^k ln(v_i) f(v_i)}{\sum_{i=1}^n v_i^k f(v_i)} - \frac{\sum_{i=1}^n ln(v_i) f(v_i)}{f(v \geq 0)} \right]^{-1} \tag{11.10}$$

$$c = \left[\frac{1}{f(v \geq 0)} \sum_{i=1}^n (v_i)^k f(v_i) \right]^{1/k} \tag{11.11}$$

where v_i is the wind speed central to bin i, n is the number of bins. $f(v_i)$ represents the Weibull frequency for the wind speed range within bin i, and $f(v \geq 0)$ is the probability for wind speed equal to or exceeding zero [22].

11.2.2.4 Least Square Method

The LSM is commonly applied in engineering and mathematics problems that are often not thought of as an estimation problem. A linear relation between the two variables has been assumed and, after some exclusive calculation for minimizing the relationship, the expression can be expressed as in Eqs. (11.12) and (11.13) [14]:

$$k = \frac{n \sum_{i=1}^{n} ln\, v \times ln[-ln\{1 - F(v)\}] - \sum_{i=1}^{n} ln\, v \times \sum_{i=1}^{n} ln[-ln\{1 - F(v)\}]}{n \sum_{i=1}^{n} ln\, v^2 - \left\{ \sum_{i=1}^{n} ln\, v \right\}^2}$$

(11.12)

$$c = exp\left[\frac{k \sum_{i=1}^{n} ln\, v - \sum_{i=1}^{n} ln[-ln\{1 - F(v)\}]}{nk} \right]$$

(11.13)

11.2.2.5 Power Density Method

The shape factor and scale factor can be determined by Eqs. (11.15) and (11.16) through this method. First, the energy pattern factor E_{pf} is computed, which is defined as a ratio between mean of cubic wind speed to cube of mean wind speed as in Eq. (11.14) [12,14,28]:

$$E_{pf} = \frac{\frac{1}{n} \sum_{i=1}^{n} v_i^3}{\left(\frac{1}{n} \sum_{i=1}^{n} v_i \right)^3} = \frac{\overline{v^3}}{(\overline{v})^3} = \frac{\Gamma\left(1 + \frac{3}{k}\right)}{\Gamma^3\left(1 + \frac{1}{k}\right)}$$

(11.14)

Once the energy pattern factor is calculated, the shape factor and scale factor can be estimated from the following formulae [12,28]:

$$k = 1 + \frac{3.69}{E_{pf}^2}$$

(11.15)

$$c = \frac{\overline{v}}{\Gamma\left(1 + \frac{1}{k}\right)}$$

(11.16)

11.2.2.6 Equivalent Energy Method

The estimation of the parameter k may be obtained from an estimator of least squares. By this method, the parameters k and c are determined using Eqs. (11.17) and (11.18) [28]:

$$\sum_{i=1}^{n} \left[W_{vi} - e^{-\left[\frac{(v_i - 1)\left\{ \Gamma\left(1 + \frac{3}{k}\right) \right\}^{1/3}}{(v_m^3)^{1/3}} \right]^k} + e^{-\left[\frac{(v_i)\left\{ \Gamma\left(1 + \frac{3}{k}\right) \right\}^{1/3}}{(v_m^3)^{1/3}} \right]^k} \right]^2 = \sum_{i=1}^{n} (\varepsilon_{vi})^2$$

(11.17)

$$c = \left(\frac{v_m^3}{\Gamma\left(1 + \frac{3}{k}\right)} \right)^{1/3} \tag{11.18}$$

where W_{vi} be the observed frequency of the wind speed, n is the number of intervals of the histogram of speed, v_i is the value of the upper limit of the ith speed interval, v_m^3 is the mean of the cubic wind speed, and ε_{vi} is the error of the approximation [14].

11.2.2.7 Graphical Method

The Weibull plot is constructed in such a way that the cumulative Weibull distribution becomes a straight line, with the shape factor k as its slope [28]. Taking logarithms on both sides of Eq. (11.1), it can be rewritten as:

$$ln[-ln\{1 - F(v)\}] = k\,ln(v) - k\,ln(c) \tag{11.19}$$

Alternatively, the horizontal axis of the Weibull plot now becomes v while $ln(1 - F(v))^{-1}$ is placed on the vertical axis. The result is a straight line with slope k. For $v = c$, one finds $F(v) = 1 - e^{-1} = 0.632$ and this gives an estimation of the value of c by drawing a horizontal line at $F(v) = 0.632$. The intersection point with the Weibull line gives the value of c [14].

11.2.2.8 Empirical Method

In the EM, Weibull shape factor k and scale factor c can be estimated using the following equations:

$$k = \left(\frac{\sigma}{\bar{v}}\right)^{-1.086} \tag{11.20}$$

$$c = \frac{\bar{v}}{\Gamma\left(1 + \frac{1}{k}\right)} \tag{11.21}$$

where $\bar{v}$ and σ are mean wind speed and standard deviation of wind speed, respectively, for any specified periods of time and can be calculated using Eqs. (11.22) and (11.23), respectively [12]:

$$\bar{v} = \frac{1}{n} \sum_{i=1}^{n} v_i \tag{11.22}$$

$$\sigma = \left[\left(\frac{1}{n-1} \sum_{i=1}^{n} (v - \bar{v})^2 \right) \right]^{1/2} \tag{11.23}$$

11.2.3 Statistical Error Analysis

To find the best method for the analysis, some statistical parameters are used to analyze the efficiency of the aforementioned methods [12,28]. The tests listed as follows can be used to achieve this goal:

1. Relative percentage error, $RPE = \left(\frac{x_{i,w} - y_{i,m}}{y_{i,m}}\right) \times 100\%$;

2. Mean percentage error, $MPE = \frac{1}{N} \sum_{i=1}^{N} \left(\frac{x_{i,w} - y_{i,m}}{y_{i,m}}\right) \times 100\%$;

3. Mean absolute percentage error, $MAPE = \frac{1}{N} \sum_{i=1}^{N} \left|\frac{x_{i,w} - y_{i,m}}{y_{i,m}}\right| \times 100\%$;

4. Root mean square error, $RMSE = \left[\frac{1}{N} \sum_{i=1}^{N} \left(y_{i,m} - x_{i,w}\right)^2\right]^{1/2}$;

5. Chi-square error, $\chi^2 = \frac{\sum_{i=1}^{N} \left(y_{i,m} - x_{i,w}\right)^2}{x_{i,w}}$;

6. Analysis of variance or efficiency of the method, $R^2 = \frac{\sum_{i=1}^{N} \left(y_{i,m} - z_{i,\bar{v}}\right)^2 - \sum_{i=1}^{N} \left(y_{i,m} - x_{i,w}\right)^2}{\sum_{i=1}^{N} \left(y_{i,m} - z_{i,\bar{v}}\right)^2}$

where N is the number of observations, $y_{i,m}$ is the frequency of observation or the ith calculated value from measured data, $x_{i,w}$ is the frequency of Weibull or the ith calculated value from Weibull distribution, and $z_{i,\bar{v}}$ is the mean of the ith calculated value from measured data. RPE shows the percentage deviation between the calculated values from the Weibull distribution and the calculated values from measured data. MPE shows the average of percentage deviation between the calculated values from the Weibull distribution and the calculated values from measured data, and MAPE shows the absolute average of percentage deviation between the calculated values from the Weibull distribution and the calculated values from measured data. Best results are obtained when these values are close to zero. R^2 determined the linear relationship between the calculated values from the Weibull distribution and the calculated values from measured data. The ideal value of R^2 is equal to 1 [12,14].

The 10-min time series wind speed data has been collected by a DL9210 anemometer with two sensors. The chip socket varies from one to two in different meters but contains left socket 32 KB and right socket 32 KB in each anemometer. Every meter has its own identification serial number, that is, 1946, 1936, 1955, and 195, respectively installed in wind monitoring stations at 20 different wind sites by the United Nations Development Programme (UNDP) wind project in Bangladesh. In this work, data for only four wind monitoring stations have been statistically analyzed. The study has been designed in a comprehensive way aiming at systematic observation on wind regimes in different suitable locations over a long period using the best Weibull distribution methods.

11.3 RESULTS AND DISCUSSION

Table 11.2 shows the complete summary of the collected wind speed data of the selected sites including monthly mean wind speed, the number of observations in hours, maximum and minimum wind speed, standard deviation and coefficient of variation (COV) at 10 m height, and also annual mean values. The *COV* demonstrates the mutability of the wind speeds and is the ratio of mean standard deviation to mean wind speed [32]. Measurement of wind speed data follows a standard practice of measurement at 10 and 20 m heights. Hence, 10 m-height data has been used for this analysis. It can be seen that the wind speeds are higher from April to September for all sites. It can also be noted that there is a wide variation of wind speed in the coastal areas of WM Station-I and Station-II throughout the year. The maximum monthly mean wind speed was found to be 8 m/s in April at WM Station-II and the minimum was 1.30 m/s in November at WM Station-IV. The WM Station-II wind speed shows better results in every month than the other sites. Hence, WM Station-II is relatively the best prospective windy site for extraction of energy for power generation, but is not totally suitable throughout the year.

Fig. 11.1 shows the wind duration curves in a single graph from which the reader can get a good idea about a site at a glance. Each point of this curve shows the number of hours in a year at which the corresponding velocity occurs, and for which higher velocities are found. Another aim of this curve is to allow easy visualization of which site is the most promising one. From Fig. 11.1, WM Station-II clearly shows better performance at a velocity above 5.5 m/s (which is the minimum requirement for power generation) than other sites [33]. More studies have been undertaken on the selected sites to obtain greater detail of the wind characteristics [34]. The study aimed to provide guidelines to the decision makers assist in determining whether the site has good potential for a particular type of wind machine precisely.

Tables 11.3—11.10 show the result of the statistical analysis for estimating Weibull parameters by eight different methods, namely MOM, MLM, MMLM, LSM, PDM, EEM, GM, and EM. The calculation process led to obtaining better methods for Weibull distribution using error analysis using RPE, MPE, MAPE, MADE, MSE, SEM, RMSE, χ^2, and R^2. The monthly variations (January to December) of Weibull shape factor, scale factor, and error analysis results for Station-I are presented in Tables 11.3 and 11.4, respectively. The value of k and c varies for the different methods. For Station-I, EM results are more relevant than other methods because the minimum error occurs in this method (see Table 11.4). In this method, the minimum value of k is 1.32 found in November and the maximum of 5.40 in March. For Station-II, LSM provides better results compared with other methods because of a minimum number of errors (for example, $\chi^2 = 0.000001$) and efficiency of the method R^2 is 0.9995. The best results are

Table 11.2 Monthly Mean Wind Speed, Number of Observations, Maximum Wind Speed, Minimum Wind Speed, Standard Deviation, and Coefficient of Variation of 10-m Height Wind Speed Data for the Selected Wind Monitoring Stations

Wind Parameter	Month of the Year												Annual Mean
	Jan	Feb	Mar	Apr	May	Jun	Jul	Aug	Sep	Oct	Nov	Dec	
Station-I													
H = 10 m													
No. obs.(h)	744	672	744	720	744	720	744	744	720	744	720	744	8760
v_{mean} (m/s)	2.01	1.9	2.93	2.41	2.93	4.23	4.34	4.74	2.48	2.16	1.73	1.82	2.81
v_{max} (m/s)	3.89	3.44	4.52	4.11	5.04	10.59	7.79	10.75	3.8	3.84	5.15	3.22	5.51
v_{min} (m/s)	0.49	0.56	2.06	0.68	1.73	1.7	2.21	1.59	1.02	0.97	0.29	0.59	1.16
σ (m/s)	0.82	0.7	0.62	0.85	0.82	1.98	1.29	2.92	0.81	0.9	1.34	0.71	1.15
COV (%)	40.8	36.84	21.16	35.27	27.99	46.81	29.72	61.6	32.66	41.67	77.46	39.01	40.92
Station-II													
No. obs.(h)	744	672	744	720	744	720	744	744	720	744	720	744	8760
v_{mean} (m/s)	3.3	2.75	3.85	8.34	2.28	3.68	5.44	4.45	5.18	4.63	4.02	2.56	4.21
v_{max} (m/s)	6.31	5.1	6.31	32.97	4.81	7.77	9.23	10.75	9.77	7.26	6.79	7.13	9.52
v_{min} (m/s)	1.39	1.49	1.38	3.14	0.41	1.75	1.91	1.59	2.33	2.24	1.61	0.87	1.68
σ (m/s)	1.42	1.08	1.41	7.13	1.12	1.33	2.63	3.03	2.16	1.18	1.45	1.84	2.15
COV (%)	43.03	39.27	36.62	85.49	49.12	36.14	48.35	68.09	41.7	25.49	36.07	71.88	48.44
Station-III													
No. obs.(h)	744	672	744	720	744	720	744	744	720	744	720	744	8760
v_{mean} (m/s)	2.85	3.45	3.88	4.12	3.44	4.67	3.36	3.1	3.19	2.38	2.5	2.37	3.28
v_{max} (m/s)	3.86	5.3	5.06	5.51	6.36	8.11	4.88	4.64	5.11	3.05	3.64	3.52	4.92
v_{min} (m/s)	1.78	1.93	2.68	2.81	1.66	2.54	2.27	1.89	2.17	1.25	1.51	1.44	1.99
σ (m/s)	0.56	0.93	0.66	0.68	1.29	1.9	0.55	0.55	0.81	0.42	0.44	0.53	0.78
COV (%)	19.65	26.96	17.01	16.5	22.15	30.08	16.37	17.74	25.39	17.65	17.6	22.36	22.95
Station-IV													
No. obs.(h)	744	672	744	720	744	720	744	744	720	744	720	744	8760
v_{mean} (m/s)	1.54	1.43	2.23	2.75	2.89	2.56	2.42	2.56	2.17	2.03	1.3	1.76	2.14
v_{max} (m/s)	2.14	2.02	4	4.54	4.41	4.15	4.34	4.11	3.8	5.71	2.82	14.59	4.72
v_{min} (m/s)	0.85	0.92	1.27	1.7	2.05	1.04	1.02	1.26	0.81	0.69	0.8	0.78	1.1
σ (m/s)	0.33	0.25	0.62	0.84	0.64	0.77	0.72	0.66	0.76	1.42	0.49	2.43	0.83
COV (%)	21.43	17.48	27.8	30.55	22.15	30.08	29.75	25.78	35.02	69.95	37.69	138.0	40.48

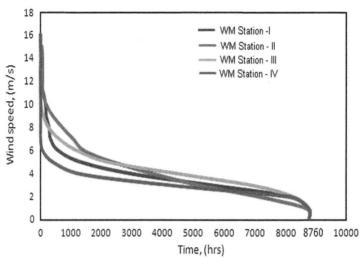

Figure 11.1 Wind duration curve for the selected wind monitoring stations.

obtained when these values (χ^2 and R^2) are close to zero and unity, respectively. On the other hand, MOM provides better results for both Station-III (see Tables 11.7 and 11.8), and Station-IV (see Tables 11.9 and 11.10). The selection of methods was done by considering minimum error and maximum efficiency, respectively. Nine statistical tools have been used and four decimal places considered of each value during numerical iteration methods. One test is enough to rank the methods, but for more precise analysis, the authors used more tools that helped to verify the conclusions about the best method. In this statistical research work, it has been found that EM, LSM, and MOM are more efficient methods to determine Weibull factors. To further justify this outcome, more analysis has been carried out on curve fitting of the methods as is briefly discussed in the following sections.

Figs. 11.2—11.5 show the curve fitting of WPDF $f(v)$, and cumulative distribution function $F(v)$ versus frequency distribution of mean wind speed. The analysis has been carried out based on calculated Weibull shape factors and scale factors using the eight numerical methods shown in Tables 11.3, 11.5, 11.7, and 11.9. The main objective of this graphical representation is to verify how the curves match with the velocity histograms and provide an idea of which method yields the best fit to the yearly measured wind speed data. As found from the figures, the wind data is well characterized by the WPDF and CDF for each station. The calculated values of WPDF and CDF curves by EM (for Station-I), LSM (for Station-II), and MOM (for both Station-III and Station-IV) closely fitted with the measured values. It was also found from Figs. 11.4 and 11.5 that the PDM and MMLM curves also closely fitted with measured data for Station-III and Station-IV, respectively.

Table 11.3 Estimation of Weibull Parameters by Eight Statistical Methods at WM Station-I

Weibull Parameters' Estimation by Eight Statistical Methods

Month	MOM		MLM		MMLM		LSM		PDM		EEM		GM		EM	
	K (−)	C (m/s)	K (−)	C (m/s)	K (−)	C (m/s)	K (−)	C (m/s)	K (−)	C (m/s)	K (−)	C (m/s)	K (−)	C (m/s)	K (−)	C (m/s)
January	2.64	2.26	2.67	2.26	2.65	2.26	2.59	2.25	2.62	2.26	2.65	1.97	2.54	2.23	2.65	2.27
February	2.95	2.13	2.96	2.13	2.89	2.13	3.47	2.18	2.82	2.13	2.96	1.90	3.98	2.22	2.96	2.13
March	5.43	3.17	4.88	3.18	4.37	3.21	4.26	3.03	3.86	3.24	5.42	3.05	3.08	2.88	5.40	3.18
April	3.10	2.70	3.17	2.69	3.07	2.70	3.28	3.05	2.96	2.70	3.10	2.42	3.46	3.39	3.10	2.70
May	3.99	3.23	3.77	3.24	3.60	3.26	3.70	3.26	3.42	3.27	3.99	3.01	3.41	3.29	3.99	3.23
June	2.27	4.77	2.29	4.79	2.24	4.78	2.22	4.78	2.18	4.77	2.28	4.01	2.17	4.79	2.28	4.78
July	3.74	4.81	3.47	4.81	3.36	4.83	3.81	4.85	3.25	4.84	3.74	4.44	3.88	4.88	3.73	4.81
August	1.68	5.31	1.78	5.38	1.73	5.35	1.62	5.41	1.68	5.31	1.69	4.02	1.56	5.50	1.69	5.31
September	3.37	2.76	3.57	2.76	3.36	2.77	2.92	2.83	3.15	2.77	3.37	2.52	2.46	2.89	3.37	2.76
October	2.58	2.43	2.64	2.44	2.59	2.44	2.98	2.50	2.54	2.43	2.59	2.11	3.37	2.57	2.59	2.43
November	1.31	1.88	1.44	1.92	1.39	1.90	1.61	1.86	1.33	1.88	1.32	1.26	1.90	1.84	1.32	1.88
December	2.77	2.04	2.85	2.04	2.82	2.04	2.76	2.03	2.78	2.04	2.78	1.80	2.75	2.01	2.78	2.04

Table 11.4 Statistical Error Analysis and Efficiency of the Eight Methods at WM Station-I

Statistical Methods	Statistical Test Efficiency Methods								
	RPE (%)	MPE (%)	MAPE (%)	MADE	MSE	SEM	RMSE	χ^2	R^2
Method of moments (MOM)	−0.8132	−0.2094	1.2146	31.1092	1646.8857	12.2359	1.0493	0.000005	0.9993
Maximum likelihood method (MLM)	−2.2948	−1.6807	2.1149	30.6204	1550.6257	11.8729	1.7475	0.000015	0.9982
Modified maximum likelihood method (MMLM)	−0.4711	0.3725	2.4839	31.075	1596.7349	12.0481	1.2416	0.000007	0.9991
Least square method (LSM)	3.9869	−0.2311	12.8314	33.6275	2010.9202	13.5208	5.6107	0.000093	0.9810
Power density method, (PDM)	1.9027	2.6304	3.0006	31.7642	1677.3173	12.3484	1.1333	0.000005	0.9992
Equivalent energy method (EEM)	−41.2906	−33.5732	33.5732	16.4946	391.8598	5.9686	26.0473	0.053021	0.5911
Graphical method (GM)	12.2819	8.3836	20.2376	35.9458	2433.1911	14.8728	11.1243	0.000251	0.9254
Empirical method (EM)	−0.9921	−0.2766	1.0637	31.0258	1628.9655	12.1691	0.884	0.000004	0.9995

Table 11.5 Estimation of Weibull Parameters by Eight Statistical Methods at WM Station-II

Weibull Parameters' Estimation by Eight Statistical Methods

Month	MOM		MLM		MMLM		LSM		PDM		EEM		GM		EM	
	K (−)	C (m/s)	K (−)	C (m/s)	K (−)	C (m/s)	K (−)	C (m/s)	K (−)	C (m/s)	K (−)	C (m/s)	K (−)	C (m/s)	K (−)	C (m/s)
January	2.49	3.72	2.58	3.74	2.54	3.73	2.84	3.55	2.5	3.72	2.5	3.2	3.18	3.37	2.5	3.72
February	2.47	2.81	2.68	3.11	2.27	2.96	3.00	2.84	1.86	2.80	2.48	2.41	3.52	2.87	2.48	2.81
March	2.97	4.31	3.15	4.31	3.05	4.32	3.42	4.06	2.94	4.32	2.98	3.85	3.87	3.81	2.98	4.31
April	1.17	8.81	1.43	9.31	1.30	9.05	1.07	8.36	1.16	8.79	1.18	5.5	0.97	7.91	1.19	8.84
May	2.15	2.57	2.21	2.58	2.19	2.58	2.65	2.63	2.17	2.57	2.16	2.13	3.15	2.69	2.16	2.57
June	3.02	4.12	2.95	4.12	2.89	4.13	2.97	4.13	2.83	4.13	3.02	3.68	2.91	4.14	3.02	4.12
July	2.19	6.14	2.3	6.17	2.31	6.16	2.29	6.16	2.31	6.14	2.2	5.1	2.39	6.17	2.2	6.14
August	1.5	4.93	1.64	5.02	1.58	4.98	1.44	5.01	1.52	4.94	1.51	3.55	1.37	5.08	1.52	4.94
September	2.58	5.83	2.65	5.85	2.61	5.84	2.88	5.53	2.56	5.83	2.59	4.13	3.17	5.23	2.59	5.83
October	4.43	5.08	4.31	5.08	3.96	5.11	4.56	5.11	3.6	5.14	4.42	4.79	4.68	5.13	4.41	5.08
November	2.92	4.36	3.19	4.5	2.91	4.44	2.86	4.44	2.62	4.38	2.92	3.87	2.8	4.51	2.92	4.36
December	1.42	2.82	1.58	2.88	1.52	2.85	1.83	2.88	1.45	2.82	1.43	1.97	2.23	2.94	1.43	2.82

Table 11.6 Statistical Error Analysis and Efficiency of Eight Methods at WM Station-II

Statistical Methods	RPE (%)	MPE(%)	MAPE(%)	MADE	MSE	SEM	RMSE	χ^2	R^2
Method of moments (MOM)	−7.5778	−1.1118	3.9582	207.0625	142,632.764	113.871	54.2612	0.000002	0.9984
Maximum likelihood method (MLM)	−24.7283	−2.8898	5.9824	151.0161	74,386.8983	82.2341	165.0128	0.000059	0.8522
Modified maximum likelihood method (MMLM)	−20.3861	−4.1248	7.6942	171.0706	96,616.1445	93.7192	126.0633	0.000020	0.9138
Least square method (LSM)	−1.0937	−7.2315	10.971	235.4961	183,946.244	129.3151	9.4032	0.000001	0.9995
Power density method (PDM)	−5.7016	1.8394	3.629	212.9264	150,516.329	116.9756	43.0584	0.000001	0.9899
Equivalent energy method (EEM)	−68.5189	−44.9993	44.9993	51.6631	7537.3722	26.1766	369.0416	0.028767	0.2609
Graphical method (GM)	7.4656	−9.0951	19.2139	267.4043	236,854.509	146.7386	61.6462	0.000001	0.9794
Empirical method (EM)	−9.8272	−1.7216	3.9564	199.7758	132882.089	109.9099	68.1175	0.000003	0.9748

Statistical Test Efficiency Methods

Table 11.7 Estimation of Weibull Parameters by Eight Statistical Methods at WM Station-III

Weibull Parameters' Estimation by Eight Statistical Methods

Month	MOM		MLM		MMLM		LSM		PDM		EEM		GM		EM	
	K (–)	C (m/s)	K (–)	C (m/s)	K (–)	C (m/s)	K (–)	C (m/s)	K (–)	C (m/s)	K (–)	C (m/s)	K (–)	C (m/s)	K (–)	C (m/s)
January	5.89	3.07	5.97	3.08	4.98	3.12	5.32	3.05	3.98	3.15	5.87	2.97	4.75	3.03	5.85	3.08
February	4.16	3.8	4.09	3.8	3.79	3.82	4.86	3.71	3.49	3.84	4.16	3.56	5.55	3.61	4.15	3.8
March	6.9	4.15	6.68	4.15	5.42	4.21	6.71	4.17	4.16	4.27	6.88	4.04	6.51	4.19	6.85	4.16
April	7.13	4.4	6.64	4.41	5.4	4.48	7.61	4.43	4.15	4.54	7.1	4.29	8.09	4.45	7.07	4.4
May	2.9	3.86	2.92	3.87	2.85	3.87	2.9	3.91	2.78	3.87	2.9	3.42	2.9	3.95	2.9	3.86
June	2.65	5.25	2.73	5.27	2.68	5.27	2.94	4.98	2.62	5.26	2.66	4.58	3.23	4.71	2.66	5.26
July	7.2	3.58	6.3	3.59	5.24	3.65	7.46	3.55	4.18	3.70	7.17	3.5	7.71	3.51	7.14	3.59
August	6.59	3.33	5.82	3.33	4.96	3.37	6.67	3.31	4.10	3.41	6.57	3.23	6.74	3.29	6.54	3.32
September	4.44	3.5	4.05	3.5	3.81	3.52	3.79	3.41	3.57	3.54	4.44	3.3	3.14	3.32	4.43	3.5
October	6.63	2.55	7.45	2.54	5.79	2.58	6.73	2.53	4.13	2.62	6.61	2.48	6.83	2.5	6.58	2.55
November	6.65	2.68	6.15	2.67	5.15	2.71	6.95	2.65	4.15	2.75	6.63	2.6	7.25	2.61	6.6	2.68
December	5.11	2.58	4.9	2.58	4.35	2.6	4.98	2.62	3.79	2.62	5.1	2.46	4.84	2.65	5.09	2.58

Table 11.8 Statistical Error Analysis and Efficiency of Eight Methods at WM Station-III

Statistical Methods	Statistical Test Efficiency Methods								
	RPE (%)	MPE(%)	MAPE(%)	MADE	MSE	SEM	RMSE	χ^2	R^2
Method of moments (MOM)	2.4023	2.0336	2.2579	15.3363	462.5151	6.4844	1.2554	0.000088	0.9964
Maximum likelihood method (MLM)	2.6342	2.2542	2.3708	15.3321	459.8674	6.4658	1.2522	0.000089	0.9964
Modified maximum likelihood method (MMLM)	5.9878	6.2702	6.2702	15.5988	469.8373	6.5355	2.1232	0.000245	0.9896
Least square method (LSM)	−3.2972	−0.7471	3.6676	13.8725	318.2328	5.3787	4.1939	0.002084	0.9594
Power density method (PDM)	10.4475	11.7156	11.7156	15.9721	480.9736	6.6125	3.4916	0.000632	0.9718
Equivalent energy method (EEM)	−16.3847	−12.2801	12.2801	11.0511	214.374	4.4146	8.6524	0.019548	0.8271
Graphical method (GM)	−8.512	−3.9882	8.2775	12.9344	238.7709	4.659	8.1589	0.014011	0.8463
Empirical method (EM)	2.6877	2.2188	2.328	15.3996	467.9113	6.5221	1.3587	0.000101	0.9957

Table 11.9 Estimation of Weibull Parameters by Eight Statistical Methods at WM Station-IV
Weibull Parameters' Estimation by Eight Statistical Methods

Month	MOM		MLM		MMLM		LSM		PDM		EEM		GM		EM	
	K (–)	C (m/s)	K (–)	C (m/s)	K (–)	C (m/s)	K (–)	C (m/s)	K (–)	C (m/s)	K (–)	C (m/s)	K (–)	C (m/s)	K (–)	C (m/s)
January	5.35	1.67	5.39	1.66	4.67	1.68	5.24	1.68	3.94	1.7	4.65	1.6	5.13	1.69	5.33	1.67
February	6.70	1.53	6.24	1.54	5.15	1.56	6.61	1.6	4.05	1.58	5.38	1.49	6.52	1.66	6.65	1.53
March	4.02	2.46	3.75	2.46	3.58	2.47	3.54	2.49	3.40	2.48	3.71	2.29	3.05	2.51	4.02	2.46
April	3.63	3.05	3.51	3.06	3.35	3.07	3.53	3.02	3.19	3.07	3.41	2.81	3.43	2.99	3.63	3.05
May	5.16	3.14	4.67	3.15	4.21	3.18	3.84	3.0	3.75	3.20	4.46	3.00	2.52	2.86	5.14	3.14
June	3.69	2.84	3.74	2.83	3.53	2.84	3.42	2.86	3.31	2.85	3.50	2.62	3.14	2.88	3.69	2.83
July	3.73	2.68	3.59	2.69	3.42	2.7	4.15	2.64	3.25	2.7	3.49	2.48	4.57	2.6	3.73	2.68
August	4.37	2.81	4.34	2.81	3.96	2.83	4.07	2.83	3.58	2.84	3.98	2.65	3.77	2.84	4.36	2.81
September	3.12	2.42	3.11	2.43	3.02	2.43	3.67	2.25	2.93	2.43	3.03	2.18	4.22	2.07	3.12	2.42
October	1.46	2.24	1.62	2.29	1.53	2.27	1.62	2.37	1.43	2.24	1.45	1.58	1.77	2.50	1.47	2.24
November	2.88	1.46	2.71	1.46	2.68	1.46	2.11	1.45	2.65	1.46	2.77	1.28	1.34	1.43	2.89	1.46
December	0.69	1.37	1.16	1.88	1.09	1.83	0.77	1.47	1.01	1.77	0.85	0.77	0.84	1.56	0.70	1.39

Table 11.10 Statistical Error Analysis and Efficiency of Eight Methods at WM Station-IV

Statistical Methods	RPE (%)	MPE(%)	MAPE(%)	MADE	MSE	SEM	RMSE	χ^2	R^2
					Statistical Test Efficiency Methods				
Method of moments (MOM)	1.0497	1.2124	2.4472	8.9758	29.9706	4.767	0.4885	0.000046	0.999
Maximum likelihood method, (MLM)	−29.088	−5.3344	9.8601	4.3226	25.3831	1.5191	13.7701	3.531541	0.206
Modified maximum likelihood method (MMLM)	−25.9329	−2.0712	11.5641	4.5971	28.5061	1.6098	13.2977	2.611319	0.2596
Least square method (LSM)	−15.1265	−1.1901	12.4107	6.3833	92.2864	2.8965	6.6473	0.062258	0.815
Power density method (PDM)	−21.8312	1.8288	13.3391	4.9615	33.6887	1.75	12.453	1.639698	0.3507
Equivalent energy method (EEM)	−50.6705	−27.1572	27.4446	3.5342	16.8853	1.239	17.3273	12.63645	−0.2571
Graphical method (GM)	−19.5862	7.9049	27.4026	5.6058	51.9159	2.1725	9.6477	0.414408	0.6103
Empirical method (EM)	−0.9587	0.6971	2.3997	8.6167	224.3201	4.5158	0.56	0.000075	0.9987

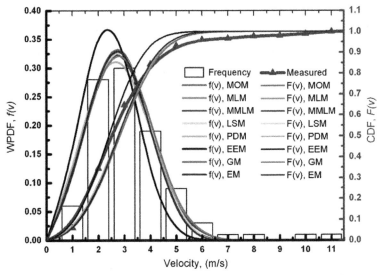

Figure 11.2 Weibull probability density function $f(v)$, cumulative distribution function $F(v)$ versus frequency distribution of mean wind speed—WM Station-I.

Tables 11.11–11.14 summarize the monthly variation of measured and calculated total wind power for each station. In these tables, the calculated wind power by different methods is compared with measured wind power. Table 11.11 shows that the calculated wind power by EM is closer with the measured power for Station-I. For Station-II, the

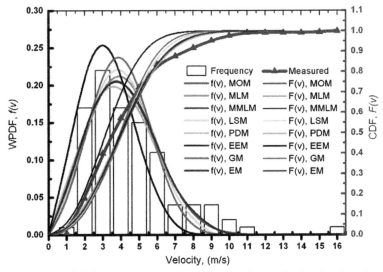

Figure 11.3 Weibull probability density function $f(v)$, cumulative distribution function $F(v)$ versus frequency distribution of mean wind speed—WM Station-II.

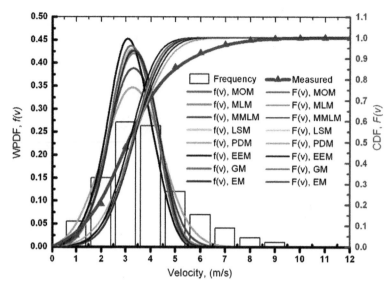

Figure 11.4 Weibull probability density function $f(v)$, cumulative distribution function $F(v)$ versus frequency distribution of mean wind speed—WM Station-III.

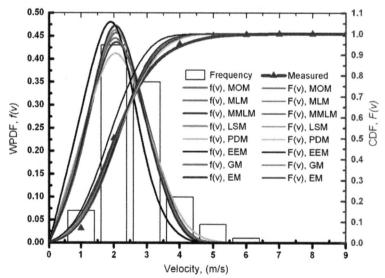

Figure 11.5 Weibull probability density function $f(v)$, cumulative distribution function $F(v)$ versus frequency distribution of mean wind speed—WM Station-IV.

Table 11.11 Monthly Mean Wind Power Based on Measured Data and Weibull Distribution Methods—WM Station-I

Month	Measured P (W/m²)	Calculated Power by Weibull Methods in W/m²							
		MOM	MLM	MMLM	LSM	PDM	EEM	GM	EM
January	7.36	7.39	7.32	7.36	7.37	7.43	4.87	7.25	7.45
February	5.86	5.84	5.82	5.9	5.9	5.96	4.13	6.03	5.82
March	16.86	16.71	16.98	17.7	14.92	18.58	14.88	13.92	16.87
April	11.24	11.38	11.15	11.42	16.02	11.56	8.19	21.7	11.38
May	18.33	18.09	18.5	19.0	18.9	19.51	14.64	19.87	18.09
June	78.45	75.00	75.49	76.43	76.87	77.85	44.56	78.84	75.47
July	61.1	60.52	61.69	63.06	61.91	64.18	47.6	62.73	60.52
August	144.45	145.36	139.36	141.51	161.77	145.36	62.58	180.77	144.22
September	11.68	11.78	11.6	11.91	13.43	12.17	8.97	15.73	11.78
October	9.13	9.07	9.09	9.19	9.19	9.17	5.94	9.53	9.07
November	10.3	10.46	9.05	9.47	3.62	10.15	3.09	5.20	10.25
December	5.23	5.30	5.22	5.25	5.24	5.30	3.64	5.09	5.30

Table 11.12 Monthly Mean Wind Power Based on Measured Data and Weibull Distribution Methods—WM Station-II

Month	Measured P (W/m²)	Calculated Power by Weibull Methods in W/m²							
		MOM	MLM	MMLM	LSM	PDM	EEM	GM	EM
January	33.81	34.03	33.83	33.91	27.57	34.03	21.66	22.41	34.03
February	17.32	14.75	19.08	18.38	13.74	18.97	9.31	13.42	14.75
March	46.38	47.43	46.28	47.19	37.71	47.99	33.8	30.22	47.43
April	1609.28	1421.52	1037.78	1175.72	1604.69	1460.34	338.18	1812.73	1373.44
May	12.28	12.32	12.15	12.22	11.29	12.16	6.97	11.13	12.24
June	41.35	40.76	41.29	41.96	41.38	42.33	29.04	42.06	40.76
July	158.07	164.69	160.12	159.34	160.3	157.8	93.84	156.28	163.75
August	137.5	139.99	127.03	85.64	158.37	137.04	51.8	183.78	137.04
September	120.94	124.98	124.4	125.74	100.73	125.68	44.43	81.94	124.98
October	69.24	69.54	69.84	72.03	70.54	74.74	58.29	71.14	69.54
November	51.05	49.78	52.74	52.57	53.04	53.46	34.81	56.13	49.78
December	28.78	29.95	26.28	27.12	21.2	28.84	10.12	18.41	29.68

calculated wind power by LSM is closely tied with the measured wind power (Table 11.12). According to this table, the maximum and minimum values of wind speed and wind power have been observed in April and February with values of 8.34 m/s and 2.75 m/s and with power of 1609.28 W/m² and 17.32 W/m², respectively. On the other hand, from Tables 11.13 and 11.14, the calculated wind power is matched with measured wind power for Station-III and Station-IV, respectively. From Fig. 11.1, it was found that Station-II is more prospective compared with other stations. From

Table 11.13 Monthly Mean Wind Power Based on Measured Data and Weibull Distribution Methods—WM Station-III

Month	Measured P (W/m²)	Calculated Power by Weibull Methods in W/m²							
		MOM	MLM	MMLM	LSM	PDM	EEM	GM	EM
January	15.46	15.4	15.53	16.29	15.14	17.23	13.94	14.97	15.55
February	27.11	30.06	30.13	31.07	27.45	32.24	24.72	25.07	30.06
March	37.23	37.36	37.36	39.14	37.89	41.94	34.47	38.45	37.63
April	42.87	44.16	44.46	46.8	45.17	49.99	40.93	45.84	44.16
May	34.2	34.03	34.3	34.6	35.37	35.08	23.67	36.47	34.03
June	87.12	89.92	89.58	90.52	72.88	91.37	59.7	59.42	90.43
July	23.86	23.75	23.95	25.29	23.18	27.01	22.19	22.43	23.95
August	18.97	19.11	19.13	19.99	18.76	21.19	17.44	18.43	18.94
September	22.02	22.7	23	23.71	21.55	24.48	19.03	19.14	22.7
October	8.58	8.61	8.52	8.93	8.41	9.64	7.92	8.11	8.61
November	9.71	10.11	10.00	10.55	9.78	11.26	9.23	9.35	10.11
December	9.22	9.22	9.25	9.60	9.68	10.06	8.00	10.04	9.22

Table 11.14 Monthly Mean Wind Power Based on Measured Data and Weibull Distribution Methods—WM Station-IV

Month	Measured P (W/m²)	Calculated Power by Weibull Methods in W/m²							
		MOM	MLM	MMLM	LSM	PDM	EEM	GM	EM
January	2.45	2.49	2.44	2.56	2.53	2.72	2.21	2.58	2.49
February	1.74	1.9	1.94	2.03	2.18	2.17	1.77	2.43	1.9
March	8.11	8.07	8.18	8.38	8.61	8.59	6.62	9.26	8.07
April	15.27	15.6	15.86	16.29	15.24	16.52	12.4	15.36	15.6
May	16.32	16.11	16.41	17.11	14.62	17.82	14.24	14.98	16.11
June	12.00	12.52	12.35	12.67	13.07	13.07	9.99	13.75	12.39
July	10.59	10.47	10.72	10.98	9.81	11.14	8.46	9.26	10.47
August	11.72	11.76	11.76	12.19	12.14	12.61	9.99	12.46	11.76
September	8.00	8.16	8.26	8.36	6.24	8.47	6.06	4.73	8.16
October	14.37	13.8	12.31	13.22	13.65	14.47	4.94	14.03	13.8
November	1.89	1.88	1.94	1.95	2.27	1.96	1.29	4.37	1.88
December	62.35	63.78	14.7	16.33	39.52	19.29	3.33	29.32	60.6

Table 11.12, it is clear that more wind power is available in Station-II compared with other stations.

Figs. 11.6—11.9 illustrate the monthly variation of energy density in the wind for each station, respectively. From Fig. 11.6, the wind energy from June to August is higher than other months. The calculated wind energy using EM coincides with the measured values. The energy estimation of wind regimes by the Weibull-based approach for Station-II is presented in Fig. 11.7. The figure shows that a pick of energy intensity in

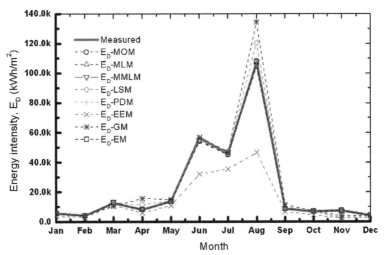

Figure 11.6 Monthly total energy intensity based on measured data and Weibull distribution methods—WM Station-I.

April which implies irregular wind behavior in that month. This is due to the wind gust identified on that time. Figs. 11.8 and 11.9 show the mean monthly variation of available wind energy for Station-III and Station-IV, respectively. The available wind energy can be used for irrigation purposes. From the previous discussion, Station-II is a better prospective wind energy site compared with other sites and LSM is the more accurate

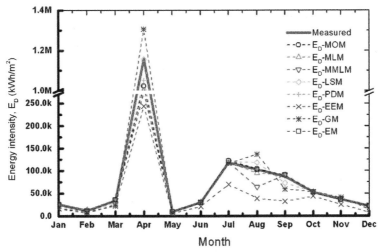

Figure 11.7 Monthly total energy intensity based on measured data and Weibull distribution methods—WM Station-II.

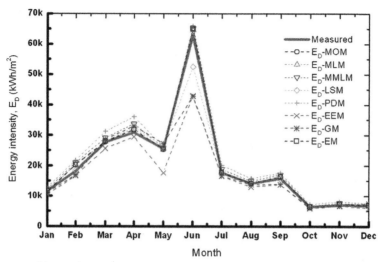

Figure 11.8 Monthly total energy intensity based on measured data and Weibull distribution methods—WM Station-III.

method for that selected site. While the effectiveness of methods for calculating Weibull distribution are site independent, their capability mainly depend on the wind characteristics of a site. It can be clearly noticed that gusty wind behavior has been identified at the selected site in April where the value of COV is 85.49% (Table 11.1) which was also observed in the velocity profile of that month in Fig. 11.10.

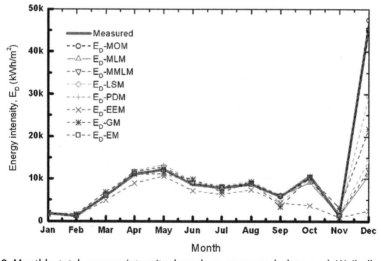

Figure 11.9 Monthly total energy intensity based on measured data and Weibull distribution methods—WM Station-IV.

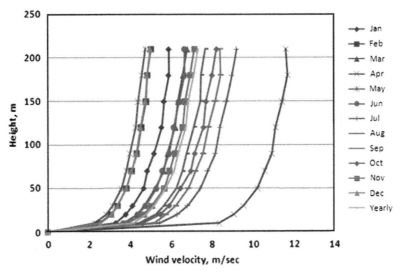

Figure 11.10 Monthly mean velocity profile at WM Station-II.

Fig. 11.10 shows the monthly mean velocity profile of the selected site. In every month, wind velocity varies up to 8 m/s except April. The average wind velocity was exceeded by more than 10 m/s in April. The common phenomenon is an increase in the wind velocity with the increase of height and this becomes free stream velocity above 100 m from the ground for each month. The analysis of velocity profile is important to determine the desirable height before installing wind turbines. Hence, the complete analysis can help to predict the more prospective windy site with minimum uncertainty and risk for power generation.

11.4 CONCLUSIONS

The study analyzed the wind energy potential at four sites using eight numerical methods of Weibull distribution. Extensive analysis was undertaken into the error and efficiency of each method. The following conclusions can be drawn from the study:

- The study found that MOM, LSM, and EM present better performance than other methods. The calculated results for wind power and wind energy are very close to the measured values with negligible error.
- Station-II is a more prospective windy site compared with the other sites. The behavior of the wind at this site was also analyzed and unusual wind behavior in April was found.
- The study found that the Weibull methods are site independent, but are dependent on the behavior of the wind pattern. Each and every site has a suitable Weibull

distribution method. There is no generalized Weibull distribution method that is a suitable fit for every site.
- Further study is recommended of long-term wind data analysis for the selected site. Investigation is also needed into a suitable design of wind turbine with computational fluid dynamics (CFD) analysis of the designated wind turbine for the selected site.

ACKNOWLEDGMENTS

The authors would like to acknowledge Mr. Tim McSweeney, Adjunct Research Fellow, Higher Education Division at Central Queensland University, Australia for his contribution in English proof read of this chapter.

NOMENCLATURE

$f(v)$	Weibull probability density function
$F(v)$	Cumulative distribution function
K	Dimensionless shape parameter.
C	Scale parameter (m/s)
COV	Coefficient of variation
v	Mean wind speed in m/s
v_i	Random sample of wind speed central to bin i
P	Total power in (W/m^2)
N	Number of sample or bin
$f(v_i)$	Weibull frequency for wind speed ranging within bin i
$f(v \geq 0)$	Probability for wind speed ≥ 0
W_{vi}	Observed frequency of the wind speed
v_m^3	Mean of the cubic wind speed (m/s)
ε_{vi}	Error of the approximation
E_{pf}	Energy pattern factor
z_{ref}	Reference height (m)
k_h	Weibull shape factor at desired height
c_h	Weibull scale factor at desired height (m/s)
v_h	Wind speed at desired height (m/s)
P_h	Power at desired height (W/m^2)
N	Total number of observations
$y_{i,m}$	i^{th} calculated value from measured data
$x_{i,w}$	ith calculated value from Weibull distribution
$z_{i,y}$	Mean of ith calculated value from measured data
R^2	Analysis of variance

GREEK LETTERS

Γ	Gamma function
Σ	Standard deviation of wind speed (m/s)
χ^2	Chi-square error

REFERENCES

[1] Ezzat MF, Dincer I. Energy and exergy analyses of a new geothermal—solar energy based system. Solar energy 2016;134:95—106.

[2] Segurado R, Madeira JFA, Costa M, Duić N, Carvalho MG. Optimization of a wind powered desalination and pumped hydro storage system. Applied Energy 2016;177:487—99.

[3] Ooba M, Fujii M, Hayashi K. Geospatial distribution of ecosystem services and biomass energy potential in eastern Japan. Journal of Cleaner Production 2016;130:35—44.

[4] Azad AK, Rasul MG, Khan MMK, Sharma SC, Hazrat MA. Prospect of biofuels as an alternative transport fuel in Australia. Renewable and Sustainable Energy Reviews 2015;43:331—51.

[5] Azad AK, Rasul MG, Khan MMK, Sharma SC, Islam R. Prospect of Moringa seed oil as a sustainable biodiesel fuel in Australia: a review. Procedia Engineering 2015;105:601—6.

[6] Azad AK, Rasul MG, Khan MMK, Sharma SC, Bhuiya MMK. Study on Australian energy policy, socio-economic, and environment issues. Journal of Renewable and Sustainable Energy 2015;7:063131 (1—20).

[7] Azad AK, Rasul MG, Khan MMK, Omri A, Bhuiya MMK, Ali MH. Modelling of renewable energy economy in Australia. Energy Procedia 2014:1902—6.

[8] Azad AK, Rasul MG, Khan MMK, Ahasan T, Ahmed SF. Energy scenario: production, consumption and prospect of renewable energy in Australia. Journal of Power and Energy Engineering 2014;2(4): 19—25.

[9] Hsieh C-H, Dai C-F. The analysis of offshore islands wind characteristics in Taiwan by Hilbert—Huang transform. Journal of Wind Engineering and Industrial Aerodynamics 2012;107:160—8.

[10] Kwon S-D. Uncertainty analysis of wind energy potential assessment. Applied Energy 2010;87(3): 856—65.

[11] Adaramola M, Paul S, Oyedepo S. Assessment of electricity generation and energy cost of wind energy conversion systems in north-central Nigeria. Energy Conversion and Management 2011; 52(12):3363—8.

[12] Mohammadi K, Mostafaeipour A. Using different methods for comprehensive study of wind turbine utilization in Zarrineh, Iran. Energy Conversion and Management 2013;65:463—70.

[13] Belu R, Koracin D. Wind characteristics and wind energy potential in western Nevada. Renewable Energy 2009;34(10):2246—51.

[14] Azad AK, Rasul MG, Yusaf T. Statistical diagnosis of the best Weibull methods for wind power assessment for agricultural applications. Energies 2014;7(5):3056—85.

[15] Azad AK, Rasul MG, Islam R, Shishir IR. Analysis of wind energy prospect for power generation by three Weibull distribution methods. Energy Procedia 2015;75:722—7.

[16] Morgan EC, Lackner M, Vogel RM, Baise LG. Probability distributions for offshore wind speeds. Energy Conversion and Management 2011;52(1):15—26.

[17] Vincent CL, Pinson P, Giebela G. Wind fluctuations over the north Sea. International Journal of Climatology 2011;31(11):1584—95.

[18] Azad AK, Rasul MG, Alam MM, Uddin SMA, Mondal SK. Assessment of wind energy potential by Weibull distribution in isolated islands. In: IEEE Xplore, 3rd International Conference on the developments in renewable energy technology (ICDRET), Dhaka, Bangladesh; 2014. p. 1—6.

[19] Ucar A, Balo F. Evaluation of wind energy potential and electricity generation at six locations in Turkey. Applied Energy 2009;86(10):1864—72.

[20] Shata AA, Hanitsch R. Evaluation of wind energy potential and electricity generation on the coast of Mediterranean Sea in Egypt. Renewable Energy 2006;31(8):1183—202.

[21] Azad AK, Alam MM. A statistical tools for clear energy: Weibull's distribution for Potentiality analysis of wind energy. International Journal of Advanced Renewable Energy Research (IJARER) 2012;1(5):240—7.

[22] Chang TP. Performance comparison of six numerical methods in estimating Weibull parameters for wind energy application. Applied Energy 2011;88(1):272—82.

[23] Azad AK, Alam MM. Determination of wind gust factor at windy areas of Bangladesh. In: 13th Asian Congress of fluid mechanics, Dhaka, Bangladesh; 2010. p. 521—4.

[24] Alam MM, Azad AK. Analysis of Weibull parameters for the three most prospective wind sites of Bangladesh. In: 9th International Conference on mechanical engineering, Dhaka, Bangladesh; 2009. p. 1—6.

[25] Azad AK, Saha M. Weibull's analysis of wind power potential at coastal sites in Kuakata, Bangladesh. International Journal of Energy Machinery 2011;4(1):36—45.

[26] Akpinar E, Akpinar S. Statistical analysis of wind energy potential on the basis of the Weibull and Rayleigh distributions for Agin-Elazig, Turkey. Proceedings of the Institution of Mechanical Engineers, Part A: Journal of Power and Energy 2004;218(8):557—65.

[27] Celik AN. A statistical analysis of wind power density based on the Weibull and Rayleigh models at the southern region of Turkey. Renewable Energy 2004;29(4):593—604.

[28] Rocha PAC, de Sousa RC, de Andrade CF, da Silva MEV. Comparison of seven numerical methods for determining Weibull parameters for wind energy generation in the northeast region of Brazil. Applied Energy 2012;89(1):395—400.

[29] Odo F, Offiah S, Ugwuoke P. Weibull distribution-based model for prediction of wind potential in Enugu, Nigeria. Pelagia Research Library. Advances in Applied Science Research 2012;3:1202—8.

[30] Ramírez P, Carta JA. The use of wind probability distributions derived from the maximum entropy principle in the analysis of wind energy. A case study. Energy Conversion and Management 2006; 47(15):2564—77.

[31] Azad AK, Rasul MG, Alam MM, Uddin SMA, Mondal SK. Analysis of wind energy conversion system using Weibull distribution. Procedia Engineering 2014;90:725—32.

[32] Azad AK, Saha M. Wind power: a renewable alternative source of green energy. International Journal of Basic and Applied Science 2012;1(2):193—9.

[33] Azad AK. Wind power: the available source of clear energy at the coastal belt of Bangladesh. Research Journal of Physical and Applied Science Vol 2012;1(1):1—6.

[34] Azad AK, Alam MM. Analysis of wind power potential in Sandwip Sea belt of Bangladesh. In: 2nd International Conference on the Developments in renewable energy Technology, Dhaka, Bangladesh; 2012. p. 143—6.

Biodiesel for Sustainable Development

CHAPTER TWELVE

Prospect of the Legume Tree *Pongamia pinnata* as a Clean and Sustainable Biodiesel Feedstock

A. Indrasumunar, P.M. Gresshoff and P.T. Scott
The University of Queensland, Brisbane, QLD, Australia

12.1 INTRODUCTION

The demand for reliable and affordable sources of energy across the globe is continuing to rise in parallel with a growing world population and a growing upper and middle class in previously poor socioeconomic populations. Currently over 7 billion people populate earth, with that number expected to rise to more than 9 billion by 2050. This is taking place at the same time that the long-term supply of conventional energy sources from the 20th century is being questioned along with the environmental impact of a fossil fuel—based economy. While the debate continues as to the reliable future supply of easily accessible fossil fuels, it is clear that the capacity of many countries to meet domestic demands for crude oil is not sustainable, a situation that is commonly referred to as "Peak Oil."

Australia is one such country, with about 85% of its oil supply imported [1], the vast majority from Singapore via the Middle East. Similarly, while the adverse environmental impacts of continued fossil fuel consumption and its contribution to climate change is overwhelmingly accepted by the scientific community, there is still some debate among politicians and the lay community regarding the proportional role of anthropogenic activities toward global climate change. Nonetheless, the atmospheric chemistry surrounding fossil fuel mining, exploration, and consumption is understood. The predominant concern is with the generation of the so-called greenhouse gases (GHGs; i.e., CO_2, CH_4, and NO_x) upon the combustion of fossil fuels. Despite these issues, exploration and mining continues unabated and with the short-term price of oil relatively low, perhaps artificially adjusted.

It is important to note at this time that much of the discussion around renewable energy among government policy makers and regulators is with technologies such as solar, hydro, and wind power that support, in part at least, the demand for electricity. The majority of demand for energy is at this stage still only met by energy-dense and readily storable liquid fuels. In light of the recognition of these impending issues, the search for

Clean Energy for Sustainable Development
ISBN 978-0-12-805423-9, http://dx.doi.org/10.1016/B978-0-12-805423-9.00012-0

environmentally sustainable biological sources of fuel has been underway for more than a decade now. The underlying assumption behind the search for suitable biofuel feedstocks has been the purported positive environmental impacts, primarily the "carbon neutral" effect, on atmospheric GHG concentrations.

Initially, biofuel feedstocks were sought from food crops such as soybean (*Glycine max*), corn (*Zea mays*), sugarcane (*Saccharum officinarum*), and rapeseed (*Brassica napus*) and were exploited for their already well-established farming systems and the ease with which their biomass could be converted into the two dominant forms of biofuel—ethanol and biodiesel. Ethanol is formed primarily by the fermentation of starch- or sugar-rich crops, such as corn, while biodiesel can be formed by the transesterification of plant seed oil of oil-rich crops, such as soybean.

The initial enthusiasm for these first-generation biofuel crops has subsequently been met with arguments that food crops and the land historically used for the cultivation of food crops should not be redirected to the production of biofuels, the so-called food versus fuel debate. In addition, the positive environmental impacts of these crops have been questioned by some [2]. Importantly, while the benefit of biofuels over fossil fuels has focused on the carbon cycle, little emphasis has been placed on the nitrogen cycle. This is particularly relevant when the economic, energetic, and environmental costs of nitrogen fertilizer production and application are taken into consideration [3]. Therefore, legumes are well placed to be examined as potential biofuel feedstocks with their ability to form symbiotic associations with soil bacteria, collectively known as rhizobia, which leads to the formation of root nodules and biological fixation of atmospheric nitrogen. One such legume under consideration and the subject of this chapter is the tree legume *Pongamia pinnata* (also called *Millettia pinnata*; hereafter referred to simply as pongamia).

12.2 PONGAMIA AS PROSPECTIVE FEEDSTOCK CANDIDATE

Pongamia has gained attention as a strong candidate for sustainable biodiesel and aviation fuel production due to its large oil-rich seeds (40—50% by volume). The oil is nonedible, and the seed cake by-product (10—20%) following oil extraction has nutritional properties indicative of potential as an animal feed supplement [4,5]. Starch, seed pods, and indigestible fiber from seeds can be used for electricity cogeneration or fermentation (Fig. 12.1).

Pongamia is native to southern and southeast Asia, as well as northern Australia. It is a fast growing nonfood legume, capable of symbiotic biological nitrogen fixation (BNF), a process absent in other more well established biodiesel feedstocks (e.g., canola, mustards, oil palm, and jatropha). Importantly, pongamia has the ability to grow well on low agriculturally productive soils typically characterized by low water availability, low nutrient content, and high salinity [6—8]. Studies carried out by the Centre for Integrative Legume Research at the University of Queensland demonstrated that the growth

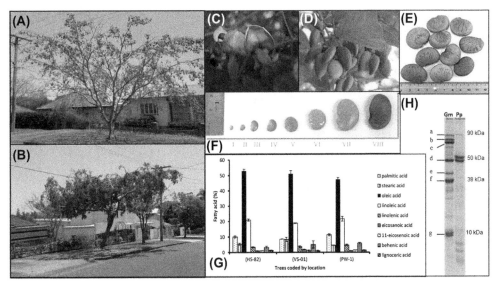

Figure 12.1 The biological properties of pongamia. (A, B) Street trees in Brisbane, Australia, with abundant pods. Note the variability of canopy structure and leaf burst caused by genetic heterogeneity. (C) Pongamia flowers, pea-like, about 6–8 mm in length. (D) Pods of pongamia. (E) Mature pongamia seeds. Each is about 2–3 g dry weight. (F) Pongamia seed development: 10–12 months from fertilization to harvest of mature seed. (G) Pongamia seed oil analysis of three trees from different locations. Oil composition is broadly stable across trees and location. (H) Pongamia seed cake proteins. Left lane: soybean (Gm); right: pongamia (Pp). (a) lipoxygenase (90 kDa); (b) 7Sα prime (70 kDa); (c) 7Sα (66.4 kDa); (d) 7Sβ conglycinin (51 kDa); (e) 11SAβ (41 kDa); (f) 11SA1a, A1b, A2 (38.5 kDa); (g) sB1a, B1, B2, B3, 11SB4 (10 kDa). The genes encoding the dominant pongamia seed storage proteins (50 and 52 kDa) were cloned and sequenced and shown to be similar to the gene encoding the low-quality seed storage protein 7S β-conglycinin of soybean.

performance of pongamia was equivalent to that of saltbush and rescue grass in saline soils [at 18 dS per meter (dS/m); [9]]. In contrast, in parallel experiments soybean perished at 5 dS/m. Unlike jatropha, another emerging biodiesel feedstock, pongamia is not listed as an invasive species in Australia and is not toxic to humans and animals [10]. In Australia, pongamia is currently grown predominantly in trial plantations in western Australia, the northern Territory, and Queensland [11].

Pongamia has wide diversity in both genotypes and phenotypes. It exhibits fast growth and high seed production potential. Within 1 year, 50 cm pongamia saplings planted as a field trial at the UQ Gatton campus reached 3 m in height, up to 2000 g dry weight biomass shoot weight, and 6% of the trees had produced seed-bearing pods. The plant is estimated to sequester 25 ton CO_2/ha/annum. Seeds can be mechanically harvested with a vibrating tree shaker. Trees bearing around 100,000 seeds have been noted. It is estimated that a tree can stay in full production for more than 35 years (note: 100-year old specimens are known in the Brisbane region). Based on conservative estimates from trees found in Brisbane, oil yields of 5 t/ha/annum are attainable. This compares well with soybean (0.8 t/ha/annum), canola (1.5 t/ha/annum), and oil palm (5 t/ha/annum). At present we estimate that a 20% diesel replacement with biodiesel for all of Australia would require 200 plantations of 6 × 6 km each. Planting of diverse germplasm using different elite clones is recommended to avoid potential disease problems arising from monoculture.

12.2.1 Pongamia and Nitrogen Fixation

Most biofuel feedstocks, including canola (*B. napus*), sugarcane (*S. officinarum* L.), sweet sorghum [*Sorghum bicolor* (L.) Moench], maize (*Z. mays* L.), and woody trees, such as eucalypts (*Eucalyptus globulus* Labill.) and willows (*Salix* spp.), require nitrogen fertilizer for their growth. The production and application of nitrogen fertilizers represent a large economic and energetic burden as costs have increased due to a dependence on fossil fuel and natural gas. Moreover, the application of nitrogenous fertilizer to crops results in resident soil bacteria producing NO_x, a powerful GHG, possessing global warming potential 296 times that of CO_2. This adverse scenario makes the supply of nitrogen to biofuel feedstocks a key issue when considering their sustainability on economic as well as ecological criteria [12,13]. In contrast, legumes are capable of forming symbiotic relationships with nitrogen-fixing rhizobia, which are housed in specialized root organs called nodules [14]. The use of perennial plants, such as pongamia, that are capable of symbiotic nitrogen fixation is a good strategy to sustain nitrogen for a more productive and diverse agroecosystem [15,16]. Pongamia as a legume can play an important role in sustainable agroforestry by fixing its own nitrogen from the atmosphere, minimizing the need for added nitrogenous fertilizers. It also enters into symbiosis with phosphate-mobilizing mycorrhizae [17].

Pongamia nodules have been reported to be determinate in nature with a spherical morphology [18,19]; however, Samuel et al. [20] reported that the "determinate-like" nodules progressed to "indeterminate-like" structures through activation of new cell divisions in the nodules of older plants. Therefore, older pongamia trees will exhibit a combination of spherical and coralloid nodules (Fig. 12.2A). This observation is consistent with a previous report by Sprent and Parsons [21] that most tree legumes do tend to have woody indeterminate nodules.

Despite being a perennial legume, the pongamia root symbiosis is very similar to that seen in annual legumes. Autoregulation of nodulation (AON) and nitrate (NO_3) inhibition that are common in annual crop legumes were also displayed in pongamia [20]. Using split root experimental methods, Samuel et al. [20] showed that the initiation of late nodulation events is developmentally suppressed by the first-formed nodules. Rhizobia inoculation of one portion of the root system systemically suppressed nodule formation on a later-inoculated and physically isolated root. Likewise, seedlings

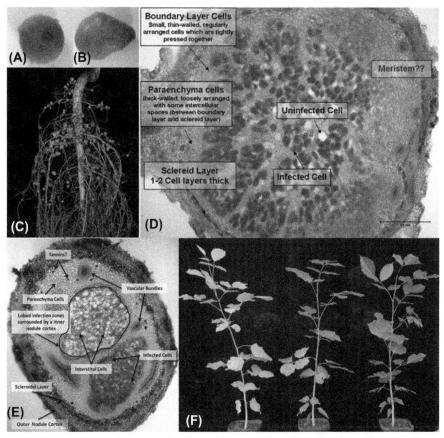

Figure 12.2 Nodulation and nitrogen fixation of pongamia. (A, B) Nodules consist of both spherical and coralloid nodules. (C) AON on pongamia nodulation. (D) Effective nodules of *B. japonicum* strain CB1809. (E) Ineffective nodule of ineffective rhizobia strain. (F) Rhizobia inoculation significantly improve the growth of pongamia in soil: *yellowing* (light gray in print version) leaves on uninoculated pongamia, *green* (gray in print version) and healthy leaves on inoculated pongamia. *(C) and (D) taken from Samuel S, Scott PT, Gresshoff PM. Nodulation in the legume biofuel feedstock tree* Pongamia pinnata. *Agriculture Research 2013;2:207—14.*

inoculated at planting are characterized by significant nodule formation on the upper portions of the root system but no nodules on the lower parts of the root (i.e., crown nodulation; Fig. 12.2B).

Unlike many other legumes that form functional nodules in association with just one specific or select few strains of rhizobia, pongamia can form nodules with several strains of both *Bradyrhizobium* and *Rhizobium* [22—24]. However, the strains of rhizobia that infect pongamia vary in their efficacy for nodulation and nitrogen fixation. Therefore, selection of superior strains of rhizobia for pongamia is very important, as it will help to promote growth and potentially increase yields of oil-rich seeds. Toward this end, our laboratory [20] tested a wide range of bacterial strains from Australia and India, and established *Bradyrhizobium japonicum* strains CB1809 and USDA 110 as the best inocula tested. The nodules produced by these strains were larger and more extensively and uniformly filled with zones of infected bacteroids (Fig. 12.2C). In contrast, the nodules produced by less-effective strains had several infection zones with variable bacterial occupancy (Fig. 12.2D). The efficacy of pongamia nodules was demonstrated using the acetylene reduction assay, where C_2H_2 (acetylene) serves as a substrate for the bacterial encoded nitrogenase, and its reduction to C_2H_4 (ethylene) is quantified by gas chromatography. Recently, we successfully isolated two new strains of rhizobia that are more efficacious than *B. japonicum* strains CB1809 and USDA110 in vermiculite (Phoebe Nemenzo-Callica, unpublished data, 2015), and in soil (Fig. 12.2E). Moreover, these new strains were also more effective than *B. japonicum* strain CB1809 on pongamia grown in degraded mined soils (A. Indrasumunar, unpublished data, 2015).

12.2.2 Pongamia and Degraded/Marginal Lands

Currently, most biodiesel is produced from food crops growing on fertile land, for example, soybean in the United States and rapeseed in Europe [25]. Due to food security concerns, future biodiesel production should be produced from crop feedstocks that can be grown on marginal/degraded land and less profitable arable crop lands in order to ensure the establishment of a biodiesel industry that will not compete for land with other food crops. Marginal/degraded lands are usually associated with soil and water limitations and other environmental stresses (e.g., salinity and acidity) that require the selection of plant species adapted to such stresses.

In Australia, there are over 1 million km^2 (100 million hectares) of marginal land, 15% of the total land area (7,687,147 km^2) [26]. In addition, land clearing in areas such as the Murray—Darling River basin has resulted in dryland salinity problems. Affected lands are currently being reclaimed by planting salt-tolerant tree legumes, such as *Acacia* spp. [27]. Such marginal lands are excellent options for planting salt-tolerant biofuel crops, such as pongamia [28,29].

Pongamia is highly desirable as it has been reported to be highly tolerant to salinity (10 dS/m) [30] and drought (survived 4 months) without rain in Brisbane during 2007—08 drought [13], and can be grown in various soil textures (stony, sandy, and clayey). It can grow in humid subtropical environments with annual rainfall ranging between 500 and 2500 mm. Pongamia can survive maximum temperatures exceeding 45°C. Cuttings and saplings survived 65°C in a glasshouse when the temperature control unit failed during the January 2011 Brisbane flood, though ample water was available [13]. Although not considered as "frost tolerant," pongamia can survive and recover from frost events [13].

12.2.3 Pongamia and Salinity

Salinity is a serious threat to agriculture in arid and semiarid regions [31,32]. Nearly 40% of the global land surface can be categorized as having potential salinity problems [33,34]. Salinity is a measure of the content of salts in soil or water. It is measured by the electrical conductivity (EC) of a solution or saturation extract of soil. Soils are considered saline when their EC exceeds 4 dS/m, and water exceeding 4.7 dS/m is unsuitable for the irrigation of most crop species [35]. However, 62% of pongamia trees survived in soil with EC values varying from 10 to 12 dS/m [7], and 13% could survive salinity values as high as 19 dS/m [8]. We have also shown that salinity of 20 dS/m did not adversely affect growth (shoot and root fresh weight, number of leaflets, root length, and plant height) of 12 week-old pongamia seedlings over an 8-week period of exposure. In addition, nodule number was not affected by salinity of up to 20 dS/m, but nodule mass and nitrogen fixation started to decrease at 4 dS/m. Efforts are required to improve nodulation and nitrogen fixation of pongamia on saline soils if it is to be a successful biofuel feedstock crop. Improvement of nodulation and nitrogen fixation in saline conditions have been shown in wattle (*Acacia ampliceps* Maslin) when inoculated with salt-tolerant strains of rhizobia [27,36]. Therefore, it is considered highly probable to improve nodulation and nitrogen fixation of pongamia in saline conditions by inoculation with salt-tolerant rhizobia [9]. Selection of rhizobia strains that are tolerant to environment stress is currently being conducted in our laboratory. We found two strains of rhizobia tolerant to higher levels of salinity than *B. japonicum* strain CB1809. These strains were able to grow on Yeast Mannitol Broth (YMB) medium containing 100 mM NaCl, while *B. japonicum* strain CB1809 was only able to grow on YMB medium containing 50 mM NaCl.

12.2.4 Pongamia and Drought

The increasing frequency and intensity of dry periods result in the consecutive occurrence of drought in Australia and other parts of the world [37,38]. Drought is a major abiotic stress limiting plant production in many countries. In 2006, more than three-quarters of Australia, encompassing 38% of the agricultural land, was affected by drought [9]. In addition, according to the Commonwealth Scientific and Industrial Research Organization, it is predicted that by 2030, rainfall in major capitals could drop by 15%. Pongamia has been reported as drought tolerant, possibly due to its dense network of lateral roots and thick, long taproot. Research in our laboratory supports the claim of this drought tolerance. Preliminary experiments indicate that seedlings are capable of withstanding extensive periods of water deprivation (25 days to 55% relative water content) without significantly affecting growth and biomass production. In the context of extensive drought periods, the vast Australian landscape should not provide an impediment to cultivation of pongamia [39]. However, further research on the effects of drought on seed and biomass production are still needed.

12.2.5 Pongamia and Mine Spoils

In addition to its ability to grow on poor quality soils for biofuel feedstock, it is worth considering the value of pongamia as a long-lived perennial plant for rehabilitation of mine spoil sites. An example of such an environment requiring rehabilitation is coal mine overburden and spoil. To date, the common practice in the rehabilitation of coal mine sites in Australia has been the planting of *Eucalyptus* spp., *Acacia* spp., or native grasses. This has been practiced at the Meandu Mine, Queensland (Stanwell Corp.) for many years. We have developed a new approach to establish an integrative rehabilitation planting scheme that contributes the restorative attributes of legume trees such as the *A.* spp., but also incorporates the planting of an economically valuable crop with multiple products and outcomes (i.e., biofuel, biochar, carbon sequestration, and carbon farming).

This new approach has some advantages. Firstly, mine rehabilitation to date has involved plantings that aim to restore any disturbed site to a landscape that resembles as close as possible undisturbed native vegetation. In this approach we integrated the capacity for rehabilitation of pongamia with its potential for the production of biomass (i.e., seed oil and associated by-products) of economic value. In promoting the value of pongamia it is worth noting that with the exception of soybean, all proposed biofuel feedstocks throughout the world to date have been nonlegume plant species. As such this approach addresses the neglected issue of the costs associated with nitrogen inputs to biofuel production through the exploitation of a perennial legume and the associated developmental process of nodulation and nitrogen fixation. Secondly, this project aims to enhance the growth performance, through improvements in nodulation and nitrogen fixation of a tree that is native to Australia and well suited to the role of a dedicated bioenergy crop. The planting of a perennial legume on coal mine spoil has an objective that directly targets the issue of long-term sustainability with both environmental and economic benefits via an agricultural biotechnology approach to complement and support the well-established coal-mining industry through restoration of disturbed mine sites.

It is widely acknowledged that pongamia can ameliorate and rehabilitate poor quality soils [29,40]. We have also shown that pongamia was successfully established in degraded mined soils (Fig. 12.3). More than 95% of pongamia seedlings survived well in the harsh environment of mine spoil. More evaluation on tree survival, growth rate, flowering, and seed production are still needed to measure the success of this mine rehabilitation. With this promising result, the potential of pongamia for mine rehabilitation needs to be extended to other mine sites.

12.3 PONGAMIA IMPROVEMENT PROGRAM

The proposal that pongamia be considered as a future bioenergy crop has primarily been on the basis of observations of individual trees growing in forests or as street

Figure 12.3 Pongamia establishment for rehabilitation and reforestation of mine spoils. Biodiesel production is seen as an additional outcome. (A) Soil and planting preparation, sculptured soil was intended to increase water retention. (B) Pongamia saplings at planting time, saplings were raised in nurseries for 10 months to get established saplings of 90 cm. (C) Most of trees dropped their leaves during winter. (D) One-year old pongamia grew well in degraded mined soil. Some trees will produce seeds within 3—4 years after planting, but in small, commercially insufficient quantity. Commercial use is expected to start in year 5 after trees' establishment.

landscape specimens. While there is a long history in the Indian subcontinent and southeast Asia of the exploitation of pongamia by humans (e.g., heating fuel and traditional medicine; [18]), pongamia is yet to undergo the comprehensive domestication and selection that is normally associated with the development of modern dominant crop species, and so future deployment of plantations will require extensive genetic improvement of currently available germplasm. Importantly, pongamia adopts a predominantly outcrossing reproductive strategy [41], which presents both challenges and opportunities. The challenges revolve around the propagation of any elite germplasm generated in a domestication program. Propagation of pongamia is likely to be through methods such as rooted cuttings, grafted saplings, and tissue culture regeneration ([42—44]; Fig. 12.4). Rooted cuttings and grafted saplings are generated through essentially low technologies, but are labor-intensive and time-consuming. Tissue culture regeneration, if reliable methods are available, can generate the large numbers of saplings that are required for broad acre plantations. Unfortunately, while it is our experience that

Figure 12.4 Phenotypic and genotypic diversity of pongamia, and its clonal propagation. (A, B) The diversity of pongamia in spring foliage regrowth in trees from a field site near Roma, Queensland. (C) A selection of pongamia seedpods sourced from trees in and around the Brisbane area. (D) Variation in morphology of leaves from two Brisbane street trees. (E) Silver-stained polyacrylamide gel of PCR products amplified with pongamia inter-simple sequence repeat markers. (F) Clonal propagation from immature cotyledons of pongamia through subsequent developmental stages (G—L). (M) Pongamia tree derived from tissue culture grew well in the field.

regeneration of whole plants is relatively easy from cotyledons, it seems to be much more difficult to regenerate whole plants through tissue culture from somatic tissue explant.

The opportunities of an outcrossing mode of reproduction are the many and varied phenotypes derived from seed-borne progeny germplasm. Phenotypic variation has been observed in traits relevant to tree architecture, annual yield of seed, timing of flowering, seed oil content, and composition. With respect to clonal propagation and phenotypic variation it has been our experience that some germplasm is more amenable to propagation than others. This may present another challenge, particularly if germplasm exhibiting traits desirable for the agronomy of biofuel production is recalcitrant to clonal propagation. Nonetheless, the wide genotypic and phenotypic diversity should provide a germplasm pool that will enable the selection of elite planting material for extensive plantations. These plantations will most likely incorporate a mix of genotypes that will avoid the potential disease and problems that have seriously affected monoculture crops in the past.

Any future pongamia domestication program should be supported with appropriate genetic and genomic tools that help to characterize and define any elite germplasm. To date, a complete genome sequence of pongamia is yet to be constructed. However, the complete sequences of the mitochondrial and chloroplast genomes have been annotated [30]. A range of molecular marker technologies have been explored for future application in marker-assisted genetic improvement. These technologies have included RAPD, AFLP, ISSR, and SSR methods [41,45–48]. While they have been demonstrated to differentiate germplasm, they are yet to positively link markers with relevant phenotypes.

12.4 QUALITY ANALYSIS AND ADVANTAGES OF PONGAMIA OIL FOR BIODIESEL

According to the American Society of Testing and Material (ASTM), biodiesel is referred to as monoalkyl esters of long-chain fatty acids (fatty acid methyl esters, FAMEs) derived from renewable biological sources, such as vegetable oils or animal fats. The oil can be converted to biodiesel (i.e., FAMEs) by transesterification with CH_3OH (methanol) in the presence of KOH [49] or NaOH [50] as catalyst. This reaction also produces glycerol, a low-value by-product for industry. The resulting biodiesel is quite similar to petroleum-based diesel fuel in its physicochemical characteristics and can be blended in any proportion with petroleum diesel to create a stable biodiesel blend [51]. It is well recognized as the best fuel substitute in diesel engines because its raw materials are renewable, it is biodegradable and more environmentally friendly than petroleum diesel, and it can be directly used in the compression ignition engines without significant modification of existing engines [52,53].

Many plants have emerged as sources of raw material for biodiesel, including canola and Indian mustard (*B. napus* and *Brassica juncea*), camelina (*Camelina sativa*; another mustard related to canola), soybean (*G. max*), oil palm (*Elaeis guineensis*), and sunflower (*Helianthus annuus*). However, it is not economically feasible to use food-grade vegetable oils to produce biodiesel because of the surge in feedstocks price as a result of competition between food and fuels. Therefore, the search for biodiesel feedstocks is focused on low-cost nonedible oils sources, and one of the most suitable species is pongamia due to favorable properties that include high oil recovery and quality of oil.

Pongamia produces an abundant supply of oil-bearing seeds with yields of more than 100 kg of seeds per tree [54]. The seeds of pongamia comprise 40–50% oil, the composition of which is dominated by oleic acid (C18:1), a fatty acid (FA) that is highly desirable for biodiesel production. It has been reported that biodiesel quality was affected by FA composition of the corresponding feedstock [54,55]. Different plant species produce oils with varying content and composition (Table 12.1). As shown in Table 12.1, pongamia is the best choice because it contains a high proportion of oleic acid but low levels of undesirable saturated and polyunsaturated fatty acids (PUFAs).

Table 12.1 Major Components of Oil of Several Plants Currently Used as Feedstock for Biodiesel Production

Plant	Oil Yield (L/ha per annum)	Percent Oleic Acid ($C_{18:1}$)	Percent Palmitic Acid ($C_{16:0}$)	Percent Stearic Acid ($C_{18:0}$)	Source or Reference(s)[a]
Corn	172	30.5—43.0	7.0—13.0	2.5—3.0	Dantas et al. (2007)
Soybean	446	22.0—30.8	2.3—11.0	2.4—6.0	Hildebrand et al. (2008)
Canola	1196	55.0—63.0	4.0—5.0	1.0—2.0	Masser (2009)
Jatropha curcas L.	1892	34.3—45.8	13.4—15.3	3.7—9.8	Becker and Makkar (2008)
Palm oil	5950	38.2—43.5	41.0—47.0	3.7—5.6	Sarin et al. (2007)
Algae[b]	59,000	1.7—14.3	3.7—40.0	0.6—6.0	Hu et al. (2008)
Tallow	NA	26.0—50.0	25.0—37.0	14.0—29.0	Canakci and Sanli (2008)
Pongamia	3000—6000	52—57	8—12	7—11	Akoh et al. (2007), Mamilla et al. (2008), CILR

[a]These references are representatives of an extensive list in the publicly available scientific literature.
[b]Algal yield represents extrapolations from smaller volume trials with multiple species. These projected yields are yet to be demonstrated on a commercial scale.
Biswas B, Scott PT, Gresshoff PM. Tree legumes as feedstock for sustainable biofuel production: opportunities and challenges. Journal of Plant Physiology 2011;168:1877—84; Biswas B, Kazakoff SH, Jiang Q, Samuel S, Gresshoff PM, Scott PT. Genetic and genomic analysis of the tree legume *Pongamia pinnata* as a feedstock for biofuel. Plant Genome 2013;6(3). http://dx.doi.org/10.3835/plantgenome2013.05.0015.

Monounsaturated oleic acid is preferred over other saturated FAs for biodiesel production because of its relatively low cloud and pour point temperatures. The pour point of a fuel indicates the lowest temperature at which the oil can still flow, whereas the cloud point determines the temperature at which the dissolved solids in the oil will precipitate from the liquid [56]. The pour and cloud points of pongamia FAMEs are 2.1°C and 8.3°C, respectively, consistent with the presence of saturated oils (palmitic acid and stearic acid; Table 12.1). Lower cloud points are more desirable for engine performance and the cloud point of pongamia FAMEs is lower than biodiesel derived from other sources, such as oil palm (*E. guineensis* Jacq.; 10°C) and beef (*Bos taurus*) tallow (13°C) [57], but higher than biodiesel made from edible oil of soybean (−1°C), rapeseed (*B. napus* L.) (−7°C), and sunflower (*H. annuus* L.) (1°C).

Engine performance tests using pongamia FAMEs showed that blends up to 40% by volume with mineral diesel were successful in reducing exhaust emissions of CO, smoke density, and NO_x without sacrificing the power output (torque, brake power, and brake thermal efficiency), and a reduction in brake-specific fuel consumption [58]. However, as the concentration of pongamia FAMEs in the blend was increased, deterioration in viscosity, cloud, and pour points (important for cold weather performance) of the fuel was detected. At low operating temperature, fuel may thicken and might not flow properly

Table 12.2 Biodiesel (B100) Fuel Quality Standard Based on ASTM D6751-09

Property	Method	Limit
Flash point	D93	>93°C
Water and sediment	D2709	<0.05% vol
Kinematic viscosity, 40°C	D445	1.9–6.0 mm^2/s
Sulfated ash	D874	<0.02% mass
Sulfur S 15 grade	D5453	<0.0015 ppm
Sulfur S 500 grade	D5453	<0.05 ppm
Copper strip corrosion	D130	<3
Cetane number	D613	>47
Cloud point[a]	D2500	Report to customer
Carbon residue	D4530	<0.05% mass
Acid number	D664	<0.05 mg KOH/g
Free glycerine	D6584	0.02% mass
Total glycerine	D6584	0.24% mass
Phosphorus content	D4951	<10 ppm
Vacuum distillation end point	D1160	<360°C
Cold soak filtration	D6751	<360 s

[a]The US ASTM D 6751 and the European prEN 14214 biodiesel standard do not set a limit for cloud point, however, the standard states that the cloud point of biodiesel is generally higher than diesel and should be taken into consideration when blending. The cold filter plug point is considered a more accurate test of biodiesel cold weather performance (Tyson, 2001 [61]).

affecting the performance of fuel lines, fuel pumps, and injectors [52]. Therefore, improvement of oil composition is needed to make pongamia FAMEs suitable for markets in all climatic regions of the world [9].

Pongamia biodiesel must meet a set of criteria (Table 12.2) to achieve ASTM D6751-09 quality standard. Biodiesel is generally considered to be a compatible substitute for all or part of mineral diesel fuel, but some key physicochemical parameters need to be considered when considering pure biodiesel replacing conventional diesel or when it is blended with conventional diesel.

A comparison between pongamia biodiesel and mineral diesel is presented in Table 12.3 [50,52,59]. It shows that in general the properties of pongamia FAMEs are comparable to mineral diesel. The fuel properties—including viscosity, density, flash point, fire point, and calorific value—of the transesterified product (biodiesel) compare well with accepted biodiesel standards. The viscosity of pongamia biodiesel is close to that of mineral diesel, with the calorific value about 12% less than that of mineral diesel. It has higher flash point than mineral diesel, and hence is safe to transport and store. In addition, biodiesel from pongamia oil shows no corrosion on piston metal and piston liners, whereas biodiesel from *Jatropha curcas* has slight corrosive effects on piston liners [60].

Table 12.3 Comparison of Pongamia Biodiesel and Mineral Diesel

Oil Properties	Pongamia Biodiesel			Mineral Diesel		
	Imran et al. [59] [a]	Mamilla et al., [50]	Bobade and Khyade [52] [b]	Imran et al. [59] [a]	Mamilla et al., [50]	Bobade and Khyade [52] [b]
Density at 15°C	0.88	0.89	0.86	0.84	0.84	0.84
Kinematic viscosity @40°C	8.53	5.6	4.78	6.06	3.8	2.98
Cloud point (°C)	—	—	6	—	—	−16
Pour point (°C)	−4	—	—	−2	—	—
Flash point, (°C)	158	217	144	70	56	74
Fire point (°C)	—	223	—	—	63	—
Heating value (kJ/kg)	—	36120	—	—	42800	—
Acid value (mg KOH/g)	0.42	—	0.42	0.34	—	0.35
Sulfur content (% mass)	0.02	—	—	0.91	—	—
Cetane number	58.22	—	41.7	51	—	49.0
Calorific value (kcal/kg)	—	—	3700	—	—	4285
Specific gravity	—	0.876	—	—	0.85	—
Water content (%)	Trace amount	—	0.02	0	—	0.02
Carbon residue (%)	0.39	—	0.005	—	—	0.01
Ash content (wt%)	0.003	—	0.005	—	—	0.02

[a]Imran et al. [59] used NaOH as catalyst.
[b]Bobade and Khyade [52] used KOH as catalyst.

Table 12.4 Co-ignition Characteristics of Pongamia Biodiesel With Mineral Diesel

Diesel Type	Ratio	Duration (min)	Load Applied (kW)	Observation
Mineral diesel:pongamia biodiesel	9:1	12 min	2130	Running smoothly
Mineral diesel:pongamia biodiesel	8:2	11 min	2130	Running smoothly
Mineral diesel:pongamia biodiesel	7:3	12 min 30 s	2130	Running more smoothly
Mineral diesel:pongamia biodiesel	6:4	10 min 25 s	2130	Running smoothly
Mineral diesel:pongamia biodiesel	5:5	9 min 32 s	2130	Running smoothly
Mineral diesel	100%	11 min 28 s	2130	Running smoothly
Pongamia biodiesel	100%	8 min 10 s	2130	Running smoothly

Imran HM, Khan AH, Islam MS, Niher RS, Sujan A, Chowdhury AMS. Utilization of Karanja (*Pongamia pinnata*) as a major raw material for the production of biodiesel. Dhaka University Journal of Science 2012;60(2):203–7.

Co-ignition of pongamia biodiesel with mineral diesel was evaluated by Imran et al. ([59]; Table 12.4). They found that pongamia biodiesel can run diesel engines smoothly, but the performance was better when it was mixed with mineral diesel. The best performance was achieved when mineral diesel and pongamia biodiesel were mixed at the ratio of 7:3.

Using pongamia biodiesel has several advantages [50], such as (1) runs in any conventional diesel engine; (2) no need for engine conversion or modification; (3) can be stored anywhere that petroleum diesel fuel is stored; (4) reduces carbon dioxide emissions by up to 100%; (5) can be used alone or mixed in any amount with petroleum diesel fuel; (6) easy to handle because it is biodegradable and nontoxic; (7) safe to transport because it has a high flash point of about 150°C compared to the flash point of petroleum diesel fuel, which is 70°C; and (8) similar fuel mileage, auto ignition, power output, and engine torque to petroleum diesel fuel.

12.5 CONCLUSION

The demand for food, fuel, and fiber continues to increase due to an increasing world population. With the declining reserves of fossil fuels, there is a strong demand for the development of future fuel that is economically viable and environmentally sustainable. It has become apparent that biofuels are destined to make a substantial contribution to the future energy demands. Pongamia has immense potential as a biofuel feedstock for several reasons. (1) pongamia is drought- and saline tolerant; it can grow on marginal lands that are globally abundant and not suitable for most food crops, (2) as a legume, pongamia does not require supplemental nitrogen fertilizers, thereby increasing sustainability, (3) pongamia also enters into a symbiosis with mycorrhizal fungi, leading to reduced phosphorus demands, (4) pongamia oil has excellent properties; this gives assurance to mixtures of pongamia oil with other liquid fuels (e.g., a B20 mix containing 20% pongamia-derived biodiesel) against engine congestion, malfunction, or low effectiveness, (5) it has the potential to yield a number of commercially viable by-products that result from the biofuel production process, such as an animal feed supplement arising from the residual seed cake following oil extraction, a source of combustible energy from the waste seedpods, and biochar that could be produced from any or all the components of the waste biomass.

Like fossil fuel, biofuel also releases carbon dioxide when combusted. However, in contrast to fossil fuel, that CO_2 is assimilated by the natural process of photosynthesis, leading to the synthesis of sucrose and subsequently FAs, as part of plant oil. Therefore, overall CO_2 emissions to the atmosphere are neutral, compared to the negative effects of burning coal, gas, or oil that was deposited after CO_2 capture millions of years ago. In addition, pongamia biodiesel is not plagued by undesired emission components, making it a prospective fuel of the future. To realize the eminent potential of this pongamia biodiesel, an optimistic and far-reaching investment and associate research support are needed.

As of 2016, the decrease in the current price of crude oil over recent years will not foster the necessary investment into renewable biofuels. However, we hope that the investors will recognize that the supply of liquid fossil fuel is limited and certainly will run out in the foreseeable future. We understand that biofuels, even those derived from the most promising plant crops, cannot supply in its entirety the global energy need. However, with proper management, they can be part of a diverse energy sources spectrum. The value of biofuels is especially valuable using feedstock such as pongamia, which has biological attributes lessening the energy-dependent inputs (N fertilizer).

There are scientific challenges to make pongamia a successful future biofuel feedstock. Research and development programs are needed to make this undomesticated and unimproved species a reliable and predictable source of oil and other valuable by-products [13]. Genetic and genomics tools in combination with traditional plant breeding should build a strong scientific foundation for genetic improvement of pongamia as a productive bioenergy crop.

ACKNOWLEDGMENTS

We thank the Australian Research Council for ARC Linkage grants LP120200562 (in partnership with BioEnergy Plantation Australia, and Stanwell Corporation). ARC and UQ also provided funds through the ARC Centre of Excellence for Integrative Legume Research grant. Also thanked are TerViva (USA), the Brisbane City Council, and the Global Change Institute as well as the Queensland Smart Futures grant.

REFERENCES

[1] Blackburn J. Australia's liquid fuel security: a report for NRMA motoring and services. 2013. http://www.mynrma.com.au/media/Fuel_Security_Report.pdf.
[2] Smith KA, Searchinger TD. Crop-based biofuels and associated environmental concerns. Global Change Biology Bioenergy 2012;4(5):479–84.

[3] Rockström J, Steffen W, Noone K, Persson Å Chapin FS, Lambin EF, Lenton TM, et al. A safe operating space for humanity. Nature 2009;461:472—5.

[4] Konwar BK, Banerjee GC. Deoiled karanja cake (*Pongamia glabra* Vent.) a new feed ingredient in cattle ration. Indian Veterinary Journal 1987;64:500—4.

[5] Ramana DBV, Singh S, Solanki KR, Negi AS. Nutritive evaluation of some nitrogen and non-nitrogen fixing multipurpose tree species. Animal Feed Science and Technology 2000;88:103—11.

[6] Daniel JN. *Pongamia pinnata* — a nitrogen fixing tree for oilseed. NFT Highlights, Nitrogen Fixing Tree Association (NFTA), FACT Net Winrock International; 1997. p. 97—103.

[7] Patil SG, Hebbara M, Devarnavadagi SB. Screening of multipurpose trees for saline vertisols and their bioameliorative effects. Annual Arid Zone 1996;35:57—60.

[8] Tomar OS. GuptaRK, Performance of some forest tree species in saline soils under shallow and saline water-table conditions. Plant and Soil 1985;87:329—35.

[9] Biswas B, Kazakoff SH, Jiang Q, Samuel S, Gresshoff PM, Scott PT. Genetic and genomic analysis of the tree legume *Pongamia pinnata* as a feedstock for biofuel. Plant Genome 2013;6(3). http://dx.doi.org/10.3835/plantgenome2013.05.0015.

[10] Low T, Booth C. The weedy truth about biofuels. Melbourne, Australia: The Invasive Species Council; December 2015. viewed 10, https://invasives.org.au/files/2014/02/isc_biofuels_revised_march08.pdf.

[11] Graham P, Braid A, Brinsmead T, Haritos V, Herr A, O'Connell D, et al. Flight path to sustainable aviation — towards establishing a sustainable aviation fuels industry in Australia and New Zealand. The sustainable aviation fuel Road Map, CSIRO energy transformed flagship. 2011.

[12] Jensen ES, Peoples MB, Boddey RM, Gresshoff PM, Hauggaard-Nielsen H, Alves BJR, et al. Legumes for mitigation of climate change and provision of feedstocks for biofuels and biorefineries. Agronomy for Sustainable Development 2012;32:329—64.

[13] Murphy HT, O'Connell DA, Seaton G, Raison RJ, Rodriguez LC, Braid AL, et al. A common view of the opportunities, challenges and research actions for Pongamia in Australia. Bioenergy Research 2012;5:778—99.

[14] Ferguson BJ, Indrasumunar A, Hayashi S, Lin M-H, Lin Y-H, Reid DE, et al. Molecular analysis of legume nodule development and autoregulation. Journal of Integrative Plant Biology 2010;52:61—76.

[15] Peoples MB, Brockwell J, Herridge DF, Rochester IJ, Alves BJR, Urquiaga S, et al. The contributions of nitrogen-fixing crop legumes to the productivity of agricultural systems. Symbiosis 2009;48:1—17.

[16] Schulze J, Temple G, Temple SJ, Beschow H, Vance CP. Nitrogen fixation by white lupin under phosphorus deficiency. Annals of Botany 2006;98:731—40.

[17] Pindi PK, Reddy RS, Reddy SM. Effect of inoculations of AMF and other bioinoculants on the growth of two agroforestry tree seedlings. Advances in Plant Sciences 2010;23:121—3.

[18] Scott PT, Pregelj L, Chen N, Hadler JS, Djordjevic MA, Gresshoff PM. *Pongamia pinnata*: an untapped resource for the biofuels industry of the future. Bioenergy Research 2008;1:2—11.

[19] Biswas B, Scott PT, Gresshoff PM. Tree legumes as feedstock for sustainable biofuel production: opportunities and challenges. Journal of Plant Physiology 2011;168:1877—84.

[20] Samuel S, Scott PT, Gresshoff PM. Nodulation in the legume biofuel feedstock tree *Pongamia pinnata*. Agriculture Research 2013;2:207—14.

[21] Sprent JI, Parsons R. Nitrogen fixation in legume and nonlegume trees. Field Crops Research 2000;65:183—96.

[22] Arpiwi NL, Yan G, Barbour EL, Plummer JA. Genetic diversity, seed traits and salinity tolerance of *Millettia pinnata* (L.) Panigrahi, a biodiesel tree. Genetic Resources and Crop Evolution 2013;60:677—92.

[23] Kesari V, Ramesh AM, Rangan L. *Rhizobium pongamiae* sp. nov. from root nodules of *Pongamia pinnata*. Biomed Research International 2013. http://dx.doi.org/10.1155/2013/165198.

[24] Rasul A, Amalraj ELD, Kumar GP, Grover M, Venkateswarlu B. Characterisation of rhizobial isolates nodulating Milletia pinnata in India. FEMS Microbiology Letter 2012;336:148—58.

[25] Tan DKY, Odeh IOA, Ancev T. Potential biodiesel crops for marginal land in Australia. The Regional Institute Online Publishing; December 2015. viewed 10, http://www.regional.org.au/au/asa/2012/climate-change/8049_tandk.htm.

[26] Bureau of Rural Sciences. Land use mapping at catchment scale: principles, procedures and definitions. edition 2. Canberra, Australia: Department of Agriculture, Food and Fisheries; 2002.

[27] Zou N, Dart PJ, Marcar NE. Interaction of salinity and rhizobial strain on growth and N_2-fixation by Acacia ampliceps. Soil Biology and Biochemistry 1995;27:409—13.

[28] Farine DR, Connell DAO, Raison JR, May BH, O'Connor MH, Crawford DF, et al. An assessment of biomass for electricity and biofuel, and for greenhouse gas emission reduction in Australia. Global Change Biology Bioenergy 2012;4:148—75.

[29] Odeh I, Tan D, Ancev T. Potential suitability and viability of selected biodiesel crops in Australian marginal agricultural lands under current and future climates. Bioenergy Research 2011;4:165—79.

[30] Kazakoff SH, Imelfort M, Edwards D, Koehorst J, Biswas B, Batley J, et al. Capturing the biofuel wellhead and powerhouse: the chloroplast and mitochondrial genomes of the leguminous feedstock tree *Pongamia pinnata*. PLoS One 2012;7:1687—99.

[31] Singh K, Yadav JSP. Effect of soil salinity and sodicity on seedling growth and mineral composition of *Pongamia pinnata*. The Indian Forester 1999;125:618—22.

[32] Hussain N, Sarwar G, Schmeisky H, Al-Rawahy S, Ahmad M. Salinity and drought management in legume crops. In: Yadav SS, McNeil DL, Redden & SA Patil R, editors. Climate change and management of cool season grain legumes. Dordrecht, Netherlands: Springer Science and Business Media; 2010. p. 171—92.

[33] Cordovilla MP, Ocana A, Ligero F, Lluch C. The effect of salinity on N_2 fixation and assimilation in *Vicia faba*. Journal of Experimental Botany 1994;45:1483—8.

[34] Manchanda G, Garg N. Salinity and its effects on the functional biology of legumes. Acta Physiologiae Plantarum 2008;30:595—618.

[35] Myers BA, West DW, Callinan L, Hunter CC. Long term effects of saline irrigation on the yield and growth of mature Williams pear trees. Irrigation Science 1995;16:35—46.

[36] Bala N, Sharma PK. Lakshminarayana. Nodulation and nitrogen fixation by salinity-tolerant rhizobia in symbiosis with tree legumes. Agriculture Ecosystems and Environment 1990;33:33—46.

[37] Hu Y, Schmidhalter U. Drought and salinity: a comparison of their effects on the mineral nutrition of plants. Journal of Plant Nutrition and Soil Science 2005;168:541—9.

[38] Hennesy K, Fawcett R, Kirono D, Mpelasoka F, Jones D, Bathols J, et al. Commonwealth scientific and industrial research organization and Bureau of meteorology report. Canberra, Australia: Australian Government, Bureau of Meteorology; 2008.

[39] Kazakoff SH, Gresshoff PM, Scott PT. *Pongamia pinnata*, a sustainable feedstock for biodiesel production. In: Halford NG, Karp A, editors. Energy crops. Cambridge, UK: Royal Society for Chemistry; 2011. p. 233—58.

[40] Kesari V, Rangan L. Development of *Pongamia pinnata* as an alternative biofuel crop — current status and scope of plantations in India. Journal of Crop Science and Biotechnology 2010;13(3):127—37.

[41] Jiang Q, Yen S-H, Stiller J, Edwards D, Scott PT, Gresshoff PM. Genetic, biochemical, and morphological diversity of the legume biofuel tree *Pongamia pinnata*. Journal of Plant Genome Sciences 2012;1(3):54—67.

[42] Sujatha K, Hazra S. Micropropagation of mature *pongamia pinnata* Pierre. In Vitro Cellular & Developmental Biology — Plant 2007;43(6):608—13.

[43] Swamy SL, Gadekar KP, Mishra A. Vegetative propagation of *Pongamia pinnata* stem cuttings. Plant Archives 2008;8(2):837—40.

[44] Mukta N, Sreevalli Y. Propagation techniques, evaluation and improvement of the biodiesel plant, *Pongamia pinnata* (L.) Pierre. A Review. Industrial Crops and Products 2010;31:1—12.

[45] Sujatha K, Rajwade AV, Gupta VS, Hazra S. Assessment of *Pongamia pinnata* (L.) — a biodiesel producing tree species using ISSR markers. Current Science 2010;99(10):1327—9.

[46] Kesari V, Rangan L. Genetic diversity analysis by RAPD markers in candidate plus trees of *Pongamia pinnata*, a promising source of bioenergy. Biomass and Bioenergy 2011;35:3123—8.

[47] Pavithra HR, Shivanna MB, Chandrika K, Prasanna KT, Gowda B. Genetic analysis of *Pongamia pinnata* (L.) Pierre populations using AFLP markers. Tree Genetics & Genomes 2014;10:173—88.

[48] Sharma SS, Islam MA, Negi MS, Tripathi SB. Isolation and characterization of a first set of nine polymorphic microsatellite loci in *Pongamia pinnata* (Fabaceae). Journal of Genetics 2014;93:e70—4.

[49] Bora DK, Baruah DC. Assessment of tree seed oil biodiesel: a comparative review based on biodiesel of a locally available tree seed. Renewable and Sustainable Energy Reviews 2012;16:1616−29.

[50] Mamilla VR, Mallikarjun MV, Rao GLN. Preparation of biodiesel from Karanja oil. International Journal of Energy Engineering 2008;1(2):94−100.

[51] Agarwal AK. Biofuels (alcohols and biodiesel) applications as fuels for internal combustion engines. Journal of Progress in Energy and Combustion Science 2007;33:233−71.

[52] Bobade SN, Khyade VB. Preparation of methyl ester (biodiesel) from karanja (*Pongamia pinnata*) oil. Research Journal of Chemical Science 2012;2(8):43−50.

[53] Soto MN, Calderon MH, Fajardo CAG, Castillo MAS, Garcia TV, Andrade IC. Biodiesel current technology: ultrasonic process a realistic industrial application. In: Fang Z, editor. Biodiesel feedstocks. Production and Applications. Tech; 2013. p. 177−207. http://dx.doi.org/10.5772/45895.

[54] Azam MM, Waris A, Nahar NM. Prospects and potential of fatty acid methyl esters of some non-traditional seed oils for use as biodiesel in India. Biomass and Bioenergy 2005;9(1):293−302.

[55] Imahara H, Minami E, Saka S. Thermodynamic study on cloud point of biodiesel with its fatty acid composition. Fuel 2006;85:1666−70.

[56] Joshi RM, Pegg MJ. Flow properties of biodiesel fuel blends at low temperatures. Fuel 2007;86:143−51.

[57] Sarin R, Sharma M, Sinharay S, Malhotra RK. Jatropha-Palm biodiesel blends: an optimum mix for Asia. Fuel 2007;86:1365−71.

[58] Raheman H, Phadatare AG. Diesel engine emissions and performance from blends of Karanja methyl ester and diesel. Biomass and Bioenergy 2004;27:393−7.

[59] Imran HM, Khan AH, Islam MS, Niher RS, Sujan A, Chowdhury AMS. Utilization of Karanja (*Pongamia pinnata*) as a major raw material for the production of biodiesel. Dhaka University Journal of Science 2012;60(2):203−7.

[60] Haas MJ, Scott KM. Combined nonenzymatic-enzymatic method for the synthesis of simple alkyl fatty acid esters from soapstock. Journal of the American Oil Chemists Society 1996;73:1393−401.

[61] Tyson SK. Biodiesel handling and use guidelines. National Renewable Energy Laboratory Report. Colorado: Golden; 2001.

CHAPTER THIRTEEN

Biodiesel From Queensland Bush Nut (*Macadamia integrifolia*)

A.K. Azad, M.G. Rasul, M.M.K. Khan and S.C. Sharma
Central Queensland University, Rockhampton, QLD, Australia

13.1 INTRODUCTION

Macadamia (*Macadamia integrifolia*) is an Australian native species inhabiting the central eastern part of Australia, particularly in New South Wales and Queensland [1,2]. It is also referred to as "bush nut." The taxonomical classification, common English names, worldwide distributions, and a distribution map for Australia are presented in Table 13.1. There are some common names, such as Queensland nut, poppel nut, bauple nut, and boppel nut, for the fruit of the plant which is now cultivated in Australia, Hawaii, Southern California, Florida, Cuba, Brazil, Kenya, Indonesia, Taiwan, and East Africa. Fig. 13.1 presents the life cycle of *Macadamia integrifolia* where different stages (baby plant, adult plant, flowering time, green fruit, and matured fruits and kernels) have been shown. After germination, the baby plant grows very well. The adult plant is an evergreen and attractive tree that can grow 12−20 m high [3]. For this reason, it is also used as an ornamental tree. After 1−2 years of growth, long pendulous clusters of white flowers come in spring, and fruit grows to about 2.5−3.5 cm in diameter [4]. An adult

Table 13.1 Taxonomical Classification, Common Names, and Distribution of Macadamia in Australia

Taxonomical Classification		Common English Names	World Distribution	
Kingdom	*Plantae*	Australian nut	Australia	
Phylum	*Charophyta*	Macadamia nut	Hawaii	
Class	*Equisetopsida*	Queensland nut	California	
Subclass	*Magnoliidae*	Bush nut	Florida	
Superorder	*Proteanae*	Poppel nut	Cuba, Brazil	
Order	*Proteales*	Bauple nut	Indonesia	
Family	*Proteaceae*	Bopple nut	Taiwan	
Genus	*Macadamia*		Kenya	
Species	*integrifolia*		East Africa	Distribution map in Australia

Figure 13.1 Life cycle of macadamia, and its different uses.

plant reaches full fruiting potential in about 5—7 years after planting. The average fruit yields are 8—13 kg/tree (about a quarter of which is kernel) and 230 gal/acre of oil was recorded in previous studies [3]. Australia is the world's largest macadamia-producing country. Some research into potential alternative fuel uses of macadamia nuts can be found in the literature. For example, Kartal et al. [5] studied recovery of tar oil from macadamia nut shells by slow pyrolysis. But the most oil-containing part is the kernel which contains about 73% (n-hexane extraction) edible oil. It could be one of the prospective sources of alternative fuel (such as biodiesel) [6]. For instance, Knothe [7] investigated fatty acid composition of the macadamia oil and found 10 fatty acids, which could be one of the potential sources of biodiesel.

Biodiesels are biodegradable, ecofriendly, and a source of renewable energy mainly produced from biological resources, such as vegetable oil, animal fats, or waste oils [8,9]. The biodiesels were derived from different sources, such as edible food crops and vegetable oils [10,11]—food crops including rice, wheat, barley, potato wastes, and sugar beets and edible vegetable oils including soybean oil [12,13], sunflower oil [14], corn oil [15], olive oil [16,17], palm oil [18,19], coconut oil [17], rapeseed oil [20], and mustard oil [21—23]. The advanced biodiesels were obtained from a wide array of feedstocks [24] and animal fat [10,25]—the nonedible feedstocks, such as animal fats [26,27], nonfood crops [28,29], and waste cooking oil [30,31] and nonedible oils. Some of the nonedible oil sources were studied in literature, for example, *Jatropha curcas* [32–34], cotton seed [35,36], *Pongamia glabra* [18,37], beauty leaf [38,39], karanja [40], and castor oil [41,42]. Some studies reported that biodiesel was obtained from microalgal biomass. It has a very distinctive growth yield compared with classical lignocellulosic biomass [10,43]. The biodiesels are the alternative fuel that can be used by blending up to 20% with petroleum fuel without any modification to compression ignition engines [44,45]. The use of biodiesel has a lot of benefits, such as it is nontoxic, noncorrosive, and has no sulfur content. It emits less greenhouse gases. Research during 2013–15 has progressively investigated different feedstocks and the sources of biodiesel [2,46]. After feedstock selection, one of the important stages is to identify suitable oil extraction techniques. These include mechanical extraction (for instance, cold press), chemical extraction (for example, n-hexane extraction), and thermal extraction (i.e., pyrolysis). Each method has some merits and demerits. In this work, oil has been extracted by the n-hexane method due to its high oil yield.

In this study, the kernels have been extracted manually from the fruits. The moisture has been removed by continuous drying at 60°C for 18 h in an electric dryer. After drying, the kernels were crushed by a grinding machine. Oil was then extracted from the grated kernel by the n-hexane method. The refined oil was converted to biodiesel utilizing an alkali catalyst transesterification reaction. The associated steps for biodiesel conversion (such as glycerin removal, excess methanol removal, washing with demineralized water, drying, and so on) have been carefully performed to qualify the biodiesel under the ASTM D6751 standard. The Fatty Acid Composition (FAC) has been measured by gas chromatography (GC) under the AOCS Ce 1a-13 standard method. The physical behavior of the fuel from winter to summer has been analyzed for varying temperatures from 10°C to 40°C. The fuel properties have been tested using appropriate ASTM and EN standards. Finally, mass and energy balances have been undertaken for the entire process of oil extraction to biodiesel conversion for macadamia oil.

13.2 MATERIALS AND METHODS

13.2.1 Oil Extraction

13.2.1.1 Seed Collection and Kernel Recovering

The macadamia seeds were collected from an Australian native plant supplier in Central Queensland, Australia. The seeds were dried at 65–70°C temperature for 72 h to remove moisture. After drying, the seeds were manually crushed to remove the shell (waste husk) in order to extract the kernels. This is the initial stage of oil extraction. The whole kernels were crushed before uniform drying to properly remove the moisture.

13.2.1.2 Moisture Control and Kernel Drying

Moisture control of the kernel is a very important stage of oil extraction because oil yield depends on moisture content. It has been reported in the literature that an optimum moisture content of about 15% can produce high oil yield using mechanical and chemical extraction [47,48]. In this study, the extracted kernels were dried carefully in a microprocessor based temperature-controlled incubator (IM550) at 60°C [49]. The drying process was closely monitored and the weight loss (using a digital balance with 0.1% tolerance) was observed every 2 h. The 2 h interval data was recorded and plotted (weight vs. drying time) to obtain the optimum moisture control point (Fig. 13.2). Fig. 13.2 shows that, after 16 h drying, kernel weight loss was negligible and the curve remains at an almost constant value. The optimum moisture content was found to be 15.8%. The trend of the drying curve follows the equation, $y = 0.1362x^2 - 5.2403x + 247.71$ where the value of goodness-of-fit is $R^2 = 0.993$. A total of 24 h data was recorded and analyzed to determine the required period for drying the kernels before undertaking oil extraction (18 h for this study). Finally, dried kernels were crushed again to a smaller size (about 0.6–0.8 mm) to extract oil by the n-hexane method.

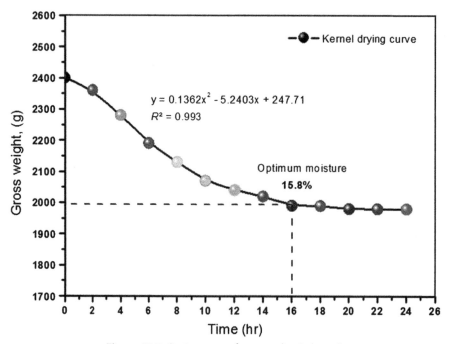

Figure 13.2 Drying curve for macadamia kernel.

Seed   Kernels   Crushing and drying   n-hexane method   Macadamia oil

Figure 13.3 Oil extraction by n-hexane method from macadamia seed.

13.2.1.3 Oil Extraction (n-Hexane Method)

Chemical extraction by the n-hexane method is one of the efficient methods to extract oil from the grated kernel. Fig. 13.3 shows a brief summary of the oil extraction technique where the dried kernels were crushed again to make finely grated particles for maximizing particle contact surface area [50]. Then the kernel samples (about 200 g each) were put into conical flasks and a 98% n-hexane solution was added in a 2:1 ratio (mL of n-hexane : grams of crushed kernels) for the initial extraction [51]. The flask was initially shaken for proper mixing with the solution, then the closed flask was placed in an orbital shaker machine at 250 rpm for 18—20 h shaking [47].

After shaking, the oil—chemical mixture was kept stationary for 5—10 h to allow the settling down of the solids from the liquid. Table 13.2 summarizes the total oil extraction process, conditions, and extracted oil percentage. It can be seen from the table that, after the initial extraction number 1, the n-hexane:kernel ratio was modified to 1:1 for extraction numbers 2 to 7 with 18—20 h of shaking time at a constant frequency of 250 rpm. The n-hexane: kernel ratio has been changed from 2:1 to 1:1 because about one-third oil has been extracted after first extraction. Variable settling down times were recorded for this experiment. The separated oil—chemical mixture was kept in the fume hood to remove unreacted n-hexane. It is to be noted that unreacted n-hexane can be recovered and recycled for large-scale production. The study found 73.07% oil yield

Table 13.2 Summary of n-Hexane Oil Extraction for 1 kg Macadamia Kernel

Extraction Number	n-Hexane and Kernel Ratio	Shaking Frequency (rpm)	Shaking Time (h)	Settling Down Time (h)	Oil from Each Extraction (g)	Total Oil Yield for All Extractions (%)
Extraction 1	2:1	250	20	10	282.97	73.07
Extraction 2	1:1	250	19	11	206.26	
Extraction 3	1:1	250	18	8	122.72	
Extraction 4	1:1	250	18	10	46.72	
Extraction 5	1:1	250	19	6	37.54	
Extraction 6	1:1	250	18	7	20.01	
Extraction 7	1:1	250	18	3	14.56	

from macadamia kernels. After removing n–hexane, the virgin oil was collected and stored at room temperature. Finally, the properties of the vegetable oil were tested and the results are briefly discussed in the following sections.

13.2.2 Biodiesel Conversion Reaction

The conversion of vegetable oil into biodiesel by chemical reaction is the most sensitive step in the fuel-processing technique. Transesterification reaction is one of the most efficient techniques to remove glycerin from the vegetable oil [52]. The conversion takes place in the presence of an alkali or acid catalytic with methanol and crude vegetable oil by maintaining a standard reaction environment [53]. Due to the higher density and viscosity of the macadamia oil, as discussed earlier, the study selected this technique (transesterification) to convert triglycerides into methyl esters to satisfy the ASTM D6751 biodiesel standard [54,55]. The study evaluated the effects of four variables on transesterification reaction (Fig. 13.4) conversion rate (i.e., yield of methyl esters). These variables were the methanol:oil molar ratio, the catalyst concentration, reaction time, and reaction concentration. For this purpose, a small quantity of oil (100 mL from each batch) was experimented by varying the methanol:oil molar ratio from 5.5 to 6.5, the catalyst concentration from 0.5 to 1.5, the reaction temperature from 50°C to 65°C, and the reaction time from 30 to 90 min (Table 13.3). The maximum reaction temperature was limited to the 64.7°C boiling point of methanol. The conversion rate or yield of methyl esters was determined by gas chromatography (using EN 14103 standard). Initially, conversion experiments were conducted using $NaOCH_3$ catalyst with standard reaction conditions. Unfortunately, this catalyst was not well suited and failed several times. Subsequently, $KOCH_3$ was selected as the alkaline catalyst for this conversion and satisfactory results were obtained, which are presented in Table 13.4. From the table, maximum conversion efficiency was achieved for 6:1 methanol:oil molar ratio with 1% KOH catalyst reacted for 1 h at 60°C.

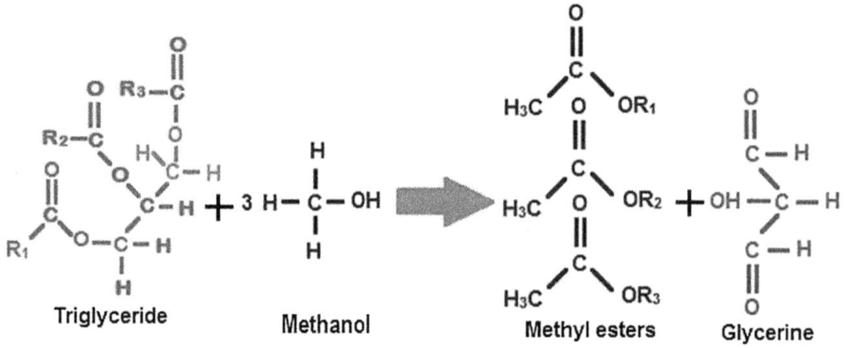

Figure 13.4 Transesterification reaction for biodiesel conversion.

Table 13.3 The Effect of Reaction Variables on Transesterification Reaction for Macadamia Vegetable Oil Conversion

Variables	Order	Molar Ratio	Catalyst Concentration (%)	Temperature (°C)	Reaction Time (min)	Observed Methyl Esters (wt%)
Change of molar ratio	1	5.5:1	1.00	60	60	86.35 ± 0.88
	2	6:1	0.50	55	30	87.30 ± 1.25
	3	6.5:1	1.00	60	60	81.00 ± 1.40
Change of % of catalyst	4	6:1	0.50	60	60	83.50 ± 1.41
	5	6:1	1.0	60	90	91.20 ± 1.15
	6	6:1	1.50	60	60	91.73 ± 1.74
Change of reaction temperature	7	6:1	1.0	50	60	91.00 ± 1.26
	8	6:1	1.0	50	60	82.00 ± 1.52
	9	6:1	1.0	65	60	89.00 ± 1.02
Change of reaction time	10	6:1	1.0	60	30	81.00 ± 1.56
	11	6:1	1.0	60	60	98.21 ± 1.32
	12	6:1	1.0	60	90	95.30 ± 1.61

13.2.3 Determination of Fuel Properties

The properties such as density, viscosity, calorific value (CV), flash point, poor point, cloud point, acid value, carbon residue, cold filter plugging point (CFPP), and so on were measured using the ASTM standards. The fatty acid compositions were measured using AOCS Ce 1a-13 standard method. The behavior of the fuel, that is, variation of density and viscosity at different temperatures (from 10°C to 40°C) were investigated. In this study, some properties, such as saponification number (SN), iodine value (IV), and cetane number (CN), were calculated numerically using the following Eqs. (13.1)–(13.3) [56–58]:

$$\text{Saponification number, SN} = \sum \left(\frac{560 \times A_i}{MW_i} \right) \tag{13.1}$$

$$\text{Iodine value, IV} = \sum \left(\frac{254 \times D \times A_i}{MW_i} \right) \tag{13.2}$$

$$\text{Cetane number, CN} = 46.3 + \left(\frac{5458}{SN} \right) - (0.225 \times IV) \tag{13.3}$$

where A_i indicates the percentage of each fatty acid component, MW_i is the molecular mass of each component (Table 13.5), and D is the number of double bonds present in component.

Table 13.4 Macadamia Vegetable Oil Properties

Properties	Unit	Macadamia Oil	Test Method
Density (at 15°C)	kg/m^3	914	ASTM D1298
Viscosity (at 40°C)	mm^2/s	35.85	ASTM D445
Calorific value	MJ/kg	38.12	ASTM D240
Cetane number	—	55.48	ASTM D613
Flash point	°C	135	ASTM D93
Pour point	°C	−3	ASTM D97
Cloud point	°C	−0.63	ASTM D2500
Acid value	mg KOH/g	0.26	ASTM D664
Free fatty acid	%	0.04	—
Cold filter plugging point	°C	3.41	ASTM D6371
Iodine value	—	79.15	ASTM D4607
Saponification number	—	201	ASTM D94-07
Oxidation stability	h	53.86	ASTM D2274
Refractive index	—	1.470	—
Moisture	%	0.00	—
Arsenic	mg/kg	<0.02	—
Cadmium	mg/kg	<0.01	—
Mercury	mg/kg	<0.01	—
Lead	mg/kg	<0.01	—

13.2.4 Mass and Energy Balance Calculations

The analysis of mass and energy balances for macadamia oil extraction, biodiesel conversion, and the total process was carried out according to the input (feedstock) and output (final products) variables. It is very important to analyze total production loss to optimize the entire process. Mass balance is calculated numerically based on input mass of the feedstock and output mass of the products plus waste. On the other hand, the energy distribution is calculated based on input energy to the system (mass of feedstock × gross heating value) and output energy from the system. Total output energy is the summation of the energy (energy = mass × gross heating value) of the individual products, by-products, and waste. The mass and energy balances are calculated using the following equations [59]:

Mass balance

$$\% \text{ of mass recovery, } M_{recovery} = \frac{W_{output}}{W_{input}} \times 100\% \tag{13.4}$$

$$\% \text{ of mass loss} = 100\% - \sum M_{recovery} \tag{13.5}$$

Table 13.5 Fatty Acid Composition of Macadamia Biodiesel (Under AOCS Ce 1a-13 Standard)

Fatty Acid Name	Structure	Chemical Name	Relative Content (%vol.)
Lauric acid	C12:0	Dodecanoic acid	0.06
Myristic acid	C14:0	Tetradecanoic acid	0.58
Palmitic acid	C16:0	Hexadecanoic acid	8.25
Palmitoleic acid	C16:1	(9Z)-Hexadec-9-enoic acid	15.39
Margaric acid	C17:0	Hexadecanoic acid	0.03
Ginkgolic acid	C17:1	*cis*-10-Heptadecenoic acid	0.08
Stearic acid	C18:0	Octadecanoic acid	3.55
Oleic acid	C18:1	(9Z)-Octadecenoic acid	61.09
Linoleic acid	C18:2	*cis*-9,12-Octadecadienoic acid	1.86
Linolenic acid	C18:3	Methyl linolenate	0.11
Arachidic acid	C20:0	Eicosanoic acid	2.94
Eicosenoic acid	C20:1	(9Z)-9-Icosenoic acid	2.55
Eicosadienoic acid	C20:2	*cis*-11-14-Eicosadienoic acid	0.06
Eicosatrienoic acid	C20:3	*cis*-11,14,17-Eicosatrienoic acid	0.03
Eicosapentaenoic acid	C20:5	5,8,11,14,17-Eicosapentaenoic	0.97
Behenic acid	C22:0	1-Docosanoic acid	0.04
Erucic acid	C22:1	(Z)-Docos-13-enoic acid	0.16
Gadolenic acid	C22:2	*cis*-13,16-Docosadienoic acid	0.13
Docosahexaenoic acid	C22:6	Cervonic acid	0.02
Tricosylic acid	C23:0	Tricosanoic acid	0.06
Lignoceric acid	C24:0	Tetracosanoic acid	0.08
Nervonic acid	C24:1	(Z)-Tetracos-15-enoic acid	0.13

Energy balance

$$\text{\% of energy recovery, } E_{recovery} = \frac{W_{outtput} \times HV_{output}}{W_{input} \times HV_{input}} \times 100\% \tag{13.6}$$

$$\text{\% of mass loss} = 100\% - \sum E_{recovery} \tag{13.7}$$

where W_{input} is the mass (kg) of input feedstock; W_{output} denotes total mass (kg) of output products (i.e., oil, biodiesel, and glycerin); $\sum M_{recovery}$ is the summation of mass recovery from all output products; HV_{output} is the heating value (MJ/kg) of the product and by-product; HV_{input} represents the heating value (MJ/kg) of the vegetable oil; and $\sum E_{recovery}$ is the summation of energy recovery from all output products.

13.3 BIODIESEL CONVERSION STEPS

Fig. 13.5 shows the steps associated with the vegetable oil to biodiesel conversion. The setup was prepared in an automatically controlled chemical fume hood (model:

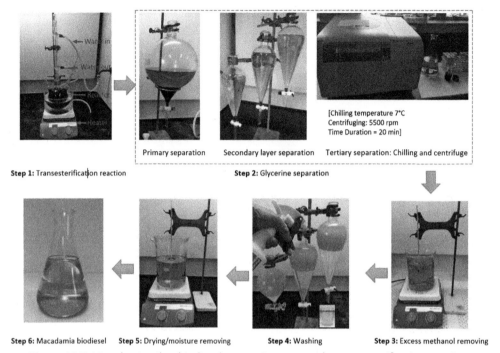

Step 1: Transesterification reaction   **Step 2:** Glycerine separation

Primary separation   Secondary layer separation   Tertiary separation: Chilling and centrifuge

[Chilling temperature 7°C
Centrifuging: 5500 rpm
Time Duration = 20 min]

Step 6: Macadamia biodiesel   **Step 5:** Drying/moisture removing   **Step 4:** Washing   **Step 3:** Excess methanol removing

Figure 13.5 Macadamia oil to biodiesel conversion process by transesterification reaction.

Dynasafe MK3) in a chemical and biomedical engineering laboratory. A three-necked round-bottomed glass reactor (model: FR1L/3s/22A, capacity 1000 mL) equipped with Liebig (model: C1/12/25/SC, 250 mm effective length and 380 mm overall length), thermometer (range $0-150°C$) and a flat bottom water bath (capacity 2.5 L) on a magnetic stirrer (model: IKA C-MAG HS7) was used for the biodiesel conversion. After designing the required reaction, the conversion process was conducted in seven batches (750 mL in each batch) to minimize the risk from any reaction failures. In Step 1, the vegetable oil was heated to $60°C$. The potassium methoxide ($KOCH_3$) solution was prepared separately at the rate of 6:1 M ratio of methanol:oil. Then, the $KOCH_3$ solution was slowly added into the preheated oil. The circulating water was started before adding the chemicals, and the reactor was then immediately closed. The oil–chemical mixture was heated by the hot plate (at $60°C$) and stirred by the magnetic stirrer (at 750 rpm) to maintain the reaction conditions. After completion of the reaction, the mixture was put into a separating funnel and cooled at room temperature in the fume hood system. In Step 2, the glycerin was separated in three stages from the oil. After $5-6$ h, primary glycerin was separated from the bottom of the funnel. The remaining methyl esters (biodiesel) were separated in three equal parts in small separating funnels.

After 18 h, secondary glycerin in these was separated from the biodiesel. Finally, the biodiesel was chilled at 7°C and centrifuged at 5500 rpm for 20 min using a centrifuge machine (model: ThermoFisher D-37520, year 2011). In Step 3, the extracted methyl esters (upper layer) were heated at 80°C for 30−40 min to remove unreacted methanol. The unreacted catalyst was removed by washing several times with warm demineralized water (DM) in separating funnels in Step 4. After removing the waste water, the biodiesel was dried at 110°C in Step 5 to remove moisture and residual water particles. In Step 6, the final product (biodiesel) was stored in a closed container to avoid oxidation of the fuel with free oxygen in the air. Finally, the properties of the biodiesel were experimentally tested and are discussed in the next section.

13.4 RESULTS AND DISCUSSIONS

13.4.1 Properties of Macadamia Vegetable Oil

Table 13.4 summarizes the physiochemical properties of the vegetable oil. The density and viscosity of the oil are higher than the acceptable limits for fuel under ASTM D1298 and ASTM D445, respectively. The density of 914 kg/m^3 was measured at 15°C and the kinematic viscosity of 35.85 mm^2/s was measured at 40°C in accordance with the temperatures specified in the relevant standards. The range limit for density is 860−900 kg/m^3 (ASTM D1298) and that for viscosity is 3.5−5.0 mm^2/s (ASTM D445) as a biofuel. These higher values for the vegetable oil make it unacceptable as a fuel in a diesel engine. The oil must therefore undergo a process that can make its properties fit within the specified limits. Other properties, such as flash point, cloud point, acid value, CN, and IV, are within the limits of the relevant standards. The analysis found very negligible amounts of arsenic (<0.02 mg/kg), cadmium (<0.01 mg/kg), mercury (<0.01 mg/kg), and lead (<0.01 mg/kg) in the fuel. Finally, the behavior of the biodiesel at different temperatures is discussed in the next section.

13.4.2 Physical Behavior of Macadamia Biodiesel

The properties and behavior of the macadamia biodiesel were analyzed after conversion and are discussed briefly in this section. The physical behavior means the variation of physical properties such as density or viscosity with temperature. It is important to analyze how the biodiesel will behave during the different seasons (various temperature ranges) of a year. According to the Bureau of Meteorology (BoM), Australia, the average temperature ranges are 8−15°C in winter, 15−23°C in autumn, 22−30°C in spring, and 30−40°C in summer. The study analyzed the variation of density and viscosity for those particular temperatures. Figs. 13.6 and 13.7show the variations of density and kinematic viscosity of macadamia biodiesel across that complete range of temperatures. In Fig. 13.6, the trend of density fluctuation from 10°C to 40°C is presented and compared with the similar trend for ultra-low sulfur diesel (ULSD). At 15°C

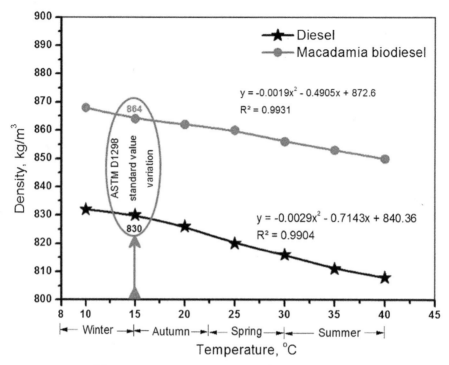

Figure 13.6 Variation of density at various temperatures (reflecting the different seasons in Australia).

temperature under ASTM D1298 standard, 830 kg/m^3 and 864 kg/m^3 are recorded for diesel and biodiesel, respectively. Density decreases with the increase of temperature for both fuels. The macadamia biodiesel curve follows the second order polynomial equation $y = -0.0019x^2 - 0.4905x + 872.6$ where the coefficient of determination (R^2), also called goodness of fit, is $R^2 = 0.9931$. On the other hand, the diesel curve follows the equation $y = -0.0029x^2 - 0.7143x + 840.36$ where the value of R^2 is 0.9904. So the derived biodiesel shows a behavior that remains within the acceptable range of the ASTM D6751 standard.

The kinematic viscosity (mm^2/s) is a very important fuel property for biodiesel. In this study, ARES (model no. BHP150) rheometer was used to measure the kinematic viscosity at various temperatures. The biodiesel and diesel follow the second order polynomial equations of $y = 0.0019x^2 - 0.3323x + 14.817$ where $R^2 = 0.9998$, and $y = 0.0006x^2 - 0.0562x + 5.389$ where $R^2 = 0.9999$, respectively. The converging trend of the curves as temperature increases is clearly evident. At low temperature, the viscosity of the biodiesel is much higher than for diesel fuel. It decreases with the increase of temperature and the curves almost meet (diesel 4.10 mm^2/s and biodiesel 4.50 mm^2/s) with each other at 40°C under the ASTM D445 standard. The biodiesel

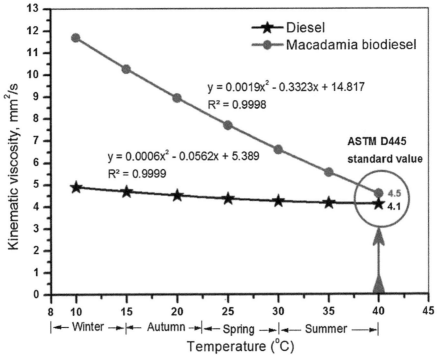

Figure 13.7 Variation of kinematic viscosity at various temperatures (reflecting the different seasons in Australia).

behaves very similarly to diesel in summer, but its viscosity is much higher than that of diesel in winter. The kinematic viscosity remains within the specified range of both the American ASTM D6751 and European EN 14214 standards. More analysis is needed on other fuel properties of the biodiesel, such as carbon structure, geometry, carbon bonds, and so on.

13.4.3 Fatty Acid Methyl Esters

The analysis of the fatty acid methyl esters (FAMEs) of biodiesel is very important to analyze the quality and structure of the fuel. In this analysis, the macadamia sample is separated on a capillary gas—liquid chromatography column having a highly polar stationary phase, according to carbon chain length, geometry, and position of the double bonds in the fuel. The FAMEs were analyzed using a SHIMADZU GC-2010 gas chromatograph equipped with capillary columns of specification 0.25 mm × 25 m inner diameter and 0.25 μm film thickness. The GC test was conducted under the European EN14103 standard and a summary of the results is presented in Table 13.5. The test results show that there are 22 fatty acids present in the macadamia biodiesel. Among them, only nine fatty acids contain a significant percentage by volume (%vol.), namely myristic acid (C14:0) 0.58%, palmitic acid (C16:0) 8.25%, palmitoleic acid (C16:1) 15.39%, stearic acid (C18:0) 3.55%, oleic acid (C18:1) 61.09%, linoleic acid (C18:2) 1.86%, arachidic acid (C20:0) 2.94%, eicosenoic acid (C20:1) 2.55%, and eicosapentaenoic acid (C20:5) 0.97%. The analysis also recorded some other fatty acids in small amounts, for instance: lauric acid (C12:0) 0.06%, margaric acid (C17:0) 0.03%, ginkgolic acid (C17:1) 0.08%, linolenic acid (C18:3) 0.11%, eicosadienoic acid (C20:0) 0.06%, eicosatrienoic acid (C20:3) 0.03%, behenic acid (C22:0) 0.04%, erucic acid (C22:1) 0.16%, gadolenic acid (C22:2) 0.13%, docosahexaenoic acid (C22:6) 0.02%, tricosylic acid (C23:0) 0.06%, lignoceric acid (C24:0) 0.08%, and nervonic acid (C24:1) 0.13%. The total amount of ester contents is 98.21%vol. The other fuel properties measured are discussed in the next section.

13.4.4 Fuel Properties

The fuel properties of the biodiesel were investigated according to relevant ASTM standards and compared with ULSD. It is important to undertake these investigations before use in a diesel engine. Table 13.6 summarizes the measured fuel properties. The density and kinematic viscosity at standard temperatures are close to ULSD and within

Table 13.6 Physiochemical Fuel Properties of Macadamia Biodiesel

Properties	Unit	Test Method	ULSD	Macadamia Biodiesel	ASTM Standard
Density (at 15°C temp.)	kg/m^3	ASTM D1298	866	868	860−900
Viscosity (at 40°C temp.)	mm^2/s	ASTM D445	4.10	4.57	3.5−5.0
Calorific value	MJ/kg	ASTM D240	42.7	39.880	—
Cetane number	—	—	44	56.91	—
Flash point	°C	ASTM D93	60	135	Minimum 100
Pour point	°C	ASTM D97	−15	−3	−15 to −16
Cloud point	°C	ASTM D2500	−8.6	6	−3 to −12
Auto-ignition temperature	°C	ASTM E659	256.9	319.56	
Acid value	mg KOH/g	ASTM D664	Max 0.5	0.15	Maximum 0.5
Carbon residue	m/m	ASTM D4530	0.01	0.01	—
Cold filter plugging point	°C	EN116	−3.0	1.62	0, <−15 winter
Iodine value	—	—	—	76.44	—
Saponification number	—	—	—	196.23	—
Oxidation stability	h	EN14112	5.0	2.6	Minimum 6
Total ester content	%	EN 14103		98.21	—

the acceptable ranges of the ASTM D1298 and ASTM D445 standards, respectively. The variation of density and viscosity with temperature was discussed earlier. A Parr 6400 automatic isoperibolic calorimeter (0.0001°C temperature resolution with 0.1% precision class instrument) was used to measure the CV of both ULSD and the biodiesel. The results show that the CV of ULSD is 42.7 MJ/kg and that of macadamia biodiesel is 39.88 MJ/kg. Hence, the produced biodiesel contains 6.6% less energy compared to ULSD. Neither the ASTM nor EN standards have any specific limit for CV, although it is reported in the literature that the minimum heating value of biodiesel is 35.0 MJ/kg [7]. Biodiesel is safer than diesel fuel due to the higher flash point which is the temperature at which the fuel can ignite when mixed with air. This study used a flash point test under the ASTM D93 standard (minimum 100°C for biodiesel) and found flash points of 135°C for macadamia biodiesel and 60°C for ULSD. Pour point is another indication of the physical behavior of fuel because it is the minimum temperature at which the fuel becomes semisolid and loses its flow characteristics [60]. The study found pour points of −15°C for ULSD and −3°C for biodiesel using the pour point test (TLC30) under the ASTM D97 standard. The cloud points were measured by a cloud point analyzer (PSA-70Xi) maintaining the ASTM D 5773 standard and found 6°C for biodiesel and −8.9°C for ULSD which means that, below these temperatures, bio-wax and wax formed a cloudy appearance, respectively.

Other fuel properties, such as auto-ignition temperature, acid value, carbon residue, and oxidation stability, were measured for both ULSD and biodiesel. Auto-ignition temperature indicates the safety of fuel. It is the temperature in which the fuel starts burning spontaneously. The study measured auto-ignition temperatures for both fuels under the ASTM E659 standard and found 256.9°C for ULSD and 319.56°C for biodiesel. Macadamia biodiesel is therefore safer than diesel for processing, handling, and storing. The measured acid value for biodiesel is 0.15 mg KOH/g which means that 0.15 mg of KOH is needed to neutralize the organic acids present per gram of macadamia biodiesel. It also represents the free fatty acids (FFA) present in the biodiesel [61]. The carbon residue for both fuels was found to be 0.01 m/m under the ASTM D4530 micro method. The CFPP is the parameter to describe the physical behavior of the fuel. CFPP indicates the estimated lowest temperature that the fuel gives trouble-free flow under winter condition. The study found CFPP at 1.62°C for biodiesel and −3°C for ULSD under ASTM D6371 test conditions. The SN, IV, and CN are calculated numerically using Eqs. (13.1), (13.2), and (13.3), respectively. The results show a higher CN of 56.91 for biodiesel compared with 44 for ULSD which indicates the ignition quality of the fuel. The higher CN value indicates lower ignition delay [62]. The SN and IVs for the biodiesel were found to be 196.23 and 76.44, respectively. The oxidation stability was measured under the ASTM D2274 standard and found to be 5.0 h for ULSD and 2.5 h for biodiesel. Finally, mass and energy balance for total process is discussed in the following section.

13.4.5 Mass and Energy Balances

The results of the mass and energy balances calculated for the oil extraction and biodiesel conversion processes are presented in Fig. 13.8. For mass balance, the dry kernel is considered as the input feedstock and the output is calculated based on both crude vegetable oil and cake material. On the other hand, the mass balance for the biodiesel conversion process is examined based on crude vegetable oil as the input variable and biodiesel (finished product) and glycerin (by-product) as the output variables. The process losses are calculated by subtraction of output variables from input variable. Fig. 13.8 shows that 73.08% (vegetable oil) mass conversion efficiency was obtained from the n-hexane oil extraction method. From this method, 21.83% cake materials and only a 5.09% processing loss were recorded in this study. For biodiesel conversion, 58.71% of the total mass was converted to biodiesel by the transesterification reaction. The study found 12.53% of the total mass was converted to glycerin as a by-product, which can be used as a raw material in the cosmetics industry. The calculated process loss is 1.94% for the vegetable oil to biodiesel conversion. Hence, total mass recoveries of 58.71% biodiesel, 12.53% glycerin, and 21.83% cake materials were found for the entire process. The calculated total loss is 7.03% (including 5.09% extraction loss and 1.94% conversion loss). Fig. 13.9 shows the energy distribution for the biodiesel conversion process which has been calculated based on the conversion masses and associated heating values (i.e., 39.88 MJ/kg for biodiesel, 21.06 MJ/kg for glycerin, and 38.12 MJ/kg for vegetable oil). The study identified that 84% energy was recovered as biodiesel, 10% as glycerin, and 6% energy loss for the biodiesel conversion process.

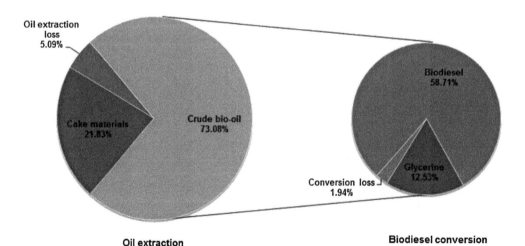

Figure 13.8 Mass balance for oil extraction and biodiesel conversion processes.

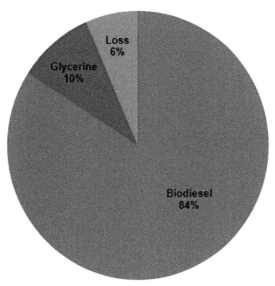

Figure 13.9 Energy balance for biodiesel conversion process.

13.5 CONCLUSIONS

The study concluded that macadamia is an oil-enriched and desirable source of bioenergy. The vegetable oil was extracted by the n-hexane method and converted to biodiesel by thermochemical reaction (transesterification in the presence of potassium methoxide catalyst). The study obtained a 98.21% conversion efficiency of the biodiesel. Both vegetable oil and biodiesel properties were measured under relevant ASTM standards and compared with ULSD. The FAME compositions were analyzed under the EN14103 standard and found 61.09%vol. oleic acid (C18:0), 15.39%vol. palmitoleic acid (C16:1), and other fatty acids present in the biodiesel. The study analyzed the physical behavior of the biodiesel for temperatures typical of the winter, autumn, spring, and summer seasons in Australia. The fuel shows very similar viscosity behavior with ULSD during summer and similar density trends throughout the seasons. Other fuel properties, such as CV, flash point, pour point, cloud point, CN, acid value, oxidation stability, and so on, were measured for the biodiesel under corresponding ASTM standards, and values found were 39.88 MJ/kg (CV), 137°C (flash point), 56.9 (CN), and 0.15 mg KOH/g (acid value). The results show that the measured physiochemical fuel properties are within the acceptable ranges of the ASTM D6751 standard. Mass and energy balances were conducted for both the oil extraction and biodiesel conversion processes. The study found 73.08% mass recovery as crude vegetable oil and 58.71% of total mass recovered as biodiesel. On the other hand, 84% energy was recovered as biodiesel and 10% as glycerin in the energy distribution. The analysis shows that a small mass loss leads to a large energy loss in the biodiesel conversion system. Further study is needed on engine performance, emissions, and combustion of the macadamia biodiesel as well as corrosion and tribo-corrosion tests on a diesel engine before commercial application of the biodiesel could be considered.

ACKNOWLEDGMENTS

This work was conducted under the Strategic Research Scholarship funded by the Central Queensland University, Australia. The authors would like to acknowledge Mr. Tim McSweeney, Adjunct Research Fellow, Higher Education Division at Central Queensland University, Australia, for his contribution in English proofread of this article.

LIST OF ABBREVIATIONS

ASTM	American Society for Testing and Materials
BoM	Bureau of Meteorology
CN	Cetane number
EN	European Standard
FAC	Fatty acid composition
FAMEs	Fatty acid methyl esters
GC	Gas chromatography
IV	Iodine value
K_2CO_3	Potassium carbonate
$KOCH_3$	Potassium methoxide
KOH	Potassium hydroxide
$NaOCH_3$	Sodium methoxide
NaOH	Sodium hydroxide
SN	Saponification number
ULSD	Ultra-low sulfur diesel

REFERENCES

[1] McFadyen L, Robertson D, Sedgley M, Kristiansen P, Olesen T. Effects of girdling on fruit abscission, yield and shoot growth in macadamia. Scientia Horticulturae 2013;164:172–7.
[2] Azad AK, Rasul MG, Khan MMK, Sharma SC, Hazrat MA. Prospect of biofuels as an alternative transport fuel in Australia. Renewable and Sustainable Energy Reviews 2015;43:331–51.
[3] Howlett BG, Nelson WR, Pattemore DE, Gee M. Pollination of macadamia: review and opportunities for improving yields. Scientia Horticulturae 2015;197:411–9.
[4] Walton DA, Wallace HM. The effect of mechanical dehuskers on the quality of macadamia kernels when dehusking macadamia fruit at differing harvest moisture contents. Scientia Horticulturae 2015;182:119–23.
[5] Nami Kartal S, Terzi E, Kose C, Hofmeyr J, Imamura Y. Efficacy of tar oil recovered during slow pyrolysis of macadamia nut shells. International Biodeterioration & Biodegradation 2011;65(2):369–73.
[6] Azad AK, Rasul MG, Khan MMK, Sharma SC, Bhuiya MMK. Study on Australian energy policy, socio-economic, and environment issues. Journal of Renewable and Sustainable Energy 2015;7:1–20. 063131.

[7] Knothe G. Biodiesel derived from a model oil enriched in palmitoleic acid, macadamia nut oil. Energy & Fuels 2010;24(3):2098−103.

[8] Corré WJ, Conijn JG, Meesters KPH, Bos HL. Accounting for co-products in energy use, greenhouse gas emission savings and land use of biodiesel production from vegetable oils. Industrial Crops and Products 2016;80:220−7.

[9] Górnaś P, Rudzińska M. Seeds recovered from industry by-products of nine fruit species with a high potential utility as a source of unconventional oil for biodiesel and cosmetic and pharmaceutical sectors. Industrial Crops and Products 2016;83:329−38.

[10] Lee RA, Lavoie J-M. From first-to third-generation biofuels: challenges of producing a commodity from a biomass of increasing complexity. Animal Frontiers 2013;3(2):6−11.

[11] Azad AK, Ameer Uddin SM, Alam MM. A comprehensive study of DI diesel engine performance with vegetable oil: an alternative bio-fuel source of energy. International Journal of Automotive and Mechanical Engineering 2012;5:576−86.

[12] Kouzu M, Kasuno T, Tajika M, Sugimoto Y, Yamanaka S, Hidaka J. Calcium oxide as a solid base catalyst for transesterification of soybean oil and its application to biodiesel production. Fuel 2008;87(12):2798−806.

[13] Balat M. Potential alternatives to edible oils for biodiesel production − a review of current work. Energy Conversion and Management 2011;52(2):1479−92.

[14] Hoekman SK, Broch A, Robbins C, Ceniceros E, Natarajan M. Review of biodiesel composition, properties, and specifications. Renewable and Sustainable Energy Reviews 2012;16(1):143−69.

[15] Fukuda H, Kondo A, Noda H. Biodiesel fuel production by transesterification of oils. Journal of Bioscience and Bioengineering 2001;92(5):405−16.

[16] Dorado MP, Ballesteros E, Arnal JM, Gómez J, López FJ. Exhaust emissions from a Diesel engine fueled with transesterified waste olive oil. Fuel 2003;82(11):1311−5.

[17] Demirbas A. Biodiesel production from vegetable oils via catalytic and non-catalytic supercritical methanol transesterification methods. Progress in Energy and Combustion Science 2005;31(5−6):466−87.

[18] Sarin R, Kumar R, Srivastav B, Puri SK, Tuli DK, Malhotra RK, et al. Biodiesel surrogates: achieving performance demands. Bioresource Technology 2009;100(12):3022−8.

[19] Demirbas A. Importance of biodiesel as transportation fuel. Energy Policy 2007;35(9):4661−70.

[20] Saka S, Kusdiana D. Biodiesel fuel from rapeseed oil as prepared in supercritical methanol. Fuel 2001;80(2):225−31.

[21] Amir Uddin MA, Azad AK. Diesel engine performance study for bio-fuel: vegetable oil, a alternative source of fuel. International Journal of Energy Machinery 2012;5(1):8−17.

[22] Ameer Uddin SM, Azad AK, Alam MM, Ahamed JU. Performance comparison of DI diesel engine by using esterified mustard oil and pure mustard oil blends with diesel. In: The 2nd International Conference on mechanical engineering and renewable energy, Chittagong, Bangladesh; 2013.

[23] Uddin SMA, Azad AK, Alam MM, Ahamed JU. Performance of a diesel engine run with mustard-kerosene blends. Procedia Engineering 2015;105:698−704.

[24] Hazrat MA, Rasul MG, Khan MMK, Azad AK, Bhuiya MMK. Utilization of polymer wastes as transport fuel resources-a recent development. Energy Procedia 2014;61:1681−5.

[25] Demirbas MF. Biorefineries for biofuel upgrading: a critical review. Applied Energy 2009;86:S151−61.

[26] Canakci M, Van Gerpen J. Biodiesel production from oils and fats with high free fatty acids. Transactions-American Society of Agricultural Engineers 2001;44(6):1429−36.

[27] Alcantara R, Amores J, Canoira L, Fidalgo E, Franco MJ, Navarro A. Catalytic production of biodiesel from soy-bean oil, used frying oil and tallow. Biomass and Bioenergy 2000;18(6):515−27.

[28] Azad AK, Rasul MG, Khan MMK, Sharma SC. In: Review of non-edible biofuel resources in Australia for second generation (2G) biofuel conversion in International Green Energy Conference, Tainjin, China; 2014. p. 867−78.

[29] Azad AK, Rasul MG, Khan MMK, Sharma SC, Mofijur M, Bhuiya MMK. Prospects, feedstocks and challenges of biodiesel production from beauty leaf oil and castor oil: a nonedible oil sources in Australia. Renewable and Sustainable Energy Reviews 2016;61:302−18.

[30] Zhang Y, Dubé MA, McLean DD, Kates M. Biodiesel production from waste cooking oil: 1. Process design and technological assessment. Bioresource Technology 2003;89(1):1—16.

[31] Vasudevan PT, Briggs M. Biodiesel production—current state of the art and challenges. Journal of Industrial Microbiology & Biotechnology 2008;35(5):421—30.

[32] Azad AK, Prince MRI. Preparation of bio-diesel from jatropha curcus seeds oil: performance and emission study of a 4-stroke single cylinder biodiesel fueled engine. International Journal of Energy and Technology 2012;4(22):1—6.

[33] Kumar N, Singh AS, Kumari S, Reddy MP. Biotechnological approaches for the genetic improvement of *Jatropha curcas* L.: a biodiesel plant. Industrial Crops and Products 2015;76:817—28.

[34] Reyes-Trejo B, Guerra-Ramírez D, Zuleta-Prada H, Cuevas-Sánchez JA, Reyes L, Reyes-Chumacero A, et al. *Annona diversifolia* seed oil as a promising non-edible feedstock for biodiesel production. Industrial Crops and Products 2014;52:400—4.

[35] Barnwal BK, Sharma MP. Prospects of biodiesel production from vegetable oils in India. Renewable and Sustainable Energy Reviews 2005;9(4):363—78.

[36] Meneghetti SMP, Meneghetti MR, Serra TM, Barbosa DC, Wolf CR. Biodiesel production from vegetable oil mixtures: cottonseed, soybean, and castor oils. Energy & Fuels 2007;21(6):3746—7.

[37] Sarin A, Arora R, Singh N, Sarin R, Sharma M, Malhotra R. Effect of metal contaminants and antioxidants on the oxidation stability of the methyl ester of Pongamia. Journal of the American Oil Chemists' Society 2010;87(5):567—72.

[38] Habibullah M, Masjuki HH, Kalam MA, Zulkifli NWM, Masum BM, Arslan A, et al. Friction and wear characteristics of *Calophyllum inophyllum* biodiesel. Industrial Crops and Products 2015;76:188—97.

[39] Mosarof MH, Kalam MA, Masjuki HH, Alabdulkarem A, Habibullah M, Arslan A, et al. Assessment of friction and wear characteristics of *Calophyllum inophyllum* and palm biodiesel. Industrial Crops and Products 2016;83:470—83.

[40] Mondal SK, Ferdous K, Uddin MR, Khan MR, Islam MA, Azad AK. Preparation and characterization of biodiesel from karanja oil by using silica gel reactor. in 1st international e-conference of energies. Switzerland 2014:14—31.

[41] Silitonga AS, Masjuki HH, Ong HC, Yusaf T, Kusumo F, Mahlia TMI. Synthesis and optimization of *Hevea brasiliensis* and *Ricinus communis* as feedstock for biodiesel production: a comparative study. Industrial Crops and Products 2016;85:274—86.

[42] Armendáriz J, Lapuerta M, Zavala F, García-Zambrano E, del Carmen Ojeda M. Evaluation of eleven genotypes of castor oil plant (*Ricinus communis* L.) for the production of biodiesel. Industrial Crops and Products 2015;77:484—90.

[43] Brennan L, Owende P. Biofuels from microalgae—a review of technologies for production, processing, and extractions of biofuels and co-products. Renewable and Sustainable Energy Reviews 2010;14(2):557—77.

[44] Bhuiya MMK, Rasul MG, Khan MMK, Ashwath N, Azad AK, Hazrat MA. Prospects of 2nd generation biodiesel as a sustainable fuel — Part 2: properties, performance and emission characteristics. Renewable and Sustainable Energy Reviews 2016;55:1129—46.

[45] Mofijur M, Rasul MG, Hyde J, Azad AK, Mamat R, Bhuiya MMK. Role of biofuel and their binary (diesel—biodiesel) and ternary (ethanol—biodiesel—diesel) blends on internal combustion engines emission reduction. Renewable and Sustainable Energy Reviews 2016;53:265—78.

[46] Azad AK, Rasul MG, Khan MMK, Omri A, Bhuiya MMK, Ali MH. Modelling of renewable energy economy in Australia. Energy Procedia 2014:1902—6.

[47] Bhuiya M, Rasul M, Khan M, Ashwath N, Azad A, Mofijur M. Optimisation of oil extraction process from Australian native beauty leaf seed (*Calophyllum inophyllum*). Energy Procedia 2015;75:56—61.

[48] Jahirul MI, Brown JR, Senadeera W, Ashwath N, Laing C, Leski-Taylor J, et al. Optimisation of bio-oil extraction process from beauty leaf (*Calophyllum inophyllum*) oil seed as a second generation biodiesel source. Procedia Engineering 2013;56:619—24.

[49] Bhuiya M, Rasul M, Khan M, Ashwath N, Azad A. Prospects of 2nd generation biodiesel as a sustainable fuel—Part: 1 selection of feedstocks, oil extraction techniques and conversion technologies. Renewable and Sustainable Energy Reviews 2015;55:1109—28.

[50] Azad AK, Rasul MG, Khan MMK, Sharma SC, Islam R. Prospect of Moringa seed oil as a sustainable biodiesel fuel in Australia: a review. Procedia Engineering 2015;105:601–6.

[51] Bhuiya MMK, Rasul MG, Khan MMK, Ashwath N, Azad AK, Hazrat MA. Second generation biodiesel: potential alternative to-edible oil-derived biodiesel. Energy Procedia 2014:1969–72.

[52] Azad AK, Uddin SMA. Performance study of a diesel engine by first generation bio-fuel blends with fossil fuel: an experimental study. Journal of Renewable and Sustainable Energy 2013;5(1):1–12. 013118.

[53] Taufiqurrahmi N, Bhatia S. Catalytic cracking of edible and non-edible oils for the production of biofuels. Energy & Environmental Science 2011;4(4):1087–112.

[54] Ghaly AE, Dave D, Brooks M, Budge S. Production of biodiesel by enzymatic transesterification: review. American Journal of Biochemistry and Biotechnology 2010;6(2):54.

[55] Azad AK, Rasul MG, Giannangelo B. Diesel engine performance and emission study using soybean biodiesel blends with fossil diesel. In: The Seventh International exergy, energy and environment symposium, Valenciennes, France; 2015. p. 1–10.

[56] Rizwanul Fattah IM, Masjuki HH, Kalam MA, Wakil MA, Rashedul HK, Abedin MJ. Performance and emission characteristics of a CI engine fueled with *Cocos nucifera* and *Jatropha curcas* B20 blends accompanying antioxidants. Industrial Crops and Products 2014;57:132–40.

[57] Mofijur M, Masjuki HH, Kalam MA, Atabani AE, Arbab MI, Cheng SF, et al. Properties and use of *Moringa oleifera* biodiesel and diesel fuel blends in a multi-cylinder diesel engine. Energy Conversion and Management 2014;82:169–76.

[58] Rizwanul Fattah IM, Masjuki HH, Kalam MA, Mofijur M, Abedin MJ. Effect of antioxidant on the performance and emission characteristics of a diesel engine fueled with palm biodiesel blends. Energy Conversion and Management 2014;79:265–72.

[59] Kongkasawan J, Capareda SC. Jatropha oil refining process and biodiesel conversion: mass and energy balance. International Energy Journal 2012;13(4):1–4.

[60] Azad AK, Rasul MG, Giannangelo B, Islam R. Comparative study of diesel engine performance and emission with soybean and waste oil biodiesel fuels. International Journal of Automotive and Mechanical Engineering 2015;12:2866–81.

[61] Demirbas A. Progress and recent trends in biodiesel fuels. Energy Conversion and Management 2009;50(1):14–34.

[62] Azad AK, Rasul MG, Khan MMK, Sharma SC, Bhuiya MMK. Recent development of biodiesel combustion strategies and modelling for compression ignition engines. Renewable and Sustainable Energy Reviews 2016;56:1068–86.

CHAPTER FOURTEEN

Assessment of Physical, Chemical, and Tribological Properties of Different Biodiesel Fuels

M. Mofijur[1], M.G. Rasul[1], N.M.S. Hassan[1], H.H. Masjuki[2], M.A. Kalam[2] and H.M. Mahmudul[3]

[1]Central Queensland University, Rockhampton, QLD, Australia
[2]University of Malaya, Kuala Lumpur, Malaysia
[3]University Malaysia Pahang, Pahang, Malaysia

14.1 INTRODUCTION

Biodiesel is a renewable energy source that offers some benefits including the reduction in greenhouse gas emission and pollutants, increasing energy diversity and economic security [1]. Biodiesel is considered as a promising alternative fuel for transportation sector as it has similar fuel properties with diesel fuel [2]. It also can be blended with diesel fuel at any percentage and can be used for power generation in a diesel engine without the change of existing infrastructure. The source-to-wheel carbon dioxide emission analysis of pure biodiesel fuel shows that it reduces 60% CO emission to the environment compared to conventional fossil fuel [3]. In this context, different countries have set their target and mandate to use biodiesel fuel in the transportation sector. For example, the European Union has a target to use 10% biodiesel and China has a target to use 10.6−12 million biodiesel by 2020. Similarly, the Australian government has also set a target to use 20% biodiesel by the year of 2020 [4]. As a result, worldwide biodiesel production has increased too. According to the BP statistics 2015, 10.3% global biodiesel production increased in 2014 compared to the year of 2004.

Biodiesel consists of long-chain alkyl ester and is produced from vegetable oils, animal fats, or waste cooking oils through transesterification reaction [5,6]. In transesterification reaction, vegetable oils are reacted with alcohol (usually methanol) in the presence of a catalyst (commonly used, KOH and NaOH). Through the reversible reaction, triglycerides are converted into monoglycerides, and glycerin is obtained as a by-product. The transesterification reaction is shown in Fig. 14.1.

One of the outstanding credit of biodiesel compared to other biofuels is that a wide range of biodiesel feedstocks are available around the world [7]. Most of the countries use the source for biodiesel production that is readily available in their country. For example,

Clean Energy for Sustainable Development
ISBN 978-0-12-805423-9, http://dx.doi.org/10.1016/B978-0-12-805423-9.00014-4

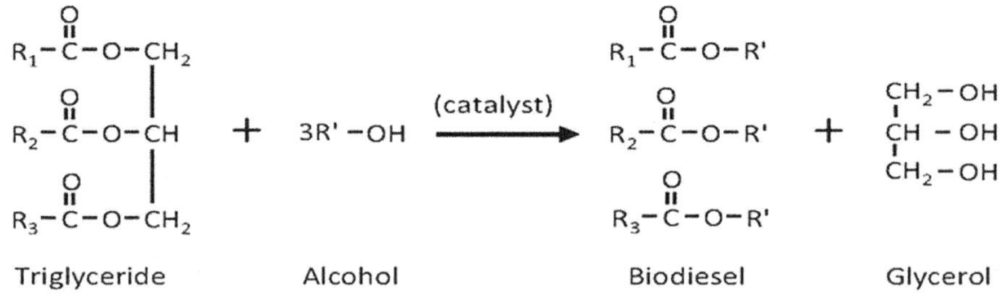

Figure 14.1 The transesterification reaction.

rapeseed oil is used widely in Europe, whereas soybean oil is used in the United States [3]. Similarly, Malaysia is a top palm oil producer country [8], that is why palm oil is widely used in Malaysia for biodiesel production.

Biodiesel is usually characterized by some physical and chemical properties including the fatty acid compositions. The properties of biodiesel, either physical or chemical properties, vary from source to source, country of origin, production process, reaction time, methanol used, temperature, and speed [9]. For this reason characterization of biodiesel is important before using as an alternative to diesel fuel. To this aim, both United States and the European Union have issued standard specification (namely ASTM D6751 by US and EN 14214 by EU) [10] that should be met by the produced biodiesel before being used in the diesel engines. Recently, in the 2010s, the governments of Australia [11], Malaysia [12], Korea [12], and some other countries also have issued their specification. Biodiesel has some limitations that it is vulnerable to oxidation, and it has poor cold temperature properties which could be tackled by using additives [13].

The selection of biodiesel source is very important as feedstock alone consists of 75% total production cost. The study on the fuel properties is also crucial as the suitability of biodiesel depends on the fatty acid composition and fuel properties. The fatty acid composition also influences many fuel properties. A research work has been done to study the effect of five edible and nonedible biodiesel feedstocks (macadamia, beauty leaf, palm, jatropha, and moringa oils) on the physical and chemical properties. Also, the correlation between composition and fuel properties of biodiesel has been established. Finally, to validate the data and choose the better alternative among them based on 15 criteria (fuel properties), namely kinematic viscosity (KV), dynamivc viscosity (DV), density (D), flash point (FP), calorific value (CV), cetane number (CN), iodine value (IV), saponification value (SV), cloud point (CP), pour point (PP), acid value (AV), oxidation stability (OS), cold filter plugging point (CFPP), polyunsaturated fatty acid (PUFA), and monounsaturated fatty acid (MUFA) using a PROMETHEE-GAIA multicriteria decision analysis software.

14.2 MATERIALS

In this study, five types of fuel, namely palm, jatropha, moringa, macadamia, and beauty leaf oils have been selected. Palm and Jatropha oils were supplied by forest research institute, Malaysia. Macadamia oil was purchased from Coles, whereas moringa and beauty leaf oils were collected from a colleague through personal communication. The detailed properties of crude oils are presented in Table 14.1. All other reagents, methanol, and 150 mm filter paper were available in the chemical laboratory, Central Queensland University (North Rockhampton, Australia).

14.2.1 Biodiesel Production

The viscosity of vegetable oils (as shown in Table 14.1) are 10–15 times higher than fossil diesel fuel. The direct use of higher viscous fuel in the internal combustion engine causes some problems including wear in the injectors, fuel pump, and engine deposition and blocking the fuel system which consequently affect the fuel spray and combustion process inside the cylinder [14]. Therefore, crude oils could be passed through a chemical process called transesterification process to reduce the viscosity. Transesterification process is very popular process among all the other biodiesel conversion process due to simplicity, reliability, low cost, and the fuel quality. One of the problems of this conversion process is the formation of soap with the reaction of higher free fatty acid (FFA) and catalyst [15]. To avoid soap formation, FFA should be reduced through the

Table 14.1 Properties of Crude Oils Used in This Study

Properties	Units	Standards	Macadamia Oil	Palm Oil	Jatropha Oil	Moringa Oil	Beauty Leaf Oil
Dynamic viscosity	mPa.s	ASTM D445	35.23	36.30	31.52	38.90	48.73
Kinematic viscosity at 40°C	mm^2/s	ASTM D445	39.22	40.40	34.93	43.33	52.13
Density at 15°C	kg/m^3	ASTM D4052	898.60	898.4	902.5	897.5	922.2
Flash point	°C	ASTM D93	167.5	165	220	268.5	195.5
Pour point	°C	ASTM D97	8	9	−3	11	8
Cloud point	°C	ASTM D2500	0	8	−2	10	8
Calorific value	MJ/kg	ASTM D240	39.89	39.44	38.66	38.05	38.51
Acid value	mg KOH/ g oil	ASTM D664	4	3.47	10.7	8.62	40

preesterification process using sulfuric acid [16]. Fig. 14.2 shows the schematic diagram of two-step biodiesel production process.

In this study, a small-scale laboratory reactor (water jacketed), as shown in Fig. 14.3A, 1 L in size equipped with reflux condenser, thermometer, and water-circulated bath was used to produce biodiesel from vegetable oils. The FFAs of crude moringa, jatropha, and beauty leaf oils were found to be higher (above two), which indicates that esterification is

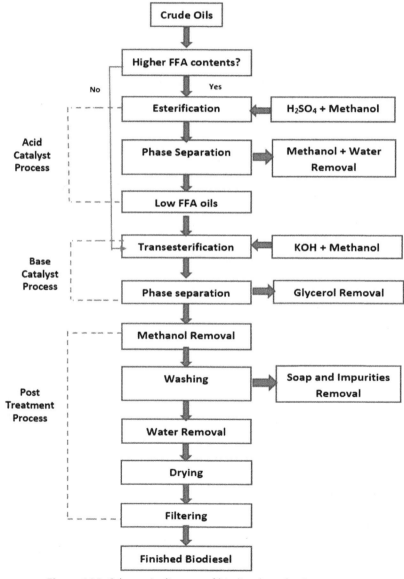

Figure 14.2 Schematic diagram of biodiesel production process.

(A)   **(B)**   **(C)**   **(D)**

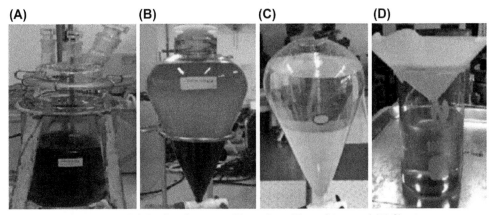

Figure 14.3 (A) Biodiesel reactor, (B) settling, (C) washing, and (D) filtering.

necessary to produce biodiesel from jatropha, moringa, and beauty leaf oils. The rest of the vegetable oils including palm and macadamia oil was processed only through the transesterification process.

In the esterification process, the molar ratio of methanol to refined beauty leaf, jatropha, and moringa oils were maintained at 12:1 (50% v/v). 1% (v/v) of sulfuric acid (H_2SO_4) was added to the preheated oils at 60°C for 3 h under 600 rpm stirring speed in a glass reactor. On completion of this reaction, the products were poured into a separating funnel to separate the excess alcohol, sulfuric acid, and impurities present in the upper layer. The lower layer was separated and heated at 95°C for 1 h to remove methanol and water from the esterified oil.

In the transesterification process, esterified/preheated oil was reacted with 6:1 M ratio of methanol to oil in the presence of 1% (w/w) of KOH catalyst. The reaction was maintained at 60°C for 2 h at 800 rpm. After completion of the reaction, the mixture was poured into a separation funnel for 14 h to be cooled and settled, and then glycerol was separated from biodiesel (Fig. 14.3B). The upper part of the funnel contained biodiesel, and the lower part contained glycerin along with excess methanol and impurities. The biodiesel was collected, and the glycerin was drawn off. The produced biodiesel was then heated at 65°C to remove remaining methanol. Then the biodiesel was washed using warm distilled water to remove entire impurities (Fig. 14.3C). Finally, the washed biodiesel was heated at 95°C for 1 h to remove water and then dried using Na_2SO_4. Then the biodiesel was filtered through a filter paper (Fig. 14.3D), and the final product was collected and stored for characterization.

14.2.2 Determination of Fatty Acid Composition

The fatty acid composition was determined using gas chromatography (GC 7890A, Agilent Technologies) equipped with a flame ionization detector. The capillary column was 30 m in length, with a film thickness of 0.25 μm, and with an internal diameter of 0.25 mm. The carrier gas He was supplied at 20 mL/min speed. The temperature was 100°C hold for 0 min and 10°C/min to 250°C hold for 5 min. The injector and detector temperatures were 250°C. The split ratio of injector ratio was 50:1 and the volume were 0.3 μL. In this test, 0.25 g of each sample was diluted with 5 mL n-heptane to analyze the fatty acid composition test.

14.2.3 Fuel Properties

In this study, all the fuel properties were measured according to the ASTM standards. Table 14.2 shows the list of equipment used in this study. CN, IV, SV, degree of unsaturation (DU), and long-chain saturated factor (LCSF) were calculated from the fatty acid profile of all biodiesel using following equations as described by [16–20]:

$$CN = 46.3 + (5458/SV) - (0.225 \cdot IV) \tag{14.1}$$

$$SV = \sum (560 \cdot A_i)/M_{wi} \tag{14.2}$$

Table 14.2 Equipment Used in This Study

Property	Equipment	Standard Method	Accuracy
Kinematic viscosity	NVB classic (Norma Lab, France)	ASTM D445	± 0.01 mm^2/s
Density	DM40 LiquiPhysics density meter (Mettler Toledo, Switzerland)	ASTM D127	± 0.1 kg/m^3
Flash point	NPM 440 Pensky-martens flash point tester (Norma Lab, France)	ASTM D93	± 0.1°C
Cloud and pour point	NTE 450 cloud and pour point tester (Norma Lab, France)	ASTM D2500	± 0.1°C
Higher heating value	IKA C 2000 calorimeter, United Kingdom	ASTM D240	± 0.001 MJ/kg
Acid number	Automation titration rondo 20 (Mettler Toledo, Switzerland)	ASTMD664 and EN 14111	± 0.001 mg KOH/g
Oxidation stability, 110°C	873 Rancimat (Metrohm, Switzerland)	EN 14112	± 0.01 h

$$IV = \sum (254 \cdot A_i \cdot D)/M_{wi} \tag{14.3}$$

$$LCSF = 0.1(C16:0, wt\%) + 0.5(C18:0\ wt\%) + 1(C20:0\ wt\%) + 1.5(C22:0\ wt\%)$$
$$+ 2.0(C24:0\ wt\%)$$

$$\tag{14.4}$$

$$DU = \sum (Monounsaturated\ Fatty\ Acid + 2\ Polyunsaturated\ Fatty\ Acid) \tag{14.5}$$

where A_i is the percentage of each component; D is the number of double bonds; and M_w is the molecular mass of each component.

In this study, a multicriteria decision analysis tool named PROMETHEE-GAIA was used to find out the best fuel that has the suitable chemical composition to ensure the compliance with standard biodiesel properties. The critical parameters of biodiesel fuel properties, such as D, KV, CN, IV, OS, and CFPP, are depended on oil nature. Therefore, data on properties of two more commercial biodiesel were obtained from literature including sunflower biodiesel (SBD) and rapeseed biodiesel (RBD) to validate the data obtained from produced biodiesels. The better alternative fuel among these biodiesels for diesel engine application was selected based on 15 criteria (fuel properties), namely KV, DV, D, FP, CV, CN, IV, SV, CP, PP, AV, OS, CFPP, PUFA, and MUFA using a multi-criteria decision analysis software. PROMETHEE-GAIA software was used for multi-criteria decision analysis because of their rational decision vector that stretches toward the preferred solution compared to other multicriteria decision software [20]. The name of the biodiesels was set to "action" and fuel properties were set to "criteria." The preference function was set as minimum (i.e., lower values preferred for good biodiesel) or maximum (i.e., higher values preferred for good biodiesel), and weighting was considered equal for all criteria in this analysis. Table 14.3 shows the variables and preference used in PROMETHEE-GAIA analysis.

14.2.4 Equipment for Tribological Study

In this study, a four-ball tester was used to evaluate the wear and friction characteristics of biodiesel. Schematic diagram of four-ball tester is shown in Fig. 14.4. The ball used was made from stainless steel, and three balls were stationary, and one ball was rotating in the steel cup. At least 10 mL of tested fuel was poured into the steel cup so that steel cup is filled minimum 3 mm and three stationary balls are fully dipped. The four-ball tester was connected to a computer to obtain the frictional data through Winducom 2008 software. The test was conducted according to the ASTM D2596 and D2783 methods and at a variable load of 40 and 80 kg. The total test run time was 300 s at a constant speed of 1800 rpm. The wear scar diameter (WSD) of tested ball was measured before removing the steel ball from the cup.

Table 14.3 The Variables and Preference Used in PROMETHEE-GAIA Analysis

Variables	Preference for PROMETHEE-GAIA
KV	Min
D	Min
HHV	Max
OS	Max
AV	Min
FP	Max
CP	Min
CFPP	Min
CN	Max
IV	Min
DU	Min
MUFA	Min
PUFA	Min

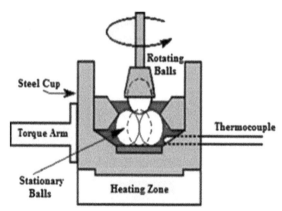

Figure 14.4 Schematic diagram of four-ball tester.

14.2.5 Determination of the Coefficient of Friction

The coefficient of friction was calculated by the multiplication of the mean friction torque and spring constant. A load cell was used to measure the frictional torque in this experiment. The frictional torque can be expressed as:

$$T = \frac{\mu \times 3 \times W \times r}{\sqrt{6}} \tag{14.6}$$

$$\approx \mu = \frac{T \times \sqrt{6}}{3 \times W \times r} \tag{14.7}$$

where μ = coefficient of friction; T = frictional torque (kg-mm); W = applied load (N); and r = distance from the center of the contact surface on the lower balls to the axis of rotation.

14.2.6 Determination of Flash Temperature Parameter

The flash temperature parameter (FTP) is a single number that is used to express the critical flash temperature at which a sample will fail under given conditions. The FTP was calculated for both loads. For the conditions used in the four-ball test, the following relationship was used:

$$\text{FTP} = \frac{W}{d^{1.4}} \qquad (14.8)$$

where W = load (kg) and d = mean wear scar diameter (mm).

14.3 RESULTS AND DISCUSSION

14.3.1 Fatty Acid Profile of Biodiesels

A systematic analysis of the fatty acid composition and comparable fuel properties are important to select the best species for biodiesel production. Table 14.4 shows the fatty acid profile of produced biodiesel, namely macadamia biodiesel (MaBD), palm biodiesel (PBD), jatropha biodiesel (JBD), moringa biodiesel (MoBD), and beauty leaf biodiesel (BBD). These fatty acids have a direct impact on the chemical and physical properties of biofuel. All the biodiesel samples contain saturated fatty acid, MUFA, and PUFA. The total saturated fatty acids of MaBD, JBD, PBD, MoBD, and BBD were found to be 15.80%, 22.6%, 44.6%, 18.6%, and 33.4%, respectively, whereas total unsaturated fatty acids of MaBD, JBD, PBD, MoBD, and BBD were found to be 84.20%, 77.4%, 55.4%, 81.4%, and 66.6%, respectively. Oleic acid (18:1) was the predominant fatty acid in all the biodiesel samples (i.e., macadamia (61.3%), jatropha (44.6%), palm (43.4%), moringa (74.1%), and beauty leaf (38.2%)) followed by the palmitoleic acid (C16:1, 16.2%) for MaBD, linoleic acid (C18:2, 31.9%) for JBD, palmitic acid (16:0, 40.3%) for PBD, palmitic acid (16:0, 7.9%) for MoBD, and linoleic acid (C18:2, 27.6%) for BBD. Among all the biodiesel fuels, JBD has the highest DU (109.60), whereas PBD has the lowest DU (67.80). Also, the DU of BBD, MoBD, and MaBD is 94.50, 85.70, and 84.50, respectively. Unlike the DU, JBD has the lowest LCSF (5.26), and BBD has the highest LCSF (11.64).

14.3.2 Analysis of Fuel Properties of Biodiesel Samples

Characterization of biodiesel is needed to check the quality of the fuels as every fuel must meet the quality standard before being considered as an automobile fuel [21]. Therefore,

Table 14.4 Fatty Acid Compositions of Biodiesel Fuels

Fatty Acids	Molecular Weight	Structure	MaBD (Wt%) [18]	JBD (Wt%)	PBD (Wt%)	MoBD (Wt%)	BBD (Wt%)
Lauric	200	12:0	0.100.6	0.1	0	0	0
Myristic acid	228	14:0	0.6	0.1	0	0.1	0
Palmitic	256	16:0	7.9	14.6	40.3	7.9	14.9
Palmitoleic	254	16.1	16.2	0.6	0	1.7	0.2
Stearic	284	18:0	3.2	7.6	4.1	5.5	17.2
Oleic	282	18:1	61.3	44.6	43.4	74.1	38.2
Linoleic	280	18:2	2.1	31.9	12.2	4.1	27.6
Linolenic	278	18:3	0.1	0.3	–	0.2	0.3
Arachidic	312	20:0	2.7	0.3	–	2.3	0.9
Eicosenoic	310	20:1	2.6	–	–	1.3	0.3
Behenic	340	22:0	0.9	–	–	2.8	0.3
Erucic	338	22:1	0.3	–	–	–	0
Lignoceric	368	24:0	0.4	–	–	–	0.1
Total saturated fatty acid			15.80	22.6	44.6	18.6	33.4
Total monounsaturated fatty acid			80.40	45.2	43.4	77.1	38.7
Total polyunsaturated fatty acid			2.20	32.2	12	4.3	27.9
Others			1.6	0	0	0	0
Degree of unsaturation (DU)			84.50	109.60	67.80	85.70	94.50
Long-chain saturated factor (LCSF)			7.24	5.26	6	10.04	11.64

the produced biodiesel from any source must meet the recognized worldwide standards to be used as IC engine fuel. Currently, global biodiesel standards are ASTM D6751 and EN14212 [22,23]. Some other countries also have identified their standards [11,14,24]. The properties of biodiesel fuel are varied from feedstock to feedstock, and it also depends on the quality, origin of the source, and biodiesel production techniques [25]. Therefore, to evaluate the impact of a fatty acid profile on the quality of fuel, the physical and chemical properties were calculated and measured. Table 14.5 shows the summary of fuel properties compared with commercial biodiesel.

14.3.2.1 Density

Density is an important property of any fuel, which affects the engine performance characteristics directly. Density influences the fuel atomization efficiency and combustion characteristics [26,27]. Density also leads to engine oil sludge problems. In this study, an Anton Paar automatic viscometer (SVM 3000) was used to measure the density (kg/cm^3) of the fuel according to ASTM D7042. The standard range of the density of biodiesel fuel is 3.5–5 in EN and Australia, respectively, but there is no specification according to US standards. The densities of MaBD, PBD, JBD, MoBD, and BBD were found to be 859.2, 858.9, 865.7, 869.6, and 868.7 kg/m^3, respectively. All the values are within the limit of EN14214 and Australian standards.

Table 14.5 Properties of Produced Biodiesel Compared With Other Commercial Biodiesel

Properties	Unit	MaBD [18]	PBD	JBD	MoBD	BBD	SBD[a]	RBD[a]	ASTM D6751	EN 14214	Australian Standards
Kinematic viscosity at 40°C	mm²/s	4.46	4.63	4.73	5.05	5.68	4.2	4.4	1.9–6	3.5–5	3.5–5
Density at 15°C	kg/m³	859.2	858.9	865.7	869.6	868.7	880	877	—	860–900	860–900
Higher heating value	MJ/kg	38.21	40.91	39.82	40.05	39.38	41.26	41.55	—	—	—
Oxidation stability	h	3.35	5.16	3.02	4.45	3.58	1.88	3.09	3 min	6 min	—
Acid value	mg KOH/g	0.07	0.05	0.05	0.05	0.34	0.15	0.16	0.5 max	0.5 max	0.8
Flash point	°C	178.5	182.5	184.5	180.5	141.5	177	176	130 min	120 min	120 min
Pour point	°C	0	11	3	19	7	—	—	Report	Report	Report
Cloud point	°C	8	10	3	19	7	—	—	Report	Report	Report
CFPP	°C	8	11	10	18	8	−3	−10	Report	Report	Report
Cetane number	—	56	59	51	56	54	44.90	61.08	47 min	51 min	51 min
Iodine number	—	77.50	61	99	77.50	86	119.47	109	—	—	—
Saponification value	—	199	206	202	199	201	214	139	—	—	—

[a]Denotes data from literature [11].

14.3.2.2 Flash Point

FP is the safety measure of fuel for storage. It is the point at which fuels are flammable [28]. According to the Australian standards and EN14214 standards, the minimum FP temperature for biodiesel should be 120°C whereas it should be 130°C according to US standards. To measure the FP value of the fuel according to the ASTM D93 method, an HFP 380 Pensky-Martens FP analyzer was used. Table 14.5 shows the FP temperature of different biodiesel fuel. All the biodiesel have a higher FP than diesel fuel which indicates that biodiesel is safer to transport and store [14,29−33]. Among all the biodiesels, JBD has the maximum FP (184.5°C), and BBD has the minimum FP temperature (141.5°C). All the values are within the specified limits.

14.3.2.3 Viscosity

Viscosity affects the fuel drop size, the jet penetration, quality of atomization, spray characteristics, and the combustion quality [18]. Very high or low viscosity of fuel affects the engine. For example, if the viscosity is very low, then it will not provide enough lubrication which will increase wear and leakage. Higher viscous fuel will form a larger droplet during injection which affects combustion quality thus leading to higher exhaust emission [15]. An Anton Paar automatic viscometer (SVM 3000) was used to measure the viscosity (mm^2/s) of the fuel according to ASTM D445. The standard range of KV of biodiesel fuel is 1.9−6 in the United States and 3.5−5 in EN and Australia. Table 14.5 shows the KV of all biodiesel fuel. BBD has a highest KV (5.68 mm^2/s) followed by the moringa (5.05 mm^2/s), jatropha (4.73 mm^2/s), palm (4.63 mm^2/s), and macadamia (4.46 mm^2/s) biodiesels, respectively. All the KV values are within the US, EN, and Australian standards.

14.3.2.4 Cold Flow Properties

Cold flow properties indicate the low-temperature operation ability of any fuel and reflect their cold weather [34]. The CP is defined as the temperature of a liquid specimen when the smallest observable cluster of wax crystals first appears upon cooling under prescribed conditions [35]. The CFPP is defined as the temperature at which the test filter starts to plug due to fuel components that begin to gel. This causes major operability problems. The cold-temperature properties of biodiesel should be reported according to the Australian, European, and US standards although the limits are not specified. An automatic NTE 450 (Norma Lab, France) CP tester and an automatic NTE 450 (Norma Lab, France) CFPP tester were used to measure the CP and CFPP of all biodiesel fuel samples according to the ASTM D2500 and ASTM D6371 methods, respectively. Table 14.5 shows the cold weather performance of different biodiesel fuels used in this study. The CP and CFPP of MaBD, PBD, JBD, MoBD, and BBD were found to be 8°C and 8°C, 10°C and 11°C, 3°C and 10°C, 19°C and 18°C, and 7°C and 8°C, respectively.

14.3.2.5 Cetane Number

It is a dimensionless number that describes the ignition quality of fuel under a fixed condition [36]. This is one of the parameters that is considered for selection of the biodiesel. Fuels having higher CN play a role to start engine rapidly and make smooth combustion in the engine [13,37]. But, fuel with lower CN affects the combustion characteristics thus emitting higher HC and PM emissions. Australian and European biodiesel standards limit the CN to a minimum value of 51 whereas ASTM standard limits it to a minimum value of 47. Table 14.5 indicates that PBD has good ignition quality followed by the moringa (56), macadamia (56), beauty leaf (54), and jatropha (51) biodiesels. Though the fatty acid composition of both macadamia and moringa is different, they have same ignition quality. The reason of highest CN of PBD could be attributed to the higher saturated fatty acid composition and biodiesel. The lowest CN of JBD may be due to the contents of higher linoleic acid. However, all the results are within the specified limits and meet the standard specification.

14.3.2.6 Higher Heating Value

The heat of combustion or the CV of a fuel blend is another very important property to determine its suitability as an alternative to diesel fuel [38]. The higher heating value of a fuel mixture influences the power output of an engine directly. In this study, the heating value of all the fuel samples was determined using IKA C 2000 calorimeter. The higher heating value (HVV) of the fuel sample used in this study is shown in Table 14.5. PBD has the highest heating value (40.91 MJ/kg) followed by the MoBD (40.05 MJ/kg), JBD (39.82 MJ/kg), BBD (39.38 MJ/kg), and MaBD (38.21 MJ/kg) respectively. The HHVs of all the fuel samples are close to each other and do not have much variation. There is no specified limit on HHV in all US, EN, and Australian standards.

14.3.2.7 Oxidation Stability

Biodiesel produced from vegetable oils is considered more vulnerable to oxidation at high temperature and contact of air, because of bearing the double bond molecules in the FFA [39]. Biodiesels show less oxidative stability compared with petroleum diesel due to their different chemical composition, and this is one of the major issues that limit the widespread use of biodiesel as a fuel in automobile engines [40]. According to the US and EN standards, the OS period should be minimum 3 and 6 h, respectively. Australia also has set the standard that is same as the EN standards (minimum 6 h). The biodiesel and its blend stability were measured by induction period. OS of samples was evaluated with commercial appliance Rancimat 743 applying accelerated oxidation test (Rancimat test) specified in EN 14112. All the biodiesel listed in Table 14.5 meet the US standards but fail to meet the EN and Australian standards. The OS values of MaBD, PBD, JBD, MoBD, and BBD were found to be 3.35, 5.16, 3.02, 4.45, and 3.58 h, respectively.

14.3.3 Effect of Fatty Acid Composition on Fuel Properties

From the previous section, it has been clear that biodiesel properties from a different source are not same. The fatty acid composition has a significant influence on the properties of biodiesel fuel [34]. The following section discusses the effect of fatty acid composition on the main fuel properties.

CN is associated with the unsaturated fatty acid composition of the biodiesel. Fig. 14.5 shows the correlation between the DU and CN. It is clear that cetane decreases with increasing DU, and it linearly fits with DU with the R^2 value of 0.9965. The following equation has been developed to predict the CN:

Cetane Number $= -0.1927 \times$ Degree of Unsaturation $+ 72.242$, $R^2 = 0.9965$

As discussed in the previous section that iodine number is the value that indicates the double bond in biodiesel, therefore, it's also related to the DU of biodiesel. The correlation between the DU and the IV is shown in Fig. 14.6. It is seen that IV increases with the DU that indicates higher the unsaturation in the fuel will have higher IV. IV linearly fits with the DU with an R^2 value of 0.9983. The following equation has been developed to predict the iodine number:

Iodine Number $= 0.9076 \times$ Degree of Unsaturation $- 0.0527$, $R^2 = 0.9983$

Biodiesel is very likely to be oxidized unless an antioxidant is used. OS is highly related to the unsaturated fatty acid [41]. Fig. 14.7 shows the correlation between DU and OS. It is seen that OS has a linear relationship with DU with a higher R^2 value of 0.9587. The OS value decreases with the DU. The level of unsaturation of most of the biodiesel fuel is high, therefore OS is poor. The following equation has been developed to predict the OS:

Oxidation Stability $= -0.0541 \times$ DU $+ 8.9285$, $R^2 = 0.7044$

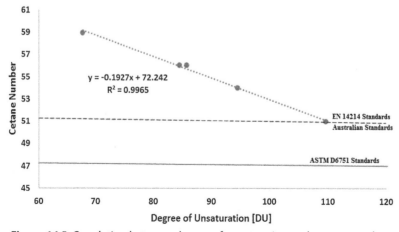

Figure 14.5 Correlation between degree of unsaturation and cetane number.

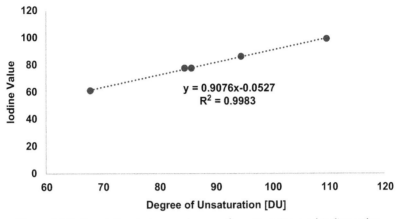

Figure 14.6 Correlation between degree of unsaturation and iodine value.

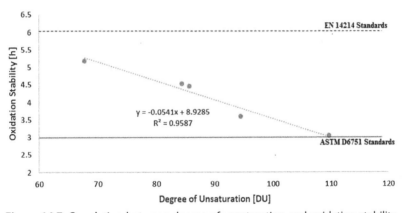

Figure 14.7 Correlation between degree of unsaturation and oxidation stability.

CFPP is an important parameter that indicates the low-temperature application capability of biodiesel which mainly depends on the LCSF. The impact of the unsaturated fatty acid composition is negligible in this case. Therefore, Fig. 14.8 shows the correlation between CFPP and LCSF. It is seen that CFPP increases with the LCSF, and correlation between them are linear with a high R^2 value of 0.9985. The following equation has been developed to predict the CFPP of biodiesel fuel:

$$\textbf{CFPP} = \textbf{3.1028} \times \textbf{Long Chain Saturated Factor} - \textbf{16.109}, \ \textbf{R}^2 = \textbf{0.9985}$$

14.3.4 Validation of Biodiesel Properties

Analysis of fuel properties and correlation analysis described in the previous section indicate that properties of all the biodiesel fuel are not same. Different biodiesels have

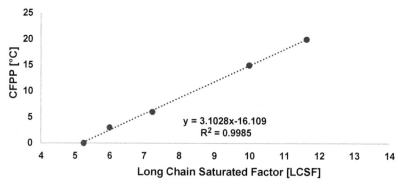

Figure 14.8 Correlation between long-chain saturated fatty acid and CFPP.

different properties, chemical composition, the DU, and LCSF. For example, MaBD has higher unsaturated fatty acid whereas PBD has higher saturated fatty acid composition. Therefore, it is necessary to assess and compare the biodiesel properties with the commercial biodiesel. In this study, a multicriteria decision software PROMETHEE-GAIA was used to evaluate and validate the data that stretches toward the preferred solution.

In GAIA plane, the different biodiesel fuels are far away from each other, which means that biodiesels from different sources have different properties. The criteria that lie near to (± 45 degrees) are correlated while the criteria that lie in opposite direction (135–225 degrees) are anticorrelated. Also, those lying in the orthogonal direction have no or minimal influence [42]. An axis represents each criterion. The direction and length of the criteria indicate their impact on the decision vector. For example, FP in Fig. 14.9 has little effect on decision vector. The choice of decision vector represents the best fuel, and the furthest criteria toward the selection vector are the most ideal [11]. In Fig. 14.9, it is seen that PBD was the farthest position from the center in the decision vector plane, thus it gave the highest ranking. The positions of macadamia and moringa are closer in the GAIA plane, which indicates that their properties are closer to each other. Although the positions of JBD and BBD are far away from the decision vector, they are positioned closer to each other which also indicates that their properties are also similar.

Table 14.6 shows the PROMETHEE II complete ranking result (phi value). The phi value is the net flow score that could be either positive or negative depending on the angular distance from the decision vector and distance from the center [43]. It is evident from Fig. 14.9 that based on all of the measured data, the highest ranked biodiesel source is palm oil and the lowest ranked source is beauty leaf oil. Among the produced biodiesels, the best biodiesel feedstocks are palm, macadamia, and moringa oils. The phi values of macadamia, rapeseed, and moringa are closer (Fig. 14.9), which indicate that their criteria values are closer to each other. The result of this analysis shows the ability to compare produced biodiesel with commercial RBD.

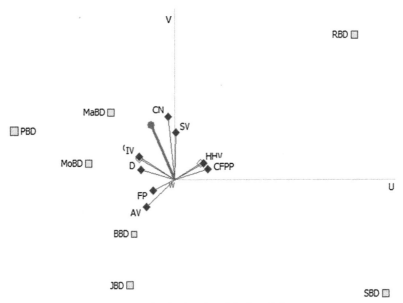

Figure 14.9 GAIA plot for biodiesel fuel and decision vector.

14.3.5 Study of Tribological Characteristics

Lubricity of an engine fuel is an imperative parameter to determine the engine life. Fuel with higher lubricity extends the engine life. It also reduces the energy consumption by reducing the friction between the moving parts. It has been reported that biodiesel exhibits better lubricity than diesel fuel, and due to this behavior, biodiesel can be used as an additive to improve the lubricity of conventional fossil fuel. According to the PROMETHEE-GAIA study in the last section, it was found that PBD is the best alternative fuel among these five biodiesels. In this section, the tribological behavior of only PBD has been studied and compared with diesel fuel.

Table 14.6 Corresponding Ranking and Phi Values of Biodiesel Fuels

Ranking	Biodiesel Samples	Phi Value
1	PBD	0.3667
2	MaBD	0.1667
3	RBD	0.1000
4	MoBD	0.0167
5	JBD	−0.1000
6	BBD	−0.1833
7	SBD	−0.3667

14.3.5.1 Friction Behavior

Frictions are highly dependent on load and temperature. Initially, the friction between metal contact surfaces is not stable which is known as a run period. After a few seconds, the condition becomes stable which is called steady-state condition [44]. In this study, coefficient of friction was calculated at both run time and steady-state conditions by changing load. Fig. 14.10 shows the coefficient of friction of both fuels at both (run time and steady-state) conditions and both (40 and 80 kg) loads. In the run period condition (Fig. 14.10A and B), diesel showed higher COF than PBD fuel. During the run period, maximum COF for PBD fuel was found to be 0.15 at 1.5 s and 0.49 at 2 s, whereas for diesel it was found to be 0.39 at 4 s and 0.52 at 3.8 s for 40 and 80 kg load conditions, respectively. The biodiesel showed better friction performance than diesel fuel due to the presence of ester molecules that are more efficient in protecting scuffing behavior [45]. In the steady-state condition, diesel fuel also shows higher COF than PBD for both 40 and 80 kg loads as presented in Fig. 14.10C and D. The COF for PBD was found to be 0.08 at 200 s for 40 kg load, and it remains almost same up to 209 s. In contrast, COF for PBD varied over the time for 80 kg load condition. In both (run-in and steady-state conditions), highest COF was found for diesel fuel in both loads.

14.3.5.2 Wear Scar Diameter and Flash Temperature Parameter

The effect of load on wear scar diameter of PBD and diesel fuel is shown in Fig. 14.11A. It is seen that the diameter is increased with the load for both fuels which could be attributed to the variation of contact surface pressure. In all load condition, PBD showed 15.5% and 54% lower WSD than diesel fuel at 40 and 80 kg load, respectively. This may be due to the presence of trace element (removal of the metallic soap film generated at high load) in biodiesel that helps to improve the lubricity of biodiesel fuel [46].

FTP is an important parameter that also indicates the lubricity of fuel samples. Higher FTP value of sample is the indication of better lubricity performance, and lower FTP value is the inverse indication of lubricity [47]. The FTP of diesel and PBD is shown in Fig. 14.11B. It is seen that FTP decreases with load, and it has the inverse character of WSD. PBD has higher FTP than diesel fuel in both loads. The maximum FTP was found for PBD at 40 kg load condition.

14.4 CONCLUSIONS

In this study, biodiesel was produced from five sources obtained locally or internationally to assess their relevant physical and chemical properties. Moreover, the properties were correlated with the composition of biodiesel, providing linear best-fit curves that will help researchers or designers in simulation work. Summary of the findings of this study are:

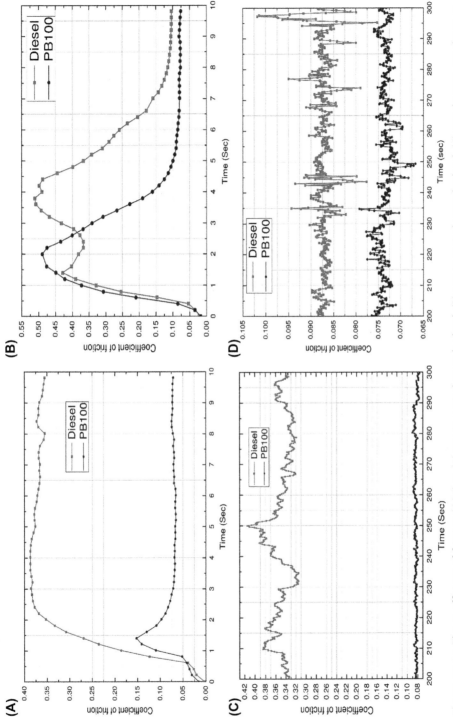

Figure 14.10 The coefficient of friction (COF) in (A) run-in-period at 40 kg load, (B) run-in-period at 40 kg load, (C) steady-state condition at 40 kg load, and (D) steady-state condition at 80 kg load.

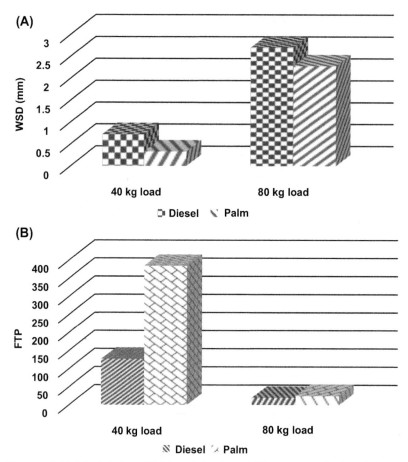

Figure 14.11 (A) Variation of WSD with loads and (B) variation of FTP with loads.

- The fuel properties of all five biodiesels are within the specified limit of ASTM D6751 and EN 14211 standards.
- A good agreement between the composition and fuel properties of biodiesels was observed. The DU was found to correlate linearly with the CN, IV, and OS. A high statistical correlation ($R^2 = 0.9985$) was also established between the long-chain saturated fatty acid and CFPP.
- The multicriteria decision analysis using PROMETHEE-GAIA software indicated that PBD could be a better alternative for diesel engine application compared much with other commercial biodiesel.
- The tribological study shows that PBD has better wear and friction performance than diesel fuel.

REFERENCES

[1] Azad AK, Rasul MG, Khan MMK, et al. Prospect of biofuels as an alternative transport fuel in Australia. Renewable and Sustainable Energy Reviews 2015;43:331—51.

[2] Mofijur M, Rasul MG, Hyde J, et al. Role of biofuel and their binary (diesel—biodiesel) and ternary (ethanol—biodiesel—diesel) blends on internal combustion engines emission reduction. Renewable and Sustainable Energy Reviews 2016;53:265—78.

[3] Giakoumis EG. A statistical investigation of biodiesel physical and chemical properties, and their correlation with the degree of unsaturation. Renewable Energy 2013;50:858—78.

[4] Mofijur M, Masjuki HH, Kalam MA, et al. Energy scenario and biofuel policies and targets in ASEAN countries. Renewable and Sustainable Energy Reviews 2015;46:51—61.

[5] Hassan NMS, Rasul MG, Harch CA. Modelling and experimental investigation of engine performance and emissions fuelled with biodiesel produced from Australian Beauty Leaf Tree. Fuel 2015;150:625—35.

[6] Rashed MM, Kalam MA, Masjuki HH, et al. Performance and emission characteristics of a diesel engine fueled with palm, jatropha, and moringa oil methyl ester. Industrial Crops and Products 2016;79:70—6.

[7] Atabani AE, Silitonga AS, Badruddin IA, et al. A comprehensive review on biodiesel as an alternative energy resource and its characteristics. Renewable and Sustainable Energy Reviews 2012;16:2070—93.

[8] Mofijur M, Masjuki HH, Kalam MA, et al. Palm oil methyl ester and its emulsions effect on lubricant performance and engine components wear. Energy Procedia 2012;14:1748—53.

[9] Ali OM, Mamat R, Abdullah NR, Abdullah AA. Analysis of blended fuel properties and engine performance with palm biodiesel—diesel blended fuel. Renewable Energy 2016;86:59—67.

[10] Silitonga AS, Masjuki HH, Mahlia TMI, et al. *Schleichera oleosa* L. oil as feedstock for biodiesel production. Fuel 2015;156:63—70.

[11] Jahirul MI, Brown RJ, Senadeera W, et al. Physiochemical assessment of beauty leaf (*Calophyllum inophyllum*) as a second-generation biodiesel feedstock. Energy Reports 2015;1:204—15.

[12] Kalam MA, Masjuki HH. Testing palm biodiesel and NPAA additives to control NO_x and CO while improving efficiency in diesel engines. Biomass and Bioenergy 2008;32:1116—22.

[13] Rizwanul Fattah IM, Masjuki HH, Kalam MA, et al. Experimental investigation of performance and regulated emissions of a diesel engine with *Calophyllum inophyllum* biodiesel blends accompanied by oxidation inhibitors. Energy Conversion and Management 2014;83:232—40.

[14] Jayed MH, Masjuki HH, Kalam MA, et al. Prospects of dedicated biodiesel engine vehicles in Malaysia and Indonesia. Renewable and Sustainable Energy Reviews 2011;15:220—35.

[15] Rao PV. Experimental investigations on the influence of properties of jatropha biodiesel on performance, combustion, and emission characteristics of a DI-CI engine. World Academy of Science, Engineering and Technology 2011;75:855—68.

[16] Mofijur M, Masjuki HH, Kalam MA, et al. Properties and use of *Moringa oleifera* biodiesel and diesel fuel blends in a multi-cylinder diesel engine. Energy Conversion and Management 2014;82:169—76.

[17] Devan PK, Mahalakshmi NV. Study of the performance, emission and combustion characteristics of a diesel engine using poon oil-based fuels. Fuel Processing Technology 2009;90:513—9.

[18] Mofijur M, Rasul M, Hassan N, Hyde J. Prospects of biodiesel production from macadamia oil as an alternative fuel for diesel engines. Energies 2016;9:403.

[19] Wang L-B, Yu H-Y, He X-H, Liu R-Y. Influence of fatty acid composition of woody biodiesel plants on the fuel properties. Journal of Fuel Chemistry and Technology 2012;40:397—404.

[20] Islam MA, Brown RJ, Brooks PR, et al. Investigation of the effects of the fatty acid profile on fuel properties using a multi-criteria decision analysis. Energy Conversion and Management 2015;98:340—7.

[21] Atabani AE, Mahlia TMI, Anjum Badruddin I, et al. Investigation of physical and chemical properties of potential edible and non-edible feedstocks for biodiesel production, a comparative analysis. Renewable and Sustainable Energy Reviews 2013;21:749—55.

[22] Zhang H, Zhou Q, Chang F, et al. Production and fuel properties of biodiesel from *Firmiana platanifolia* L.f. as a potential non-food oil source. Industrial Crops and Products 2015;76:768—71.

[23] Sanjid A, Masjuki HH, Kalam MA, et al. Characterization and prediction of blend properties and evaluation of engine performance and emission parameters of a CI engine operated with various biodiesel blends. RSC Advances 2015;5:13246–55.

[24] Jayed MH, Masjuki HH, Saidur R, et al. Environmental aspects and challenges of oilseed produced biodiesel in Southeast Asia. Renewable and Sustainable Energy Reviews 2009;13:2452–62.

[25] Hathurusingha S. Retraction notice to "Periodic variation in kernel oil content and fatty acid profiles of *Calophyllum inophyllum* L.: a medicinal plant in northern Australia". Industrial Crops and Products May 2011;33(3):775–8. 2013;46:268.

[26] Rahman MM, Hassan MH, Kalam MA, et al. Performance and emission analysis of *Jatropha curcas* and *Moringa oleifera* methyl ester fuel blends in a multi-cylinder diesel engine. Journal of Cleaner Production 2014;65:304–10.

[27] Atabani AE, Mofijur M, Masjuki HH, et al. Effect of *Croton megalocarpus*, *Calophyllum inophyllum*, *Moringa oleifera*, palm and coconut biodiesel–diesel blending on their physico-chemical properties. Industrial Crops and Products 2014;60:130–7.

[28] Arbab MI, Masjuki HH, Varman M, et al. Fuel properties, engine performance and emission characteristic of common biodiesels as a renewable and sustainable source of fuel. Renewable and Sustainable Energy Reviews 2013;22:133–47.

[29] Atadashi IM, Aroua MK, Abdul Aziz AR, Sulaiman NMN. Production of biodiesel using high free fatty acid feedstocks. Renewable and Sustainable Energy Reviews 2012;16:3275–85.

[30] Atadashi IM, Aroua MK, Aziz AA. High-quality biodiesel and its diesel engine application: a review. Renewable and Sustainable Energy Reviews 2010;14:1999–2008.

[31] Dwivedi G, Sharma MP. Prospects of biodiesel from Pongamia in India. Renewable and Sustainable Energy Reviews 2014;32:114–22.

[32] Jain S, Sharma MP. Prospects of biodiesel from Jatropha in India: a review. Renewable and Sustainable Energy Reviews 2010;14:763–71.

[33] Lim S, Teong LK. Recent trends, opportunities, and challenges of biodiesel in Malaysia: an overview. Renewable and Sustainable Energy Reviews 2010;14:938–54.

[34] Atabani AE, Silitonga AS, Ong HC, Mahlia TMI, Masjuki HH, Badruddina IA, et al. Non-edible vegetable oils: a critical evaluation of oil extraction, fatty acid compositions, biodiesel production, characteristics, engine performance and emissions production. Renewable and Sustainable Energy Reviews 2013;18:211–45.

[35] Wakil MA, Kalam MA, Masjuki HH, et al. Influence of biodiesel blending on physicochemical properties and importance of mathematical model for predicting the properties of biodiesel blend. Energy Conversion and Management 2015;94:51–67.

[36] Habibullah M, Masjuki HH, Kalam MA, et al. Potential of biodiesel as a renewable energy source in Bangladesh. Renewable and Sustainable Energy Reviews 2015;50:819–34.

[37] Rizwanul Fattah IM, Masjuki HH, Liaquat AM, et al. Impact of various biodiesel fuels obtained from edible and non-edible oils on engine exhaust gas and noise emissions. Renewable and Sustainable Energy Reviews 2013;18:552–67.

[38] Silitonga AS, Masjuki HH, Mahlia TMI, et al. Overview properties of biodiesel diesel blends from edible and non-edible feedstock. Renewable and Sustainable Energy Reviews 2013;22:346–60.

[39] Mofijur M, Masjuki HH, Kalam MA, et al. Prospects of biodiesel from Jatropha in Malaysia. Renewable and Sustainable Energy Reviews 2012;16:5007–20.

[40] Hoekman SK, Broch A, Robbins C, et al. Review of biodiesel composition, properties, and specifications. Renewable and Sustainable Energy Reviews 2012;16:143–69.

[41] Karmakar A, Karmakar S, Mukherjee S. Properties of various plants and animals feedstocks for biodiesel production. Bioresource Technology 2010;101:7201–10.

[42] Islam M, Magnusson M, Brown R, et al. Microalgal species selection for biodiesel production based on fuel properties derived from fatty acid profiles. Energies 2013;6:5676.

[43] Figueira J, Greco S, Ehrgott M. Multiple criteria decision analysis: state of the art surveys. Springer; 2005.

[44] Mosarof MH, Kalam MA, et al. Assessment of friction and wear characteristics of Calophyllum inophyllum and palm biodiesel. Industrial Crops and Products 2016;83:470–83.

[45] Habibullah M, Masjuki HH, Kalam MA, et al. Friction and wear characteristics of *Calophyllum inophyllum* biodiesel. Industrial Crops and Products 2015;76:188—97.

[46] Jayadas NH, Prabhakaran Nair K, A G. Tribological evaluation of coconut oil as an environment-friendly lubricant. Tribology International 2007;40:350—4.

[47] Habibullah M, Masjuki HH, Kalam MA, et al. Tribological characteristics of *Calophyllum inophyllum*—based TMP (Trimethylolpropane) ester as energy-saving and biodegradable lubricant. Tribology Transactions 2015;58:1002—11.

CHAPTER FIFTEEN

Biodiesel Production Through Chemical and Biochemical Transesterification: Trends, Technicalities, and Future Perspectives

M.W. Mumtaz[1,2], A. Adnan[2], H. Mukhtar[2], U. Rashid[3] and M. Danish[1]

[1]University of Gujrat, Gujrat, Pakistan
[2]Government College University Lahore, Lahore, Pakistan
[3]Universiti Putra Malaysia (UPM), Serdang, Selangor, Malaysia

15.1 BACKGROUND

The scientific community of the world is striving for appropriate alternative solution of current energy crises due to the depletion of fossil-based energy resources like natural gas, petroleum, and coal. Although these fossil-based fuels are satisfying the energy requirements of the world, the utilization of these nonrenewable fossil-based fuels are frightening the whole world with their environmental implications including global warming and ozone depletion [1—3]. Therefore sustainable and clean source of energy is the fundamental need of the time. With regard to energy discrepancies, compared with the developed world, the situation is worse in developing countries. Therefore to cope with these drastic energy crises and to trim down the dependence on nonrenewable fossil-based petroleum fuels, the use of alternative renewable energy resources may be considered as one of the imperative option. In this scenario, biofuels such as biodiesel has emerged as sustainable alternative fuel, which is renewable, technically compatible with conventional diesel fuel, nontoxic, biodegradable, and environment friendly [4—8].

15.2 BIODIESEL AS SUSTAINABLE FUEL

Biodiesel has gained much acceptance worldwide as alternative to the fossil-based conventional diesel because it is a renewable source of energy with favorable energy balance, biodegradable and ecofriendly nature, excellent fuel properties, and nontoxic profile [9—15]. Moreover, biodiesel reduces carbon emissions providing cleaner and sustainable environment with lower carbon monoxide (CO) and particulate matter (PM) emissions on combustion compared with diesel [16—20]. Due to clean emission profile, ease in usage along with many other advantages, biodiesel is well recognizing as alternative fuel. Therefore use of biodiesel offers great opportunity to escalate economic development along with energy security by minimizing dependence on the fossil-based nonrenewable fuel sources.

15.3 STRATEGIES TO MINIMIZE VISCOSITY OF VEGETABLE OIL

The research for an alternative to diesel fuel started with the use of vegetable oils directly without any modification in diesel engines. Although lack of sulfur, high ash point, and safe storage are considerable advantages of vegetable oils over petroleum diesel [21], however, their high viscosity, polyunsaturation character, low volatility, poor fuel atomization, incomplete combustion, insufficient mixture formation inside the combustion chamber, formation of carbon deposits, piston ring sticking, injector coking, long ignition delay, and reduced cold starting misfire are the main constraints for direct use of vegetable oils as an alternative to petroleum-based fossil fuels in diesel engine [19,22,23].

Studies revealed that, certain processes such as pyrolysis, microemulsification, and transesterification can be used to overcome the deficiencies associated with direct use of vegetable oil as a fuel in diesel engines.

15.3.1 Microemulsification and Pyrolysis

Vegetable oil/diesel blending as alternative biodiesel fuel has been recognized as one of the promising techniques effectively being used to reduce viscosity of vegetable oils. This technique demonstrates the feasibility of biodiesel fuel production through the vegetable oil extraction utilizing "diesel-based reverse micellar microemulsions" as extraction solvent [24]. Microemulsion of vegetable oil has been proved a means for lowering the viscosity of the vegetable oils but associated disadvantages are heavy carbon deposits, uneven injector needle sticking, and incomplete combustion [25].

On the other hand, pyrolysis has attracted the researchers with its effective capacity of producing liquid fuels. Pyrolysis is a process by which thermal decomposition of the biomass is carried out in the absence of oxygen. Enormous amount of research has been conducted for the production of biofuels via thermochemical conversion of biomass [26]. Products resulting from pyrolysis of vegetable oils usually exhibit advantages like low viscosity; acceptable levels of sulfur, water, and sediments; satisfactory copper corrosion values; and high cetane number; however, some disadvantages regarding ash contents, carbon residues, and pour points render them unacceptable [25].

15.3.2 Transesterification

Transesterification gained much acceptance in recent years for the conversion of vegetable oils into products with technically more compatible fuel properties. Transesterification is an imperative process for biodiesel production, as it can reduce the viscosity of the feedstock/vegetable oils to a level closer to the conventional fossil-based

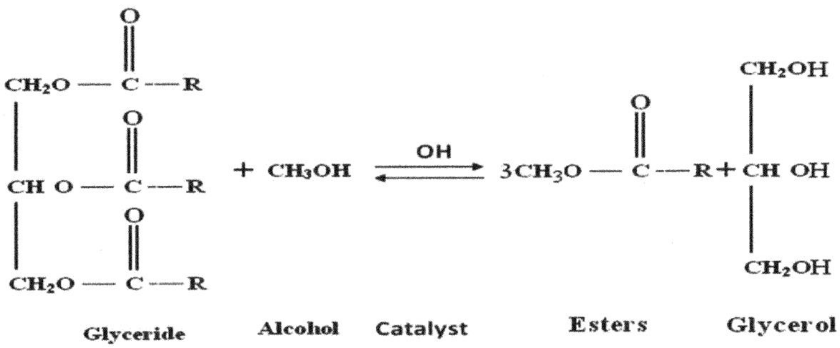

Scheme 15.1 Transesterification process.

$$\mathrm{TG + ROH} \underset{k^{-1}}{\overset{k^{1}}{\rightleftharpoons}} \mathrm{DG + FAME}$$

$$\mathrm{DG + ROH} \underset{k^{-2}}{\overset{k^{2}}{\rightleftharpoons}} \mathrm{MG + FAME}$$

$$\mathrm{MG + ROH} \underset{k^{-3}}{\overset{k^{3}}{\rightleftharpoons}} \mathrm{GL + FAME}$$

Scheme 15.2 Overall transesterification reaction.

diesel oil [20]. Transesterification represents an important group of organic reactions during which interchange of the alkoxy moiety results in the transformation of one ester into another as per Scheme 15.1. Transesterification is an equilibrium reaction describing the alcoholysis of carboxylic esters usually performed in the presence of conventional catalyst (e.g., NaOH and KOH) for valuable acceleration of the equilibrium adjustment to achieve higher yields of esters [7,19,27].

Chemically vegetable oils are triglyceride molecules with structural differences in their glycerol bound alkyl moiety. Transesterification of these triglyceride molecules with short-chain alcohols in the presence of suitable catalyst results in fatty acid methyl esters and glycerol (Scheme 15.1) [28]; a sequence of three consecutive reversible reactions [29] illustrates the overall transesterification process as described in Scheme 15.2.

15.4 FEEDSTOCK FOR BIODIESEL PRODUCTION

For the production of biodiesel, lots of feedstock is available viz., edible/nonedible/waste cooking oils, animal fats, and algal oils. Geographical distribution and price of the feedstock are the main factors to be considered during their selection for biodiesel production [30–32].

Plant-derived oils are considered to be the most efficient feedstock/raw materials for biodiesel production due to their inexhaustible, biodegradable, nontoxic, renewable, and ecofriendly nature [33]. On the bases of regional soil fertility and environmental conditions, the availability of these oils differs significantly in various countries [34].

Numerous edible oils have been investigated by various researchers as potential raw material/feedstock for the production of biodiesel. Bueso et al. [32] investigated palm oil along with jatropha oil as feedstock for the lipase-catalyzed production of biodiesel. Azócar et al. [35] reported the use of waste frying oil for the optimized production of biodiesel. Selmi and Thomas [12] investigated sunflower oil as feedstock for biodiesel production in solvent-free medium. Dossat et al. [36] used sunflower oil for biodiesel production, Antolin et al. [37] also explored sunflower oil as potential feedstock for biodiesel production, and Granados et al. [38] evaluated sunflower oil as feedstock for biodiesel production using activated calcium oxide as catalyst for transesterification.

Watanabe et al. [39] transesterified soybean oil to biodiesel; Xie and Huang [40] synthesized biodiesel using soybean oil as raw material. Ryan et al. [41] investigated methanolysis of soybean oil; Yan et al. [42] used soybean oil as feedstock for the production of biodiesel using tert-amyl alcohol as solvent. Palm oil also gained acceptance as potential feedstock for biodiesel as explored by researchers worldwide. Darnoko and Cheryan [43] used palm oil as raw material and transesterified into biodiesel in a batch reactor; Oliveira et al. [44] investigated transesterification of palm oil into biodiesel in the presence of n-hexane; Jitputti et al. [45] executed transesterification experiments on palm kernel oil and coconut oil and synthesized biodiesel.

Kose et al. [46] investigated cottonseed oil as feedstock for the production of cottonseed oil fatty acid methyl esters; Meneghetti et al. [47] synthesized castor and cottonseed oil ethyl esters using castor and cottonseed oils as feedstock; Hem et al. [48] used cottonseed oil and synthesized biodiesel; Mohammed et al. [49] reported castor oil as potential raw material for biodiesel. Kulkarni et al. [50] reported the feasibility of canola oil as feedstock for biodiesel production. In the past few years rapeseed has also been well explored by the researchers as suitable raw material for the production of biodiesel. Korus et al. [51] reported biodiesel production from rape seed oil; Li et al. [52] investigated the possibility of rapeseed oil as raw material for synthesizing biodiesel; Wang et al. [53] also reported rapeseed oil as potential feedstock transesterified into biodiesel using NaOH as catalyst.

Some nonedible oils have also been studied as fruitful raw materials for biodiesel production [2]. Bueso et al. [32] and Lee et al. [54] used jatropha oils for the production of biodiesel through enzymatic transesterification. Narwal et al. [55] reported the use of nonedible castor oil for the production of biodiesel. Steinke et al. [56] evaluated the feasibility of crambe oil and camelina oil as raw material for biodiesel production; Dorado et al. [57] and Gemma et al. [58] reported transesterification of *Brassica carinata* oil;

Ghadge and Raheman [59] explored mahua oil for biodiesel synthesis; Karmee and Chadha [60] also used crude *Pongamia pinnata* as potential raw material for biodiesel production; Ramadhas et al. [61] reported the use of rubber seed oil as raw material for biodiesel production; Meher et al. [62] transesterified *P. pinnata* oil for biodiesel production; Veljkovic et al. [63] carried out transesterification of tobacco seed oil; Sharma et al. [64] investigated karanja oil as potential feedstock for biodiesel production; Sahoo et al. [65] reported polanga seed oil as raw material for biodiesel production; Tiwari et al. [66] evaluated jatropha oil as potential feedstock for the production of biodiesel; Yang et al. [3] used *Camelina sativa* oil as feedstock for biodiesel production.

Obviously, use of vegetable oils as feedstock for the production of biodiesel is much expensive; therefore waste cooking oils and animal fats gained much attention because of their low cost. Zhang et al. [67] used waste cooking oil for biodiesel production, Cetinkaya and Karaosmanoglu [68] executed optimized transesterification of used cooking oil, Leung and Guo [69] reported biodiesel production from used frying oil, Canakci [70] investigated restaurant waste lipids as biodiesel raw material, and Wang et al. [71] executed two-step transesterification of waste cooking oil for biodiesel preparation. In other studies, Uzun et al. [72], Al-Hamamre et al. [73], and Atapour et al. [8] reported the use of used/waste frying oil as feedstock for biodiesel. Ma et al. [74] reported the use of beef tallow as potential feedstock for biodiesel production, whereas Tashtoush et al. [75] investigated transesterification of waste animal fat for biodiesel production. Besides the low cost of waste cooking oil, presence of water and free fatty acid (FFA) results in soap formation as a consequence of saponification reaction with alkali catalysts. Therefore additional steps are required to remove water, FFAs, and even soap resulting from side reactions. The attention is also focused by researchers toward the use of cheap and renewable feedstock, i.e., microalgal oil for biodiesel production. Heterotrophic fermentation and substrate feeding are the methods mostly used for the production of algal oil. Ginzburg [76] explored halophilic algae as raw material for biodiesel production, Miao and Wu [77] reported the use of microalgal oil as feedstock for the production of biodiesel, and Xu et al. [78] investigated microalga *Chlorella protothecoides* as potential feedstock for the production of high-quality biodiesel.

15.5 CHEMICAL TRANSESTERIFICATION REACTIONS

Chemically biodiesel can be synthesized either by acid-catalyzed or base-catalyzed transesterification of feedstock.

15.5.1 Acid-Catalyzed Transesterification Reactions

Acid-catalyzed transesterification reactions are mostly carried out by Bronsted acids, preferentially sulfuric, hydrochloric, and sulfonic acids [27,79]. The mechanism of acid esterification is described in Scheme 15.3.

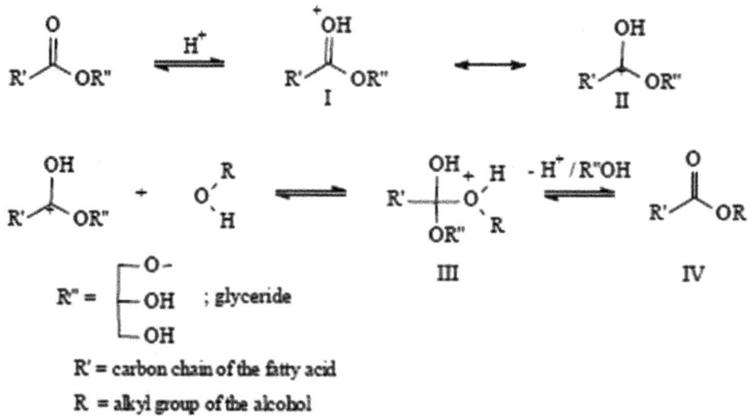

R'—C(=O)—OR" ⇌ H+ R'—C(+OH)—OR" (I) ⟷ R'—C(OH)—OR" (II)

R'—C(OH)—OR" + O(R)(H) ⇌ R'—C(OH+,OR",O—R,H) (III) ⇌ -H+/R"OH R'—C(=O)—OR (IV)

R" = [—O— / —OH / —OH] ; glyceride

R' = carbon chain of the fatty acid

R = alkyl group of the alcohol

Scheme 15.3 Mechanism of acid esterification for monoglyceride can be extended to diglycerides and triglycerides, respectively.

Scheme 15.3 illustrates that carbocation II results from carbonyl group protonation of the ester as a first step followed by nucleophilic attack of an alcohol producing a tetrahedral intermediate III, which after eliminating glycerol molecule results in a new ester IV and catalyst H$^+$ is regenerated [27]. Presence of water may decrease the alkyl ester yield due to the formation of the carboxylic acids by reaction with carbocation II; therefore competitive carboxylic acid formation can be avoided using water-free feedstock. Although high yields of the alkyl esters can be achieved using acid esterification, certain disadvantages, i.e., slow reaction speed, high temperature requirement, and difficult glycerol recovery render it unfit for use [27,79]. Mostly acid esterification is recommended as a prestep for biodiesel production via base-catalyzed transesterification where an acid value lesser than 2.0—4.0 mg KOH/g is required, which can be easily achieved by acid transesterification of the feedstock [61,64,80,81].

Chemical transesterification reactions catalyzed by acids are highly beneficial for the feedstock with higher FFA content. Acid-catalyzed transesterification results in better yield, although the reaction is time consuming, slow, and requires higher temperature conditions. Acid esterification is well accepted as a prestep for the base-catalyzed transesterification reactions to esterify FFAs if higher than 2%. The acid-catalyzed transesterification therefore helps in reducing the levels of FFAs to a level compatible with alkaline transesterification. Freedman et al. [11] described the sensitivity of base-catalyzed transesterification reactions toward the purity level of the raw material/feedstock used for the production of biodiesel. Michael et al. [82] revealed that, for acid esterification of feedstock with higher FFA, sulfuric acid is a more effective catalyst compared with the others including hydrochloric acid, formic acid, nitric acid, and acetic acid. Inadequacy of base-catalyzed transesterification reactions for the vegetable oils with high FFA content is also reported by Canakci and Gerpan [83]. Ramadhas et al. [61] reported the reduction of

FFA content of rubber seed oil to almost 2% with acid esterification. Veljkovic et al. [63] described H_2SO_4-catalyzed esterification of tobacco seed oil and the initial FFA level, i.e., about 35% was reduced to lesser than 2%. Reduction in the FFA content from 2.535% to 0.95% is also reported by Sharma and Singh [84] using H_2SO_4-catalyzed esterification of karanja oil. Sahoo et al. [65] performed acid esterification of polanga oil using H_2SO_4 as a prestep before performing alkaline transesterification, and successfully reduced the FFA level to about 2.0%. Reduction in FFA from 14% to <01% is also reported by Tiwari et al. [66] using acid-catalyzed esterification, whereas Ghadge and Raheman [59] reported the reduction of FFA from 19% to <01% of mahua oil by using two step pretreatment process. In another study, Wang et al. [71] executed acid esterification of waste cooking oil and reported FFA reduction from 37.96% to 1.055%. In another study, Banani et al. [85] performed acid-catalyzed esterification using H_2SO_4 of waste frying oil and reported reduction in FFA under 2%.

The cited literature clearly revealed that acid esterification is an imperative approach to achieving FFA levels in feedstocks feasible for alkaline transesterification.

15.5.2 Base-Catalyzed Transesterification Reactions

Alkaline-catalyzed transesterification reactions are much faster than acid-catalyzed esterification reactions and have gained much attention [19,25,72,73]. Substantially anhydrous feedstock with the least FFA content gives the best results regarding ester yield using base-catalyzed transesterification. Because of the less corrosive nature of alkaline catalysts compared with acid catalysts, at industrial scale alkaline-catalyzed transesterification is usually preferred. The most commonly employed alkaline catalysts are sodium and potassium hydroxides and alkoxides. These catalysts are well accepted for industrial-scale biodiesel production, because these are cheap and easy to transport and store. Comparatively, sodium and potassium methoxides are preferably being used to catalyze continuous flow processes for the production of biodiesel [19,25,72].

Debora et al. [86] reported the transesterification of soybean oil using NaOH as alkaline catalyst. The effect of NaOH levels (0.5—1.5 wt/wt%) on percentage yield of biodiesel was investigated. An inverse relationship was depicted between catalyst concentration and the percentage biodiesel yield. In another study, biodiesel production from fish oil via alkaline-catalyzed transesterification was carried out by Roberto et al. [87] and 98% fish oil biodiesel yield was achieved using 0.8% C_2H_5ONa. Veljkovic et al. [63] reported 91% biodiesel yield from tobacco seed oil via base-catalyzed transesterification reaction catalyzed by 1.0% KOH. Leung and Guo [69] employed NaOH, KOH, and CH_3ONa to catalyze transesterification reactions using used frying oil as feedstock. They reported that the highest biodiesel yield was achieved by using CH_3ONa as a catalyst; however, lesser concentration of NaOH was utilized compared with CH_3ONa and KOH under similar reaction conditions [69]. In another study, Li et al. [88] performed transesterification of soybean oil using KOH as alkaline catalyst and reported 96% biodiesel yield, when transesterification reactions were carried out at reaction temperature (45°C), methanol to oil ratio (4.5:1), cosolvent dichloromethane (4.0%), reaction time (2.0 h), and (1.0 wt.%) potassium hydroxide. Atapour et al. [8] achieved 92.05% biodiesel yield from used frying oil through NaOH-catalyzed transesterification, Yang et al. [3] reported the optimized biodiesel production by using response surface methodology (RSM). The authors achieved optimal yield of fatty acid methyl ester (FAME) (98.9%) using reaction conditions viz., reaction temperature (38.7°C), molar ratio of methanol/oil (7.7), reaction time (40 min), and catalyst concentration (1.5 wt.%). In one study, RSM was employed to achieve optimal safflower oil methyl ester yield (94.69%) by performing NaOH-catalyzed transesterification of safflower oil [20].

Ellis et al. [89] and Srivastava and Verma [90] investigated the catalytical behavior of potassium methoxide ($KOCH_3$) and $NaOCH_3$ toward transesterification for biodiesel production, respectively. Of KOH and NaOH, the former is proved to be a better catalyst than NaOH as potassium soaps are much softer than sodium soaps causing the least blockage of the separatory funnel as described by Sharma and Singh [64] and Tiwari et al. [66]. Ramadhas et al. [61] reported maximum conversion of triglycerides to biodiesel via alkaline transesterification using 0.5% NaOH, whereas beyond this level of NaOH, emulsion and gel formation was observed. Sahoo et al. [65] reported maximum biodiesel yield from polanga seed oil using alkaline transesterification catalyzed by 1.5 KOH after acid esterification. Meher et al. [62] investigated KOH-catalyzed transesterification of karanja oil and reported maximum biodiesel yield using 1.0% KOH.

The only disadvantage associated with base-catalyzed transesterification is additional purification requirements for biodiesel and glycerol for the removal of the basic catalyst. To avoid this drawback researchers have also investigated heterogeneous catalysts for their potential to catalyze transesterification [91,92]. Simple filtration can separate heterogeneous catalysts from the end products and thus can be reused. Some important heterogeneous catalysts investigated by the scientific community to catalyze transesterification reactions include Hb-Zeolite, ZnO and montmorillonite K-10, TiO_2/ZrO_2, Al_2O_3/ZrO_2, zeolites, alkaline earth oxides, hydrotalcites, $CaTiO_3$, KF/Al_2O_3 $CaMnO_3$, metal-loaded MgAl, KF/Ca—Mg—Al hydrotalcite, ion-exchange resins, dolomites, vanadyl phosphate, sodium aluminate, double-layered hydroxide, $SnCl_2$, CaO, and MgO [92].

15.5.3 Mechanism for Base-Catalyzed Transesterification Reactions

Base-catalyzed transesterification of vegetable oil starts with the reaction of alcohol with alkaline catalyst resulting in the generation of alkoxide along with protonated catalyst (Eq. 15.1), a tetrahedral intermediate is then formed as a result of nucleophilic attack of alkoxide on carbonyl moiety of triglyceride (Eq. 15.2) followed by alkyl ester production and anion generation corresponding to diglyceride (Eq. 15.3), and

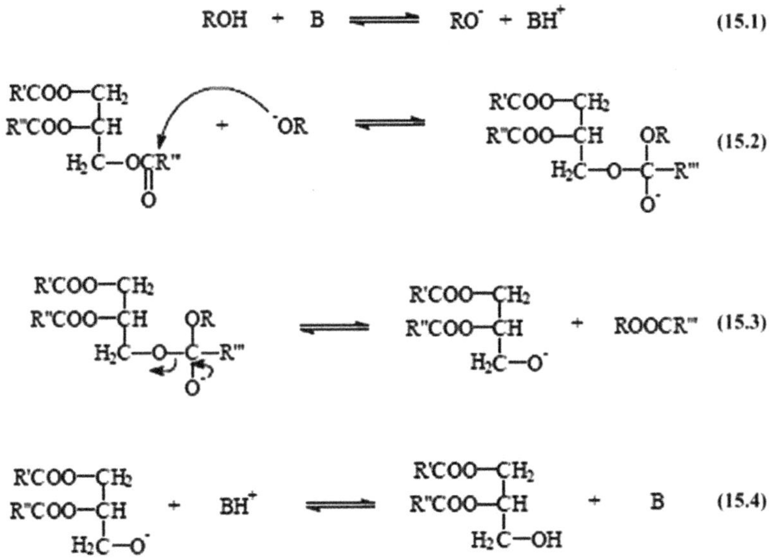

$$ROH + B \rightleftharpoons RO^- + BH^+ \qquad (15.1)$$

$$(15.2)$$

$$(15.3)$$

$$(15.4)$$

Scheme 15.4 Base-catalyzed transesterification.

deprotonation of the catalyst regenerates the active species (Eq. 15.4) ready for staring a new catalytic cycle [19,93]. The mechanism of base-catalyzed transesterification is shown in Scheme 15.4.

15.6 BIOCHEMICAL/ENZYMATIC TRANSESTERIFICATION REACTIONS

Compared with the use of acids or alkalis to catalyze transesterification reactions for biodiesel production, enzymes with significant advantages are attracting researchers. The advantages associated with the use of enzymes include enzyme specificity, reuse ability, mild reaction conditions requirement, efficiency improvement by genetic engineering, whole-cell immobilization, capacity to accept multiple substrates, being natural, and their thermal stability to catalyze green reactions [25,94–98]. Whereas, enzyme based catalysts reaction system is time consuming as compared to conventional catalyst reaction process.

Lipases extracted from different microbial strains have been utilized as biocatalysts for the production of biodiesel by researchers. Lipase-catalyzed biodiesel production from sunflower oil has been reported by Mittlebach [99]. Bueso et al. [32] investigated palm and jatropha oils as potential feedstock for lipase-catalyzed production of biodiesel. Narwal et al. [55] investigated the *Bacillus aerius* lipase as a potential biocatalyst for the production of castor oil–based biodiesel. Lee et al. [54] used jatropha oils for the production of biodiesel through enzymatic transesterification. Mangas-Sánchez and Adlercreutz [100] presented a new and simple but efficient method "aqueous/organic two-phase lipase-catalyzed system" for biodiesel production and reported the successful conversion of triacylglycerol oils (with high FFA level) into ethyl esters using *Thermomyces lanuginosus* lipase. In another study, Cesarini et al. [101] explored a novel lipase for the production of FAMEs from crude soybean oil in water-containing systems. Azócar et al. [35] described lipase-catalyzed process for biodiesel production using waste frying oil in an anhydrous system with enzyme reutilization. Ciudad et al. [102] reported the use of *Rhizopus oryzae* (the whole-cell catalyst) for biodiesel production using a three-phase bioreactor system.

The lipases from *Pseudomonas fluorescens, Candida* sp., and *Mucor miehei* were employed and of these lipase derived from *P. fluorescens* was reported to have superior characteristics when compared with the lipases extracted from *M. miehei* and *Candida* sp. Transesterification was performed both in the presence of a solvent and in a solvent-free system. Iso et al. [103] reported the use of lipase from *P. fluorescens* to catalyze transesterification reactions for biodiesel production. Soumanou and Bornscheuer [94] described the use of lipase from *P. fluorescens* as well as from *Rhizomucor miehei* to catalyze transesterification reactions. Matsumoto et al. [104] reported the application of lipase from *R. oryzae* for biodiesel synthesis. Lipase based on *Candida rugosa* is reported by Chen and Wu [105] as an important enzyme catalyzing transesterification reactions, whereas Xu et al. [106] reported the use of lipase from *T. lanuginosus* for the same purpose. Lipase from *Candida antarctica* commercially available as NOVOZYME-435 is the most widely used enzyme for the synthesis of biodiesel as described by Lai et al. [97]. The only disadvantage associated with the use of lipase is the inactivation of enzyme by methanol during transesterification reactions. This problem can be removed by adding methanol stepwise in the reaction mixture as Shimada et al. [107] reported 90% waste cooking oil—based biodiesel using this protocol of stepwise methanol addition.

High stability and repeated use of immobilized lipases were revealed to be the superior characteristics compared with free lipase for biodiesel production. Fiber cloth, porous kaolinite, macroporous resins, hydrotalcite, and silica gel are among the most widely used immobilization carriers as reported by Refs. [103,108—110]. Different immobilization methods have been reported by different researchers, but of these the most appropriate method was found to be lipase entrapment on sol—gel matrices with hydrophobic nature and lipase adsorption on hydrophobic carriers like polypropylene, whereas lyophilized powders and immobilized preparations are the recognized commercially available lipases.

In the recent years researchers have investigated different reaction systems for conducting the best lipase-catalyzed transesterification reactions. Solvent-free system, organic solvent medium with hydrophobic nature, hydrophilic reaction medium, and ionic liquid medium are among the well considered reaction systems for biodiesel production [108].

15.7 RESPONSE SURFACE METHODOLOGY AS IMPERATIVE TOOL FOR BIODIESEL OPTIMIZATION

RSM is gaining much importance these days regarding the optimization of various processes including biodiesel production. This methodology normally utilizes statistical and mathematical techniques for the said purpose [19,111,112]. RSM is recognized as an imperative technique involved in the optimization of specific response influenced by variables. Hence, modeling of the experimental response of interest to get the optimal results is the main objective associated with the emphatic use of RSM, although its usage has now been extended up to the modeling and optimization of numerical experiments as well [19,113]. So, the RSM is a highly valuable tool for the development, optimization, and improvement of the response variables and it can be mathematically expressed as:

$$y = f(x_1, x_2) + e$$

where "y" is the response of interest and depends on "x_1 and x_2" independent variables and "e" is experimental error, which shows the influence of countable error if any on the response of interest [111].

In the current scenario, both the first- and second-order response surface models are in use. The first-order model can be employed for approximating response-based function, based upon response usually represented by linear function of variables having independent nature. The simplest first-order model can be expressed as:

$$y = \beta_0 + \beta_1 x_1 + \beta_1 x_2 + e$$

where β_o, β_1, and β_2 represent the regression coefficient.

Comparatively, the second-order model usually represents the approximating function with two variables. In addition to the first-order interaction model terms, it also considers all quadratic and cross-product terms [19,111]. A second-order model can mathematically be described as:

$$y = \beta_0 + \beta_1 x_1 + \beta_1 x_2 + \beta_{11} x_1^2 + \beta_{22} x_2^2 + \beta_{12} x_1 x_2 + e$$

Central Composite Design (CCD) is the most acceptable design that attracted the scientific community worldwide in the recent years for process optimization [19,113]. CCD is characterized either by a full factorial design with two levels (2^k) or with fractional factorial designs (2^{k-f}) having several design points. This design comprises factorial points n_f, axial points n_a, and central points n_c. All possible regression parameters may be measured with CCD on the basis of collective results of design points. Moreover, CCD is considered to be rotatable provided the variance depends on response (of interest) from design center rather than direction. In general, the purpose of RSM is topographical understanding of the response surfaces to find optimal response [19,111].

RSM has been employed by numerous researchers to optimize the reaction parameters involved in the biodiesel production process, e.g., methanolysis of *B. carinata* oil was optimized using RSM for biodiesel production by Gemma et al. [58]. Hem et al. [48] reported the optimized biodiesel production from cottonseed oil; CCD comprising eight factorial points, six axial points, and six replicated center was employed for the optimization of reaction temperature, catalyst concentration, and ethanol to oil molar ratio. The authors reported 98% yield of biodiesel under the reaction conditions, i.e., catalyst concentration (1.07%), ethanol to oil molar ratio (20:1), and reaction temperature (25°C). Kuan et al. [114] applied RSM in combination with Box-Behnken design for the optimized biodiesel production from soybean oil or waste cooking oil. They optimized the levels of different reaction parameters and found that the optimal biodiesel yield was procured when the reaction was performed at 43.6°C using substrate molar ratio (4.3%). Sangaletti et al. [115] also reported the use of RSM to optimize various reaction parameters involved in the production of biodiesel. The optimal levels of reaction variables were depicted to be temperature 40°C, 1:4.5 oil to ethanol molar ratio, and 9.5% catalyst concentration for 24 h resulting in 85.4% biodiesel yield [115]. In another study, Mumtaz et al. [19] described the use of RSM and reported that maximum sunflower oil–based biodiesel yield was obtained when reactions were conducted using (0.75%) KOH/NaOH and (6:1) methanol to oil molar ratio at reaction temperature (45°C) for 60 min.

Tiwari et al. [66] optimized various reaction parameters for pretreatment and transesterification of jatropha oil using RSM. In another study, Ghadge and Raheman [59] applied the same technique for optimized reduction of FFAs in *Madhuca indica* seed oil. Domingos et al. [116] optimized the reaction variables for ethanolysis of *Raphanus sativus* seed oil using RSM. RSM has also been applied by Bouaid et al. [117] and Chen et al. [118] for the optimization of transesterification conditions to produce biodiesel using jojoba oil and acid oil as feedstocks. Vicente et al. used factorial design along with RSM for optimization of methanolysis of sunflower [119]. Wu et al. [120] investigated the optimization of reaction parameters (lipase concentration, time, temperature, and molar ratio of various reactants) of lipase-catalyzed transesterification of grease and reported 85.4% biodiesel yield at optimized reaction conditions, i.e., 13.7 wt% lipase-PS-30, 38.4°C, 2.47 h, and 1:6.6 M ratio. Rodrigues et al. [121] reported the optimization of soybean oil–based biodiesel production and RSM was employed for the evaluation of impact of various reaction variables including catalyst concentration and alcohol:oil molar ratio on the transformation of soybean oil to biodiesel. The investigator procured 91.8% biodiesel yield using alcohol to oil molar ratio of 10.2 when the reaction was performed for 30 min.

15.8 ANALYTICAL METHODS FOR BIODIESEL CHARACTERIZATION

Different analytical methods have been reported by the researchers for monitoring and compositional characterization of biodiesel. Among these, the most emphasized are chromatographic and spectroscopic methods of analysis. Thin layer chromatography equipped with flame ionization detector (TLC-FID), high-performance liquid chromatography (HPLC), gas chromatography using flame ionization and mass spectrometric detections (GC-FID and GC−MS), and gel permeation chromatography (GPC) are among the most extensively reported chromatographic methods, whereas among the spectroscopic methods Fourier transform infrared spectroscopy (FTIR) and nuclear magnetic resonance spectroscopy (NMR) are potentially utilized methods.

TLC/FID is the first reported chromatographic technique employed for transesterification reactions, whereas Trathnigg and Mittelbach reported HPLC as an analytical method using isocratic solvent system for the characterization of methyl esters along with mono-, di-, and triglycerides [122]. Foglia and Jones [123] characterized reaction mixtures resulting from enzyme (lipase)-catalyzed transesterification using HPLC equipped with evaporative light scattering detector (ELSD). Holcapek et al. [124] reported an extensive study related to the characterization of biodiesel using reversed phase HPLC with different detection systems like UV detection, ELSD, and atmospheric pressure chemical ionization mass spectrometry. Dimmig et al. [125] employed [13]C−NMR for the characterization of rapeseed oil−based biodiesel. Sunflower oil biodiesel was characterized using GPC and GC by Madras et al. [126]. Sangaletti et al. [115] analyzed fatty acid ethyl esters using GC−MS. Mumtaz et al. [19] analyzed fatty acid methyl esters of sunflower oil−based biodiesel with GC−MS and FTIR. Li et al. [88] investigated the biodiesel resulting from the transesterification of soybean oil with GC-FID and FTIR.

Karmee and Chadha [60] used GC-FID for characterizing biodiesel derived from *P. pinnata* oil. In another study, Foroutan et al. [127] employed GC-FID for the investigation of fatty acid ester composition of sunflower oil and palm kernel oil. Veljkovic et al. [63] and Sahoo et al. [65] reported the use of HPLC for characterization of biodiesel resulting from transesterification of tobacco seed oil and polanga seed oil, respectively. Meher et al. [62] used HPLC, GC, and ^{1}H- NMR for the characterization of polanga seed oil−based biodiesel synthesized by base-catalyzed transesterification. Leung and Guo [69] employed TLC and GC-FID for characterization of biodiesel. Xu et al. [78] employed GC−MS as an analytical tool for characterization of biodiesel synthesized by transesterification of microalgal oil. Soybean oil biodiesel was characterized by Furuta et al. [128] using GPC, GC−MS, and X-ray diffractometer. Biodiesel resulting from transesterification of different vegetable oils including jatropha, pongamia, sunflower, soybean, and palm oil was characterized by Sarin et al. [129]. Wang et al. [53] reported the use of

GC-FID for waste cooking oil–based biodiesel characterization. Granados et al. [38] also reported the use of GC for biodiesel characterization. Correa and Arbilla [130] employed size exclusion chromatography and FTIR for the analysis of biodiesel.

15.9 FUEL PROPERTIES AND EXHAUST EMISSIONS OF BIODIESEL

Commercial applications of biodiesel depend much on the fuel quality of biodiesel as determined by standard protocols. For the assessment of the fuel quality different countries throughout the world have defined guideline values for different fuel characteristics. Of these, contributions of Austria, France, Italy, the Czech Republic, Germany, and United States are acknowledgeable. These guideline values not only serve as highly valuable selection criteria for customers so that they can purchase high-quality fuel but also provide standard tools for safety risk and environmental pollution assessment [131].

Specified measurement procedures and uniform standards were formulated by the European Committee for standardization, which were mandated in 1997 (for biodiesel consisting of FAMEs). The resulting standards for fuel quality, i.e., EN 14214 then comes into action in 2004. Now EN 14214 is validated all over the European Union member states instead of their own national fuel specifications. The United States developed its own standards for fuel quality, i.e., ASTM D 6751. The fuel characteristics covered by these fuel quality standards include viscosity, density, sulfur content, carbon residue, flash point, cetane number, free glycerol, acid number, methanol, and ester and water content [122,131].

Diesel engine exhaust with higher levels of CO, NO_x, and PM is described as carcinogenic to human beings by International Agency on Cancer Research [16]. Health risks like lung cancer (50%) associated with long-term exposure of diesel engine exhaust have already been established [17]. Researchers and environmentalists are therefore focusing on the development of safer and environment-friendly alternative fuels. Although different countries are trying to develop treatment strategies for the simultaneous reduction of PM, CO, and NO_x emissions, still the use of ecofriendly and renewable biodiesel instead of conventional fossil-based diesel fuel is being recommended worldwide to give best results regarding cleaner environment. Biodiesel has several merits over conventional diesel fuel and can also be used in diesel engine without major modifications [132,133].

It has been established experimentally that the use of biodiesel instead of conventional diesel or in blended form results in reduced levels of CO and PM emissions but higher nitrogen oxide (NO_x) emissions [133–135]. Graboski and McCormick observed a 12% increase in NO_x emissions on combustion of soy biodiesel (100%) when compared with that of petroleum diesel [135]. However, 20% soy biodiesel blend with diesel resulted in only 2–4% increase in NO_x emissions when compared with diesel. The higher NO_x emissions even at a small level have a negative effect on biodiesel usage [135]. In his study, Mustafa et al. [136] revealed that the NO_x emissions can be reduced to acceptable levels by using high-oleic soy biodiesel instead of normal soy biodiesel. Mumtaz et al. [133] reported decreased levels of PM and CO in exhaust emissions of biodiesel blends (POB-5 to POB-100) compared with conventional diesel. The percentage change in PM and CO emission levels ranged from −6.2 to −58.4% and −2.1 to −68.7%, respectively, compared with conventional diesel. The carcinogenic effects of diesel exhaust are mainly attributed to the inhalation of the soot particles, as many carcinogenic agents such as polycyclic aromatic hydrocarbons are known to reside or be absorbed as organic phase over the PM surface. PM can readily be inhaled due to their median dynamic diameter (0.1−0.3 μm) leading to their deposition in the alveolar region of lungs [18,137,138].

15.10 FUTURE PERSPECTIVES OF BIODIESEL PRODUCTION

Biodiesel has been proved to be a potential candidate for future energy requirements, hence sustaining the energy security. Its usage as alternative fuel not only will minimize the fossil fuel dependence but also will provide the means of economic benefits and employment. Its ecofriendly nature with the least greenhouse gas emissions is still another advantage pertaining to its sustainability.

Although biodiesel production creates significant impact to cope with the future energy discrepancies, focus should be on its large-scale production and reducing costs. Therefore future studies should be more focused on making the biodiesel production process more cost-effective and one of the possible options may be value-added use of its by-product, i.e., glycerol leading to improved environmental and economic viability.

15.11 CONCLUSION AND RECOMMENDATIONS

From the earlier discussion we conclude that biodiesel is an alternative energy source for recent worldwide energy crises. Transesterification has advantages over pyrolysis and microemulsion. Biodiesel has variable synthetic routes, usually synthesized by chemical or biochemical transesterification of vegetable oils/animal fat. Base-catalyzed transesterification reactions give higher yield of biodiesel than acid catalysts. However, enzymatic transesterification is much better than chemical transesterification because of enzyme specificity, efficiency, reusability, and thermal stability. If we consider environmental and health issues related to exhaust of petrodiesel and biodiesel, then biodiesel is more environment and health friendly. Although a lot of research has been done on biodiesel synthesis, the cost of biodiesel is still a question because it is somewhat higher than petrodiesel. Therefore future studies should be more focused on making the biodiesel production process more cost-effective either by exploring the novel and cheap feedstocks or by shortening the reaction time using different solvo-thermal techniques or by synthesizing the reusable and stable heterogeneous catalysts.

REFERENCES

[1] Tickell J. In: From the fryer to the fuel tank: the complete guide to using vegetable oil as an alternative fuel. 3rd ed. Tickell Energy Consultants; 2003.

[2] Bankovic-Ilic IB, Stamenkovic OS, Veljkovic VB. Biodiesel production from non-edible plant oils. Renewable and Sustainable Energy Review 2012;16:3621—47.

[3] Yang J, Corscadden K, He QS, Caldwell C. The optimization of alkali-catalyzed biodiesel production from *Camelina sativa* oil using a response surface methodology. Journal of Bioprocess and Biotechnology 2015;5(7):1—8.

[4] Ma F, Clements LD, Hanna MA. The effect of mixing on transesterification of beef tallow. Bioresource Technology 1999;69:289—93.

[5] Knothe G, Steidley KV. Kinematic viscosity of biodiesel fuel components and related compounds. Influence of compound structure and comparison to petrodiesel fuel components. Fuel 2005;84:1059—65.

[6] Xie W, Peng H, Chen L. Transesterification of soybean oil catalyzed by potassium loaded on alumina as a solid base catalyst. Applied Catalysis A: General 2006;300:67—74.

[7] Rashid U, Anwar F. Production of biodiesel through base-catalyzed transesterification of safflower oil using an optimized protocol. Energy and Fuels 2008;22:1306—12.

[8] Atapour M, Kariminia H-R, Moslehabadi PM. Optimization of biodiesel production by alkali-catalyzed transesterification of used frying oil. Process Safety and Environmental Protection 2014;92(2):179—85.

[9] Bartholomov D. Vegetable oil fuel. Journal of the American Oil Chemists' Society 1981;58:286—8.

[10] Clark J, Wangner L, Schrock MD, Piennaar PG. Methyl and ethyl esters as renewable fuels for diesel engines. Journal of the American Oil Chemists' Society 1984;61:1632.

[11] Freedman B, Pryde EH, Mounts TL. Variables affecting the yields of fatty esters from transesterified vegetable oils. Journal of the American Oil Chemists' Society 1984;61:1638.

[12] Selmi B, Thomas D. Immobilized lipase-catalyzed ethanolysis of sunflower oil in a solvent-free medium. Journal of the American Oil Chemists' Society 1998;75:691.

[13] De BK, Bhattacharyya DK, Bandhu C. Enzymatic synthesis of fatty alcohol esters alcoholysis. Journal of the American Oil Chemists' Society 1999;76:451.

[14] Ma F, Hanna MA. Biodiesel production: a review. Bioresource Technology 1999;70:1—15.

[15] Kurki A, Hill A, Morris M. Biodiesel: the sustainability dimensions. ATTRA Publications; 2006. p. 1—12.

[16] International Agency for Research on Cancer. Evaluation of carcinogenic risks to humans: diesel and gasoline engine exhausts and some nitroarenes. IARC Monographs 1989;46:41—185.

[17] Mauderly JL. Toxicological and epidemiological evidence for health risks from inhaled diesel engine emissions. Environmental Health Perspectives 1994;102:165—71.

[18] Jürgen K, Jürgen B, Olaf S, Axel M, Gerhard K. Exhaust emissions and health effects of particulate matter from agricultural tractors operating on rapeseed oil methyl ester. Journal of the American Oil Chemists' Society 2002;79:7.

[19] Mumtaz MW, Adnan A, Anwar F, Mukhtar H, Raza MA, Ahmad F, et al. Response surface methodology: an emphatic tool for optimized biodiesel production using rice bran and sunflower oils. Energies 2012;5:3307—28.

[20] Math MC, Chandrashekhara KN. Optimization of alkali catalyzed transesterification of safflower oil for production of biodiesel. Journal of Engineering 2016;8928673:7.

[21] Giannelos PN, Sxizas S, Lois E, Zannikos F, Anastopoulos G. Physical, chemical and fuel related properties of tomato seed oil for evaluating its direct use in diesel engines. Industrial Crops and Products 2005;22:193—9.

[22] Knothe G, Dunn RO, Bagby MO. Biodiesel: the use of vegetable oils and their derivatives as alternative diesel fuels. Oil Chemical Research, National Center for Agricultural Utilization Research, Agricultural Research Service, U.S. Department of Agriculture; 1996.

[23] Agarwal AK. Biofuels (alcohols and biodiesel) applications as fuels for internal combustion engines. Progress in Energy and Combustion Science 2007;33:233—71.

[24] Nguyen T, Do L, Sabatini DA. Biodiesel production via peanut oil extraction using diesel-based reverse-micellar microemulsions. Fuel 2010;89:2285—91.

[25] Fukuda H, Kondo A, Noda H. Biodiesel fuel production by transesterification of oils. Journal of Biosciences and Bioengineering 2001;92:405−16.

[26] Jahirul MI, Rasul MG, Chowdhury AA, Ashwath N. Biofuels production through biomass pyrolysis—a technological review. Energies 2012;5:4952−5001.

[27] Ulf S, Ricardo S, Rogério MV. Transesterification of vegetable oils: a review. Journal of Brazilian Chemical Society 1998;9:199−210.

[28] Freedman B, Butterfield R, Pryde E. Transesterification kinetics of soybean oil. Journal of the American Oil Chemists' Society 1986;63:1375−80.

[29] Aksoy HA, Becerik I, Karaosmanoglu F, Yatmaz HC, Civelekoglu H. Utilization prospects of Turkish Raisin seed oil as an alternative engine fuel. Fuel 1990;69:600−3.

[30] Abdulla R, Chan ES, Ravindra P. Biodiesel production from *Jatropha curcas*: a critical review. Critical Reviews in Biotechnoogy 2011;31:53−64.

[31] Christopher LP, Kumar H, Zambare VP. Enzymatic biodiesel: challenges and opportunities. Applied Energy 2014;119:497−520.

[32] Bueso F, Moreno L, Cedeño M, Manzanarez K. Lipase-catalyzed biodiesel production and quality with *Jatropha curcas* oil: exploring its potential for Central America. Journal of Biological Engineering 2015;9:12.

[33] Casimir CA, Shu-Wei C, Guan-Chiun L, Jei-Fu S. Enzymatic approach to biodiesel production. Journal of Agriculture and Food Chemistry 2007;55:8995−9005.

[34] Pinto AC, Guarieiro LLN, Rezende MJC, Ribeiro NM, Torres EA, Lopes WA, et al. Biodiesel: an overview. Journal of the Brazilian Chemical Society 2005;16:1313−30.

[35] Azócar L, Navia R, Beroiz L, Jeison D, Ciudad G. Enzymatic biodiesel production kinetics using co-solvent and an anhydrous medium: a strategy to improve lipase performance in a semi-continuous reactor. New Biotechology 2014;31:422−9.

[36] Dossat V, Combes D, Marty A. Continuous enzymatic transesterification of high oleic sunflower oil in a packed bed reactor: influence of the glycerol production. Enzyme Microbiology and Technology 1999;25:194−200.

[37] Antolin G, Tinaut FV, Briceňo Y. Optimisation of biodiesel production by sunflower oil transesterification. Bioresource Technology 2002;83:111−4.

[38] Granados ML, Poves MDZ, Alonso DM, Mariscal R, Galisteo FC, Tost RM. Biodiesel from sunflower oil by using activated calcium oxide. Applied Catalysis B 2007;73:317−26.

[39] Watanabe Y, Shimada Y, Sugihara A. Conversion of degummed soybean oil to biodiesel fuel with immobilized *Candida Antarctica* lipase. Journal of Molecular Catalysis 2002;17:151−5.

[40] Xie W, Huang X. Synthesis of biodiesel from soybean oil using heterogeneous KF/ZnO catalyst. Catalysis Letter 2006;107:53−9.

[41] Ryan D, Samir KK, David GB. Kinetic parameters of a homogeneous transmethylation of soybean oil. Journal of the American Oil Chemists' Society 2008;85:271−6.

[42] Yan Z, Jing Q, Xin N, Li-Min Z, Bo J, Zhi-Yan H. Lipase-catalyzed transesterification of soybean oil for biodiesel production in tert-amyl alcohol. World Journal of Microbiology and Biotechnology 2009;25:41−6.

[43] Darnoko D, Cheryan M. Kinetics of palm oil transesterification in a batch reactor. Journal of the American Oil Chemists' Society 2000;77:1263−7.

[44] Oliveira D, Oliveira JD. Enzymatic alcoholysis of palm kernel oil in *n*-hexane and SC−CO_2. Journal of Supercritical Fluids 2001;19:141−8.

[45] Jitputti J, Kitiyanan B, Rangsunvigit P, Bunyakiat K, Attanatho L, Jenvanitpanjakul P. Transesterification of crude palm kernel oil and crude coconut oil by different solid catalysts. Chemical Engineering Journal 2006;116:61−6.

[46] Kose O, Tuter M, Aksoy HA. Immobilized *Candida antarctica* lipase catalyzed alcoholysis of cotton seed oil in a solvent-free medium. Bioresource Technology 2002;83:125−9.

[47] Meneghetti SMP, Meneghetti MR, Wolf CR, Silva EC, Lima GES, Coimbra MA, et al. Ethanolysis of castor and cottonseed oil: a systematic study using classical catalysts. Journal of the American Oil Chemists' Society 2006;83:819−22.

[48] Hem CJ, Joe T, Terry W. Optimization of cottonseed oil ethanolysis to produce biodiesel high in gossypol content. Journal of the American Oil Chemists' Society 2008;85:357−63.

[49] Mohammed H, Chakrab A, Rafiq A. Transesterification studies on castor oil as a first step towards its use in biodiesel production. Pakistan Journal of Botany 2008;40:1153—7.

[50] Kulkarni MG, Dalai AK, Bakhshi NN. Transesterification of canola oil in mixed methanol/ethanol system and use of esters as lubricity additive. Bioresource Technology 2007;98:2027—33.

[51] Korus RA, Hoffman DS, Bam N, Peterson CL, Drown DC. Transesterification process to manufacture ethyl ester of rape oil. First Biomass Conference of the Americas, Burlington 1993;2:815—22.

[52] Li LL, Du W, Liu DH. Lipase-catalyzed transesterification of rapeseed oils for biodiesel production with a novel organic solvent as the reaction medium. Journal of Molecular Catalysis B 2006;43:58—62.

[53] Wang L, He H, Xie Z, Yang J, Zhu S. Transesterification of the crude oil of rapeseed with NaOH in supercritical and subcritical methanol. Fuel Processing Technology 2007;88:477—81.

[54] Lee JH, Kim SB, Yoo HY, Suh YJ, Kang GB, Jang WI. Biodiesel production by enzymatic process using Jatropha oil and waste soybean oil. Biotechnology and Bioprocess Engineering 2013;8:703—8.

[55] Narwal SK, Saun NK, Dogra P, Chauhan G, Gupta R. Production and characterization of biodiesel using nonedible castor oil by immobilized lipase from *Bacillus aerius*. BioMed Research International 2015:1—6.

[56] Steinke G, Kirchhoff R, Mukherjee KD. Lipase-catalyzed alcoholysis of crambe oil and camelina oil for the preparation of long-chain esters. Journal of the American Oil Chemists' Society 2015;77:361—6.

[57] Dorado MP, Ballesteros E, López FJ. Optimization of alkalicatalyzed transesterification of *Brassica carinata* oil for biodiesel production. Energy and Fuels 2004;18:77—83.

[58] Gemma V, Mercedes M, José A. Optimization of *Brassica carinata* oil methanolysis for biodiesel production. Journal of the American Oil Chemists' Society 2005;82:899—904.

[59] Ghadge SV, Raheman H. Biodiesel production from mahua (*Madhuca indica*) oil having high free fatty acids. Biomass and Bioenergy 2005;28:601—5.

[60] Karmee SK, Chadha A. Preparation of biodiesel from crude oil of *Pongamia pinnata*. Bioresource Technology 2005;96:1425—9.

[61] Ramadhas AS, Jayaraj S, Muraleedharan C. Biodiesel production from high FFA rubber seed oil. Fuel 2005;84:335—40.

[62] Meher LC, Dharmagadda VSS, Naik SN. Optimization of alkalicatalyzed transesterification of *Pongamia pinnata* oil for production of biodiesel. Bioresource Technology 2006;97:1392—7.

[63] Veljkovic VB, Lakicevic SH, Stamenkovic OS, Todorovic ZB, Lazic ML. Biodiesel production from tobacco (*Nicotiana tabacum* L.) seed oil with a high content of free fatty acids. Fuel 2006;85:2671—5.

[64] Sharma YC, Singh B. Development of biodiesel from karanja, a tree found in rural India. Fuel 2007;87:1740—2.

[65] Sahoo PK, Das LM, Babu MKG, Naik SN. Biodiesel development from high acid value polanga seed oil and performance evaluation in a CI engine. Fuel 2007;86:448—54.

[66] Tiwari AK, Kumar A, Raheman H. Biodiesel production from jatropha oil (*Jatropha curcas*) with high free fatty acids: an optimized process. Biomass and Bioenergy 2007;31:569—75.

[67] Zhang Y, Dube MA, McLean DD. Biodiesel production from waste cooking oil: 1. Process design and technological assessment. Bioresource Technology 2003;89:1—16.

[68] Cetinkaya M, Karaosmanoglu F. Optimization of base-catalyzed transesterification reaction of used cooking oil. Energy Fuel 2004;8(6):1888—95.

[69] Leung DYC, Guo Y. Transesterification of neat and used frying oil: optimization for biodiesel production. Fuel Processing and Technology 2006;87:883—90.

[70] Canakci M. The potential of restaurant waste lipids as biodiesel feedstocks. Bioresource Technology 2007;98:183—90.

[71] Wang Y, Ou S, Liu P, Zhang Z. Preparation of biodiesel from waste cooking oil via two-step catalyzed process. Energy Conversion & Management 2007;48:184—8.

[72] Uzun BB, Kılıç M, Özbay N, Pütün AE, Pütün E. Biodiesel production from waste frying oils: optimization of reaction parameters and determination of fuel properties. Energy 2012;44:347—51.

[73] Al-Hamamre Z, Yamin J. Parametric study of the alkali catalyzed transesterification of waste frying oil for Biodiesel production. Energy Conversion & Management 2014;79:246—54.

[74] Ma F, Clements LD, Hanna MA. The effects of catalyst, free fatty acids, and water on transesterification of beef tallow. Transactions of the American Society of Agricultural Engineers 1998;41:1261—4.

[75] Tashtoush GM, Al-Widyan MI, Al-Shyoukh AO. Experimental study on evaluation and optimization of conversion of waste animal fat into biodiesel. Energy Conversation & Management 2004;45:2697—711.

[76] Ginzburg BZ. Liquid fuel (oil) from halophilic algae: a renewable source of non-polluting energy. Renewable Energy 1993;3:249—52.

[77] Miao X, Wu Q. Biodiesel production from heterotrophic microalgal oil. Bioresource Technology 2006;97:841—6.

[78] Xu H, Miao X, Wu Q. High quality biodiesel production from a microalga *Chlorella protothecoides* by heterotrophic growth in fermenters. Journal of Biotechnology 2006;126:499—500.

[79] Pathak S. Acid catalyzed transesterification. Journal of Chemical & Pharmaceutical Research 2015;7(3):1780—6.

[80] Canakci M, Gerpen JV. Biodiesel production via acid catalysis. Transactions of the American Society of Agricultural Engineers 1999;42:1203—10.

[81] Diserio M, Tesser R, Ferrara A, Santacesaria E. Heterogeneous basic catalysts for the transesterification and the polycondensation reactions in PET production from DMT. Journal of Molecular Catalysis A 2004;212:251—7.

[82] Michael J, Goff NS, Bauer SL, William RS, Galen JS. Acid-catalyzed alcoholysis of soybean oil. Journal of the American Oil Chemists' Society 2004;81:415—20.

[83] Canakci M, Gerpan JV. Biodiesel production from oils and fats with high free fatty acids. Transactions of the American Society of Agricultural Engineers 2001;44:1429—36.

[84] Sharma YC, Singh B, Upadhyay SN. Advancements in development and characterization of biodiesel: a review. Fuel 2008;87:2355—73.

[85] Banani R, Youssef S, Bezzarga M, Abderrabb M. Waste frying oil with high levels of free fatty acids as one of the prominent sources of biodiesel production. Journal of Materials & Environmental Science 2015;6(4):1178—85.

[86] Debora DO, Marco DL, Carina F, Clarissa DR, Joao PB, Nadia L, et al. Optimization of alkaline transesterification of soybean oil and castor oil for biodiesel production. Applied Biochemistry & Biotechnology 2005;121—124:553—60.

[87] Roberto EA, Mircea V, Adam MB, Jaroslav AK, Colin JB. Transesterification of fish oil to produce fatty acid ethyl esters using ultrasonic energy. Journal of the American Oil Chemists' Society 2007;84:1045—52.

[88] Li Y, Qiu F, Yang D, Sun P, Li X. Transesterification of soybean oil and analysis of bioproduct. Food & Bioproducts Processing 2012;90:135—40.

[89] Ellis N, Guan F, Chen T, Poon C. Monitoring biodiesel production (transesterification) using in situ viscometer. Chemical Engineering Journal 2007;138:200—6.

[90] Srivastava PK, Verma M. Methyl ester of karanja oil as alternative renewable source energy. Fuel 2008;87:1673—7.

[91] Farooq M, Ramli A, Naeem A. Biodiesel production from low FFA waste cooking oil using heterogeneous catalyst derived from chicken bones. Renewable Energy 2015;76:362—8.

[92] El-Gendy NS, Deriase SF, Hamdy A, Abdallah RI. Statistical optimization of biodiesel production from sunflower waste cooking oil using basic heterogeneous biocatalyst prepared from eggshells. Egyptian Journal of Petroleum 2015;24(1):37—48.

[93] Guthrie JP. Concerted mechanism for alcoholysis of esters: an examination of the requirements. Journal of the American Oil Chemists' Society 1991;113:3941—9.

[94] Soumanou MM, Bornscheuer UT. Lipase-catalyzed alcoholysis of vegetable oils. European Journal of Lipid Science & Technology 2003;105:656—60.

[95] De OD, Di LM, Faccio C, Dalla RC, Bender JP, Lipke N, et al. Optimization of enzymatic production of biodiesel from castor oil in organic solvent medium. Applied Biochemistry & Biotechnology 2004;113—116:771—80.

[96] Chang HM, Liao HF, Lee CC, Shieh CJ. Optimized synthesis of lipase-catalyzed biodiesel by Novozym 435. Journal of Chemical Technology & Biotechnology 2005;80:307—12.

[97] Lai CC, Zullaikah S, Vali SR, Ju YH. Lipase-catalyzed production of biodiesel from rice bran oil. Journal of Chemical Technology & Biotechnology 2005;80:331—7.

[98] Noureddini H, Gao X, Philkana RS. Immobilized *Pseudomonas cepacia* lipase for biodiesel fuel production from soybean oil. Bioresource Technology 2005;96:769—77.

[99] Mittlebach M. Lipase catalyzed alcoholysis of sunflower oil. Journal of the American Oil Chemists' Society 1990;67:168—70.

[100] Mangas-Sánchez J, Adlercreutz P. Highly efficient enzymatic biodiesel production promoted by particle-induced emulsification. Biotechology & Biofuel 2015;8:58.

[101] Cesarini S, Diaz P, Nielsen PM. Exploring a new soluble lipase from FAMEs production in water-containing systems using crude soybean oil as feedstock. Process Biochemistry 2013;48:484—7.

[102] Ciudad G, Reyes I, Jorquera M, Azócar L, Wick L, Navia R. Novel three-phase bioreactor concept for fatty acid alkyl ester production using *R. oryzae* as whole cell catalyst. World Journal of Microbiology & Biotechnology 2011;27:2505—12.

[103] Iso M, Chen BX, Eguchi M, Kudo T, Shrestha S. Production of biodiesel fuel from triglycerides and alcohol using immobilized lipase. Journal of Molecular Catalysis Enzyme 2001;16:53—8.

[104] Matsumoto T, Takahashi S, Kaieda M, Ueda M, Tanaka A, Fukuda H, et al. Yeast whole-cell biocatalysts constructed by intracellular overproduction of *Rhizopus oryzae* lipase is applicable to biodiesel fuel production. Applied Microbiology & Biotechnology 2001;57:515—20.

[105] Chen JW, Wu WT. Regeneration of immobilized *Candida antarctica* lipase for transesterification. Journal of Biosciences & Bioengineering 2003;92:231—7.

[106] Xu YY, Du W, Zeng J, Liu DH. Conversion of soybean oil to biodiesel fuel using lipozyme TL 1M in a solvent-free medium. Biocatalysis & Biotransformation 2004;22:45—8.

[107] Shimada Y, Watanabe Y, Sugihara A, Tominaga Y. Enzymatic alcoholysis for biodiesel fuel production and application of the reaction to oil processing. Journal of Molecular Catalysis 2002;17:133—42.

[108] Du W, Xu YY, Liu DH, Li ZB. Study on acyl migration in immobilized lipozyme TL-catalyzed transesterification of soybean oil for biodiesel production. Journal of Molecular Catalysis B: Enzymatic 2005;37:68—71.

[109] Orcaire O, Buisson P, Pierre AC. Application of silica aerogel encapsulated lipases in the synthesis of biodiesel by transesterification reactions. Journal of Molecular Catalysis 2006;42:106—13.

[110] Yagiz F, Kazan D, Akin AN. Biodiesel production from waste oils by using lipase immobilized on hydrotalcite and zeolites. Chemical Engineering Journal 2007;134:262—7.

[111] Raymond HM, Douglas MC. Response surface methodology process and product optimization using designed experiments. 2nd ed. John Wiley & Sons, Inc.; 2002.

[112] Montgomery DC. Design and analysis of experiments: response surface method and designs. New Jersey: John Wiley and Sons, Inc.; 2005.

[113] Box GEP, Norman RD. Empirical model-building and response surfaces. Wiley; 1987.

[114] Kuan I-C, Lee C-C, Tsai B-H, Lee S-L, Lee W-T, Yu C-Y. Optimizing the production of biodiesel using lipase entrapped in biomimetic silica. Energies 2013;6:2052—64.

[115] Sangaletti N, Cea M, Bismara Regitano-d'Arce MA, Ferreira de Souza Vieira TM, Navia R. Enzymatic transesterification of soybean ethanolic miscella for biodiesel production. Journal of Chemical Technology & Biotechnology 2013;88:2098—106.

[116] Domingos AK, Saad EB, Wilhelm HM, Ramos LP. Optimization of the ethanolysis of *Raphanus sativus* (L. Var.) crude oil applying the response surface methodology. Bioresource Technology 2008;99:1837—45.

[117] Bouaid A, Bajo L, Martinez M, Aracil J. Optimization of biodiesel production from jojoba oil. Transactions of the Institution of Chemical Engineers 2007;85:378—82.

[118] Chen X, Du W, Liu D. Response surface optimization of biocatalytic biodiesel production with acid oil. Journal of Biochemical Engineering 2008;40:423—9.

[119] Vicente G, Coteron A, Martinez M, Aracil J. Application of the factorial design of experiments and response surface methodology to optimize biodiesel production. Industrial Crops & Products 1998;8:29—35.

[120] Wu WH, Foglia TA, Marmer WN, Phillips JG. Optimizing production of ethyl esters of grease using 95% ethanol by response surface methodology. Journal of the American Oil Chemists' Society 1999;76:517−21.

[121] Rodrigues S, Mazzone LCA, Santos FFP, Cruz MGA, Fernandes FAN. Optimization of the production of ethyl esters by ultrasound assisted reaction of soybean oil and ethanol. Brazilian Journal of Chemical Engineering 2009;26:361−6.

[122] Trathnigg B, Mittelbach M. Analysis of triglyceride methanolysis mixtures using isocratic HPLC with density detection. Journal of Liquid Chromatography 1990;13:95−105.

[123] Foglia TA, Jones KC. Quantitation of neutral lipid mixtures using high performance liquid chromatography with light scattering detection. Journal of Liquid Chromatography and Related Technologies 1997;20:1829−38.

[124] Holcapek M, Jandera P, Fischer J, Prokes B. Analytical monitoring of the production of biodiesel by high−performance liquid chromatography with various detection methods. Journal of Chromatography A 1999;858:13−31.

[125] Dimmig T, Radig W, Knoll C, Dittmar T. ^{13}C−NMR spectroscopic determination of the conversion and reaction kinetics of transesterification of triacylglycerols to methyl esters. Chemical Technology (Leipzig) 1999;51:326−9.

[126] Madras G, Kolluru C, Kumar R. Synthesis of biodiesel in supercritical fluids. Fuel 2004;83:2029−33.

[127] Foroutan R, Naeimi B, Khamisipour GR, Mohebbi GH, Dobaradaran S, Ghodrati S, et al. Biodiesel production by base-catalyzed trans-esterification of sunflower and date seed oils using methanol: Optimization of parameters. Journal of Chemical & Pharmaceutical Research 2015;7(4):1187−93.

[128] Furuta S, Matsuhashi H, Arata K. Biodiesel fuel production with solid amorphous-zirconia catalysis in fixed bed reactor. Biomass & Bioenergy 2006;30:870−3.

[129] Sarin R, Sharma M, Sinharay S, Malhotra RK. Jatropha-palm biodiesel blends: an optimum mix for Asia. Fuel 2007;86:1365−71.

[130] Correa SM, Arbilla G. Carbonyl emissions in diesel and biodiesel exhaust. Atmospheric Environment 2008;42:769−75.

[131] Rashid U, Anwar F, Arif M. Optimization of base catalytic methanolysis of sunflower (*Helianthus annuus*) seed oil for biodiesel production by using response surface methodology. Industrial & Engineering Chemistry Research 2009.

[132] Doo SB, Young CH. The effect of biodiesel and ultralow sulfur diesel fuels on emissions in 11,000 cc heavy-duty diesel engine. Journal of Mechanical Science & Technology 2005;19:870−6.

[133] Mumtaz MW, Mukhtar H, Anwar F, Saari N. RSM based optimization of chemical and enzymatic transesterification of palm oil: Biodiesel production and assessment of exhaust emission levels. The Scientific World Journal 2014;1−11.

[134] American Biofuels Association Information Resources (ABAIR). Biodiesel: a technology performance and regulatory overview. Jefferson City, MO: National Soy Diesel Development Board; 1994.

[135] Graboski MS, McCormick RL. Combustion of fat and vegetable oil derived fuels in diesel engines. Progress in Energy and Combustion Science 1998;24:125−64.

[136] Mustafa ET, Paul SW, Jon HVG, Thomas EC. Exhaust emissions from an engine fueled with biodiesel from high-oleic soybeans. Journal of the American Oil Chemists' Society 2007;84:865−9.

[137] Scheepers PTJ, Bos RP. Combustion of diesel fuel from a toxicological perspective, II. Toxicity. International Archives of Occupational and Environmental Health 1992;64:163−77.

[138] Scheepers PTJ, Bos RP. Combustion of diesel fuel from a toxicological perspective, I. Origin of incomplete combustion products. International Archives of Occupational and Environmental Health 1992;64:149−61.

CHAPTER SIXTEEN

Mesoporous Catalysts for Biodiesel Production: A New Approach

S. Soltani[1], U. Rashid[1,2], S.I. Al-Resayes[2] and I.A. Nehdi[2]
[1]Universiti Putra Malaysia (UPM), Serdang, Selangor, Malaysia
[2]King Saud University, Riyadh, Saudi Arabia

16.1 INTRODUCTION

Energy is a principal necessity for human existence on the earth. The energy crisis has become the main concern for the human beings. This concern is strongly associated to the availability of natural sources of power (i.e., coal and natural gases) which are depleting at rapid rates. These natural sources have been hugely consumed for transportation, industrial, and domestic sectors. Apart from these facts, the increasing consumption of petroleum has been endangering planet earth and more particularly human health [1,2].

However, the world's most energy needs are met mainly by petrochemical sources; the depletion of these sources has warned human communities to discover alternative sources of energy. The increasing trend on the number of researches on biodiesel production might be a true evidence that biodiesel would be the best substitute for petroleum-based diesel. Biodiesel is a sulfur-free liquid fuel, which can enhance the world's energy security [3]. Biodiesel is generated from various feedstocks, which has comparable physical and chemical characteristics to petroleum [4]. In terms of energy generation, biodiesel is a promising source of energy which is nontoxic in nature, sustainable, and biodegradable with a low emission of greenhouse gases [5—7].

16.2 BIODIESEL

The seed oil—based fuel is referred to as ester, well known as biodiesel, normally produced in presence of a catalyst with a short-chain alcohol. Biodiesel is considered as a sulfur-free, nontoxic, and biodegradable source of energy which burns cleaner than petroleum-based fuels.

16.2.1 Triglycerides As Diesel Fuel

The history of using triglyceride (TG) as a diesel fuel comes from 1885 where Rudolph Diesel designed a diesel engine and used peanut oil as raw material. Palm, sunflower, soybean, olive, coconut, and peanut oils have been utilized majorly for biodiesel

Clean Energy for Sustainable Development
ISBN 978-0-12-805423-9, http://dx.doi.org/10.1016/B978-0-12-805423-9.00016-8

production [8]. Climate condition and geographical location are two important factors that affect the cultivation of these feedstocks. Unfortunately, the number of drawbacks, such as high viscosity, flash point, cloud point, pour point, and carbon deposition, have limited the use of edible oils in industrial applications [9]. Besides, choosing food–based vegetable oils as raw material may affect both biodiesel and food industries. Rising demand for edible oils will consequentially increase their prices in both markets [10]. Moreover, there is a need to explore nonedible oils, such as waste cooking oils (WCOs) and animal fats, which are completely unfit for human food consumption. Thereby, the cost of the production process can be substantially reduced using nonedible oils as raw material.

16.2.2 Process of Biodiesel Production

Several synthesis methods have been utilized for biodiesel production, including: (1) direct utilization of vegetable oils, (2) micro emulsions, (3) thermal cracking, (4) transesterification, and (5) supercritical methanol (detailed in Table 16.1) [11,12].

Table 16.1 Some Important Approaches to Produce Biodiesel

Type of Process	Method	Pros	Cons
Direct mixing and dilution	Blending plant oil without any type of pretreatment with petrodiesel and using directly in engine	No engine modification required	Low ignition/low cold filter plugging point/carbon deposition on engine
Micro emulsion	—	Simple process	Low stability High viscosity
Thermal cracking	Converting of an organic substance to another one by heating	Simple process/ nonpolluting	High temperature required/high cost of equipment/less purity/ nonpolluting
Transesterification	Converting of TG into alkyl esters in presence of alcohol and catalyst	Similar fuel properties to diesel/high conversion/low cost/ suitable to scale up to industrial scale	Undesired byproduct/ separation costing
Supercritical methanol	Accelerate free catalytic transesterification using supercritical fluid at high pressure/ temperature	Free catalyst/high conversion efficiency/ very short reaction time/no product purification/ simultaneously esterification and transesterification	High pressure and temperature required/high energy using/ expensive equipment

16.2.3 Transesterification

Alcoholysis known as transesterification to synthesize alcoholic ester as a result of interaction between an ester with an alcohol in the presence of a proper catalyst. The general equation of the transesterification reaction is illustrated as follows [3]:

$$
\begin{array}{l}
CH_2-OOC-R' \\[6pt]
\quad | \\[6pt]
CH-OOC-R'' \quad + \quad 3CH_3OH \quad \xrightleftharpoons{Catalyst} \\[6pt]
\quad | \\[6pt]
CH_2-OOC-R''' \\[4pt]
(Triglyceride) \qquad (Methanol)
\end{array}
\qquad
\begin{array}{l}
R'-COOCH_3 \quad CH_2-OH \\[6pt]
\quad | \qquad\qquad\quad | \\[6pt]
R''-COOCH_3 \; + \; CH-OH \\[6pt]
\quad | \qquad\qquad\quad | \\[6pt]
R'''-COOCH_3 \\[4pt]
\qquad\qquad\qquad\; CH_2-OH \\[4pt]
(Methyl\ Ester) \qquad (Glycerol)
\end{array}
$$

$$(16.1)$$

where R', R'', and R''' are referred to as hydrocarbons (oils) or free fatty acid (FFA) chains.

The main reason to transesterify vegetable oils is to enhance their combustion by reducing their high viscosity [13]. Transesterification is a chemical reaction that includes a sequence of converting TG to diglyceride, and then diglyceride to monoglyceride, and finally monoglyceride to glycerol [14]. Typically, transesterification is performed to convert TG to methyl ester (ME). The molar ratio of alcohol to oil (MeOH:oil), catalyst amount, level of FFAs, operating temperature, and reaction time are some important reaction variables that affect the conversion rate [9]. Generally, the conversion percentage of TG to glycerol is generally obtained from:

$$
Conversion\ \% = \frac{Calculated\ weight\ of\ methyl\ esters}{weight\ of\ methyl\ ester\ phase} \times 100
$$

$$
Conversion\ \% = \sum \frac{Ai}{A} \times \frac{C \times V}{W} \times 100 \tag{16.2}
$$

where $\sum A_i$ is the total maximum of methyl esters, A is the area of methyl ester, C is the strength in mg/ml of the methyl heptadecanoate, V is the volume in ml of the methyl heptadecanoate, and W is the weight in mg of the sample [15].

Biodiesel properties have to be in accordance with either ASTM D6751-07 or EN 14214:2003 standards, utilized as an ecofriendly fuel [16].

16.3 CATALYSTS

Typically, a proper catalyst should be active enough to catalyze transesterification reaction which consists of a large number of saturated fatty acids (SFAs) and long carbon chains [17−20]. Catalysts are used to maximize the biodiesel yield in a shorter reaction rate by increasing the number of active sites. Using a proper catalyst may provide better condition due to enhancing the miscibility of the reactants into alcohol.

16.3.1 Classification of Catalysts

Generally, transesterification is performed in the presence of acid or base catalysts considering the fatty acid content of feedstocks. Transesterification is a reversible reaction, therefore, catalyst dosage should be sufficient to catalyze both the forward and backward reactions [21,22]. Generally, catalysts are categorized into three major groups; homogeneous catalyst, heterogeneous catalyst, and biocatalyst, of which the subclassification is shown in Scheme 16.1 [23]. Commonly, biodiesel is produced using homogeneous base catalyst (such KOH and NaOH) or homogeneous acid catalysts (such as H_2SO_4), which are available in market but are corrosive in nature.

16.3.2 Problems With Homogeneous Catalysts

Conventionally, transesterification has been conducted using homogeneous acidic catalysts that are ironically nongreen and corrosive. Besides, loading higher MeOH:oil ratio is required to complete the reaction. Consequently, loading higher MeOH:oil ratio prolongs the transesterification reaction time. Homogeneous base catalysts (such as sodium methoxide and potassium hydroxide) are also used for catalyzing the transesterification reaction [24]. On the other hand, homogeneous base catalysts are quite sensitive to water and FFAs. Furthermore, huge amount of water is needed to separate catalyst from the final product which consequently increases the production cost of biodiesel production. Generally, the hygroscopic nature, water and soap formation, oil losses, and difficulty in separation are some undesired drawbacks of the homogenous catalysts during catalyzing the transesterification reaction. Thereby, research efforts have been switched onto heterogeneous catalysts [15,25].

16.3.3 Problems With Base and Acid Catalysts

Hana et al. [24] concluded that the base catalysts are not suitable in cases of high levels of FFAs (>1%) but according to 2012 reports, these types of catalysts can efficiently catalyze feedstocks with FFAs over 4% [26]. It is also reported that solid base catalysts have higher catalytic activity and stability than the solid acid catalysts [27]. On the other hand, acid content is an important feature of the feedstocks which increases by raising the reaction temperature [28]. As an example, the acid value of soybean oil will be increased from 0.04% to 1.51% as the temperature goes above 190°C [29]. In this case, using heterogeneous acid catalysts performs much better than heterogeneous base catalysts. Heterogeneous acid catalysts are mostly preferred than the base ones for the transesterification of feedstocks with high FFA [30]. Besides, high acid density is an important advantage of solid acid catalysts in organic reactions which significantly eases

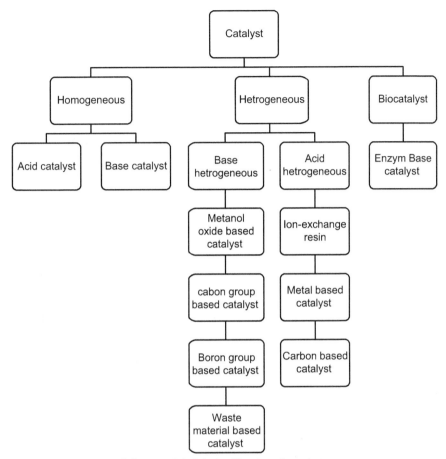

Scheme 16.1 Subclassification of catalysts.

the adsorption of reactants to the catalyst' active sites. Thereby, presence of heterogeneous catalyst can increase the catalytic activity through transesterification reaction. Solid acid catalysts can be simply detached from the reactants after filtration process.

It is observed that the base catalysts are able to complete the reaction at a lower temperature at a shorter reaction time than the acid catalysts [31,32]. Activation of solid acid catalysts is completely low at low temperatures in the transesterification reaction. However, gaining higher conversion rate requires an increase in the reaction temperature ($>170°C$), but some solid acid catalysts are not stable enough against severe reaction temperatures. On the other hand, less surface area and less porosity may decline the performance of solid acid catalysts [33]. For instance, zirconia as an inorganic catalyst has fairly high stability at high temperatures, but its small surface area and small pore size reduce its activity. In order to enhance the activity of heterogeneous acid catalysts, they can be functionalized with other types of materials with a higher surface area [34].

16.4 POROUS MATERIALS

Generally, porous materials can be classified into three major groups by their size: micropores, mesopores, and macropores, which consist of pores with diameters less than 2 nm, between 2 and 50 nm, and more than 50 nm, respectively. The macropore channel leads to the penetration of bulky TGs into the active sites of the catalyst via mass transfer [35]. In contrast, the mesopore channel leads to the enhancement of the catalyst's active sites on the surface. Indeed, the function of each macropore and mesopore channel is to provide fast transportation channels and to enhance catalyst' active sites on the surface, respectively [36].

16.4.1 Mesoporous Material Characteristics

Utilizing heterogeneous catalysts causes a slow mass transfer in transesterification reaction. This may be attributed to their small surface area which causes low contact between the catalyst and reactants. This feature prolongs the reaction time and reduces the yield of the final product. In the last few decades, zeolites have drawn considerable interest owing to their high surface area, unique porosity, and high thermal stability. On the other hand, interrupting mass transfer (while reacting with large reactant particles, especially in liquid phase) has limited their effective utilization [37]. It is important to note that catalysts with a lower pore diameter are inappropriate for biodiesel preparation because of the low penetration of big-sized particles into the pore system [14]. Therefore, it is required to enhance the diffusion rate of reactants into the mesoporous channels to overcome these limitations.

Several research works have been carried out in the field of mesoporous materials to enhance porosity with thicker pore wall [38]. Gaining a higher pore diameter and thicker pore wall results in greater thermal stability [39,40]. The thermal stability is a considerable advantage of a mesoporous material specifically in the process of catalyst fabrication which requires high temperature stability [41]. Despite the dimensions of TG (5.8 nm), methyl oleate (2.5 nm), and glycerin (0.6 nm) [15,22] particles, the pore size of the synthesized catalysts has to be large enough to let the reactants diffuse into the pore structure of the catalyst [42]. Flexibility in the pore diameter is one of the remarkable advantage of mesoporous materials which make them favorable in catalytic reactions [43].

It is reported that the catalytic activity of mesoporous catalysts is not high at low temperatures. Therefore, a higher reaction temperature is needed to activate them [44,45]. In this regard, a microwave-assisted reactor is a promising technique that is capable of conducting transesterification at a higher temperature in a short time from hours to minutes. As a matter of fact, this attractive option vastly simplifies the process of biodiesel production in a shorter reaction time, because microwave irradiation can directly penetrate into the reactants [46–48].

16.5 VARIOUS TYPES OF MESOPOROUS CATALYSTS

16.5.1 Mesoporous Silica Material

Mesoporous ordered silicate (mesoporous silica material, MSM) is a hexagonal ordered mesoporous structure material that has prominent properties, such as high porosity and thick pore wall. These inherited properties result in higher mass transfer through the active sites. High hydrothermal stability is another remarkable advantage which increases the demand for its application in many fields of studies [14]. So far, MSM has been functionalized by a number of metal oxides (such as Al_2O_3, MgO, CuO, and La_2O_3) to transform its function into a heterogeneous base catalyst [49]. It should be considered that interaction between the silicon element and strong alkaline destructs the mesoporous silicate structure. Therefore, weaker loading of the alkali has been recommended to avoid destroying of the mesopore channels [50].

16.5.2 Mesoporous Carbon Material

Mesoporous carbon material (MCM) could be a perfect candidate to enhance catalytic activity [51,52]. The unique surface area and uniform porosity are two excellent features of MCMs that lead to the absorption of long chains of FFAs and prevent the absorption of water [53]. High thermal stability is another important characteristic of MCMs, which promotes its application. In order to improve the catalytic activity of MCMs, they can be modified by grafting via electrochemistry [54,55], chemical reducing of aryl diazoniums, and by reduction of alkylation and arylation [56,57].

16.5.3 Mesoporous Metal Oxide Catalyst

A number of reports in the literature underscore the excellent performance of the metal oxides as filler in mesoporous materials. To date, a wide range of nanosized materials (such as Al_2O_3, ZnO, TiO_2, and ZrO_2) has been utilized as support elements in mesoporous catalyst preparations [58]. Coating nonmetal ions [59,60], doping of noble metal nanoparticles [61,62], and mixing different semiconductors [63,64] are some attractive approaches to modify the structural and textural properties of the mesoporous catalysts [65]. Nanotransition, high thermal stability, high firmness, and low-temperature deposition capacity are some significant aspects of mesoporous metal oxide catalysts. The combination of these properties makes them applicable in many fields, such as sensing of methanol, hydrogen production, sensors, solar cells, and biodiesel [66,67].

16.6 APPLICATION OF MESOPOROUS MATERIALS

Much attention has been focused on the application of mesoporous materials since 1990s. Some effective characteristics of mesoporous materials, such as large surface area, large and uniform pore structure, low density, and delivery capacity, have attracted the focus of the international scientific communities to utilize them for various bioapplications [68–70]. The combination of remarkable physicochemical, textural, and structural properties implies that the mesoporous solid catalysts are the most encouraging applicants in many branches of science, such as quantum dot, optic, electronic, optoelectronic, magnetic, mechanical, nanoreactor, as well as biofuel production [69,71]. Some important techniques to fabricate porous materials consist of solid-state reaction, sol–gel, coprecipitation, polymeric precursor, hydrothermal, and microwave hydrothermal.

A large number of research has been conducted on mesoporous materials to investigate the three principal considerations, including [69]:

1. synthesis of mesoporous materials for a required structure;
2. examination of textural, morphological, structural, and physicochemical characteristics;
3. functionalizing the surface of synthesized mesoporous materials for desirable applications.

In order to make the mesoporous solid materials technology successful, it is necessary to find out different compositions of the mesoporous materials with different properties for different applications [72,73].

16.7 PERFORMANCE OF THE MESOPOROUS CATALYST

16.7.1 Catalytic Activity

Based on reports, there is modification proposed which may maximize the activity of mesoporous catalysts including [42,74,75] as suggested as follows:

1. formation of a wide number of active sites with large surface area;
2. increase accessibility of the reagent particles to active sites;
3. widen the pore diameter by shortening the length of the macro-perforated channels;
4. decrease the level of the by-products and deposition of carbon within the synthesis process.

16.7.2 Thermal Stability

The catalyst stability is an important feature that significantly reduces the cost of production in industrial scale. In order to commercialize the process of biodiesel production, the stability and particularly reusability of the catalyst should be improved. A stable catalyst never gets dissolved in the reactants, so the total quantity will remain the same at the end of the reaction [76]. After several runs, the adsorbed glycerol on the active sites prohibits interaction between the active sites and reactants, which causes deactivation of the catalyst. To some degree, it can handle with decontamination of the deposited particles by thermal degradation [42]. Furthermore, the low acid strength is another factor that causes lower thermal stability. This problem can be solved with postsulfonation treatment to attach acid functional groups to the surface of the active sites [71]. In order to increase the thermal stability of a product, we can functionalize the surface of the catalyst with phosphoric acid [77,78] or sulfuric acid (SO_3/H_2SO_4) [79] functional groups. Postsulfonation is a promising approach that significantly enhances the thermal stability of the synthesized samples up to 500°C. Among the type of mesoporous catalyst, thermal stability of mesoporous carbon is quite high which is staying intact at temperatures as high as 1400°C.

16.7.3 Basicity and Acidity

Another prominent factor that has huge impact on catalyst performance is the basicity and the acidity of the catalyst. Acid (or basic) modification can be done by introducing a proper acid (or base) functional group to the surface of the sample [80]. For instance, postsulfonation method is an effective method that transforms the nature of materials by enhancing the aromatic chains and reducing the aliphatic rings in the presence of oxidized bands [81]. To examine the basic sites of the basic solid catalyst, an acidic molecular particle is required to be adsorbed on the surface of the origin basic sites, which is usually preferred to be carbon dioxide. The level of the adsorption of the CO_2 determines the quantity of the basic sites. On the other hand, to examine the acid sites of the acid solid catalyst, a basic molecular particle is required to be adsorbed on the surface of the original acid sites, which is usually preferred to be ammonia. The level of adsorption of the ammonia determines the quantity of the acid sites. It is noteworthy that the number of active sites plays a prominent role to enhance the yield of the product even more than the surface area [82]. In a later section, we discuss about preparation of base/ acid mesoporous catalyst and their catalytic activity.

16.7.4 Pore Filling

Good preservation of mesostructure is highly challenging due to pore filling process which might occur during the saturation reaction. The pore filling is a reaction that might take place as a result of sulfate species. In such cases, postcalcination should be performed to prevent pore filling [83]. Calcination is an integral part of catalyst fabrication which enhances the leaching resistance of samples via the transformation of salt into a more stable form. It also helps in activating synthesized catalyst via catalysis reaction. For instance, mesoporous CaO catalyst can simply get poisoned by hydration and carbonation in air. In this regard, postcalcination process is required to activate poisoned sample [36,84]. Thereby, postcalcination treatment is an effective way to prevent pore filling and protect the channels through the saturation–diffusion process.

16.8 THE DIFFUSION PROCESS OF THE REACTANTS INTO MESOPORE CHANNELS

The reaction between mesoporous catalyst and adsorbed reactants usually occurs in heterogeneous catalysis. In case of catalytic transesterification reaction, reactants diffuse through catalyst surface (mainly to the internal surface of pores) into porous channels. The rate of diffusion is highly associated with the accessibility of the reactants with majority of active sites through catalyst porous channels (Fig. 16.1) [85]. The chemisorption reaction occurs in order to convert the reactant (A) to product (B) (Scheme 16.2);

16.9 SURFACE MODIFICATIONS

Several treatments have been established to fulfil surface modification of nano-compositions. In general, surface modification can be classified into physical and chemical treatments.

16.9.1 Physical Treatment

Physical treatment is applied to cover nanoparticles with an appropriate surfactant to generate secondary forces including electrostatic, hydrogen, and van der Waals forces [86].

16.9.1.1 Surfactant

A surfactant is a lengthy aliphatic chain with one or more polar bonds. The main role of surfactant treatment is to adsorb polar-band surfactant onto the surface of the fillers through electrostatic interactions. It results in the formation of ionic bonds between the filler's surface and surfactants which consequently decreases the surface energy of filler's surface by making a shield around it. Furthermore, surfactant treatment can increase the surface area of the treated particles [86].

16.9.1.2 Dispersant

Discernments are also used to modify the surface area by encapsulating inorganic particles with in situ—formed polymers. Typically, the polymeric dispersing agent contains two major components: (1) a functional group (such as $-OH$, $-NH_2$, and $-COOH$) and (2) soluble polymeric chains (such as polyester, polyether, polyolefin, and so on). The function of these two components helps binding of the dispersing agent to the mesopore surface via hydroxide and electrostatic bonds. It also causes dispersion of the particles in different directions to facilitate creation of homogeneous coverage [86].

16.9.2 Chemical Treatment

Significantly, surface modification through chemical interaction occurs by applying modifier to create strong chemical bonding with the matrix. Surface modification by

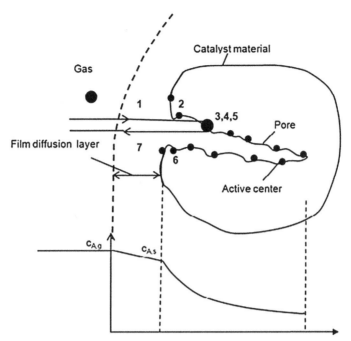

Figure 16.1 Sequential process of diffusion with heterogeneous catalyst [85].

coupling agents (such as Zn and Al) is another treatment method that provides the bonding between inorganic fillers and matrix polymer. Another advantage is the elimination of the hydrophilic properties to some extent. Furthermore, unstable nanoparticle agglomerates become more strengthened due to the transformation of nanoparticles into a nanocomposite structure. Therefore, surface modification by coupling agents is a promising approach to create a unique surface coverage onto the mesopore channels [86].

Soltani et al. [87] hydrothermally synthesized a polymeric mesoporous $ZnAl_2O_4$ catalyst, using polyethylene glycol as surfactant and D-glucose as a template. The as-prepared catalyst was further sulfonated in order to improve the hydrophobicity through catalysis reaction, and the proposed mechanism is presented in Fig. 16.2. The optimized mesoporous $SO_3H-ZnAl_2O_4$ catalyst possessed unique characteristics, such as a surface area of 352.39 m^2/g, the average pore diameter of 3.10 nm, a total pore volume of 0.13 cm^3/g, and high acid density up to 1.95 mmol/g, simultaneously. The combination of the mesostrucure and functionalization methods (in situ doping Al source and sulfonation process) resulted in high catalytic performance of synthesized catalyst, giving FAME yield of 94.65%. It was claimed that the polymerization step provided a better occasion for the sulfonic groups to scatter on the surface of the formed mixed metal oxide particles.

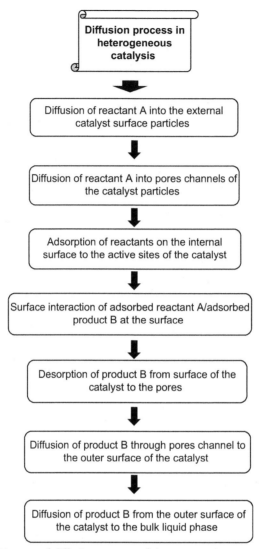

Scheme 16.2 Diagram of diffusion process of the reactant into mesopore channels.

16.10 THE EFFECT OF MESOPOROUS CATALYST ON TRANSESTERIFICATION REACTION

Different types of mesoporous catalysts were used for transesterifying TG to biodiesel. Kazemian et al. [84] functionalized an ordered mesoporous silicate catalyst (SBA-15) with cesium species (CsNO$_3$) in order to catalyze transesterification of canola oil, using pressurized batch stirred tank reactor. A high surface area of 628.15 m^2/g was obtained with a pore volume of 0.837 cm^3/g. In order to gain high conversion, the effect of different variables were examined, including different ratio of methanol-oil (20:1 and 40:1), catalyst concentration (100–200 mg), different reaction times (3–24 h), and reaction temperatures (65–135°C). The high biodiesel yield (25.35%) was obtained under optimized condition: mesoporous 2wt%-CsNO$_3$-SBA-15 catalyst concentration of 100 mg, MeOH:oil of 40:1, reaction time of 5 h, and operating temperature of 135°C.

Albuquerque et al. [41] prepared mesoporous solid basic catalyst through impregnation with different ratios of calcium acetate CaO as a support on SBA-15, using the incipient wetness method. The catalytic activity was examined through transesterification reaction. To obtain higher conversion rate, the effects of different quantities of catalyst (0.4 and 1.6 wt.%) were examined. The highest conversion (95%) was obtained by only loading 1 wt.% of the catalyst which proves high catalytic activity of the synthesized mesoporous catalyst.

In another study, [14] impregnated surface area of mesoporous SBA-15 with KOH solution. A high surface area of 539 m^2/g, average pore diameter of 5.63 nm, and pore volume of 0.63 cm^3/g were recorded, which illustrates the high potential of mesoporous K-SBA-15 catalyst. The response surface methodology was established to examine the impact of different variables on the biodiesel yield. The optimized condition found 3.91 wt.% for the catalyst concentration, 11:6 for MeOH:oil ratio, 70°C for the operating temperature, and 5 h for the reaction time. The optimum reaction condition results in 93% conversion of palm oil to biodiesel through transesterification reaction. The high catalyst activity was attributed to extremely high surface area and high porosity of the prepared catalyst, which simplified the diffusion of reagents into mesostructure.

Mesoporous MgO catalyst was prepared by Jeon et al. [36] in the presence of template-MgO, using the sol–gel method as shown in Fig. 16.3. The surface areas obtained were 32.9 and 79.6 m^2/g for nontemplate-MgO and template-MgO, respectively. It shows that the surface area of the template-MgO sample is 2.4 times larger than that of nontemplate-MgO. The basicity was determined with CO$_2$-TPD which shows that the basicity of template-MgO catalyst was 2.5 times higher than that of nontemplate-MgO catalyst. The transesterification of canola oil was performed with

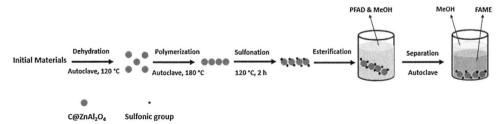

Figure 16.2 Synthesis of the sulfonated mesoporous ZnAl$_2$O$_4$ catalyst and FAME production [87]. *Copyright 2016 Elsevier Ltd.*

methanol/oil molar ratio of 20:3, catalyst loading of 3%, and operating temperature of 190°C for 2 h, using batch reactor. The rapid transportation structure of the synthesized catalyst resulted in 96.5% conversion of biodiesel through transesterification reaction.

In another study that was carried out by [49], mesoporous silica was used in MgO applying two methods: impregnated and in situ–coated method which resulted in different physical characteristics of SBA-15 catalyst. In comparison with impregnated method, less-blocked mesopores were reported via in situ-coating method which was mainly due to smooth formation of MgOx particles on SBA-15. Also, the X-ray photoelectron spectroscopy (XPS) data indicated low level of Mg particles on the surface area of the in situ–coated SBA-15 catalysts which resulted in a higher surface area and higher pore volume in comparison with the impregnated method. The catalytic activity was examined by transesterification of blended vegetable oils. The high conversion (96%) was obtained in the presence of mesoporous SBA-15-MgO catalyst through transesterification reaction.

In another research, the sol—gel method was reported by Wu et al. [42] to impregnate K_2SiO_3 into mesoporous Al-SBA-15. By increasing the amount of K_2SiO_3, the surface area of parent material was reduced because the surface area of the support (K_2SiO_3) covered through the preparation process. Besides, the basicity of the synthesized mesoporous Al-SBA-15 was also increased by introducing higher concentration of potassium compounds. In order to examine the catalytic activity, transesterification was carried out, using Jatropha oil under batch condition. The highest conversion (95%) was obtained in the presence of mesoporous Al-SBA-15-K_2SiO_3 catalyst.

Xie and Zhao [88] hydrothermally impregnated $Ca(NO_3)_2$ and $(NH_4)_6Mo_7O_{24}$ into mesoporous SBA-15 to synthesize mesoporous CaO—MoO$_3$—SBA-15. The effect of different postcalcination temperatures (623—823K) was examined to gain higher catalytic activity. It was found that catalytic activity was increased from 48.3% to 83.2% by increasing the postcalcination temperature from 623K to 823K.

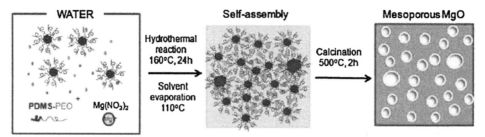

Figure 16.3 Schematic illustration of the synthesis of the mesoporous MgO catalyst using a PDMS—PEO comb-like copolymer. *Reproduced with permission from Jeon H, Kim DJ, Kim SJ, Kim JH. Synthesis of mesoporous MgO catalyst templated by a PDMS—PEO comb-like copolymer for biodiesel production. Fuel Processing Technology 2013;116:325—31. Copyright 2013 Elsevier Ltd.*

Grafting method was established by Xie and Fan [89] in order to functionalize mesoporous SBA-15 with ammonium chloride organosilane. The high surface area of $810 \ m^2/g$ and large pore diameter of 6.75 nm show the high potential of synthesized catalyst. The catalytic activity of mesoporous SBA-15-pr-NR_3OH was examined via transesterification of soybean under various reaction conditions. A quite high yield (99.4%) was obtained under the suitable reflux condition, including: catalyst concentration 2.5 wt.%, MeOH:oil ratio 12:1, and reaction time 30 min. The synthesized catalyst was recycled for several runs without even a slight drop of activity.

Xie et al. [90] synthesized SBA-15-pr-ILOH under hydrothermal condition through impregnation of 4-butyl-1,2,4-triazolium hydroxide onto SBA-15 silica. The surface area and pore size of the support were obtained at $590 \ m^2/g$ and 6.64 nm, respectively. After surface functionalization, the surface area and pore diameter were significantly dropped to $341 \ m^2/g$ and 5.58 nm, respectively. The catalytic activity of synthesized catalyst was examined through transesterification of soybean oil. It should be considered, however, that the textural properties were dropped in some extent after treatment, but the conversion (95.4 wt.%) confirmed high potential of synthesized mesoporous base catalyst.

16.11 CONCLUSION AND RECOMMENDATION

Generally, small surface area is the main concern for heterogeneous catalysts, which causes a slow mass transfer through catalysis reaction. Consequently, lower contact between catalyst' active sites and reactants prolongs the reaction time. Furthermore, catalysts with a lower pore diameter are inappropriate for biodiesel preparation because of the lower penetration of big-sized particles (FFAs) into the pore system. The large surface area, uniform pore structure, and high thermal stability are the most important characteristics of mesoporous materials, which proposed them in many applications as catalysts or catalyst support. The excellent textural properties of mesoporous catalysts enhance the catalytic activity of them due to easy accessibility of the reactants into mesopores. Up to now, many scientists have utilized mesoporous catalysts in order to accelerate the process of biodiesel production. The interactions between variables such as reaction time, temperature, the amount of methanol, and catalyst loading contribute to the great potential of biodiesel production via transesterification reaction. In order to gain prominent success in the mesoporous solid materials technology, it is necessary to find out different compositions of the mesoporous materials with different physicochemical properties for different applications. It is also recommended to utilize novel metal oxide mesoporous materials rather than mesoporous silica as alternative materials for catalyzing chemical reaction.

ACKNOWLEDGMENTS

The authors extend their appreciation to the International Scientific Partnership Program (ISPP) at King Saud University for funding this research work through ISPP# 0035.

NOMENCLATURE/ABBREVIATION

FFA   Free fatty acid
TG   Triglyceride
WCO   Waste cooking oil
ME   Methyl ester
SFA   Saturated fatty acid
MSM   Mesoporous silica material
MCM   Mesoporous carbon material
MeOH   Methanol

REFERENCES

[1] Jothiramalingam R, Wang MK. Review of recent developments in solid acid, base, and enzyme catalysts (heterogeneous) for biodiesel production via transesterification. Industrial Engineering Chemistry Research 2009;48:6162−72.

[2] Piriou B, Vaitilingom G, Veyssière B, Cuq B, Rouau X. Potential direct use of solid biomass in internal combustion engines. Progress in Energy and Combustion Science 2013;39:169−88.

[3] Motasemi F, Ani FN. A review on microwave-assisted production of biodiesel. Renewable and Sustainable Energy Reviews 2012;16:4719−33.

[4] Cantrell DG, Gillie LJ, Lee AF, Wilson K. Structure-reactivity correlations in MgAl hydrotalcite catalysts for biodiesel synthesis. Applied Catalysis A: General 2005;287:183−90.

[5] Meher L, Vidyasagar D, Naik S. Technical aspects of biodiesel production by transesterification-a review. Renewable and Sustainable Energy Reviews 2006;10:248−68.

[6] Sharma YC, Singh B. Development of biodiesel: current scenario. Renewable and Sustainable Energy Reviews 2009;13:1646−51.

[7] Lee SB, Han KH, Lee JD, Hong IK. Optimum process and energy density analysis of canola oil biodiesel synthesis. Journal of Industrial and Engineering Chemistry 2010;16:1006−10.

[8] Fukuda H, Kondo A, Noda H. Biodiesel fuel production by transesterification of oils. Journal of Bioscience and Bioengineering 2001;92:405−16.

[9] Murugesan A, Umarani C, Chinnusamy TR, Krishnan M, Subramanian R, Neduzchezhain N. Production and analysis of bio-diesel from non-edible oils—a review. Renewable and Sustainable Energy Reviews 2009;13:825−34.

[10] Kansedo J, Lee KT, Bhatia S. *Cerbera odollam* (sea mango) oil as a promising non-edible feedstock for biodiesel production. Fuel 2009;88:1148−50.

[11] Helwani Z, Othman MR, Aziz N, Kim J, Fernando WJN. Solid heterogeneous catalysts for transesterification of triglycerides with methanol: a review. Applied Catalysis A: General 2009;363:1−10.

[12] Borges ME, Díaz L. Recent developments on heterogeneous catalysts for biodiesel production by oil esterification and transesterification reactions: a review. Renewable and Sustainable Energy Reviews 2012;16:2839−49.

[13] Basha SA, Gopal KR, Jebaraj S. A review on biodiesel production, combustion, emissions and performance. Renewable and Sustainable Energy Reviews 2009;13:1628−34.

[14] Abdullah AZ, Razali N, Lee KT, et al. Optimization of mesoporous K/SBA-15 catalyzed transesterification of palm oil using response surface methodology. Fuel Processing Technology 2009;90:958−64.

[15] Granados ML, Poves MDZ, Alonso DM, Mariscal R, Galisteo FC, Moreno-Tost R. Biodiesel from sunflower oil by using activated calcium oxide. Applied Catalysis B: Environmental 2007;73:317−26.

[16] López DE, Goodwin JG, Bruce DA, Lotero E. Transesterification of triacetin with methanol on solid acid and base catalysts. Applied Catalysis A: General 2005;295:97−105.

[17] Ganduglia-Pirovano MV, Popa C, Sauer J, Abbott H, Uhl A, Baron M. Role of ceria in oxidative dehydrogenation on supported vanadia catalysts. Journal of the American Chemical Society 2010;132:2345−9.

[18] Wang R, Song B, Zhou W, Zhang Y, Hu D, Bhadury PS. A facile and feasible method to evaluate and control the quality of *Jatropha curcas* L. seed oil for biodiesel feedstock: gas chromatographic fingerprint. Applied Energy 2011;88:2064—70.

[19] Maia ECR, Borsato D, Moreira I, Spacino KR, Rodrigues PRP, Gallina AL. Study of the biodiesel B100 oxidative stability in mixture with antioxidants. Fuel Processing Technology 2011;92:1750—5.

[20] Lin C-Y, Cheng H-H. Application of mesoporous catalysts over palm-oil biodiesel for adjusting fuel properties. Energy Conversion and Management 2012;53:128—34.

[21] Liang X, Xiao H, Qi C. Efficient procedure for biodiesel synthesis from waste oils using novel solid acidic ionic liquid polymer as catalysts. Fuel Processing Technology 2013;110:109—13.

[22] Jacobson K, Gopinath R, Meher L, Dalai A. Solid acid catalyzed biodiesel production from waste cooking oil. Applied Catalysis B: Environmental 2008;85:86—91.

[23] Soltani S, Rashid U, Yunus R, Taufiq-Yap YH. Synthesis of biodiesel through catalytic transesterification of various feedstocks using fast solvothermal technology: a critical review. Catalysis Reviews: Science and Engineering 2015;57(4):1—29.

[24] Ma F, Hanna MA. Biodiesel production: a review. Bioresource Technology 1999;70:1—15.

[25] Xie W, Li H. Alumina-supported potassium iodide as a heterogeneous catalyst for biodiesel production from soybean oil. Journal of Molecular Catalysis A: Chemical 2006;255:1—9.

[26] Yücel Y. Optimization of biocatalytic biodiesel production from pomace oil using response surface methodology. Fuel Processing Technology 2012;99:97—102.

[27] Ramachandran K, Suganya T, Nagendra Gandhi N, Renganathan S. Recent developments for biodiesel production by ultrasonic assist transesterification using different heterogeneous catalyst: a review. Renewable and Sustainable Energy Reviews 2013;22:410—8.

[28] Marchetti JM, Errazu AF. Esterification of free fatty acids using sulfuric acid as catalyst in the presence of triglycerides. Biomass and Bioenergy 2008;32:892—5.

[29] Tyagi VK, Vasishtha AK. Changes in the characteristics and composition of oils during deep-fat frying. Journal of the American Oil Chemists' Society 1996;73:499—506.

[30] Peterson GR, Scarrah SW. Rapeseed oil transesterification by heterogeneous catalysis. Journal of the American Oil Chemists' Society 1984;61:1593—7.

[31] Demirbas A. Importance of biodiesel as transportation fuel. Energy Policy 2007;35:4661—70.

[32] Casas A, Fernández CM, Ramos MJ, Pérez Á RJF. Optimization of the reaction parameters for fast pseudo single-phase transesterification of sunflower oil. Fuel 2010;89:650—8.

[33] Jiang T, Zhao Q, Li M, Yin H. Preparation of mesoporous titania solid superacid and its catalytic property. Journal of Hazardous Materials 2008;159:204—9.

[34] Shao GN, Sheikh R, Hilonga A, Lee JE, Park Y-H, Kim HT. Biodiesel production by sulfated mesoporous titania—silica catalysts synthesized by the sol—gel process from less expensive precursors. Chemical Engineering Journal 2013;215—216:600—7.

[35] Woodford JJ, Dacquin J-P, Wilson K, Lee AF. Better by design: nanoengineered macroporous hydrotalcites for enhanced catalytic biodiesel production. Energy & Environmental Science 2012;5:6145—50.

[36] Jeon H, Kim DJ, Kim SJ, Kim JH. Synthesis of mesoporous MgO catalyst templated by a PDMS—PEO comb-like copolymer for biodiesel production. Fuel Processing Technology 2013;116:325—31.

[37] Taguchi A, Schüth F. Ordered mesoporous materials in catalysis. Microporous and Mesoporous Materials 2005;77:1—45.

[38] Chen A, Yu Y, Lv H, Wang Y, Shen S, Hu Y, et al. Thin-walled, mesoporous and nitrogen-doped hollow carbon spheres using ionic liquids as precursors. Journal of Materials Chemistry A 2013;1:1045—7.

[39] Antonelli DM, Ying JY. Synthesis of hexagonally packed mesoporous TiO_2 by a modified sol—gel method. Angewandte Chemie International Edition in English 1995;34:2014—7.

[40] Tian B, Liu X, Tu B, Yu C, Fan J, Wang L. Self-adjusted synthesis of ordered stable mesoporous minerals by acid-base pairs. Natural Materials 2003;2:159—63.

[41] Albuquerque MCG, Jiménez-Urbistondo I, Santamaría-González J, Mérida-Robles JM, Moreno-Tost R, Rodríguez-Castellón E. CaO supported on mesoporous silicas as basic catalysts for transesterification reactions. Applied Catalysis A: General 2008;334:35—43.

[42] Wu H, Zhang J, Liu Y, Zheng J, Wei Q. Biodiesel production from Jatropha oil using mesoporous molecular sieves supporting K_2SiO_3 as catalysts for transesterification. Fuel Processing Technology 2014;119:114–20.

[43] Lin VS-Y, Lai C-Y, Huang J, Song S-A, Xu S. Molecular recognition inside of multifunctionalized mesoporous silicas: toward selective fluorescence detection of dopamine and glucosamine. Journal of the American Chemical Society 2001;123:11510–1.

[44] Kulkarni MG, Dalai AK. Waste cooking oil an economical source for biodiesel: a review. Industrial & Engineering Chemistry Research 2006;45:2901–13.

[45] Lotero E, Goodwin Jr JG, Bruce DA, Suwannakarn K, Liu Y, Lopez DE. The catalysis of biodiesel synthesis. Catalysis 2006;19:41–83.

[46] Barnard TM, Leadbeater NE, Boucher MB, Stencel LM, Wilhite BA. Continuous-flow preparation of biodiesel using microwave heating. Energy & Fuels 2007;21:1777–81.

[47] Geuens J, Kremwner JM, Nebel BA, Schober S. Microwave-assisted catalyst-free transesterification of triglycerides with 1-butanol under supercritical conditions. Energy & Fuels 2008;22:643–5.

[48] Zuo D, Lane J, Culy D, Schultz M, Pullar A, Waxman M. Sulfonic acid functionalized mesoporous SBA-15 catalysts for biodiesel production. Applied Catalysis B: Environmental 2013;129:342–50.

[49] Li E, Rudolph V. Transesterification of vegetable oil to biodiesel over MgO-functionalized meso-porous catalysts. Energy & Fuels 2008;22:145–9.

[50] Wei YD, Zhang SG, Li GS, Yin SF, Qu ZT. Research progress on solid superbase catalysts in the last decade. Chinese Journal of Catalysis 2011;32:891–8.

[51] Kao H, TSai Y, Chao S. Functionalized mesoporous silica MCM-41 in poly(ethylene oxide)-based polymer electrolytes: NMR and conductivity studies. Solid State Ionics 2005;176:1261–70.

[52] Kim S, Park S-J. Preparation and electrochemical behaviors of polymeric composite electrolytes containing mesoporous silicate fillers. Electrochimica Acta 2007;52:3477–84.

[53] Liu R, Wang X, Zhao X, Feng P. Sulfonated ordered mesoporous carbon for catalytic preparation of biodiesel. Carbon 2008;46:1664–9.

[54] Delamar M, Hitmi R, Pinson J, Saveant JM. Covalent modification of carbon surfaces by grafting of functionalized aryl radicals produced from electrochemical reduction of diazonium salts. Journal of the American Chemical Society 1992;114:5883–4.

[55] Bahr JL, Yang J, Kosynkin DV, Bronikowshi MJ, Smalley RE, Tour JM. Functionalization of carbon nanotubes by electrochemical reduction of aryl diazonium salts: a bucky paper electrode. Journal of the American Chemical Society 2001;123:6536–42.

[56] Wildgoose GG, Leventis HC, Davis IJ, Crossley A, Lawrence NS, Jiang L, et al. Graphite powder derivatised with poly-lcysteine using "building-block" chemistry - a novel material for the extraction of heavy metal ions. Journal of Materials Chemistry 2005;14:2375–82.

[57] Stephenson JJ, Sadan AK, Higginbotham AL, Tour JM. Highly functionalized and soluble multiwalled carbon nanotubes by reductive alkylation and arylation: the billups reaction. Chemistry of Materials 2006;18:4658–61.

[58] Wang L, Yang W, Wang J, Evans DG. New nanocomposite polymer electrolyte comprising nanosized $ZnAl_2O_4$ with a mesopore network and PEO-LiClO4. Solid State Ionics 2009;180:392–7.

[59] Bu X-Z, Zhang G-K, Gao Y-Y, Yang Y-Q. Preparation and photocatalytic properties of visible light responsive N-doped TiO_2/rectorite composites. Microporous and Mesoporous Materials 2010;136:132–7.

[60] Xiao Q, Ouyang L, Gao L, Yao C. Preparation and visible light photocatalytic activity of mesoporous N, S-codoped $TiO_2(B)$ nanobelts. Applied Surface Science 2011;257:3652–6.

[61] Chiarello GL, Aguirre MH, Selli E. Hydrogen production by photocatalytic steam reforming of methanol on noble metal-modified TiO_2. Journal of Catalysis 2010;273:182–90.

[62] Alem A, Sarpoolaky H. The effect of silver doping on photocatalytic properties of titania multilayer membranes. Solid State Sciences 2010;12:1469–72.

[63] Long M, Cai W, Cai J, Zhou B, Chai X, Wu Y. Efficient photocatalytic degradation of phenol over Co_3O_4/$BiVO_4$ composite under visible light irradiation. The Journal of Physical Chemistry B 2006;110:20211–6.

[64] Yan J, Chen H, Zhang L, Jiang J. Inactivation of *Escherichia coli* on immobilized CuO/$CoFe_2O_4$-TiO_2 thin-film under simulated sunlight irradiation. Chinese Journal of Chemistry 2011;29:1133–8.

[65] Remmel J, Geerk J, Linker G, Meyer O, Smithey R, Strehlau B. Superconductivity and structure of ion irradiated LaSrCuO thin films. Physica C: Superconductivity 1990;165:212−20.

[66] Saberi A, Golestani-Fard F, Sarpoolaky H, Willert-Porada M, Gerdes T, Simon R. Chemical synthesis of nanocrystalline magnesium aluminate spinel via nitrate−citrate combustion route. Journal of Alloys and Compounds 2008;462:142−6.

[67] Kumar RT, Selvam NCS, Ragupathi C, Kennedy LJ, Vijaya JJ. Synthesis, characterization and performance of porous Sr(II)-added ZnAl$_2$O$_4$ nanomaterials for optical and catalytic applications. Powder Technology 2012;224:147−54.

[68] Ying JY, Mehnert CP, Wong MS. Synthesis and applications of supramolecular-templated mesoporous materials. Angewandte Chemie International Edition 1999;38:56−77.

[69] Wan Y, Yang H, Zhao D. "Host-Guest" chemistry in the synthesis of ordered nonsiliceous mesoporous materials. Accounts of Chemical Research 2006;39:423−32.

[70] Chen XY, Ma C. Spherical porous ZnAl$_2$O$_4$:Eu^{3+} phosphors: PEG-assisted hydrothermal growth and photoluminescence. Optical Materials 2010;32:415−21.

[71] Corma A. From microporous to mesoporous molecular sieve materials and their use in catalysis. Chemical Reviews 1997;97:2373−419.

[72] Stein A, Melde BJ, Schroden RC. Hybrid inorganic−organic mesoporous silicates−nanoscopic reactors coming of age. Advanced Materials 2000;12:1403−19.

[73] Davis ME. Ordered porous materials for emerging applications. Nature 2002;417:813−21.

[74] Zheng J, Zeng Q, Yi Y, Wang Y, Ma J, Qin B. The hierarchical effects of zeolite composites in catalysis. Catalysis Today 2011;168:124−32.

[75] Li H, He S, Ma K, Wu Q, Jiao Q, Sun K. Micro-mesoporous composite molecular sieves H-ZSM-5/MCM-41 for methanol dehydration to dimethyl ether: effect of SiO$_2$/Al$_2$O$_3$ ratio in H-ZSM-5. Applied Catalysis A: General 2013;450:152−9.

[76] Liu Q, Wang L, Wang C, Qu W, Tian Z, Ma H, et al. The effect of lanthanum doping on activity of Zn-Al spinel for transesterification. Applied Catalysis B: Environmental 2013;136−137:210−7.

[77] Schüth F. Non-siliceous mesostructured and mesoporous materials. Chemistry of Materials 2001;13:3184−95.

[78] He X, Antonelli D. Recent advances in synthesis and applications of transition metal containing mesoporous molecular sieves. Angewandte Chemie International Edition 2002;41:214−29.

[79] Inagaki S, Guan S, Ohsuna T, Terasaki O. An ordered mesoporous organosilica hybrid material with a crystal-like wall structure. Nature 2002;416:304−7.

[80] Chang B, Li Y, Guo Y, Yin H, Zhang S, Yang B. SO$_3$H-functionalized hollow mesoporous carbon sphere prepared by simultaneously achieving sulfonation and hollow structure. Journal of Porous Materials 2015;22:629−34.

[81] Fraile JM, García-Bordejé E, Pires E, Roldán L. Catalytic performance and deactivation of sulfonated hydrothermal carbon in the esterification of fatty acids: comparison with sulfonic solids of different nature. Journal of Catalysis 2015;324:107−18.

[82] León M, Díaz E, Bennici S, Vega A, Ordóñez S, Auroux A. Adsorption of CO$_2$ on hydrotalcite-derived mixed oxides: sorption mechanisms and consequences for adsorption irreversibility. Industrial & Engineering Chemistry Research 2010;49:3663−71.

[83] Yang H, Lu R, Wang L. Study of preparation and properties on solid superacid sulfated titania−silica nanomaterials. Materials Letters 2003;57:1190−6.

[84] Kazemian H, Turowec B, Siddiquee MN, Rohani S. Biodiesel production using cesium modified mesoporous ordered silica as heterogeneous base catalyst. Fuel 2013;103:719−24.

[85] Bajirao UR. Kinetics and reaction engineering aspects of syngas production by the heterogeneously catalysed reverse water gas shift reaction; 2012.

[86] Rong MZ, Zhang MQ, Ruan WH. Surface modification of nanoscale fillers for improving properties of polymer nanocomposites: a review. Materials Science and Technology 2006;22:787−96.

[87] Soltani S, Rashid U, Yunus R, Taufiq-Yap YH. Biodiesel production in the presence of sulfonated mesoporous ZnAl$_2$O$_4$ catalyst via esterification of palm fatty acid distillate (PFAD). Fuel 2016;178:253−62.

[88] Xie W, Zhao L. Heterogeneous CaO–MoO$_3$–SBA-15 catalysts for biodiesel production from soybean oil. Energy Conversion and Management 2014;79:34–42.

[89] Xie W, Fan M. Biodiesel production by transesterification using tetraalkylammonium hydroxides immobilized onto SBA-15 as a solid catalyst. Chemical Engineering Journal 2014;239:60–7.

[90] Xie W, Hu L, Yang X. Basic ionic liquid supported on mesoporous SBA-15 silica as an efficient heterogeneous catalyst for biodiesel production. Industrial & Engineering Chemistry Research 2015;54:1505–12.

CHAPTER SEVENTEEN

Edible and Nonedible Biodiesel Feedstocks: Microalgae and Future of Biodiesel

A.E. Atabani[1,2], M.M. El-Sheekh[3], G. Kumar[4] and S. Shobana[5]
[1]Erciyes University, Kayseri, Turkey
[2]BioGreen Power Arge, Erciyes Teknopark, Kayseri, Turkey
[3]Tanta University, Tanta, Egypt
[4]National Institute for Environmental Studies (NIES), Tsukuba, Japan
[5]Aditanar College of Arts and Science, Tirchendur, India

17.1 INTRODUCTION

Historically, fossil fuels have played a vital role in the global energy demand. They have been used for running vehicles, power plants, and motor engines in the transportation, agricultural, and industrial sectors, respectively. However, it is believed that climate change, acid rain, and smog are currently the most pressing global environmental problems that are attributed to burning fossil fuels. Fossil fuels are the main contributor of carbon dioxide (CO_2), nitrogen oxide (NO_x), volatile organic compounds (VOC), and hydrocarbons (HC) [1]. Declining reserves of fossil fuels, beside recognition that climate change has stemmed from growing carbon dioxide emissions, have generated the interest in promoting biofuels as one of the leading renewable energy sources. The sustainable production of biofuels is a valuable tool in stemming climate change, boosting local economies, particularly in lesser-developed parts of the world, and enhancing energy security for all [1,2].

Biodiesel is defined as monoalkyl esters of long-chain fatty oils derived from renewable lipid feedstock, such as vegetable oils or animal fats, and alcohol with or without a catalyst, for use in compression ignition (diesel) engines as given by the National Biodiesel Board, 1996 [3—7]. Biodiesel seems very interesting for several reasons; it is highly biodegradable and has minimal toxicity and can replace diesel fuel in many different applications, such as boilers and internal combustion engines, without major modifications. Furthermore, a small decrease in performances is reported and it almost emits zero emissions of sulfates, aromatic compounds, and other chemical substances that are destructive to the environment. Biodiesel produces lower emissions; possesses high flash point, better lubrication, and high cetane number; and has very close physico-chemical characteristics to those of conventional diesel fuel allowing its use either on its

Clean Energy for Sustainable Development
ISBN 978-0-12-805423-9, http://dx.doi.org/10.1016/B978-0-12-805423-9.00017-X

own (pure biodiesel, B100) or mixed with petroleum-based diesel fuel (preferred ratio 5–20%, B5–B20) with very few technical adjustments or no modification [8]. Biodiesel has been in use in many countries, such as United States, Malaysia, Indonesia, Brazil, Germany, France, Italy, and other European countries. Therefore, there is a good potential for its production and application.

Microalgae can thrive in a wide range of habitats where light and water are present including ocean, lake, soils, ice, and rivers [9]. Microalgae demonstrate a great biodiversity (between 200,000 and several millions of species) [10], which can be divided into different divisions and classes depending on their pigmentation, biological structure, life cycle, and metabolism. Algae, like other plants, during photosynthesis process, convert solar energy into chemical energy. Algae store chemical energy in the form of carbohydrates, oils, and proteins. The stored algae oil can be converted to biodiesel, which is why biodiesel is a form of solar energy. The efficiency of a particular plant depends on the conversion of solar energy into chemical energy, the better form is the biodiesel and algae are among the most photosynthetically efficient plants on the earth. Microalgae can produce different kinds of biofuels which are mainly the biomethane produced by anaerobic digestion [11], biohydrogen by photobiological process [12], bioethanol by fermentation [13], liquid oil by thermal liquefaction [14], and biodiesel [15]. Even if industrial-scale biofuels from microalgae remain at an early stage, they remain a sustainable solution as a transportation fuel. Production of biodiesel from algae is of interest in the United States and many other countries because it is a renewable energy form and can reduce energy dependence on imported petroleum. In addition, biodiesel does not compete with traditional agricultural crops for land and water [16]. Also, algae are an excellent alternative and renewable form of energy that has gained a lot of interest over the last years for production of biodiesel. Algal biodiesel may also produce 10–30 times more than oil producing crops. Furthermore, algae do not require wide land, and it can be cultivated in sweater and wastewater for the bioremediation purposes.

17.1.1 Current Challenges of Biodiesel Industry

The prolonged reliance on food-grade vegetable oils (edible oils) as feedstocks for biodiesel production has threatened the supply of edible oils to food industry and raised some doubts on the future of biodiesel industry. Moreover, they are economically not feasible because of day-to-day increase in feedstock price. Due to these factors, it is crucial to find other alternative oil feedstock to substitute edible oil in the production of biodiesel. Therefore, biodiesel extraction from various nonedible feedstocks, microalgae, and waste resources has become the hot area of research in the last 3 years. Much attention has been devoted to the application of low-cost nonconventional and nonedible feedstocks from wild plants to produce biodiesel [1,8,17,18]. Enhancing the economic aspects of biodiesel production through integrated strategies targeting different stages (i.e., upstream, mainstream, and downstream) can be seen in Ref. [19].

17.1.2 Objectives

In this review paper, the authors have tried to promote biodiesel as one of the leading renewable energy sources. Based on the challenges faced by the biodiesel industry, as mentioned in the previous section, this chapter covers the following aspects:

1. Identification of various potential nonedible oil—bearing plants, beside those bearing edible oil, for biodiesel production.
2. Physicochemical properties of crude oils.
3. Characterization and determination of physicochemical properties and fatty oil compositions of the produced biodiesel.
4. Investigation of blending opportunities to improve the final product properties.
5. Engine performance and emission analysis of various biodiesel feedstocks.
6. Impact of additives on biodiesel properties and engine performance.
7. Future of biodiesel.
8. Role of microalgae in biodiesel industry.
9. Role of algae in wastewater treatment for biodiesel production.

This chapter gathers the latest publications of authors since 2012, beside many other recent publications.

17.2 BIODIESEL FEEDSTOCKS

Globally, there are more than 350 oil-bearing crops identified as potential sources for biodiesel production [20]. The wide range of available feedstocks for biodiesel production represents one of the most significant factors that promotes biodiesel industry and makes it an important choice for policy makers worldwide. The availability of feedstock for producing biodiesel depends on some factors such as regional climate, geographical locations, local soil conditions, and agricultural practices of a country.

To consider any feedstock as a biodiesel source, the oil percentage and the yield per hectare are the important parameters. The estimated oil content and yields of many biodiesel feedstocks can be found in Refs. [8,19,21,22]. The feedstock for biodiesel production is chosen according to quality, availability in each country, physicochemical properties, and its production cost. Composition of the oil is also one of the important criteria to determine the suitability of oil as a raw material for biodiesel production [8].

Feedstock alone represents 75% of the overall biodiesel production cost [21,23]. Therefore, selecting the cheapest feedstock is vital to ensure low biodiesel production cost. In general, biodiesel feedstock can be divided into four main categories as [5,23,24]:

1. edible vegetable oils;
2. nonedible vegetable oils;

3. waste or recycled oils;

4. animal fats: tallow, yellow grease, chicken fat, and by-products from fish oil.

The use of edible oils raises many concerns, such as food versus fuel crisis, and major environmental problems, such as serious destruction of vital soil resources, deforestation, and usage of much of the available arable land. Moreover, since mid-2000s, the prices of vegetable oil plants have increased dramatically that negatively affect the economic viability of biodiesel [25–27]. Furthermore, their use is not feasible in the long term because of the expected growing gap between supply and demand of such oils in many countries. For instance, dedicating all US soybean to biodiesel production would meet only 6% of diesel demands [28].

Nonedible oils are one of the possible solutions to reduce the utilization of the edible oil for biodiesel production. Nonedible oil resources are gaining much attention because they are easily available in many parts of the world, especially waste lands that are not suitable for food crops, eliminate competition for food, reduce deforestation rate, are more environmentally friendly, produce useful by-products, and are very economical when compared with edible oils. There are many publications that have been published on nonedible oils in the past few years. For further reading, see Refs. [18,26,29–41].

More recently, microalgae have emerged to be the third generation of biodiesel feedstock. Microalgae are photosynthetic microorganisms that convert sunlight, water, and CO_2 to algal biomass, but they do it more efficiently than conventional crop plants. It represents a very promising feedstock because of its high photosynthetic efficiency to produce biomass, higher growth rates, productivity, and high oil content when compared to edible and nonedible feedstocks. Microalgae have the potential to produce an oil yield that is up to 25 times higher than the yield of oil palm and 250 times the amount of soybeans. This is because microalgae can be grown in a farm or a bioreactor. Moreover, they are easier to cultivate than many other plants. It is believed that microalgae can play an important role in solving the problem between the production of food and that of biodiesel in the near future. Moreover, among other generations of biodiesel feedstocks, microalgae appear to be the only source of renewable biodiesel that is capable of meeting the global demand for fuels and can be sustainably developed in the future. The main obstacle for the commercialization of microalgae is its high production cost due to the requirement of high oil–yielding algae strains and effective large-scale bioreactors. Recent studies indicate that algae for biodiesel production can grow on flue gas, giving opportunities in consuming greenhouse gas [23,42–44]. Januan and Ellis [3] and Lin et al. [45] show that genetically engineered plants such as poplar, switchgrass, miscanthus, and big bluestem can be considered new feedstocks for biodiesel production. These feedstocks will create new bioenergy crops that are not associated with food crops. Therefore, they are expected to represent a sustainable biodiesel feedstock in the future. However, precaution on biosafety must be considered for these feedstocks. Fig. 17.1A and B show some plant families contributing to biodiesel production and some important plant species with the highest oil content, respectively, extracted from the aerial parts as well as the roots [46].

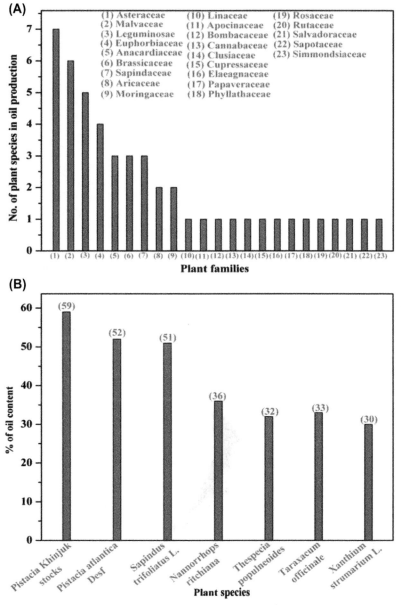

Figure 17.1 (A) Plant families contributing to biodiesel production. (B) Plant species with the highest oil content [46].

17.3 BIODIESEL RESEARCH METHODOLOGY

Fig. 17.2 has been suggested by the author to promote biodiesel research work [1,17,47—51].

17.4 OIL EXTRACTION

Mechanical pressing [52—54] and solvent extraction [39,41,52,55—60] are the most commonly used methods for commercial oil extraction from edible and nonedible feedstocks. In this subchapter, two examples of oil extraction from *Calophyllum ino-phyllum* [49] and *Pangium edule* [50] have been presented. The oil was extracted from *C. inophyllum* using mechanical extraction technique, while the oil was extracted from *P. edule* using solvent extraction technique.

17.4.1 Oil Extraction From *Calophyllum inophyllum*

The seeds of *C. inophyllum* were dried under sunlight for 2—3 days. The kernel was then separated from the shell and found to have high oil content (70%). The ideal conditions to preserve the kernel are 26—27°C and 60—70% humidity. The place in which the kernel is stored must be well ventilated and the storing period should not be too long.

Before extraction, the kernels were mixed with rice husk. This is very important to increase the oil yield during the extraction process. There are two types of pressing machines that are used to extract the oil from the kernel: hydraulic manual pressing machine and screw extruder machine. The oil extracted using pressing machine is very low and about 20—30%. Therefore, hydraulic machine was used to increase oil yield from *C. inophyllum*. The cake that remained after extraction was of high commercial and

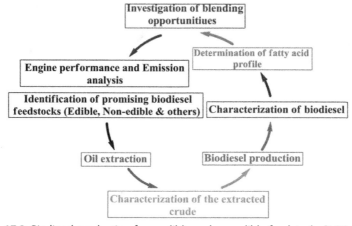

Figure 17.2 Biodiesel production from edible and nonedible feedstocks [1,17,47—51].

Figure 17.3 Pictures of *Calophyllum inophyllum* (A) tree, (B) seeds, and (C) extracted crude oils [49].

marketing value. Therefore, it can be used either for agricultural or industrial applications. Fig. 17.3A–c shows *C. inophyllum* tree, seeds, and extracted crude oils [49].

17.4.2 Oil Extraction From Crude *Pangium edule* Oil

The kernels of the *P. edule* fruits were obtained manually and cleaned before being dried overnight in an oven at 353 K. Moderate temperature for the drying of the *P. edule* kernels was used to prevent possible decomposition or oxidation of the kernels at higher temperature (>373 K) at which the properties of the extracted oil could be affected. The dried *P. edule* kernels were ground to fine particles using food processor and then dried for the second time in the oven to remove excess moisture. The oil extraction process was carried out using Soxhlet apparatus and the duration of each extraction process was set at about 4 h under temperature of 343 K (reflux temperature of the solvent). *P. edule* oil was obtained after separating the mixture of solvent and oil using rotary evaporator. The resultant *P. edule* oil was decanted mechanically to remove impurities and other components (glycosides) that may be present in the oil [50].

17.5 ANALYSIS OF PHYSICOCHEMICAL PROPERTIES

The quality of biodiesel and its feedstocks must be expressed in terms of fuel properties, such as cetane number, calorific value (kJ/kg), density (kg/m^3), kinematic viscosity (mm^2/s), cloud and pour points (°C), flash point (°C), cold filter plugging point (°C), acid value (mg KOH/g oil), ash content (%), copper corrosion, carbon residue, water content and sediment, sulfur content, glycerin (% m/m), phosphorus (mg/kg), and oxidation stability. The physical and chemical properties of biodiesel fuel basically depend on the type of feedstock and their fatty oil composition [34,45,61]. The important physicochemical properties of the crude oils and their respective methyl esters are tested according to ASTM D 6751 standard. Table 17.1 shows the findings of physicochemical properties of various edible and nonedible feedstocks.

Table 17.1 Physicochemical Properties of Various Edible and Nonedible Biodiesel Feedstocks [8,17,47,48,50,62,63]

	Property	CCIO	CCIO	CJCO	CJCO	CSFO	CSFO	CMOO	CMOO	CCMO	CCMO
1	Kinematic viscosity (mm²/s) at 40°C	55.677	71.98	48.091	47–54.8	75.913	49.7	43.4680	43.2	29.8440	64
2	Kinematic viscosity (mm²/s) at 100°C	9.5608	–	9.1039	–	13.608	–	9.0256	–	7.2891	–
3	Dynamic viscosity (mPa s) at 40°C	51.311	–	43.543	–	69.408	–	38.9970	–	27.1570	–
4	Viscosity index (VI)	165.4	–	174.10	–	184.80	–	195.20	–	224.20	–
5	Flash point (°C)	236.5	221	258.5	210–240	246.5	158	263	–	235	–
6	Cold filter plugging point (°C)	26	–	21	–	29	–	18	–	10	–
7	Density (g/cm³) at 40°C	0.9216	–	0.9055	–	0.9143	–	0.8971	–	0.9100	–
8	Acid Value (mg KOH/g oil)	41.74	44	17.63	0.92–6.16	9.49	0.36	8.62	–	12.07	3.343
9	Calorific value (kJ/kg)	38,511	39,25	38,961	37,830–42,050	39,793	39,650	N/D	–	N/D	–
10	Copper strip corrosion (3 h at 50°C)	1a	–	1a	–	1a	–	1a	–	1a	–
11	Refractive index	1.4784	–	1.4652	–	1.4651	–	1.4661	–	1.4741	–
12	Transmission (%T)	34.7	–	61.8	–	26.6	–	69.2	–	87.5	–
13	Absorbance (abs)	0.46	–	0.209	–	0.574	–	0.16	–	0.058	–
14	Oxidation stability (h at 110°C)	0.23	–	0.32	15.6	0.15	–	41.75	90.8	0.14	–
15	pH at 26°C	4.60	–	4.83	–	4.84	–	N/D	–	N/D	–

	Property	CCO	CCO	CPO	CPO	CSO	CSO	CCaO	CCaO	CPEO	CMO
1	Kinematic viscosity (mm²/s) at 40°C	27.640	27.26	41.932	44.79	31.7390	28.87	35.706	34.72	27.175	132.750
2	Kinematic viscosity (mm²/s) at 100°C	5.9404	–	8.496	–	7.6295	–	8.5180	–	6.6407	20.624
3	Dynamic viscosity (mPa s) at 40°C	25.123	–	37.731	–	28.796	–	32.286	–	24.393	122.810
4	Viscosity index (VI)	168.5	–	185.0	–	223.2	–	213.5	–	216.20	179.9
5	Flash point (°C)	264.5	–	254.5	267	280.5	254	290.5	246.5	–	192.5
6	Cold filter plugging point (°C)	22	–	23	–	13	–	15	–	–	–
7	Density (g/cm³) at 40°C	0.9089	–	0.8998	–	0.9073	–	0.9042	–	0.8976	0.9251
8	Acid Value (mg KOH/g oil)	N/D	–	N/D	–	N/D	–	N/D	–	19.62	2.08
9	Calorific value (kJ/kg)	37,806	–	39,867	–	39,579	39,600	39,751	39,700	39,523	38,682
10	Copper strip corrosion (3 h at 50°C)	1a	–	1a	–	1a	–	1a	–	1a	–
11	Refractive index	1.4545	–	1.4642	–	1.4725	–	1.471	–	1.4683	1.487

Property										
12 Transmission (%T)	91.2	—	63.2	—	65.2	—	62.9	—	86.1	89.3
13 Absorbance (abs)	0.04	—	0.199	—	0.186	—	0.202	—	0.064	0.049
14 Oxidation stability (h at 110°C)	6.93	92.23	0.08	2.7	6.09	14.1	5.64	—	0.08	0.16
15 pH at 26°C	6.71	—	N/D	—	N/D	—	5.60	—	—	—

Property	CAPO	TPSO
1 Kinematic viscosity (mm²/s) at 40°C	35.093	470.42
2 Kinematic viscosity (mm²/s) at 100°C	7.2547	—
3 Dynamic viscosity (mPa s) at 40°C	32.159	—
4 Viscosity index (VI)	177.9	190
5 Flash point (°C)	N/D	—
6 Cold filter plugging point (°C)	N/D	—
7 Density (g/cm³) at 40°C	0.9164	—
8 Acid value (mg KOH/g oil)	26.7	4.7
9 Calorific value (kJ/kg)	38,729	—
10 Copper strip corrosion (3 h at 50°C)	N/D	—
11 Refractive index	1.4789	—
12 Transmission (%T)	61.6	—
13 Absorbance (abs)	0.209	—
14 Oxidation stability (h at 110°C)	0.09	—
15 pH at 26°C	6.71	—

CAPO, Crude *Aphanamixis polystachya* oil; CCaO, Crude canola oil; CCIO, Crude *Calophyllum inophyllum* oil; CCMO, Crude *Croton megalocarpus* oil; CCO, Crude coconut oil; CJCO, Crude *Jatropha curcas* oil; CMO, Crude manketti oil; CMOO, Crude *Moringa oleifera* oil; CPEO, Crude *Pangium edule* oil; CPO, Crude palm oil; CSFO, Crude *Sterculia foetida* oil; CSO, Crude soybean oil; and TPSO, *Thevetia peruviana* seed oil.

Table 17.2 shows the physicochemical properties of the biodiesels from different edible and nonedible oils.

Moreover, Table 17.3 presents a comparison of some physicochemical properties of edible and nonedible biodiesel relative to palm oil biodiesel (PB) as the baseline fuel.

17.6 BIODIESEL PRODUCTION

Fig. 17.4 shows an example of a small-scale laboratory biodiesel production reactor. The reactor consists of 1 L jacketed glass batch reactor, reflux condenser to recover methanol, a sampling device, overhead mechanical stirrer, refrigerator, and a circulating water bath to control the reaction temperature [1]. Ref. [65] can also show another good example of schematic layout of biodiesel production setup.

Alkaline-catalyzed transesterification is the most widely used process for biodiesel production because it is very fast and yields large amount of biodiesel (Fig. 17.5).

The excess methanol is later recovered and reused. In transesterification process, the triglyceride molecule was taken after neutralizing the free fatty oils (free fatty acid, FFA), releasing the glycerin, and creating an alcohol methyl ester when methanol is used as catalyst. This is accomplished by mixing methanol with sodium hydroxide to make sodium methoxide. This liquid is then mixed into vegetable oil. The entire mixture then settles. The methyl esters or biodiesel will be on top while glycerin will be left at the bottom. The produced glycerin can be used in soap manufacture and the methyl esters are washed and filtered.

However, to use alkaline catalysts, the FFA level should be below a desired limit (ranging from less than 0.5% to less than 3%). Most of the nonedible oils have high FFA values. Therefore, transesterification with alkali–based catalyst yield a considerable amount of soap which are emulsifiers that make the separation of glycerol and ester phases very difficult. *Acid-catalyzed esterification* was found to be a good solution to this problem. However, the reaction rate was considerably less, requiring lengthy reaction periods. Therefore, the best approach to produce biodiesel from nonedible oils with high FFA values is the *acid-catalyzed esterification* process followed by *Alkaline-catalyzed transesterification* process.

Therefore, production of biodiesel can be conducted as follows [1]:

1. pretreatment process,
2. esterification process,
3. transesterification process,
4. posttreatment process.

17.6.1 Pretreatment Process

In this process, crude oil is entered into a rotary evaporator and heated to remove moisture for 1 h at 95°C under vacuum.

Table 17.2 Physicochemical Properties of the Produced Biodiesel [1,17,47,48,50,63,64]

No	Property	CIME	JCME	SME	CME	SFME	PME	COME	MOME	CMME	PEME	MME	ASTM D6751 Limit
1	Kinematic viscosity at 40°C (mm²/s)	5.5377	4.9476	4.3745	4.5281	6.3717	4.6889	3.1435	5.0735	4.0707	5.2296	8.3425	1.9–6.0.
2	Dynamic viscosity at 40°C (mPa s)	4.8599	4.2758	3.8014	3.9212	5.5916	4.0284	2.705	4.3618	3.453	4.5551	7.4067	–
3	Density at 40°C (kg/cm³)	0.8776	0.8642	0.869	0.866	0.8776	0.8591	0.8605	0.8597	0.8704	0.8710	0.8878	–
4	Oxidation stability (h at 110°C)	6.12	4.84	4.08	7.08	1.46	23.56	8.01	12.64	0.71	0.57	0.52	3 h min
5	CFPP (°C)	11	10	−3	−10	2	12	−1	18	−4	−8	5	–
6	Cloud point (°C)	12	10	1	−3	1	13	1	21	−3	−6	1	Report
7	Pour point (°C)	13	10	1	−9	2	15	−4	19	−2	−4	3	–
8	Flash point (°C)	162.5	186.5	202.5	186.5	130.5	214.5	118.5	176	164	N/D	N/D	130 min
9	Copper strip corrosion (3 h at 50°C)	1a	1a	1a	1a	1a	1a	1a	1a	1a	1a	1a	No.3 max
10	Caloric value (kJ/kg)	39,513	39,738	39,976	40,195	40,001	40,009	38,300	40,115	39,786	39,625	39,070	–
11	CCR (m/m%)	0.4069	0.0440	0.0204	0.0291	0.2911	0.0118	0.0114	0.022	0.028	N/D	N/D	0.050% mass max
12	Saponification value	201.3	201.5	201.2	199.25	N/D	206.7	272.5	196.5	200.4	200.8	201.7	–
13	Iodine value	87	105.7	137	122.1	N/D	59.9	7.8	71.224	152	119.2	149.2	–
14	Cetane number	53.8	49.6	42.6	46.2	N/D	59.2	64.6	90.1	39.3	46.7	39.8	47 min
15	Total sulfur (ppm)	4.11	3.84	0.86	0.83	7.02	1.81	0.94	N/D	N/D	N/D	N/D	15 max (S15) 500 max (S500)
16	Absorbance (abs) at WL 656.1	0.057	0.045	0.037	0.041	0.057	0.05	0.035	0.046	0.041	0.160	0.146	–
17	Transmission (%) at WL 656.1	87.7	90.3	92	91.1	87.9	89.1	92.3	90	91.1	69.2	71.5	–
18	Refractive index (RI) at 25°C	1.4574	1.4513	1.4553	1.4544	1.4557	1.4468	1.4357	1.4494	1.4569	1.4551	1.4698	–
19	Viscosity index	183.2	194.6	257.8	236.9	174.4	203.6	230.8	206.7	276.3	211.8	176	–
20	Viscosity at 100°C (mm²/s)	1.998	1.8557	1.764	1.7864	2.1954	1.7921	1.3116	1.9108	1.6781	1.9651	2.6683	–

(Continued)

Table 17.2 Physicochemical Properties of the Produced Biodiesel [1,17,47,48,50,63,64]—cont'd

No	Property	FOME	APME	ASTM D6751 Limit
1	Kinematic viscosity at 40°C (mm^2/s)	4.77	4.7177	1.9–6.0.
2	Dynamic viscosity at 40°C (mPa s)	N/D	4.1210	—
3	Density at 40°C (kg/cm^3)	0.8776	0.8735	—
4	Oxidation stability (h at 110°C)	N/D	0.16	3 h min
5	CFPP (°C)	11	5	—
6	Cloud point (°C)	N/D	8	Report
7	Pour point (°C)	N/D	8	—
8	Flash point (°C)	174.8	188.5	130 min
9	Copper strip corrosion (3 h at 50°C)	1a	N/D	No.3 max
10	Caloric value (kJ/kg)	39,954	39,960	—
11	CCR (m/m%)	0.0206	N/D	0.050% mass max
12	Saponification value	201.3	201.5	—
13	Iodine value	61	105.7	—
14	Cetane number	N/D	49.6	47 min
15	Total sulfur (ppm)	4.11	3.84	15 max (S15)
				500 max (S500)
16	Absorbance (abs) at WL 656.1	N/D	0.045	—
17	Transmission (%) at WL 656.1	N/D	90.3	—
18	Refractive index (RI) at 25°C	N/D	1.4513	—
19	Viscosity index	N/D	220.7	—
20	Viscosity at 100°C (mm^2/s)	N/D	1.8239	—

APME, Aphanamixis polystachya methyl ester; *CIME*, Calophyllum inophyllum methyl ester; *CME*, Canola methyl ester; *CMME*, Croton megalocarpus methyl ester; *COME*, Coconut oil methyl ester; *FOME*, Fish oil methyl ester; *JCME*, Jatropha curcas methyl ester; *MME*, Manketti methyl ester; *MOME*, Moringa oleifera methyl ester; *N/D*, Not determined; *N/S*, Not specified; *PEME*, Pangium edule methyl ester; *PME*, Palm methyl ester; *SFME*, Sunflower methyl ester; and *SME*, Soybean methyl ester.

Table 17.3 A Comparison of Some Physicochemical Properties of Edible and Nonedible Biodiesel Relative to Palm Oil Biodiesel (PME) [1]

		CIME	JCME	SFME	MOME	CMME	PEME	PME	SME	CME	CoME	MME
No	Property	NE1 (%)	NE2 (%)	NE3 (%)	NE4 (%)	NE5 (%)	NE7 (%)	E1 (Ref) (%)	E2 (%)	E3 (%)	E4 (%)	E5 (%)
1	Kinematic viscosity at 40°C (mm²/s)	+18.1	+5.5	+35.9	+8.2	-13.2	+11.5	0.0	-6.7	-3.4	-33.0	+77.9
2	Dynamic viscosity at 40°C (mPa s)	+20.6	+6.1	+38.8	+8.3	-14.3	+13.1	0.0	-5.6	-2.7	-32.9	+83.9
3	Density at 40°C (kg/cm³)	+2.2	+0.6	+2.2	+0.1	+1.3	+1.4	0.0	+1.2	+0.8	+0.2	+3.3
4	Oxidation stability (h at 110°C)	-74.0	-79.5	-93.8	-46.3	-97.0	-97.6	0.0	-82.7	-69.9	-66.0	-97.8
5	CFPP (°C)	-8.3	-16.7	-83.3	+50.0	-133.3	-166.7	0.0	-125.0	-183.3	-108.3	-58.3
6	Cloud point (°C)	-7.7	-23.1	-92.3	+61.5	-123.1	-146.2	0.0	-92.3	-123.1	-92.3	-92.3
7	Pour point (°C)	-13.3	-33.3	-86.7	+26.7	-113.3	-126.7	0.0	-93.3	-160.0	-126.7	-80.0
8	Flash point (°C)	-24.2	-13.1	-39.2	-17.9	-23.5	N/D	0.0	-5.6	-13.1	-44.8	N/D
9	Caloric value (kJ/kg)	-1.2	-0.7	+0.0	+0.3	-0.6%	-1.0	0.0%	-0.1%	+0.5%	-4.3%	-2.3
10	Absorbance (abs) at WL 656.1	+14.0%	-10.0%	+14.0%	-8.0%	-18.0%	+220.0	0.0%	-26.0%	-18.0%	-30.0%	+192.0
11	Transmission (%) at WL 656.1	-1.6%	+1.3%	-1.3%	+1.0%	+2.2%	-22.3	0.0%	+3.3%	+2.2%	+3.6%	-19.8
12	Refractive index (RI) at 25°C	+0.7	+0.3	+0.6	+0.2	+0.7	+0.6	0.0	+0.6	+0.5	-0.8	+1.6
13	Viscosity index	-10.0	-4.4	-14.3	+1.5	+35.7	+4.0	0.0	+26.6	+16.4	+13.4	-13.6
14	Viscosity at 100°C (mm²/s)	+11.5	+3.5	+22.5	+6.6	-6.4	+9.7	0.0	-1.6	-0.3	-26.8	+48.9

CIME, Calophyllum inophyllum methyl ester; *CME,* Canola methyl ester; *CMME, Croton megalocarpus* methyl ester; *CoME,* Coconut oil methyl ester; *E,* Edible; *JCME, Jatropha curcas* methyl ester; *MME,* Manketti methyl ester; *MOME, Moringa oleifera* methyl ester; *N/A,* Not determined; *NE,* Nonedible; *PEME, Pangium edule* methyl ester; *PME,* Palm methyl ester; *SFME,* Sunflower methyl ester; *SME,* Soybean methyl ester.

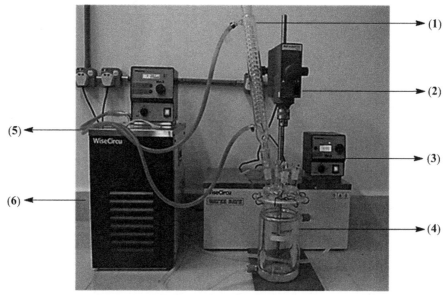

Figure 17.4 Experimental setup used to perform biodiesel production: (1) reflux condenser, (2) overhead mechanical stirrer, (3) circulating water bath, (4) jacketed glass batch tank reactor (1 L), (5) hoses, and (6) refrigerator [1].

$$
\begin{array}{ccccc}
\text{CH}_2\text{-OCOR}_1 & & & \text{CH}_2\text{-OH} & \text{CH}_3\text{-OCOR}_1 \\
| & & \xrightarrow{\text{Catalyst}} & | & \\
\text{CH-OCOR}_2 & + \ 3\text{HOCH}_3 & \rightleftharpoons & \text{CH- OH} & + \ \text{CH}_3\text{-OCOR}_2 \\
| & & & | & \\
\text{CH}_2\text{-OCOR}_3 & & & \text{CH}_2\text{-OH} & \text{CH}_3\text{-OCOR}_3 \\
\text{Triglyceride} & \text{Methanol} & & \text{Glycerol} & \text{Methyl esters} \\
\text{(Plant oil)} & \text{(Alcohol)} & & \text{(Triol)} & \text{(Biodiesel)}
\end{array}
$$

where R_1, R_2 & R_3 = Alkyl groups

Figure 17.5 Schematic representation of transesterification process [66].

17.6.2 Esterification Process

In this process, the molar ratio of methanol to crude oils with high acid values was maintained at 12:1 (50% v/v); 1% (v/v) of sulfuric acid (H_2SO_4) was added to the preheated oils at 60°C for 3 h and 400 rpm stirring speed in a glass reactor. On completion of this reaction, the products were poured into a separating funnel to separate the excess alcohol, sulfuric acid, and impurities presented in the upper layer. The lower layer was separated and entered into a rotary evaporator and heated at 95°C under vacuum conditions for 1 h to remove methanol and water from the esterified oil.

Due to the high acid value of some crude oils, the amount of methanol can be increased to reduce the acid value to less than 4 mg KOH/g oil.

17.6.3 Transesterification Process

In this process, crude oils with low acid values and esterified oils were reacted with 25% (v/v) of methanol and 1% (m/m) of potassium hydroxide (KOH) and maintained at 60°C for 2 h and 400 rpm stirring speed. After completion of the reaction, the produced biodiesel was deposited in a separation funnel for 12 h to separate glycerol from biodiesel. The lower layer that contained impurities and glycerol was drawn off.

17.6.4 Posttreatment Process

Methyl ester formed in the upper layer from the previous process was washed to remove the entrained impurities and glycerol. In this process, 50% (v/v) of distilled water at 60°C was sprayed over the surface of the ester and stirred gently. This process was repeated several times until the pH of the distilled water became neutral. The lower layer was discarded and upper layer was entered into a flask and dried using Na_2SO_4 and then further dried using rotary evaporator to make sure that biodiesel is free from methanol and water.

The conversion of crude oil to methyl ester and the purity of the produced biodiesel can be examined by the FT-IR spectroscopy [59,65,67−69].

A high-quality flow chart of production process of *P. pinnata*, as an example, can be found in Ref. [65]. Moreover, some technologies can be seen in Refs. [19,70].

17.7 DETERMINATION OF FATTY OIL COMPOSITION

To determine the fatty oil compositions of biodiesel samples (1 μL) can be injected into gas chromatography equipped with a flame ionization detector and a BPX70 capillary column of 30 m × 0.25 μm × 0.32 mm. An initial temperature of 140°C was held for 2 min, which was then increased at 8°C per minute to 165°C, 3°C per minute to 192°C, and finally 8°C per minute to 220°C. The column was held at the final temperature for another 5 min. The oven, injector, and the detector ports were set at 140°C, 240°C, and 260°C, respectively. The carrier gas was helium with column flow rate at 1.10 mL/min at a 50:1 split ratio. Table 17.4 shows an example of GC operating condition to determine fatty oil compositions [1,50].

Fatty oil composition is an important property for any biodiesel feedstock as it determines the efficiency process to produce biodiesel. The percentage and type of fatty oil composition relies mainly on the plant species and their growth conditions. The fatty oil composition and distribution of some feedstocks are generally aliphatic compounds with a carboxyl group at the end of a straight chain. The most common fatty oils are C_{16} and C_{18} acids. However, some feedstocks contain significant amounts of fatty oils other than the typical C_{16} and C_{18} oils [71]. Table 17.5 shows the chemical

Table 17.4 GC Operating Conditions to Determine Fatty Oil Composition [1,50]

Property	Specifications
Carrier gas	Helium
Linear velocity	24.4 cm/sec
Flow rate	1.10 mL/min (column flow)
Detector temperature	260.0°C
Column head pressure	56.9 kPa
Column dimension	BPX 70, 30.0 m × 0.25 μm × 0.32 mm ID
Injector column oven	240.0°C
Temperature ramp	140.0°C (hold for 2 min)
	8°C/min 165.0°C
	3°C/min 192.0°C
	8°C/min 220.0°C (hold for 5 min)

Table 17.5 The Chemical Structures of Common Fatty Oil [8,18]

Fatty Oil	Structure	Systematic Name	Chemical Structure
Lauric	(12:0)	Dodecanoic	$CH_3(CH_2)_{10}COOH$
Myristic	(14:0)	Tetradecanoic	$CH_3(CH_2)_{12}COOH$
Palmitic	(16:0)	Hexadecanoic	$CH_3(CH_2)_{14}COOH$
Palmitoleic	(16:1)	*cis*-9-Hexadecenoic	$CH_3(CH_2)5CH=CH(CH_2)_7COOH$
Stearic	(18:0)	Octadecanoic	$CH_3(CH_2)_{16}COOH$
Oleic	(18:1)	*cis*-9-Octadecenoic	$CH_3(CH_2)_7CH=CH(CH_2)_7COOH$
Linoleic	(18:2)	*cis*-9-cis-12-Octadecadienoic	$CH_3(CH_2)_4CH=CHCH_2CH$ $=CH(CH_2)_7COOH$
Linolenic	(18:3)	*cis*-9-cis-12,	$CH_3CH_2CH=CHCH_2CH$ $=CHCH_2CH=CH(CH_2)_7COOH$
Arachidic	(20:0)	Eicosanoic	$CH_3(CH_2)_{18}COOH$
Gondoic	(20:1)	*cis*-11-Eicosenoic	$CH_3(CH_2)_7CH=CH(CH_2)_9COOH$
Behenic	(22:0)	Docosanoic	$CH_3(CH_2)_{20}COOH$
Erucic	(22:1)	*cis*-13-Docosenoic	$CH_3(CH_2)_7CH=CH(CH_2)_{11}COOH$
Lignoceric	(24:0)	Tetracosanoic	$CH_3(CH_2)_{22}COOH$
Nervonic	(24:1)	*cis*-15-Tetracosenoic	$CH_3(CH_2)_7CH=CH(CH_2)_{13}COOH$
Cerotic	(26:0)	Hexacosanoic	$CH_3(CH_2)_{24}COOH$

structures of common fatty oils. The results of fatty oil composition of various methyl esters are given in Table 17.6.

17.8 PREDICTION OF PROPERTIES OF BLENDS

Polynomial curve fitting method is also a good approach to estimate the properties of other blends. This method is an attempt to describe the relationship between variable *X* as a function of available data and a response *Y*, which seeks to find a smooth curve that

Table 17.6 Fatty Oil Composition of Biodiesel [1,8,48,50,63,64]

Name of Fatty Oil	Structure	Chemical Name of Fatty Oils	JCME	JCME	JCME	JCME	PME	PME	PME	PME	PEME
Caproic	C6:0	Hexanoic	N/D	N/D	N/D	N/D	N/D	N/D	N/D	N/D	N/D
Caprylic	C8:0	Octanoic	N/D	N/D	N/D	N/D	N/D	N/D	N/D	N/D	N/D
Capric	C10:0	Decanoic	N/D	N/D	N/D	N/D	N/D	N/D	N/D	N/D	N/D
Lauric	C12:0	Dodecanoic	N/D	N/D	N/D	N/D	0.20	0.267	0.20	N/D	N/D
Myristic	C14:0	Tetradecanoic	0.10	0.10	N/D	1.40	0.90	1.434	1.10	N/D	0.10
Palmitic	C16:0	Hexadecanoic	13.0	14.2	13.23	15.6	38.6	46.13	44.0	42.6	8.30
Palmitoleic	C16:1	cis-9-Hexadecenoic	0.70	0.70	0.850	N/D	0.20	N/D	N/D	0.30	0.10
Margaric	C17:0	Heptadecanoic	N/D	0.10	N/D	N/D	N/D	N/D	N/D	N/D	N/D
Stearic	C18:0	Octadecanoic	5.80	7.00	05.40	9.70	4.40	3.684	4.50	4.40	4.00
Oleic	C18:1	cis-9-Octadecenoic	44.5	44.7	41.62	40.8	44.6	37.47	39.2	40.5	45.2
Linoleic	C18:2	cis-9,cis-12-octadecedianoic	35.4	32.8	36.99	32.1	10.5	11.02	10.1	10.1	39.3
Linolenic	C18:3	cis-6,cis-9,cis-12-octadecatrienoic	0.30	0.2	0.220	N/D	0.20	N/D	0.40	0.20	2.50
Stearidonic	C18:4	15-Octadecatetraenoic	N/D	N/D	N/D	N/D	N/D	N/D	N/D	1.10	N/D
Arachidic	C20:0	Eicosanoic	0.20	0.20	N/D	0.40	0.40	N/D	N/D	N/D	0.20
Eicosenoic	C20:1	cis-11-Eicosenoic	N/D	N/D	N/D	N/D	N/D	N/D	N/D	N/D	0.30
Eicosadienoic	C20:2	11,14-Eicosadienoic	N/D	N/D	N/D	N/D	N/D	N/D	N/D	N/D	N/D
Behenic	C22:0	Docosanoic	N/D	N/D	N/D	N/D	N/D	N/D	N/D	N/D	N/D
Erucic	C22:1	cis-13-Docosenoic	N/D	N/D	N/D	N/D	N/D	N/D	N/D	N/D	N/D
Lignoceric	C24:0	Tetracosanoic	N/D	N/D	N/D	N/D	N/D	N/D	N/D	N/D	N/D
Nervonic	C24:1	15-Tetracosenoic	N/D	N/D	N/D	N/D	N/D	N/D	N/D	N/D	N/D

Name of Fatty Oil	Structure	Chemical Name of Fatty Oils	MOME	MOME	MOME	MOME	COME	COME	COME	COME	SME	SME	SME	MME	FOME	APME
Caproic	C6:0	Hexanoic	N/D	N/D	N/D	N/D	0.50	N/D	8.90	N/D	N/D	N/D	N/D	N/D	N/D	N/D
Caprylic	C8:0	Octanoic	N/D	N/D	N/D	N/D	8.60	8.60	6.20	6.30	N/D	N/D	N/D	N/D	N/D	N/D
Capric	C10:0	Decanoic	N/D	N/D	N/D	N/D	6.40	6.40	6.00	N/D	N/D	N/D	N/D	N/D	N/D	N/D
Lauric	C12:0	Dodecanoic	N/D	N/D	N/D	N/D	47.2	49.2	48.8	48.8	N/D	N/D	N/D	N/D	N/D	N/D
Myristic	C14:0	Tetradecanoic	0.1	N/D	N/D	N/D	17.7	18.5	19.9	17.7	0.10	0.10	N/D	N/D	3.180	N/D
Palmitic	C16:0	Hexadecanoic	N/D	N/D	N/D	N/D	9.10	7.80	N/D	N/D	11.3	11.0	12	10.5	27.25	18.4
Palmitoleic	C16:1	cis-9-Hexadecenoic	6.7	7.0	9.10	8.70	8.70	0.10	0.10	N/D	N/D	N/D	N/D	N/D	1.270	0.30
Margaric	C17:0	Heptadecanoic	1.4	2.0	2.10	N/D	N/D	N/D	0.10	N/D	0.10	N/D	N/D	N/D	0.94	0.30
Stearic	C18:0	Octadecanoic	N/D	N/D	N/D	N/D	N/D	N/D	N/D	N/D	4.30	4.00	3.0	N/D	05.10	N/D
Oleic	C18:1	cis-9-Octadecenoic	6.2	4.0	2.70	3.10	3.10	2.70	3.00	4.30	22.8	23.4	23	18.2	13.23	11.8
Linoleic	C18:2	cis-9,cis-12-Octadecedianoic	71	78	79.4	6.30	6.50	6.30	4.40	6.50	22.8	23.4	23	43.9	42.06	18.3

(*Continued*)

Name of Fatty Oil	Structure	Chemical Name of Fatty Oils												
Linolenic	C18:3	cis-6,cis-9,cis-12-Octadecatrienoic	2.1	1.0	0.70	1.20	1.70	0.80	52.7	53.2	55	19.3	02.04	26.7
Stearidonic	C18:4	15-Octadecatetraenoic	0.2	N/D	0.20	N/D	N/D	N/D	7.60	7.80	6.0	N/D	0.160	23.2
Arachidic	C20:0	Eicosanoic	3.5	4.0	5.80	0.30	N/D	N/D	N/D	N/D	N/D	N/D	0.210	N/D
Eicosenoic	C20:1	cis-11-Eicosenoic	1.9	N/D	N/D	N/D	N/D	N/D	0.40	N/D	N/D	0.3	N/D	0.50
Eicosadienoic	C20:2	11,14-Eicosadienoic	N/D	N/D	N/D	N/D	N/D	N/D	0.20	N/D	N/D	N/D	N/D	0.20
Behenic	C22:0	Docosanoic	N/D	N/D	N/D	N/D	N/D	N/D	N/D	N/D	N/D	N/D	N/D	N/D
Erucic	C22:1	cis-13-Docosenoic	6.0	4.0	N/D	N/D	N/D	N/D	0.40	N/D	N/D	N/D	N/D	N/D
Lignoceric	C24:0	Tetracosanoic	0.1	N/D	N/D	N/D	N/D	N/D	N/D	N/D	N/D	N/D	N/D	N/D
Nervonic	C24:1	15-Tetracosenoic	0.8	N/D	N/D	N/D	N/D	N/D	0.10	N/D	N/D	N/D	N/D	N/D

Name of Fatty Oil	Structure	Chemical Name of Fatty Oils	CME	CME	CME	CME	CMME	CMME	CIME	CIME	CIME	CIME	TPSO
Caproic	C6:0	Hexanoic	N/D	N/D	N/D	N/D	N/D	N/D	N/D	N/D	N/D	N/D	N/D
Caprylic	C8:0	Octanoic	N/D	N/D	N/D	N/D	N/D	N/D	N/D	N/D	N/D	N/D	N/D
Capric	C10:0	Decanoic	N/D	N/D	N/D	N/D	N/D	N/D	N/D	N/D	N/D	N/D	N/D
Lauric	C12:0	Dodecanoic	N/D	N/D	N/D	N/D	N/D	N/D	N/D	N/D	N/D	N/D	N/D
Myristic	C14:0	Tetradecanoic	0.10	N/D	N/D	N/D	0.10	0.10	N/D	0.090	N/D	N/D	N/D
Palmitic	C16:0	Hexadecanoic	4.00	3.80	3.50	3.0	6.80	6.50	14.7	14.60	12.01	17.9	23.28
Palmitoleic	C16:1	cis-9-Hexadecenoic	N/D	0.30	N/D	N/D	0.10	0.10	0.30	N/D	N/D	2.50	N/D
Margaric	C17:0	Heptadecanoic	N/D	N/D	N/D	N/D	N/D	0.10	N/D	N/D	N/D	N/D	N/D
Stearic	C18:0	Octadecanoic	1.70	1.90	0.90	1.0	4.20	3.80	13.2	19.96	12.95	18.5	10.71
Oleic	C18:1	cis-9-Octadecenoic	59.9	63.9	64.1	64	11.8	11.6	46.1	37.57	34.09	42.7	43.72
Linoleic	C18:2	cis-9,cis-12-Octadecedianoic	22.6	19.0	22.3	22	71.2	72.7	24.7	26.33	38.26	13.7	19.85
Linolenic	C18:3	cis-6,cis-9,cis-12-Octadecatrienoic	9.60	9.70	8.20	8.0	4.20	3.90	0.20	0.270	0.300	2.10	N/D
Stearidonic	C18:4	15-Octadecatetraenoic	N/D	N/D	N/D	N/D	N/D	N/D	N/D	N/D	N/D	N/D	2.410
Arachidic	C20:0	Eicosanoic	0.60	0.60	N/D	N/D	0.40	N/D	0.80	N/D	N/D	N/D	N/D
Eicosenoic	C20:1	cis-11-Eicosenoic	1.20	N/D	N/D	N/D	1.00	0.90	N/D	N/D	N/D	N/D	N/D
Eicosadienoic	C20:2	11,14-Eicosadienoic	N/D	N/D	N/D	N/D	N/D	0.20	N/D	N/D	N/D	N/D	N/D
Behenic	C22:0	Docosanoic	0.30	0.40	N/D	N/D	0.10	N/D	N/D	N/D	N/D	N/D	N/D
Erucic	C22:1	cis-13-Docosenoic	N/D	N/D	N/D	N/D	N/D	N/D	N/D	N/D	N/D	N/D	N/D
Lignoceric	C24:0	Tetracosanoic	N/D	0.20	N/D	N/D	0.10	N/D	N/D	N/D	N/D	2.60	N/D
Nervonic	C24:1	15-Tetracosenoic	N/D	0.20	N/D	N/D	N/D	N/D	N/D	N/D	N/D	N/D	N/D

APME, Aphanamixis polystachya methyl ester; COME, Coconut oil methyl ester; FOME, Fish oil methyl ester; JCME, Jatropha curcas methyl ester; MME, Manketti methyl ester; MOME, Moringa oleifera methyl ester; N/D, Not detected; PEME, Pangium edule methyl ester; PME, Palm methyl ester; SME, Soybean methyl ester. CIME, Calophyllum inophyllum methyl ester; CME, Canola methyl ester; CMME, Croton megalocarpus methyl ester; TPSO, T. peruviana seed oil.

best fits the data. Mathematically, a polynomial of order k in X is expressed in the following form [1,17,50,51,72−74]:

$$Y = C_0 + C_1X + C_2X^2 + \ ..... \ + C_kX^k \qquad (17.1)$$

where X is the variable as a function of available data and Y is the predicted value. The polynomial curve fitting method has been used in several studies to predict the properties of biodiesel−diesel blends. Table 17.7 shows some mathematical equations for predicting properties for various biodiesel blends while Table 17.8 shows some mathematical equations for predicting properties of various biodiesel feedstocks.

The prediction of important physical and chemical properties of biodiesel and its blends (whether with diesel or biodiesel) is a very important factor in the design of fuel spray, atomization, and combustion and emission system for diesel engines. It is also a highly demanding parameter because research is going on with various feedstocks for biodiesel production. These equations would help to predict the property at any percentages of biodiesel in biodiesel−diesel blend. Several studies have been conducted to examine the physical and chemical properties of biodiesel biodiesel−diesel blends. The following paragraph will summarize the most important work done in this aspect. Iqbal et al. [75] optimized the blend ratio of PB with coconut biodiesel (CB) and jatropha biodiesel (JB) using MATLAB optimization tool. A linear relationship among fuel properties was considered for MATLAB coding. The resulting optimum blend ratio and the equations of the MATLAB code were used to predict the fuel property values and were compared with the experimental values of the optimum blend fuel properties. Assuming X, Y, and Z are the blend ratio of JB, PB, and CB, respectively:

For density: $0.8833X + 0.8793Y + 0.8771Z < 0.8793$

For viscosity: $4.805X + 4.663Y + 3.180Z < 4.663$

For induction time: $2.08X + 3.24Y + 5.12Z > 3$

For flash point: $202.5X + 188.5Y + 136.5Z > 160$

For calorific value: $39.839X + 39.907Y + 36.985Z > 38.5$

Relation among the ratios: $X + Y + Z = 1$

Solving the previously mentioned inequalities using MATLAB, the obtained values of X, Y, and Z are 0.23, 0.559, and 0.211, respectively. Therefore, the blending ratios of JB, PB, and CB for the JPC blend are 23%, 55.9%, and 21.1%, respectively.

Saxena et al. [76] reviewed various methods for the prediction of important thermophysical properties, such as cetane number, kinematic viscosity, density, higher heating value, flash point, cloud point, pour point, cold filter plugging point, and vapor pressure for various biodiesel feedstocks.

Sivaramakrishnan and Ravikumar [77] developed an equation to calculate cetane number of various vegetable oils and their biodiesel from their viscosity, density, flash point, and higher calorific value. They concluded that this equation gives an accuracy of 90%.

Table 17.7 Mathematical Equation for Predicting Properties for Various Biodiesel Blends

Biodiesel Blends	Property	Mathematical Equation	R^2	Variable	Reference(s)
Biodiesel-Diesel Blending					
CIME + diesel	Kinematic viscosity at 40°C	$y = 2.5379x + 3.1036$	0.9911	x is the dependent variable; x, biodiesel%	[51,88]
	Kinematic viscosity at 100°C	$y = 0.7883x + 3.2287$	0.9976		
	Density at 40°C	$y = 0.0429x + 0.8346$	0.9997		
	Viscosity index	$y = -44.242x^2 + 103.13x + 113.02$	0.9828		
	Calorific value	$y = -5798.5x + 45,323$	0.9828		
	Oxidation stability	N/D	N/D		
	Flash point	$y = 10.382x^2 + 13.54x + 69.655$	0.9888		
COME + diesel	Kinematic viscosity at 40°C	$y = 0.8337x + 3.186$	0.9902		
	Kinematic viscosity at 100°C	$y = 0.3082x + 1.2602$	0.9933		
	Density at 40°C	$y = 0.0312x + 0.8349$	0.9993		
	Viscosity index	$y = -17.235x^2 + 60.243x + 139.2$	0.9732		
	Calorific value	$y = -7328.9x + 45,232$	0.9992		
	Oxidation stability	$y = -236.64x^3 + 453.53x^2 - 382.69x + 171.04$	0.9928		
	Flash point	$y = -236.64x^3 + 453.53x^2 - 382.69x + 171.04$	0.9928		
CMME + diesel	Kinematic viscosity at 40°C	$y = 0.7205x + 3.33$	0.9727		
	Kinematic viscosity at 100°C	$y = 0.3787x + 1.2896$	0.9796		
	Density at 40°C	$y = 0.043x + 0.8271$	0.9997		
	Viscosity index	$y = -134.32x^2 + 304.89x + 91.683$	0.9835		
	Calorific value	$y = -5798.5x + 45,323$	0.9965		
	Oxidation stability	$y = 28.003x^2 - 51.792x + 25.173$	0.9848		

	Property	Equation	R^2	Ref.
	Flash point	$y = 336.3x^3 - 384.77x^2 + 156.15x + 70.097$	0.997	
MOME + diesel	Kinematic viscosity at 40°C	$y = 1.7446x + 3.2962$	0.996	
	Kinematic viscosity at 100°C	$y = 0.581x + 1.2795$	0.9895	
	Density at 40°C	$y = 0.0328x + 0.827$	0.9993	
	Viscosity index	$y = -11.088x^2 + 108.82x + 89.982$	0.9925	
	Calorific value	$y = -5363.9x + 45,277$	0.9942	
	Oxidation stability	$y = -593.44x^3 + 1106.5x^2 - 724.66x + 238.14$	0.9984	
	Flash point	$y = 169.01x^3 - 208.93x^2 + 121.23x + 68.147$	0.9925	
POME + diesel	Kinematic viscosity at 40°C	$y = 1.3841x + 3.2024$	0.9948	
	Kinematic viscosity at 100°C	$y = 0.4976x + 1.2616$	0.9941	
	Density at 40°C	$y = 0.0238x + 0.8353$	0.9979	
	Viscosity index	$y = -43.106x^2 + 106.33x + 130.68$	0.9936	
	Calorific value	$y = -5465.1x + 45100$	0.9917	
	Oxidation stability	$y = -417.44x^3 + 893.48x^2 - 641.88x + 165.28$	0.9829	
	Flash point	$y = 348.01x^3 - 366.52x^2 + 132.24x + 67.398$	0.9947	
SFME + diesel	Kinematic viscosity at 40°C	$y = 0.0257x + 3.7969$	1	[1]
	Kinematic viscosity at 100°C	$y = 0.0076x + 1.4337$	1	
	Viscosity index	$y = 0.542x + 120.2$	1	

(Continued)

Table 17.7 Mathematical Equation for Predicting Properties for Various Biodiesel Blends—cont'd

Biodiesel Blends	Property	Mathematical Equation	R^2	Variable	Reference(s)
PEME + diesel	Kinematic viscosity at 40°C	$y = 2.9943x^2 + 0.9235x + 3.2291$	0.9951	x, biodiesel%; 0% ≤ x ≥ 20%	[1,50]
	Kinematic viscosity at 100°C	$y = -1.9229x^2 + 1.0152x + 1.2473$	0.99		
	Density at 40°C	$y = 5.7333x^3 - 2.3114x^2 + 0.313x + 0.8273$	0.9914		
	Calorific value	$y = -5750x + 45236$	0.9937		
	Oxidation stability	$y = 3521x^2 - 1239.3x + 127.84$	0.9949	5% ≤ x ≥ 20%	
MME + diesel	Kinematic viscosity at 40°C	$y = 5.2543x^2 + 1. + 3.2265$	0.9977	x, biodiesel% 0% ≤ x ≥ 20%	[1,48]
	Calorific value	$y = -6312x + 45249$	0.9943		
	Density at 40°C	$y = 6x^3 - 2.3314x^2 + 0.3193x + 0.8273$	0.995		
	Viscosity index	$y = 32867x^3 - 12503x^2 + 1453.4x + 90.636$	0.9864		
MOME + diesel	Kinematic viscosity at 100°C	$y = 0.0583x + 1.6872x + 3.2997$	0.9962	x, biodiesel% 0% ≤ x ≥ 100%	[89]
	Density at 40°C	$1.014x^2 + 31.813x + 827.19$	0.9994		
	Cloud point	$-6.1772x^2 + 21.177x + 3.8462$	0.9866		
	Calorific value	$-0.0564x^2 - 5.1609x + 45.149$	0.9937		
	Pour point	$-11.538x^2 + 31.084x + 0.042$	0.9929		
	CFPP	$7.6923x^2 + 6.3077x + 4.7902$	0.9829		
	Flash point	$46.028x^2 + 26.034x + 73.296$	0.9651		
Biodiesel–Biodiesel Blending					
SFME–POME	Kinematic viscosity at 40°C	$y = -0.5159x^2 - 1.1195 + 6.3599$	0.9908	x, POME%	[1,17,88]
SFME–COME	40°C	$y = 0.9533x^2 - 4.1182x + 6.3457$	0.9981	x, COME%	
POME–CME	Cloud point	$y = 3.4286x^2 - 20.629x + 13.429$	0.9704	x, CME%	

JCME–CME		$y = -1.1429x^2 - 12.857x + 10.457$	0.979
CIME–CME		$y = -3.4286x^2 - 12.171x + 12.171$	0.9867
POME–CME	Pour point	$y = -2.2857x^2 - 20.114x + 14.114$	0.9784
JCME–CME		$y = -13.714x^2 - 6.2857x + 10.286$	0.9785
CIME–CME		$y = -13.714x^2 - 8.6857x + 13.286$	0.9972
POME–CME	Cold filter plugging	$y = -6.8571x^2 - 15.543x + 11.943$	0.9843
JCME–CME	point	$y = -6.8571x^2 - 14.743x + 10.543$	0.9639
CIME–CME		$y = -5.7143x^2 - 16.286x + 11.486$	0.9918

CIME, Calophyllum inophyllum methyl ester; CME, Canola methyl ester; CMME, Croton megalocarpus methyl ester; COME, Coconut oil methyl ester; JCME, Jatropha curcas methyl ester; MME, Manketti methyl ester; MOME, Moringa oleifera methyl ester; N/D, Not determined; PME, Palm methyl ester; PEME, Pangium edule methyl ester; and SFME, Sterclia feotida methyl ester.

Table 17.8 Mathematical Equation for Predicting Properties for Various Biodiesel Feedstocks

Blends	Property	Mathematical Equation	R^2	Reference(s)
CMME	Flash point (FP) versus	$FP = 183.95 \times (KV)^2 + 1221.6 \times (KV) + 2099.5$	0.9534	[51,88]
CIME	kinematic viscosity (KV)	$FP = 0.4884 \times (KV)^2 + 5.1448 \times (KV) + 47.913$	0.9887	
COME		$FP = 33.934 \times (KV) + 188.35 \times (KV) + 325.3$	0.9933	
POME		$FP = 74.797 \times (KV)^2 + 517.44 \times (KV) + 968.12$	0.9569	
MOME		$FP = 13.79 \times (KV)^2 + 73.438 \times (KV) + 164.68$	0.9724	
CMME	Calorific value (CV) versus	$CV = -2410.4 \times (KV)^2 + 10,323 \times (KV) + 37,233$	0.9891	
CIME	kinematic viscosity (KV)	$CV = 560.27 \times (KV)^2 - 7392.4 \times (KV) + 63,326$	0.9975	
COME		$CV = 33.934 \times (KV)^2 - 188.35 \times (KV) + 325.3$	0.9933	
POME		$CV = 1413.7 \times (KV)^2 + 15,028 \times (KV) + 79,180$	0.996	
MOME		$CV = -3063.7 \times (KV) + 55,367$	0.9912	

CIME, Calophyllum inophyllum methyl ester; CMME, Croton megalocarpus methyl ester; COME, Coconut oil methyl ester; MOME, Moringa oleifera methyl ester; PME, Palm methyl ester.

Atabani et al. [1,17] discussed the concept of biodiesel–biodiesel blending to improve the properties of some feedstocks. For instance, blending of Sterculia feotida methyl ester (SFME) and coconut methyl ester (CoME) improves the viscosity of (SFME) from 6.3717 mm²/s to 5.3349 mm²/s (3:1), 4.4912 mm²/s (1:1) and 3.879 mm²/s (1:3) respectively. Same work was done on the effect of biodiesel–biodiesel blending on cloud point, pour point, and cold filter plugging point. The properties at different biodiesel–biodiesel blend percentages were estimated using the polynomial curve fitting method. This chapter concludes that blending of edible and nonedible biodiesel feedstocks could be considered an approach to improve the properties of the final product.

Moser [78] indicated that the fuel properties of soybean methyl ester were improved through blending with canola, palm, and sunflower methyl esters to satisfy the IV (<120) and OSI (>6 h) specifications contained within EN 14214. The CFPP of Palm methyl ester was improved by up to 15°C through blending with canola methyl ester. Statistically significant relationships were elucidated between oxidation stability and iodine value, oxidation stability and saturated fatty oil methyl ester (sunflower methyl ester) content, oxidation stability and CFPP, CFPP and iodine value, and CFPP and sunflower methyl ester content. However, the only relationship of practical significance was that of CFPP versus sunflower methyl ester content when sunflower methyl ester content was greater than 12 wt%.

Oghenejoboh and Umukoro [79] indicated that blending of biodiesel from some feedstocks, such as palm, palm kernel, jatropha, and rubber oils, with diesel has resulted in an increase in the calorific value and decrease in density, cloud point, pour point, kinematic viscosity, and flash point of biodiesel. The same work was done by Krishna [80] to improve the cold flow properties of biodiesel.

Moser [74] studied blending of biodiesels prepared from field pennycress and meadowfoam seed oils with biodiesels from camelina, cottonseed, palm, and soybean oils in an effort to improve the properties of these biodiesel fuels. For example, camelina, cottonseed, and soybean oil—derived biodiesels exhibited poor oxidative stabilities but satisfactory kinematic viscosities. Field pennycress and meadowfoam seed oil methyl esters exhibited excellent cold flow properties but high kinematic viscosities. Therefore, field pennycress and meadowfoam-derived biodiesel fuels were blended with the other biodiesels to improve cold flow, oxidative stability, and viscosity deficiencies inherent to the individual fuels. Highly linear correlations were observed between blend ratio and cold flow as well as viscosity after least squares statistical regression whereas a nonlinear relationship was observed for oxidative stability. The author concluded that, biodiesel—biodiesel blending enhanced fuel properties, such as cold flow, kinematic viscosity, and oxidative stability of biodiesel.

Sivaramakrishnan and Ravikumar [81] developed an equation to predict the higher heating value of biodiesel based on its kinematic viscosity, flash point, and density with 0.949 accuracy.

A review on the physical and chemical properties and the fatty acid composition of 26 biodiesel feedstocks (including 22 edible and nonedible oils and 4 animal fats) was conducted by [82]. The author concluded that an excellent correlation exists between iodine number and the degree of unsaturation, while a small statistical correlation ($R^2 > 0.60$) was also established for cetane number, density, pour point, carbon content, number of carbon atoms, stoichiometric air—fuel ratio, and T90 distillate temperature.

Kalayasiri et al. [83] developed two empirical equations to predict the saponification number and iodine value of biodiesel based on its fatty oil composition:

$$SN = \sum \left(\frac{560 \times A_i}{MW_i} \right) \tag{17.2}$$

$$IV = \sum \left(\frac{254 \times D \times A_i}{MW_i} \right) \tag{17.3}$$

where SN, the saponification number; A_i, the percentage of each component; D, the number of double bond; MW_i, the molecular mass of each component; and IV, the iodine value.

Krisnangkura [84] illustrated a simple method to estimate the cetane number of biodiesels based on their saponification and iodine numbers. The range of the calculated values covers all the cetane numbers of vegetable oil methyl esters determined experimentally. When it was applied to individual fatty oil methyl esters from C_8 to C_{24}, a straight line parallel to that of Klopfenstein was obtained. The developed equation was as follows:

$$CN = \left(46.3 + \left(\frac{5458}{SN}\right) - (0.225 \times IV)\right) \tag{17.4}$$

where CN, the cetane number; SN, the saponification number; and IV, the iodine value.

Ramírez–Verduzco et al. [85] attempted to develop four empirical correlations that can be used to estimate the cetane number, kinematic viscosity, density, and higher heating value of biodiesels based on their molecular weight and degree of unsaturation. The estimated values have been found to be in good agreement with the experimental values, and an average absolute deviation (AAD) of 5.95%, 2.57%, 0.11%, and 0.21% for the cetane number, kinematic viscosity, density, and higher heating value, respectively, were found. Those derived equations were as follows:

$$\varnothing_i = -7.8 + 0.302 \cdot M_i - 20 \cdot N \tag{17.5}$$

$$\ln(n_i) = -12.503 + 2.496 \cdot \ln(M_i) - 0.178 \cdot N \tag{17.6}$$

$$P_i = 0.8463 + \frac{4.9}{M_i} - 0.0118 \cdot N \tag{17.7}$$

$$\delta_i = 46.19 - \frac{1794}{M_i} - 0.21 \cdot N \tag{17.8}$$

where $\varnothing_i$, the cetane number of the ith FAME; M_i, the molecular weight of the ith FAME; N, the number of double bonds in a given FAME; n_i, the kinematic viscosity at 40°C of the ith FAME in mm^2/s; P_i, the density at 20°C of the ith FAME in g/cm^3; and δ_i, the higher heating value of the ith FAME in MJ/kg.

Talebi et al. [86] developed a new software package (the BiodieselAnalyzer) that can predict 16 different properties of biodiesel based on the fatty oil methyl ester profile of the oil feedstock used in making it.

Gulum and Bilgin [87] investigated densities of corn oil biodiesel and its blends with petrodiesel. The effects of temperature (10°C, 15°C, 20°C, 30°C, and 40°C) and biodiesel percentage in blend (B5, B10, B15, B20, B50, and B75) on the densities of blends were examined. New one- and two-dimensional regression models have been developed for estimating densities at various temperatures and biodiesel percentages. Moreover, the qualities of the corn oil biodiesel and its blends were evaluated by determining the other important properties, such as flash point temperature and higher heating value. In order to predict these properties, some equations were also evaluated as a function of biodiesel percentage in blend.

17.9 ENGINE AND EMISSIONS TESTS

Biodiesel is not compatible with all engine components. Hence it is absolutely important for you to check your owner's engine manual to find out the types of fuels you can use in your engine. Some manufacturers, such as Volkswagen and ford, have taken a step in this direction and are in the process of creating a diesel car or diesel engine that is more compatible with blended biodiesel fuels. Manufacturers such as Mercedes in their BR 300, 400, 500, and 900 series engines; DAF in their LF, CF, and XF Models; Scania and MAN Trucks; and Peugeot Citroen Group have all approved certain models and engine types to be used with B30–B100 biodiesel [90].

The fuel properties of any fuel play a significant role in the engine performance. Biodiesel has a major influence on engine performance due to its higher oxygen contents, higher heat of vaporization, higher density, higher viscosity, lower heating values, and higher cetane number [91].

According to Ref. [92], the higher peak in cylinder pressure with biodiesel compared to diesel is attributed to the higher cetane number and reduced ignition delay. This is also attributed to the higher oxygen content of the biodiesel that lead to the improved combustion. Refs. [67,93,94] attributed the higher brake–specific fuel consumption (BSFC) of biodiesel compared to diesel to the lower energy contents and higher density of biodiesel fuel, while Ref. [95] attributed the reduction in brake power (BP) of biodiesel compared to diesel to the lower heating values and higher viscosities of biodiesel.

Table 17.9 gives a summary of engine performance and emissions of *Croton megalocarpus* (nonedible), coconut (edible), and *C. inophyllum* (nonedible) methyl esters relative to diesel and possible causes.

It can be seen that all biodiesel–diesel blends have lower torque, brake power, CO and HC, and higher BSFC. However, *C. megalocarpus* (nonedible) and coconut (edible) have higher NO while *C. inophyllum* (nonedible) has lower NO compared to diesel. Biodiesel is not universally compatible with every automobile ever built. Hence it is absolutely important for you to check your owner's engine manual to find out the types of fuels you can use in your engine. Also, biodiesel is not compatible with all engine components. Some manufacturers, such as Volkswagen and ford, have taken a step in this direction and are in the process of creating a diesel car or diesel engine that is more compatible with blended biodiesel fuels. Manufacturers such as Mercedes in their BR 300, 400, 500, and 900 series engines; DAF in their LF, CF, and XF Models; Scania and MAN Trucks; and Peugeot Citroen Group have all approved certain models and engine types to be used with B30-B100 biodiesel [90].

Table 17.9 Summary of Engine Performance and Emissions Test of Biodiesel–Diesel Blends of CMME, CIME, and COME (B10 and B20) Relative to Diesel [1]

	CMME 10	CMME20	CIME10	CIME20	COME10	COME20	Possible Causes
Engine Performance							
Torque	−4.05%	−5.95%	−5.04%	−6.77%	−2.7%	−4.84%	1. Lower calorific value of biodiesel 2. Higher viscosity and density of biodiesel
BP	−3.94%	−5.95%	−5.06%	−6.85%	−2.43%	−4.88%	1. Lower heating value of biodiesel 2. Higher viscosity of biodiesel
BSFC	+5.3%	+8.9%	+6.85%	+12.33%	+2.99%	+8.31%	1. Higher density of biodiesel 2. Lower energy content of biodiesel
Emissions Analysis							
CO	−8.89%	−29.04%	−16.81%	−34.49%	−15.44%	−34.72%	1. Higher oxygen content of biodiesel 2. Lower carbon and hydrogen of biodiesel
HC	−3.89%	−11.68%	−5.19%	−19.48%	−3.89%	−15.58%	For CMME and COME
NO	+7.31%	+8.06%	−0.54%	−1.55%	+1.55%	+6.16%	1. The lean air/fuel ratio, as biodiesel is an oxygenated fuel that causes higher chamber temperature by improving combustion at warmed-up condition 2. Adiabatic flame temperature can be also considered as another reason of increasing NO/NOx. Biodiesel fuel contains higher percentages of unsaturated fatty oils that have higher adiabatic flame temperature which causes higher NO/NOx emission For CIME 1. Lower combustion chamber temperatures using CIME.

CIME, Calophyllum inophyllum methyl ester; CMME, Croton megalocarpus methyl ester; COME, Coconut oil methyl ester.

17.10 IMPORTANCE OF STATISTICAL AND UNCERTAINTY ANALYSIS

Rizwanul Fattah et al. [96] indicated that a statistical analysis can be carried out by applying two-sided Student's t-test for independent variables to test for significant differences between samples set using Microsoft Excel 2013. Differences between mean values at a level of $p = 0.05$ (95% confidence level) were considered statistically significant.

On the other hand, errors and uncertainties in the experiments can arise from instrument selection, experimental condition, equipment calibration, ambient environment, observation, reading, and test planning. Uncertainty analysis in engine performance tests is required to prove the accuracy of the experiments. Percentage uncertainties of various parameters such as total fuel consumption, brake power, brake-specific fuel consumption, and brake thermal efficiency can be calculated using the percentage uncertainties of various instruments used in the experiment [97].

The relative uncertainty can be determined using the linearized approximation method of uncertainty using the following equation [96,98,99]:

$$\frac{\Delta z}{z} \approx \frac{1}{z} \sum_{i=1}^{p} \frac{\partial z}{\partial x_i} \partial x_i \qquad (17.9)$$

Using the principle of propagation of errors, the total percentage uncertainty of an experimental trial can be computed as [97]:

$$= \text{Square root of} (\text{uncertainty of tfc})^2 + (\text{uncertainty of brake power})^2$$
$$+ (\text{uncertainty of specific fuel consumption})^2$$
$$+ (\text{uncertainty of brake thermal efficiency})^2 + (\text{uncertainty of Co})^2$$
$$+ (\text{uncertainty of Hc})^2 + (\text{uncertainty of No})^2 + (\text{uncertainty of smoke})^2$$
$$+ (\text{uncertainty of EGT indicator})^2$$

$$(17.10)$$

17.11 EFFECTS OF ADDITIVES ON BIODIESEL QUALITY

Basha and Anand [100] prepared biodiesel emulsion (83% of JB, 15% of water, and 2% of surfactants (Span80 and Tween80)) by emulsification technique with the aid of a mechanical agitator. The prepared fuel is mixed with the alumina nanoparticles in the mass fractions of 25, 50, and 100 ppm with the help of an ultrasonicator. The whole investigation is carried out in a constant speed diesel engine in three phases using JB, JB emulsion fuel, and alumina nanoparticle−blended JB emulsion fuels. The experimental results revealed a substantial enhancement in the performance and a reduction in harmful emissions for the biodiesel emulsion fuels compared to those of neat biodiesel.

Kivevele and Huan [101] evaluated the effect of metal contents and antioxidant additives on the storage stability of biodiesel produced from *C. megalocarpus* and *Moringa oleifera*. The produced *C. megalocarpus* and *M. oleifera* biodiesels displayed oxidation stability of 2.5 and 5.3 h, respectively. The oxidation stability of *C. megalocarpus* biodiesel did not meet the minimum requirement prescribed in ASTM D6751 and EN 14214 standards of 3 and 6 h, respectively. However, *M. oleifera* biodiesel met the minimum requirement of ASTM D6751 but not the EN 14214. In order to improve the storage stability of *C. megalocarpus* and *M. oleifera* biodiesels, two synthetic antioxidants—1,2,3 tri-hydroxy benzene (pyrogallol, PY) and 3,4,5-tri-hydroxy-benzoic acid (propyl gallate, PG)— were doped to the biodiesel samples to examine its effectiveness. Moreover, the samples with and without antioxidant were mixed with different transition metals (Fe, Ni, Mn, Co, and Cu) to evaluate the impact of these metals on storage stability of *C. megalocarpus* and *M. oleifera* biodiesels. The samples were stored indoor for 6 months in completely closed and open translucent plastic bottles, the oxidation stability were measured every month.

It was found that 200 ppm of PY and PG was enough for *C. megalocarpus* and *M. oleifera* biodiesels to meet EN 14112 biodiesel oxidation stability. Moreover, the antioxidant PY was more effective than PG. Fe displayed least detrimental effect on oxidation stability of *C. megalocarpus* and *M. oleifera* biodiesels while Cu had great detrimental impact. The samples in the completely closed (airtight) bottles recorded higher oxidation stability compared to the ones kept in the open bottles, because the latter's exposure to air enhanced the oxidation of the samples.

Kivevele et al. [102] evaluate oxidation stability of biodiesel produced from *C. megalocarpus* oil. It was found that oxidation stability of *C. megalocarpus* biodiesel did not meet the specifications of EN 14214 (6 h). The effectiveness of three antioxidants: 1,2,3 tri-hydroxy benzene (PY), 3,4,5-tri-hydroxy benzoic acid (PG), and 2-tertbutyl-4-methoxy phenol (butylated hydroxyanisole, BHA) on oxidation stability of *C. megalocarpus* biodiesel was investigated. The article of Kivevele et al. concluded that the effectiveness of these antioxidants was in the order of PY > PG > BHA.

Agarwal et al. [103] investigated the effectiveness of five cheap and commercially available antioxidants namely 2,6-di-tertbutyl-4-methyl phenol (BHT), 2-tertbutyl-4-methoxy phenol (BHA), 2-tertbutyl hydroquinone (TBHQ), 1,2,3 tri-hydroxy benzene (PY), and 3,4,5-tri-hydroxy benzoic acid (PG) on biodiesel produced from nonedible vegetable oils of karanja, neem, and jatropha. The antioxidants were dosed in concentrations ranging from 100 to 1000 ppm. JB showed higher oxidation stability than karanja and neem-based biodiesel; however, it was still unable to meet the prevailing ASTM biodiesel oxidation stability specifications.

Optimum concentration required to meet the specifications for karanja biodiesel was: PG 500 ppm, PY 300 ppm, and BHA 500 ppm. The optimum concentration to stabilize neem biodiesel was PG 200 ppm and PY 200 ppm and for JB, it was: BHT700 ppm, PG 100 ppm, PY 100 ppm, and BHA 300 ppm. The authors concluded that the effectiveness of antioxidants investigated was: PY > PG > BHA > BHT > TBHQ, for karanja biodiesel, neem biodiesel, and JB.

Ryu [104] investigated the effects of antioxidants on the oxidation stability of biodiesel derived from soybean oil, the engine performance, and the exhaust emissions of a diesel engine. The results show that the efficiency of antioxidants is in the order TBHQ > PG > BHA > BHT > α-tocopherol. The oxidative stability of biodiesel fuel attained the 6 h quality standard with 100 ppm TBHQ and with 300 ppm PG in biodiesel fuel. Exhaust emissions were not influenced by the addition of antioxidants in biodiesel. However, the BSFC of biodiesel with antioxidants decreased more than that of biodiesel without antioxidants.

Imtenan et al. [105] investigated an improvement in emission and performance characteristics of PB—diesel blend with the help of ethanol, n-butanol, and diethyl ether as additives. The blends consisted of 80% diesel, 15% PB, and 5% additive. These additives improved the fuel blend regarding density and viscosity which in turn improved atomization and showed better combustion characteristics through higher engine brake power, lower BSFC, and higher BTE than PB—diesel blend (B20). Diethyl ether showed highest 6.25% increment of brake power, 3.28% decrement of BSFC, and about 4% increment of BTE than 20% PB—diesel blend when used as additive through its low density and viscosity profile with quite a high calorific value. Other two additives also showed interesting improvement regarding performance. All the blends with additives showed decreased NO and CO emission but HC emission showed a slight increment. The authors concluded from this experiment that, ethanol, n-butanol, and diethyl ether are quite effective regarding emission and performance even when they are used only about 5% as additive. The same authors have done similar work in Refs. [106,107]. Palash et al. [108] present an experimental study on a four-cylinder diesel engine to evaluate the performance and emission characteristics of JB blends (JB5, JB10, JB15, and JB20) with and without the addition of N,N-o-diphenyl-1,4-phenylenediamine (DPPD) antioxidant at 0.15% (m). Engine tests were conducted at engine speeds 1000—4000 rpm at an interval of 500 rpm under the full-throttle condition. The results showed that the additives reduce NO_x emissions significantly with a slight penalty in terms of engine power and brake-specific fuel consumption as well as CO and HC emissions. However, when compared to diesel combustion, the emissions of HC and CO with the addition of the DPPD additive were found to be nearly the same or lower. By the addition of 0.15% (m) DPPD additive in JB5, JB10, JB15, and JB20, the reduction in NO_x emissions were 8.03%, 3.503%, 13.65%, and 16.54%, respectively, compared to biodiesel blends without the additive. Moreover, the addition of DPPD additive to all biodiesel blend samples reduced the exhaust gas temperature (EGT). Shahabuddin et al. [109] investigated engine and emission performance of B20 with

IRGANOR NPA as a fuel corrosion inhibitor. The final result implied that the biodiesel with some additives (B20 + 1%) produces 1.73% and 9% higher brake power as compared to fuel B20 and OD, respectively; consumes 26% and 6% lower SFC as compared to fuel B20 and OD, respectively; and reduces CO, NO_x, and CO_2 emissions as compared to other fuels. Swaminathan and Sarangan [110] tested fish oil biodiesel in single-cylinder diesel engine blended with diethyl ether (DEE) at 1%, 2%, and 3% to improve the performance and reduce the emission of the engine.

The percentage reduction toward 91% CO; 62% CO_2; 92% NO_x; and 90% C_xH_y were attained when the engine was run at maximum load using fish oil biodiesel with 2% additive with EGR. The authors concluded that the optimum values were obtained with 2% of additives.

17.12 DIFFERENT TYPES OF ALGAE CULTURES FOR BIODIESEL PRODUCTION

Microalgae can be cultivated in different cultures according to the algae species and the purpose of cultivation. Microalgae cultures include cultivation in open ponds, closed ponds, photobioreactors, desert-based algae cultivation, waste water, marine algae cultivation, and cultivation next to power plants.

17.12.1 Cultivation in Open Pond

Algae can be cultivated in open ponds for large-scale production of biodiesel. The most commonly used systems include shallow big ponds (Fig. 17.6A), circular ponds, and raceway ponds (Fig. 17.6B). One of the major advantages of open ponds is that they are easier to construct and operate than most closed systems.

Figure 17.6 (A) Open pond and (B) raceway open pond for cultivation of algae [111].

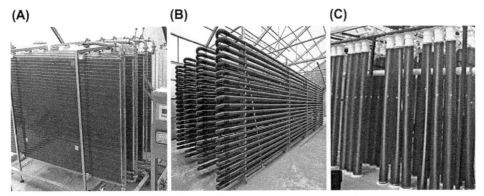

Figure 17.7 (A) Horizontal plastic plate [113], (B) tubular [114], and (C) column photobioreactors [115].

17.12.2 Photobioreactors

Photobioreactors are the most effective way to cultivate microalgae. In closed photobioreactors the culture condition parameters, such as carbon dioxide, light intensity, nutrient levels, and temperature of the culture are controlled. Under optimal conditions, microalgae are capable of doubling within hours and reaching high cell densities [112]. Closed photobioreactors have many advantages, which are making them a suitable culture environment for biofuel production.

Most closed photobioreactors are designed as plate reactors (Fig. 17.7A), tubular reactors (Fig. 17.7B), or columnar reactors (Fig. 17.7C).

17.13 ALGAE GROWTH ON WASTEWATER FOR BIODIESEL PRODUCTION

As mentioned earlier, one of the advantage of cultivation of microalgae for biodiesel production is their ability to grow on wastewater—either domestic or agricultural or industrial wastewaters. The cultivation of microalgae on wastewater to obtain biodiesel can be an economically feasible strategy with minimal environmental impacts [116]. The cultivation of microalgae species, such as *Chlamydomonas mexicana*, *Chlorella vulgaris*, *Scenedesmus obliquus*, and *Micractinium reisseri*, in wastewaters has been demonstrated previously [117,118]. Abou-Shanab et al. [119] isolated the microalgal species *M. reisseri* and *S. obliquus* from municipal wastewater mixed with agricultural drainage. *M. reisseri* was proved to grow well in municipal wastewater (influent, secondary, and tertiary effluents) which varied in nutrient concentration. *M. reisseri* showed growth rate (μ_{opt}) of 1.15, 1.04, and 1.01 1/day for the influent and the secondary and tertiary effluents, respectively. One of the advantages of algae cultivation in wastewater is their effective ability to assimilate both organic carbons and inorganic nutrients from wastewater to produce maximum algal biomass, efficient nutrient removal, and lipid accumulation which can then be converted into biodiesel [120]. In this connection, [119] found that the microalga *M. reisseri* removed 94% of phosphorus from secondary effluent of municipal wastewater and supported 40% saturated fatty oil content. The microalgae when grown on tertiary effluent supported 40% lipid content, unsaturated fatty acid content, including monounsaturated and polyunsaturated fatty acids (66%), and nitrogen removal (80%). The fatty acids of *M. reisseri* were mainly composed of palmitic, oleic, linoleic, and α-linolenic acids.

In addition to the lipid accumulation which can range from 10% to 80% of the dry cell weight [121], another advantage of using microalgae to treat wastewater is their photosynthetic CO_2 fixation, which contributes to reduce greenhouse gases [122].

17.14 MICROALGAL POTENTIAL FOR BIODIESEL PRODUCTION

Some algae are capable of accumulating lipids and fatty acids during the photosynthesis process by converting carbon dioxide in the presence of sunlight to be used as biofuel, food, and other valuable compounds. In earlier work, we screened 13 freshwater microalgae for fatty oils productivity. We found that *S. obliquus* was a promising microalga for lipid production in large-scale cultures because of its high gain of biomass production which resulted in high lipid and fatty oil productivity [123]. The idea of using microalgae as a source of biofuel is not new, but it nowadays has been taken seriously because of the rising price of petroleum more significantly and the emerging concern about burning of fossil fuels that causes global warming [124]. Gouviea and Oliveira [125] screened the microalgae *C. vulgaris*, *Spirulina maxima*, *Nannochloropsis* sp., *Neochloris oleabundans*, *S. obliquus*, and *Dunaliella tertiolecta* in order to choose the best producing alga for biofuel production in terms of quantity and quality as oil source. They found that the freshwater microalga *N. oleabundans* and marine microalga *Nannochloropsis* sp. proved to be suitable raw materials for biodiesel production, due to their high oil content (29.0% and 28.7%, respectively). El-Mekkawi et al. [126] isolated the cyanobacterium *Spirulina platensis* and green algae *S. obliquus* and *Nannochloropsis* sp. from Nile River in order to select the best alga and culture conditions for oil production. They found that maximum amount of oil production in photobioreactor was optimal when the factors affecting algal growth as light intensity, mixing, temperature, and aeration are optimal. They concluded that the achieved productivity was 0.54 g/L of algae biomass after 13 days growth with doubling time 2.8 days and specific growth rate $0.25\,d^{-1}$. In this connection, Gouviea and Oliveira [125] found that *S. obliquus* presents the most adequate fatty oil profile, namely in terms of linolenic and other polyunsaturated fatty oils. However, the microalgae *N. oleabundans*, *Nannochloropsis* sp., and *D. tertiolecta* can also be used if associated with other microalgal oils and/or vegetable oils. The effort of scientists to find out algal strains that can produce high amounts of lipids and biodiesel and at the same time can be cultivated economically on seawater and on wastewater for the dual purposes of biodiesel production and bioremediation of the wastewater is in progress. One of the most important decisions is the choice of species to use. High lipid productivity is a key desirable characteristic of a species for biodiesel production. Griffiths and Harrison [127] reviewed information available in the literature on microalgal growth rates, lipid content, and lipid productivities for 55 species of microalgae, including 17 Chlorophyta, 11 Bacillariophyta, and 5 Cyanobacteria as well as other taxa. Marine diatoms may be promising microalgae for biodiesel production, in this connection Matsumoto et al. [128] isolated and identified the marine diatom *Navicula* sp. (JPCC DA0580) and the green microalga *Chlorella* sp. (NKG400014), respectively, by 18S rDNA analysis. They studied the growth and lipid accumulation of both marine microalgal strains by changing nutrient concentrations in growth media and initial illumination intensity. They concluded that the highest productivity of fatty oil methyl ester (FAME) reached 154 mg/L/week for *Chlorella* sp. and 185 mg/L/week for *Navicula* sp. From gas chromatography/mass spectrometry analysis, it was found that FAME fraction from *Chlorella* sp. (NKG400014) mainly contained 9-12-15-octadecatrienoate (C18:3) and that from *Navicula* sp. (PCC DA0580) mainly contained methyl palmitate (C16:0) and methyl palmitoleate (C16:1) oils. Furthermore, calorimetric analysis revealed that the energy content of strain was 4233 kcal/kg for *Chlorella* sp. (NKG400014) and 6423 kcal/mg for *Navicula* sp. (JPCC DA0580), respectively. The value from *Navicula* sp. (JPCC DA0580) was equivalent to that of coal. The strains *Chlorella* sp. (NKG400014) and *Navicula* sp. (JPCC DA0580) will become promising microalgae candidates that can grow as dominant species in the open ocean toward production of biofuels.

In spite of their slow growth, the microalgae *Chlamydomonas reinhardtii, Dunaliella salina, Chlorella spp.* and *Botryococcus braunii* can contain over 60 wt% lipid, much of which is secreted into the cell wall [129]. Other important algal groups include the diatoms *Phaeodactylum tricornutum* and *Thalassiosira pseudonana* and other heterokonts including *Nannochloropsis* and *Isochrysis* sp [130].

17.15 ADVANTAGE OF BIODIESEL OVER HIGHER PLANTS

Algae from either eukaryotes or prokaryotes are wide diverse group of photosynthetic organisms and have attracted significant attention as renewable energy feedstocks due to their rapid growth rate, biomass production, and great potential to synthesize fatty acids which can be converted to biodiesel [131,132] when compared to oleaginous plants [133]. In addition, its oil content is usually between 10% and 50% of their dry cell weight and in some cases some species can produce 80% of oil under certain growth conditions [134,135].

Microalgae possess the following advantages over higher plants:

1. They grow at high rates and can synthesize and accumulate large quantities of neutral lipids (20–50% dry weight of biomass).
2. They can exceed the oil yield of the oilseed crops of all year round production because it can grow in continuous cultures for long time providing a reliable and continuous supply of oil.
3. The consumption of water by microalgae is less than that for terrestrial crops, therefore it saves freshwater resources in case of cultivation of freshwater algae species.
4. Microalgae cultures do not require herbicides or pesticides application.
5. They can reduce emissions of greenhouse gas through exploiting CO_2 from flue gases emitted from fossil fuel–fired power plants and other sources in photosynthesis process.
6. They can grow on wastewater from different sources, such as agricultural runoff, animal feed wastes, industrial, and municipal wastewaters, and have the ability for bioremediation by removal of ammonia, nitrate, phosphate, and toxic metals.
7. They can be cultivated in saline water on nonaerable land and do not compete for resources with conventional agriculture.
8. According to the microalgae species, other valuable compounds may also be extracted, which can be used in different industrial sectors, including a large range of fine chemicals and bulk products, such as polyunsaturated fatty oils, antioxidants, polysaccharides, pigments, and natural dyes.
9. The remaining algae dry weight biomass after extraction of lipids and fatty acids can be used as feed or food for animal and aquaculture nutrition because it contains high amount of proteins [136–138].
10. Microalgal biodiesel has a higher photon conversion efficiency and produces nontoxic and highly biodegradable biofuels [139].

17.16 ALGAE CULTURE CONDITIONS AND BIODIESEL PRODUCTION

The culture conditions as well as the medium components play an important role in the biodiesel production from microalgae. El-Sheekh et al. [140] studied the enhancement of fatty oil productivity from green microalga *S. obliquus* through certain modifications and optimization of the culture medium composition. They studied the effect of different concentrations of potassium nitrate, sodium bicarbonate, salinity, glycerol, and molasses of sugarcane on the production of biomass and esterified fatty acids of the alga. The results showed that low concentrations of $NaHCO_3$ (0.5 g/L) caused an increase in the biomass productivity, while it decreased fatty oil productivity at all the tested concentrations. Salinity increase in the culture medium enhanced both biomass and fatty oil productivity. The optimum NaCl concentration and sea water ratio were 0.94 g/L and 25%, respectively, which resulted in 56% and 39% increase in fatty oil productivity, respectively. The growth medium of algae must provide the inorganic elements that constitute the algal cells. Essential elements include phosphorus, nitrogen, iron, and in some cases silicon. Minimal nutritional requirements can be estimated using the approximate molecular formula of the microalgal biomass, that is, $CO_{0.48}H_{1.83}N_{0.11}P_{0.01}$, derived after [141]. El-Sheekh et al. [140] found that nitrogen deficiency increased fatty acid content by 54% over control but fatty oil productivity was decreased as a result of growth inhibition. Addition of 0.05 and 0.1 M of glycerol increased the biomass productivity by 6% and 5%, respectively, but showed no significant effect on fatty oil productivity as a result of decrease in fatty acid content. Sugarcane molasses stimulated both biomass and fatty acid content of the studied algae cells. Microalgae produce lipids, carbohydrates, and proteins during photosynthesis process, using nutrients and light. These metabolic products are tightly connected to environmental and nutrient conditions, including light intensity, temperature, pH, CO_2 concentrations, and nutrients availability [142]. Environmental and nutritional factors exert great effects on lipid content and biomass productivity of microalgae such as pH, temperature, light, nitrogen, and phosphorus. Depending on the microalgal strain, the percentage of lipid content can be increased under stress conditions, for example, low temperature or nitrogen deficiency and phosphorus in the medium [143]. Chavan et al. [143] also stated that the exact combination of different environmental factors (such as temperature, pH, photoperiod, and nitrogen) for large-scale cultivation of microalgae for biodiesel production differ from species to species. This production efficiency depends on nutrient availability in the medium and environmental conditions of the culture. Another technique to produce biodiesel from microalgae through cultivating the algae under heterotrophic conditions using cheap organic carbon source, such as sugarcane molasses or corn powder hydrolysate, instead of glucose which is relatively expensive than remnants of plants. Xu et al. [144] cultivated the green microalgae *Chlorella protothecoides* in a medium containing organic carbon source, corn powder hydrolysate, instead of glucose to reduce the cost of biodiesel production in heterotrophic culture medium fermenters. They can reach to 55% lipid under these conditions and also the result showed that cell density significantly increased under the heterotrophic condition and the highest cell concentration reached 15.5 g/L. The biodiesel produced was characterized by a high heating value of 41 MJ/kg, a density of 0.864 kg/L, and a viscosity of 5.2×10^{-4} Pa s (at 40°C).

17.16.1 Hydrogen Ion Concentration

Hydrogen ion concentration (the pH) is one of the important factors that affect growth and multiplication of algae and hence the oil and biodiesel production. Most algal growth occurs in the region of neutral pH, although optimum pH is the pH of initial culture in which an alga is adapted to grow [145]. Bartley et al. [146] found that pH of around 8 seems most beneficial for maximum growth rate and lipid accumulation of *Nannochloropsis salina* and to minimize invading organisms. However, adding buffers will not be cost-effective or realistic at a large scale. They also demonstrated that higher pH values per se do not slow *Nannochloropsis* production. Thus, the addition of CO_2 at large scales is mostly valuable for providing an inorganic carbon source for algae.

Moheimani [147] found pH 7 and 7.5 to be ideal for lipid accumulation in *Tetraselmis suecica* and *Chlorella* sp. While, Bartley et al. [146] found no significant effect of pH change on lipid accumulation, the treatment with a pH change to 8 exhibited the greatest overall accumulation (averaging 24.75% by mass) of *N. salina*. Rodolfi et al. [148] found the lipid content (% biomass) for different *Nannochloropsis* spp. to be 24.4−35.7%. The earlier results indicate that pH may not be an important stress factor that triggers increased lipid accumulation in microalgae. Acidic pH of culture media can alter nutrient uptake or induce metal toxicity and therefore have an effect on algal growth and oil production [149]. The green microalga *Chlamydomonas acidophila* and the diatom *Pinnularia braunii* accumulate storage lipids, such as triacylglycerides, under extremely acidic environment (pH 1) [150]. However, basic pH decreases membrane-associated polar lipids due to cell cycle inhibition. In basic pH conditions, membrane lipids in *Chlorella* were observed to be less unsaturated [151].

17.16.2 Light

Light is one of the factors that affect growth and lipid accumulation in photoautotrophic algae. Algae utilize light energy to convert CO_2 to organic compounds, such as sugars and lipids. Microalgae need light and dark regimes for photosynthesis. The light phase provides adenosine triphosphate (ATP) and nicotinamide adenine dinucleotide phosphate-oxidase (NADPH) [152]. The light intensity is also important to the specific algae species to get an optimal growth with maximum amount of lipids and fatty acids to be esterified to form biodiesel. Since the microalgae cultivation for the purpose of biodiesel production using photobioreactors, the average provision of light is linked to bioreactor depth, diameter, mixing, and algal cell concentration. Therefore, a larger surface area to volume ratio, may be achieved through areas of thin paneling or narrow tubing, results in higher light provision. At low light intensities, photosynthetic efficiency is highest. At high light intensities, in spite of the faster photosynthetic rate, there is a lesser amount of efficient use of absorbed light energy by algal cells [143]. Algae can grow in light/dark rhythm or in continuous light; in this connection the (16:8 h light/dark photoperiod) method was found to be suitable for outdoor cultivation in Oman because of enough solar energy throughout the year, also artificial lights can be used in cloudy winter days if it is necessary [153].

There are many factors that affect algal growth and biomass of microalgae in pho-tobioreactors. There are several suggestions to increase distribution of light through sufficient agitation, aeration, and energy demand for higher performance. In closed bioreactors, there are many ways to increase light rather than increasing the transparent surface, and bringing the algal growth to the light is to use alternative ways to bring light to the biomass layers. This can be achieved by using milli- and micro-scaled multi structure fleece of glass fibers, which can be used in light conductive structure to guide light into a compact closed reactor [154].

17.16.3 Salinity

Salinity is an important factor for the growth and biodiesel production by microalgae. Some species of microalgae require high salinity concentration in the growth medium because they were isolated from saline habitats, for example, *Dunaliella* sp. Therefore, marine microalgae species that have the ability to grow in brackish water or seawater could be considered as a good economic candidates for biodiesel production than freshwater algae since it can save freshwater. In addition, Ho et al. [155] stated that some critical biological features, such as rate of growth, lipid content, and lipid composition, are important for justifying the feasibility of using microalgae to produce lipids for biodiesel synthesis. Thus, selecting a marine alga candidate with rapid growth, high lipid content, and appropriate lipid composition is essential for high biodiesel production. Several marine microalgae have been considered as potential biodiesel producers due to their ability to tolerate high salt and their high lipid content. These species include *Chlorella sorokiniana* [156], *Nannochloropsis* sp. F&M-M24 [157], *Nannochloropsis gaditana* [158], *D. tertiolecta* ATCC30929 [159], and *C. protothecoides* [160]. Salinity stress induced physiological and biochemical changes in microalga *Scenedesmus* sp. during single stage cultivation, 33.13% lipid and 35.91% carbohydrate content were found in 400 mM NaCl-grown culture. During two-stage cultivation of unicellular microalga, salinity stress of 400 mM for 3 days resulted in 24.77% lipid (containing 74.87% neutral lipid) along with higher biomass as compared with single-stage cultivation, making it an efficient strategy to enhance biofuel production potential of *Scenedesmus* sp. [161]. El-Sheekh et al. [140] studied the effect of salinity on the production of biomass, fatty oils, and biodiesel from the microalga *S. obliquus*. They found that increase of salinity enhanced both biomass and fatty oil productivity. The optimum NaCl concentration and sea water ratio were 0.94 g/L and 25%, respectively, which resulted in 56% and 39% increase in fatty oil productivity, respectively. Bartley et al. [162] grew *N. salina* at 22 PSU until the populations were at stationary phase and then increased salinity to 34, 46, and 58 PSU. The lipid content increased significantly at these higher salinities and was highest at 34 PSU (36% dry tissue mass). Analysis of Folch extracts by FT-IR and mass spectrometry showed a monotonic increase in triglyceride content and decreased membrane lipid content with increased salinity levels.

17.16.4 Temperature

Temperature is one of the vital environmental factors that influence algal growth, cell size, biochemical composition, and nutrient requirements. Microalgae have the ability to grow in a range of temperature, depending upon strain, region, and time of the year. The optimum growth rate of most algae occurs between temperatures 20°C and 25°C for mesophilic species, such as *Plectonema calothricoides* and *Phormidium luridum*. In thermophilic algae strains, such as *Mastigocladus laminosus*, the optimum growth temperature increase for instance 55°C, and it decreases down for psychrophilic strains; for instance, 4°C for *Mastigocladus vaginatus* [163]. Microalgal growth and their metabolic activities including lipid biosynthesis are sensitive to temperature changes. In the coupled system of wastewater treatment and biodiesel production by microalgae, increasing of lipid productivity via temperature adjustment has an advantage that the composition of wastewater does not need to be changed or primarily chemically treated, making it suitable for processing and ecologically safe [164]. Converti et al. [165] studied the effect of temperature on the lipid content of the microalgae *Nannochloropsis oculata* and *C. vulgaris*, and they found that the lipid content was strongly affected by variations in growth temperature of the studied algae. They also found that the increase in temperature from 20°C to 25°C doubled the lipid content of *N. oculata* (from 7.90% to 14.92%), while temperature increase from 25°C to 30°C resulted in a decrease of the lipid content of *C. vulgaris* from 14.71% to 5.90%. Nitrogen concentration in the medium decreased by 75%, with respect to the optimal values for growth. The lipid fractions of *N. oculata* increased from 7.90% to 15.31% and of *C. vulgaris* from 5.90% to 16.41%, respectively.

17.17 FUTURE

According to the IEA perspectives, the share of renewables is expected to increase in the coming years. Biodiesel is a well-known technology and the production potential of installed plants is largely underexploited. So it seems reasonable to presume that biodiesel production will still keep growing in the near future. However, competition for feedstock supply, development of new technologies (i.e., green diesel), and concerns about biodiesel degradability and long-term effects on engines have also to be taken into consideration.

In the following section, the author reflects three options that may contribute to the future of biodiesel industry.

17.17.1 Conversion of Biodiesel—Derived Crude Glycerol to n-Butanol

Khanna et al. [166] present a research paper on the conversion of biodiesel-derived crude glycerol to n-butanol via anaerobic fermentation process using immobilized *Clostridium pasteurianum* cells. Experiments were done to optimize the growth conditions and growth medium, namely, reinforced clostridial media (RCM) and cooked meat media (CMM). Varying concentration of crude glycerol as a substrate was found to have significant effect on product profile with 25 g/L glycerol giving maximum yield of n-butanol after 120 h.

17.17.2 Conversion of Lipid From Food Waste to Biodiesel

Karmee et al. [167] reported the utilization of lipid obtained from food waste as a nonedible and low-cost resource for biodiesel production. Lipid obtained from food waste was transesterified with methanol using base (chemical catalyst) or lipase (biocatalyst) as catalysts. The maximum biodiesel yield was 100% for the base (KOH)-catalyzed transesterification at 1:10 lipid to methanol molar ratio in 2 h at 60°C, whereas Novozyme-435 yielded a 90% FAME conversion at 40°C and 1:5 lipid to methanol molar ratio in 24 h. Lipid obtained from fungal hydrolysis of food waste is found to be a suitable feedstock for biodiesel production.

Karmee and Lin [168] highlighted that lipid fraction that can be obtained after hydrolysis of food waste is considered nonedible and can be used as a potential source to produce biodiesel. Some examples of food waste areas follows:

1. rotten fruits and vegetables;
2. fish and poultry organs, intestine, meat trimmings, and other residues;
3. fruit and vegetable peelings;
4. meat, fish, shellfish shells, and bones;
5. plate scrapings and leftover of cooked food.

The routes for the conversion of food waste to biodiesel and bioethanol has been proposed in Refs. [168,169].

17.17.3 One-Step Production of Biodiesel From High Fatty Nonedible Oils

Pua et al. [170] investigated the direct production of biodiesel from high-acid value jatropha oil with solid acid catalyst derived from lignin. Solid acid catalyst was prepared from Kraft lignin by chemical activation with phosphoric acid, pyrolysis, and sulfuric acid. The reaction mixtures were loaded into an autoclave (reaction temperature: 120°C) for esterification and transesterification to biodiesel directly.

The highest catalytic activity was achieved with a 96.1% esterification rate, and the catalyst can be reused three times with little deactivation under optimized conditions. Biodiesel production from jatropha oil was studied under such conditions. It was found that 96.3% biodiesel yield from nonpretreated jatropha oil with high-acid value (12.7 mg KOH/g) could be achieved.

17.18 CONCLUSION

Biodiesel has attracted attention as a feasible renewable energy source due to its potential to reduce dependence on fossil fuels. Traditionally, biodiesel has been produced from edible oils due to their low free fatty oils. However, this has elevated some issues such as food versus fuel. Therefore, utilization of nonedible oils may decrease the cost of biodiesel. This chapter covers biodiesel production from edible and nonedible feedstocks. Several physical and chemical properties were determined. This chapter also discusses the concept of blending to improve some properties of biodiesel—diesel blend. Moreover, the effect of biodiesel—diesel blends of 10% and 20% (by volume) of some methyl esters on engine performance and emissions was evaluated. Based on the experimental study, it can be concluded that biodiesel production should be mainly considered from nonedible oil feedstocks or waste resources, such as waste cooking oil or food waste. This will significantly save the environment and reduce the cost of biodiesel production.

ACKNOWLEDGMENTS

The authors would like to acknowledge Erciyes University, Kayseri, Turkey for the financial support under FOA-2015-5817 and FOA-2015-5790 projects.

REFERENCES

[1] Atabani AE. A comprehensive analysis of edible and non-edible biodiesel feedstocks, in Department of mechanical engineering. Kuala Lumpur (Malaysia): University of Malaya; 2014.

[2] Jatrofuels. From feedstock cultivation to full market integration. 2012 [cited 2012 9th December]; Available from: http://www.jatrofuels.com/161-0-Biofuels.html#Fuel%20characteristics%20and%20advantages.

[3] Janaun J, Ellis N. Perspectives on biodiesel as a sustainable fuel. Renewable and Sustainable Energy Reviews 2010;14(4):1312—20.

[4] Shahid EM, Jamal J. Production of biodiesel: a technical review. Renewable and Sustainable Energy Reviews 2011;15(9):4732—45.

[5] Kafuku G, Mbarawa M. Biodiesel production from *Croton megalocarpus* oil and its process optimization. Fuel 2010;89(9):2556—60.

[6] Satyanarayana M, Muraleedharan C. A comparative study of vegetable oil methyl esters (biodiesels). Energy 2011;36(4):2129—37.

[7] Howell S. Biodiesel standards-an update of current activities. Society of Automotive Engineers (SAE) Technical Paper; 1997. p. 203—8.

[8] Basumatary S. Yellow oleander (*Thevetia peruviana*) seed oil biodiesel as an alternative and renewable fuel for diesel engines: a review. International Journal of ChemTech Research 2014—2015;7(6):2823—40.

[9] Deng X, Li Y, Fei X. Microalgae: a promising feedstock for biodiesel. African Journal of Microbiology Research 2009;3(13):1008—14.

[10] Natrah FMI, Yusoff FM, Shariff M, Abas F, Mariana NS. Screening of Malaysian indigenous microalgae for antioxidant properties and nutritional value. Journal of Applied Phycology 2007;19(6):711—8.

[11] Sialve B, Bernet N, Bernard O. Anaerobic digestion of microalgae as a necessary step to make microalgal biodiesel sustainable. Biotechnology Advances 2009;27(4):409—16.

[12] Kapdan IK, Kargi F. Bio-hydrogen production from waste materials. Enzyme and Microbial Technology 2006;38:569—82.

[13] Choi SP, Nguyen MT, Sim SJ. Enzymatic pretreatment of *Chlamydomonas reinhardtii* biomass for ethanol production. Bioresource Technology 2010;101:5330—6.

[14] Yu G, Zhang Y, Schideman L, Funk T, Wang Z. Distributions of carbon and nitrogen in the products from hydrothermal liquefaction of low-lipid microalgae. Energy and Environmental Science 2011;4:4587—95.

[15] Koberg M, Cohen M, Ben-Amotz A, Gedanken A. Bio-diesel production directly from the microalgae biomass of *Nannochloropsis* by microwave and ultrasound radiation. Bioresource Technology 2011;102:4265—9.

[16] Richardson JW, Outlaw JL, Alison L. The economics of microalgae oil. AgBioForum 2010;13(2):119—30.

[17] Atabani AE, Mahlia TMI, Masjuki HH, Badruddin IA, Yussof HW, Chong WT, et al. A comparative evaluation of physical and chemical properties of biodiesel synthesized from edible and non-edible oils and study on the effect of biodiesel blending. Energy 2013;58:296—304.

[18] Atabani AE, Silitonga AS, Ong HC, Mahlia TMI, Masjuki HH, Badruddin IA, et al. Non-edible vegetable oils: a critical evaluation of oil extraction, fatty acid compositions, biodiesel production, characteristics, engine performance and emissions production. Renewable and Sustainable Energy Reviews February 2013;18:211—45.

[19] Tabatabaei M, Karimi K, Horváth IS, Kumar R. Recent trends in biodiesel production. Biofuel Research Journal 2015;7:258—67.

[20] Bart JCJ, Palmeri N, Cavallaro S. Biodiesel science and technology: from soil to oil. Woodhead Publishing Limited; 2010.

[21] Atabani AE, Silitonga AS, Badruddin IA, Mahlia TMI, Masjuki HH, Mekhilef S. A comprehensive review on biodiesel as an alternative energy resource and its characteristics. Renewable and Sustainable Energy Reviews 2012;16(4):2070—93.

[22] Razon LF. Review alternative crops for biodiesel feedstock. 2009 [cited 2012 8th February]; Available from: http://leoneresources.com/downloads/alternativecropsforbiodieselfeedstock.pdf.

[23] Ahmad AL, Yasin NHM, Derek CJC, Lim JK. Microalgae as a sustainable energy source for biodiesel production: a review. Renewable and Sustainable Energy Reviews 2011;15(1):584—93.

[24] Karmee SA, Chadha A. Preparation of biodiesel from crude oil of *Pongamia pinnata*. Bioresource Technology 2005;96(13):1425—9.

[25] Balat M, Balat H. Progress in biodiesel processing. Applied Energy 2010;87(6):1815—35.

[26] Balat M. Potential alternatives to edible oils for biodiesel production — a review of current work. Energy Conversion and Management 2011;52(2):1479—92.

[27] Deng X, Fang Z, Liu YH, Yu CL. Production of biodiesel from Jatropha oil catalyzed by nanosized solid basic catalyst. Energy 2011;36(2):777—84.

[28] Chapagain BP, Yehoshua Y, Wiesman Z. Desert date (*Balanites aegyptiaca*) as an arid lands sustainable bioresource for biodiesel. Bioresource Technology 2009;100(3):1221—6.

[29] Banković-Ilić IB, Stamenković OS, Veljković VB. Biodiesel production from non-edible plant oils. Renewable and Sustainable Energy Reviews 2012;16(6):3621—47.

[30] Chhetri AB, Tango MS, Budge SM, Watts KC, Islam MR. Non-edible plant oils as new sources for biodiesel production. International Journal of Molecular Sciences 2008;9:169—80.

[31] Gui MM, Lee KT, Bhatia S. Feasibility of edible oil vs. non-edible oil vs. waste edible oil as biodiesel feedstock. Energy 2008;33(11):1646—53.

[32] Kumar A, Sharma S. Potential non-edible oil resources as biodiesel feedstock: an Indian perspective. Renewable and Sustainable Energy Reviews 2011;15(4):1791—800.

[33] Mofijur M, Atabani AE, Masjuki HH, Kalam MA, Masum BM. A study on the effects of promising edible and non-edible biodiesel feedstocks on engine performance and emissions production: a comparative evaluation. Renewable and Sustainable Energy Reviews 2013;23:391—404.

[34] Murugesan A, Umarani C, Chinnusamy TR, Krishnan M, Subramanian R, Neduzchezhain N. Production and analysis of bio-diesel from non-edible oils — a review. Renewable and Sustainable Energy Reviews 2009;13(4):825—34.

[35] Saravanan S, Nagarajan G, Rao GLN. Investigation on nonedible vegetable oil as a compression ignition engine fuel in sustaining the energy and environment. Journal of Renewable and Sustainable Energy 2010;2(013108):1—8.

[36] Sarin R, Sharma M, Khan AA. Studies on *Guizotia abyssinica* L. oil: biodiesel synthesis and process optimization. Bioresource Technology 2009;100(18):4187—92.

[37] Sarma A, Konwer D, Bordoloi PK. A comprehensive analysis of fuel properties of biodiesel from Koroch seed oil. Energy & Fuels 2005;19(2):656—7.

[38] Falasca SL, Flores N, Lamas MC, Carballo SM, Anschau A. *Crambe abyssinica*: an almost unknown crop with a promissory future to produce biodiesel in Argentina. International Journal of Hydrogen Energy 2010;35(11):5808—12.

[39] Zhang H, Zhou Q, Chang F, Pan H, Liu X-F, Li H, et al. Production and fuel properties of biodiesel from *Firmiana platanifolia* L.f. as a potential non-food oil source. Industrial Crops and Products 2015;76:768—71.

[40] Shirazi MJA, Bazgir S, Shirazi MMA. Edible oil mill effluent; a low-cost source for economizing biodiesel production: electrospun nanofibrous coalescing filtration approach. Biofuel Research Journal 2014;1:39—42.

[41] Haile H. Integrated volarization of spent coffee grounds to biofuels. Biofuel Research Journal 2014;1:65—9.

[42] Sharma YC, Singh B. Development of biodiesel: current scenario. Renewable and Sustainable Energy Reviews 2009;13(6—7):1646—51.

[43] Singh SP, Singh D. Biodiesel production through the use of different sources and characterization of oils and their esters as the substitute of diesel: a review. Renewable and Sustainable Energy Reviews 2010;14(1):200—16.

[44] Chen W, Ma L, Zhou P-P, Zhu Y-M, Wang X-P, Luo X-A, et al. A novel feedstock for biodiesel production: the application of palmitic acid from *Schizochytrium*. Energy 2015;86:128—38.

[45] Lin L, Zhou C, Saritporn V, Shen X, Dong M. Opportunities and challenges for biodiesel fuel. Applied Energy 2011;88(4):1020—31.

[46] Asif S. An overview of some potential feedstocks for biodiesel production. Islamabad (Pakistan): Faculty of Biological Sciences, Quaid-i-Azam University; 2016 [Biodiesel lab].

[47] Atabani AE, Badruddin IA, Mahlia TMI, Masjuki HH, Mofijur M, Lee KT, et al. Fuel properties of *Croton megalocarpus*, *Calophyllum inophyllum*, and *Cocos nucifera* (coconut) methyl esters and their performance in a multicylinder diesel engine. Energy Technology 2013;1(11):685—94.

[48] Atabani AE, Mofijur M, Masjuki HH, Badruddin IA, Chong WT, Cheng SF, et al. A study of production and characterization of Manketti (*Ricinodendron rautonemii*) methyl ester and its blends as a potential biodiesel feedstock. Biofuel Research Journal 2014;1:139—46.

[49] Atabani AE, Cesar A. *Calophyllum inophyllum* L. — a prospective non-edible biodiesel feedstock. Study of biodiesel production, properties, fatty acid composition, blending and engine performance. Renewable and Sustainable Energy Reviews 2014;37:644—55.

[50] Atabani AE, Badruddin IA, Masjuki HH, Chong WT, Lee KT. *Pangium edule* Reinw: a promising non-edible oil feedstock for biodiesel production. Arabian Journal for Science and Engineering 2015;40(2):583—94.

[51] Atabani AE, Mofijur M, Masjuki HH, Badruddin IA, Kalam MA, Chong WT. Effect of *Croton megalocarpus*, *Calophyllum inophyllum*, *Moringa oleifera*, palm and coconut biodiesel—diesel blending on their physico-chemical properties. Industrial Crops and Products 2014;60:130—7.

[52] Beerens P. Screw-pressing of Jatropha seeds for fueling purposes in less developed countries. Eindhoven: Eindhoven University of Technology; 2007.

[53] Evangelista RL, Cermak SC. Full-press oil extraction of *Cuphea* (PSR23) seeds. Journal of American Oil and Chemists' Society 2007;84:1169—75.

[54] Forson FK, Oduro EK, Hammond-Donkoh E. Performance of Jatropha oil blends in a diesel engine. Renewable Energy 2004;29(7):1135—45.

[55] Amarasinghe BMWPK, Gangodavilage NC. Rice bran oil extraction in Sri Lanka: data for process equipment design. Food and Bioproducts Processing 2004;82(1):54—9.

[56] Qian J, Shi H, Yun Z. Preparation of biodiesel from *Jatropha curcas* L. oil produced by two-phase solvent extraction. Bioresource Technology 2010;101(18):7025—31.

[57] Nabilah Aminah L, Leong ST, Wong YS, Ong SA, Kairulazam CK. Biodiesel production of *Garcinia Mangostana* Linn. seeds by two-phase solvent extraction and alkali-catalyzed transesterification. International Journal of Chemical Engineering and Applications 2012;4(3):92—5.

[58] Kartikaa IA, Yania M, Arionob D, Evon Ph, Rigal L. Biodiesel production from jatropha seeds: solvent extraction and in situ transesterification in a single step. Fuel 2013;106:111—7.

[59] Salaheldeen M, Aroua MK, Mariod AA, Cheng SF, Abdelrahman MA, Atabani AE. Physicochemical characterization and thermal behavior of biodiesel and biodiesel-diesel blends derived from crude *Moringa peregrina* seed oil. Energy Conversion and Management 2015;92:535—42.

[60] Azeem MW, Hanif MA, Al-Sabahi JN, Khan AA, Naz S, Ijaz A. Production of biodiesel from low priced, renewable and abundant date seed oil. Renewable Energy 2016;86:124—32.

[61] Atadashi IM, Aroua MK, Abdul Aziz A. High quality biodiesel and its diesel engine application: a review. Renewable and Sustainable Energy Reviews 2010;14(7):1999—2008.

[62] Atabani AE, Mahlia TMI, Badruddin IA, Masjuki HH, Chong WT, Lee KT. Investigation of physical and chemical properties of potential edible and non-edible feedstocks for biodiesel production, a comparative analysis. Renewable and Sustainable Energy Reviews May 2013;21:749—55.

[63] Palash SM, Masjuki HH, Kalam MA, Atabani AE, Rizwanul Fattah IM, Sanjid S. Biodiesel production, characterization, diesel engine performance, and emission characteristics of methyl esters from *Aphanamixis polystachya* oil of Bangladesh. Energy Conversion and Management 2015;91:149—57.

[64] Alptekin E, Canaki M, Sanli H. Evaluation of leather industry wastes as a feedstock for biodiesel production. Fuel 2012;95:214—20.

[65] Rao MS, Anand RB. Production characterization and working characteristics in DICIengine of Pongamia biodiesel. Ecotoxicology and Environmental Safety 2015;121:16—21.

[66] Sharif Hossain ABM, Salleh A, Nasrulhaq Boyce A, Chowdhury P, Naqiuddin M. Biodiesel fuel production from algae as renewable energy. American Journal of Biochemistry and Biotechnology 2008;4(3):250—4.

[67] Mofijur M, Masjuki HH, Kalam MA, Atabani AE. Evaluation of biodiesel blending, engine performance and emissions characteristics of *Jatropha curcas* methyl ester: Malaysian perspective. Energy 2013;55(15):879—87.

[68] Omidvarborna H, Kumar A, Kim D-S. NO_x emissions from low-temperature combustion of biodiesel made of various feedstocks and blends. Fuel Processing Technology 2015;140:113—8.

[69] Hajra B, Sultana N, Pathak AK, Guria C. Response surface method and genetic algorithm assisted optimal synthesis of biodiesel from high free fatty acid sal oil (*Shorea robusta*) using ion-exchange resin at high temperature. Journal of Environmental Chemical Engineering 2015;3:2378—92.

[70] Yunus Khan TM, Atabani AE, Badruddin IA, Badarudin A, Khayoon MS, et al. Recent scenario and technologies to utilize non-edible oils for biodiesel production. Renewable and Sustainable Energy Reviews 2014;37.

[71] Knothe G, Cermak SC, Evangelista RL. Biodiesel and renewable diesel: a comparison. Progress in Energy and Combustion Science 2010;36(3):364—73.

[72] Rizwanul Fattah IM, Kalam MA, Masjuki Hh, Wakil MA. Biodiesel production, characterization, engine performance, and emission characteristics of Malaysian Alezandrian laurel oil. RSC Advances 2014;4:17787—96.

[73] Demirbas A. Relationships derived from physical properties of vegetable oiland biodiesel fuels. Fuel 2008;87(8—9):1743—8.

[74] Moser BR. Fuel property enhancement of biodiesel fuels from common and alternative feedstocks via complementary blending. Renewable Energy 2016;85:819—25.

[75] Iqbal MI, Varman M, Masjuki HH, Kalam MA, Hossain S, Imtenan S. Tailoring fuel properties using jatropha, palm and coconut biodiesel to improve CI engine performance and emission characteristics. Journal of Cleaner Production 2015;101:262—70.

[76] Saxena P, Jawale S, Joshipura MH. A review on prediction of properties of biodiesel and blends of biodiesel. Procedia Engineering 2013;51:395—402.

[77] Sivaramakrishnan K, Ravikumar P. Determination of cetane number of biodiesel and it's influence on physical properties, vol. 7(2). Asian Research Publishing Network (ARPN); 2012. p. 205—11.

[78] Moser BR. Influence of blending canola, palm, soybean, and sunflower oil methyl esters on fuel properties of biodiesel. Energy & Fuels 2008;22(6):4301—6.

[79] Oghenejoboh KM, Umukoro PO. Comparative analysis of fuel characteristics of bio-diesel produced from selected oil-bearing seeds in Nigeria. European Journal of Scientific Research 2011;58(2):238—46.

[80] Krishna CR, Butcher T, Mahajan D, Thomassen K, Brown C. Improving cold flow properties of biodiesel. 2008 [cited 2014 19th May]; Available from: ttp://www.aertc.org/conference/AEC_Sessions%5CCopy%20of%20Session%202%5CTrack%20B-%20Renewables%5C1.%20Climate%20Change%20Biofuels%20III%5C1.%20Dr.%20CR%20Krishna%5CCR%20Krishna%20presentation.pdf.

[81] Sivaramakrishnan K, Ravikumar P. Determination of higher heating value of biodiesels. International Journal of Engineering Science and Technology (IJEST) 2011;3(11):7981—7.

[82] Giakoumis EG. A statistical investigation of biodiesel physical and chemical properties, and their correlation with the degree of unsaturation. Renewable Energy 2013;50(0):858—78.

[83] Kalayasiri P, Jayashke N, Krisnangkura K. Survey of seed oils for use as diesel fuels. Journal of American Oil Chemical Society 1996;73(4):471—4.

[84] Krisnangkura K. A simple method for estimation of cetane index of vegetable oil methyl esters. Journal of American Oil Chemical Society 1986;63(4):552—3.

[85] Ramírez-Verduzco LF, Rodríguez-Rodríguez JE, Jaramillo-Jacob AD. Predicting cetane number, kinematic viscosity, density and higher heating value of biodiesel from its fatty acid methyl ester composition. Fuel 2012;91(1):102—11.

[86] Talebi AF, Tabatabaei M, Chisti Y. Biodiesel analyzer: a user-friendly software for predicting the properties of prospective biodiesel. Biofuel Research Journal 2014;1(2):55—7.

[87] Gülüm M, Bilgin A. Density, flash point and heating value variations of corn oil biodiesel—diesel fuel blends. Fuel Processing Technology 2015;134:456—64.

[88] Wakil MA, Kalam MA, Masjuki HH, Atabani AE, Rizwanul Fattah IM. Influence of biodiesel blending on physicochemical properties and importance of mathematical model for predicting the properties of biodiesel blend. Energy Conversion and Management 2015;94:51—67.

[89] Mofijur M, Masjuki HH, Kalam MA, Rasul MG, Atabani AE, Hazrat MA, et al. Effect of biodiesel-diesel blending on physico-chemical properties of biodiesel produced from *Moringa oleifera*. Procedia Engineering 2015;105:665—9.

[90] Raj FRMS, Sahayaraj JW. A comparative study over alternative fuel (biodiesel) for environmental friendly emission. In: Recent advances in space technology services and climate change (RSTSCC), Chennai, India; 2010.

[91] Özener O, Yüksek L, Ergenç AT, Özkan M. Effects of soybean biodiesel on a DI diesel engine performance, emission and combustion characteristics. Fuel January 2012;115:875—83.

[92] Abdullah NR, Liu D, Park T, Xu HM, Wyszynski ML, Tsolakis A. Effect of variations of biodiesel blends on engine combustion and emissions. 2014 [cited 2015 3rd February]; Available from: http://www.birmingham.ac.uk/Documents/college-eps/mechanical/staff/xu/xu-biodesiel-blends.pdf.

[93] Fontaras G, Karavalakis G, Kousoulidou M, Tzamkiozis T, Ntziachristos L, Bakeas E, et al. Effects of biodiesel on passenger car fuel consumption, regulated and non-regulated pollutant emissions over legislated and real-world driving cycles. Fuel 2009;88(9):1608—17.

[94] Ng J-H, Ng HK, Gan S. Characterisation of engine-out responses from a light-duty diesel engine fuelled with palm methyl ester (PME). Applied Energy 2012;90(1):58—67.

[95] Muralidharan K, Vasudevan D, Sheeba KN. Performance, emission and combustion characteristics of biodiesel fuelled variable compression ratio engine. Energy 2011;36(8):5385—93.

[96] Rizwanul Fattah IM, Masjuki HH, Kalam MA, Wakil MA, Ashraful AM, Shahi SA. Experimental investigation of performance and regulated emissions of a diesel engine with *Calophyllum inophyllum* biodiesel blends accompanied by oxidation inhibitors. Energy Conversion and Management 2014;83:232—40.

[97] Savariraj S, Ganapathy T, Saravanan CG. Experimental investigation of performance and emission characteristics of Mahua biodiesel in diesel engine. ISRN Renewable Energy 2011;2011:1—6.

[98] Mofijur M, Masjuki HH, Kalam MA, Atabani AE, Arbab MI, Cheng SF, et al. Properties and use of *Moringa oleifera* biodiesel and diesel fuel blends in a multi-cylinder diesel engine. Energy Conversion and Management 2014;82:169–76.

[99] Mofijur M, Masjuki HH, Kalam MA, Atabani AE, Rizwanul Fattah IM, Mobarak HM. Comparative evaluation of performance and emission characteristics of *Moringa oleifera* and palm oil based biodiesel in a diesel engine. Industrial Crops and Products 2014;53:78–84.

[100] Basha JS, Anand RB. Role of nanoadditive blended biodiesel emulsion fuel on the working characteristics of a diesel engine. Journal of Renewable and Sustainable Energy 2011;3(023106):1–17.

[101] Kivevele T, Huan Z. Influence of metal contaminants and antioxidant additives on storage stability of biodiesel produced from non-edible oils of Eastern Africa origin (*Croton megalocarpus* and *Moringa oleifera* oils). Fuel 2015;158:530–7.

[102] Kivevele TT, Mbarawa MM, Bereczky A, Laza T, Madarasz J. Impact of antioxidant additives on the oxidation stability of biodiesel produced from *Croton Megalocarpus* oil. Fuel Processing Technology 2011;92:1244–8.

[103] Agarwal AK, Khurana DK, Dhar A. Improving oxidation stability of biodiesels derived from Karanja, Neem and Jatropha: step forward in the direction of commercialisation. Journal of Cleaner Production 2015;107:646–52.

[104] Ryu K. The characteristics of performance and exhaust emissions of a diesel engine using a biodiesel with antioxidants. Bioresource Technology 2010;101:S78–82.

[105] Imtenan S, Masjuki HH, Varman M, Arbab MI, Sajjad H, Rizwanul Fattah IM, et al. Emission and performance improvement analysis of biodiesel-diesel blends with additives. Procedia Engineering 2014;90:472–7.

[106] Imtenan S, Masjuki HH, Varman M, Kalam MA, Arbab MI, Sajjad H, et al. Impact of oxygenated additives to palm and jatropha biodiesel blends in the context of performance and emissions characteristics of a light duty diesel engine. Energy Conversion and Management 2014;83:149–58.

[107] Imtenan S, Masjuki HH, Varman M, Rizwanul Fattah IM, Sajjad H, Arbab MI. Effect of n-butanol and diethyl ether as oxygenated additives on combustion–emission-performance characteristics of a multiple cylinder diesel engine fuelled with diesel–jatropha biodiesel blend. Energy Conversion and Management 2015;94:84–94.

[108] Palash SM, K MA, Masjuki HH, Arbab MI, Masum BM, Sanjid A. Impacts of NO_x reducing antioxidant additive on performance and emissions of a multi-cylinder diesel engine fueled with Jatropha biodiesel blends. Energy Conversion and Management 2014;77:577–85.

[109] Shahabuddin M, et al. Effect of additive on performance of C.I. engine fuelled with bio diesel. Energy Procedia 2012;140(0):1624–9.

[110] Swaminathan C, Sarangan J. Performance and exhaust emission characteristics of a CI engine fueled with biodiesel (fish oil) with DEE as additive. Biomass and Bioenergy 2012;39(0):168–74.

[111] ROBOPLANT. Elsewhere in science: we need to talk about biofuels. 2016 [cited 2016 24th January]; Available from: https://roboplant.wordpress.com/2013/07/23/elsewhere-in-science-we-need-to-talk-about-biofuels/.

[112] Rosenberg J, Oyler G, Wilkinson L, Betenbaugh M. A green light for engineered algae: redirecting metabolism to fuel a biotechnology revolution. Biotechnology 2008;19:430–6.

[113] Wikimedia Commons. Photobioreactor PBR 500 P IGV Biotech. 2016 [cited 2016 20th January]; Available from: https://commons.wikimedia.org/wiki/File:Photobioreactor_PBR_500_P_IGV_Biotech.jpg.

[114] Oilalgae. Tubular photobioreactors. 2016 [cited 2016 31st January]; Available from: http://www.oilgae.com/algae/cult/pbr/typ/tub/tub.html.

[115] UK, A. Algae. 2016 [cited 2016 31st January]; Available from: http://www.bae.uky.edu/biofuels/algae/photo_gallery/img6.jpg.

[116] Smith AL, Stadler LB, Love NG, Skerlos SJ, Raskin L. Perspectives on anaerobic membrane bioreactor treatment of domestic wastewater: a critical review. Bioresource Technology 2012;122:149–59.

[117] Abou-Shanab RAI, Ji M-K, Kim H-C, Paeng K-J, Jeon B-H. Microalgal species growing on piggery wastewater as a valuable candidate for nutrient removal and biodiesel production. Journal of Environmental Management 2013;115:257—64.

[118] Ji MK, Kim HC, Sapireddy VR, Yun HS, Abou-Shanab RA, Choi J, et al. Simultaneous nutrient removal and lipid production from pretreated piggery wastewater by *Chlorella vulgaris*. Applied Microbiology and Biotechnology 2013;97(6):2701—10.

[119] Abou-Shanab RAI, El-Dalatony MM, EL-Sheekh MM, Min-Kyu J, Salamaa E, Kabraa AN, et al. Cultivation of a new microalga *Micractinium reisseri* in municipal wastewater for nutrient removal, biomass, lipid and fatty acid production. Biotechnology and Bioprocess Engineering 2014;19:510—8.

[120] Octavio PG, Froylan ME, de-Bashan LE, Yoav B. Heterotrophic cultures of microalgae: metabolism and potential product. Water Research 2011;45:11—36.

[121] Pittman JK, Dean AP, Osundeko O. The potential of sustainable algal biofuel production using wastewater resources. Bioresource Technology 2011;102:17—25.

[122] Van DES, Vervaeren H, Boon N. Flue gas compounds and microalgae: (Bio-) chemical interactions leading to biotechnological opportunities. Biotechnology Advances 2012;30:1405—24.

[123] Abomohra A, Wagner M, El-Sheekh M, Hanelt D. Lipid and total fatty acid productivity in photoautotrophic fresh water microalgae: screening studies towards biodiesel production. Journal of Applied Phycology 2012;25(4):931—6.

[124] Chitis Y. Biodiesel from microalgae. Biotechnology Advances 2007;25(3):294—306.

[125] Gouviea L, Oliveira AC. Microalgae as a raw material for biofuels production. Journal of Industrial Microbiology & Biotechnology 2009;36(2):269—74.

[126] El-Mekkawi S, El-Ardi O, El-Ibiari N, Moustafa T, Ismail I. Cultivation of local fresh water microalgae in closed systems. International Journal of Innovative Science, Engineering & Technology 2015;2(3):18—22.

[127] Griffiths MJ, Harrison TL. Lipid productivity as a key characteristic for choosing algal species for biodiesel production. Journal of Applied Phycology 2009;21(5):493—507.

[128] Matsumoto M, Sugiyama H, Maeda Y, Sato R, Tanaka T, Matsunaga T. Marine diatom, *Navicula* sp. strain JPCC DA0580 and marine green alga, *Chlorella* sp. strain NKG400014 as potential sources for biodiesel production. Applied Biochemistry and Biotechnology 2010;161:1—8.

[129] Metzger P, L C. *Botryococcus braunii*: a rich source for hydrocarbons and related ether lipids. Applied Microbiology and Biotechnology 2005;66:486—96.

[130] Scott SA, Davey MP, Dennis JS, Horst I, Howe CJ, Lea-Smith DJ, et al. Biodiesel from algae: challenges and prospects. Current Opinion in Biotechnology 2010;21:277—86.

[131] Rittmann BE. Opportunities for renewable bioenergy using microorganisms. Biotechnology and Bioengineering 2008;100:203—12.

[132] Groom MJ, Gray EM, Townsend PA. Biofuels and biodiversity: principles for creating better policies for biofuel production. Conservation Biology 2008;22(3):602—9.

[133] Johnson MB, Wen Z. Production of biodiesel fuel fromthe microalga *Schizochytrium limacinum* by direct transesterification of algal biomass. Energy Fuels 2009;23:5179—83.

[134] Metting FBJ. Biodiversity and application of microalgae. Journal of Industrial Microbiology 1996;17:477—89.

[135] Spolaore P, Joannis-Cassan C, Duran E, Isambert A. Commercial applications of microalgae. Journal of Bioscience and Bioengineering 2006;101:87—96.

[136] Um B-H, Kim Y-S. Review: a chance for Korea to advance algal-biodiesel technology. Journal of Industrial and Engineering Chemistry 2009;15:1—7.

[137] Brennan L, Owende P. Biofuels from microalgae-a review of technologies for production, processing, and extractions of biofuels and co-products. Renewable and Sustainable Energy Reviews 2010;14:557—77.

[138] Mata TM, Martins AA, Nidia CS. Microalgae for biodiesel production and other application: a review. Renewable and Sustainable Energy Reviews 2010;14:217—32.

[139] Schenk PM, Thomas-Hall SR, Stephens E, Marx UC, Mussgnug JH, Posten C, et al. Second generation biofuels: high-efficiency microalgae for biodiesel production. BioEnergy Research 2008;1(1):20—43.

[140] El-Sheekh M, Abomohra A, Haanelt D. Optimization of biomass and fatty acid productivity of *Scenedesmus obliquus* as a promising microalga for biodiesel production. World Journal of Microbiology and Biotechnology 2013;29:915—22.

[141] Grobbelaar JU. Algal nutrition-mineral nutrition. In: Handbook of microalgal culture: biotechnology and applied phycology. Blackwell Publishing Ltd.; 2004. p. 97—115.

[142] Mata TM, Almeidab R, Caetanoa NS. Effect of the culture nutrients on the biomass and lipid productivities of microalgae *Dunaliella tertiolecta*. Chemical Engineering 2013;32:973—8.

[143] Chavan KJ, Chouhan S, Jain S, Singh P, Yadav M, Tiwari A. Environmental factors influencing algal biodiesel production. Environmental Engineering Science 2014;31(11):602—11.

[144] Xu H, Miao X, Wu Q. High quality biodiesel production from a microalga *Chlorella prototecoides* by heterotrophic growth in fermenters. Journal of Biotechnology 2006;126(4):499—507.

[145] Hansen PJ. Effect of high pH on the growth and survival of marine phytoplankton: implications for species succession. Aquatic Microbial Ecology 2002;28(3):279—88.

[146] Bartley ML, Boeing WJ, Dungan BN, Holguin FO, Schaub T. pH effects on growth and lipid accumulation of the biofuel microalgae *Nannochloropsis salina* and invading organisms. Journal of Applied Phycology 2014;26:1431—7.

[147] Moheimani NR. Inorganic carbon and pH effect on growth and lipid productivity of *Tetraselmis suecica* and *Chlorella* sp (Chlorophyta) grown outdoor in bag photobioreactors. Journal of Applied Phycology 2013;25:387—98.

[148] Rodolfi L, C.Z G, Bassi N, Padovani G, Biondi N, Bonini G, et al. Microalgae for oil: strain selection, induction of lipid synthesis and outdoor mass cultivation in a low-cost photobioreactor. Biotechnology and Bioengineering 2008;102:100—12.

[149] Visviki I, Palladino J. Growth and cytology of *Chlamydomonas acidophila* under acidic stress. Bulletin of Environmental Contamination and Toxicology 2001;66(5):623—30.

[150] Poerschmann J, Spijkerman E, Langer U. Fatty acid patterns in *Chlamydomonas* sp. as a marker for nutritional regimes and temperature under extremely acidic conditions. Microbial Ecology 2004;48(1):78—89.

[151] Guckert JB, Cooksey KE. Triglyceride accumulation and fatty acid profile changes in *Chlorella* (Chlorophyta) during high pH-induced cell cycle inhibition. Journal of Phycology 1990;26(1):72—9.

[152] Kim HS, Weiss TL, Thapa HR, Devarenne TP, Han A. A microfluidic photobioreactor array demonstrating high-throughput screening for microalgal oil production. Lab Chip 2014;14(8):1415—25.

[153] Al-Qasmi M, Sahar Talebi S, Al-Rajhi S, Al-Barwani T. Review of effect of light on microalgae growth. In: Proceedings of the World Congress on Engineering; 2012.

[154] Posten C. Design principles of photo-bioreactors for cultivation of microalgae. Engineering in Life Sciences 2009;9(3):165—77.

[155] Ho S-H, Nakanishi A, Ye X, Chang J-S, Hara K, Hasunuma T, et al. Optimizing biodiesel production in marine *Chlamydomonas* sp. JSC4 through metabolic profiling and an innovative salinity-gradient strategy. Biotechnology for Biofuels 2014;7(97):1—16.

[156] Chen C-Y, Chang J-S, Chang H-Y, Chen T-Y, Wu J-H, Lee W-L. Enhancing microalgal oil/lipid production from *Chlorella sorokiniana* CY1 using deep-sea water supplemented cultivation medium. Biochemical Engineering Journal 2013;77:74—81.

[157] Bondioli P, Bella LD, Rivolta G, Zittelli GC, Bassi N, Rodolfi L, et al. Oil production by the marine microalgae *Nannochloropsis* sp. F&M-M24 and *Tetraselmis suecica* F&M-M33. Bioresource Technology 2012;114:567—72.

[158] San Pedro A, González-López CV, Acién FG, Molina-Grima E. Marine microalgae selection and culture conditions optimization for biodiesel production. Bioresource Technology 2013;134:353—61.

[159] Takagi M, Karseno S, Yoshida T. Effect of salt concentration on intracellular accumulation of lipids and triacylglyceride in marine microalgae Dunaliella cells. Journal of Bioscience and Bioengineering 2006;101:223—6.

[160] Campenni L, Nobre BP, Santos CA, Oliveira A, Aires-Barros M, Palavra A, et al. Carotenoid and lipid production by the autotrophic microalga *Chlorella protothecoides* under nutritional, salinity, and luminosity stress conditions. Applied Microbiology and Biotechnology 2013;97:1383—93.

[161] Pancha I, Chokshi K, Maurya R, Trivedi K, Patidar SK, Ghosh A, et al. Salinity induced oxidative stress enhanced biofuel production potential of microalgae *Scenedesmus* sp. CCNM 1077. Bioresource Technology 2015;189:341—8.

[162] Bartley ML, Boeing WJ, Corcoran A, Holguin FO, Schaub T. Effects of salinity on growth and lipid accumulation of biofuel microalga *Nannochloropsis salina* and invading organisms. Biomass and Bioenergy 2013;54:83—8.

[163] Chen C-H, Berns DS. Thermotropic properties of thermophilic, mesophilic, and psychrophilic blue-green algae. Plant Physiology 1980;66(4):596—9.

[164] Xin L, Hong-ying H, Yu-ping Z. Growth and lipid accumulation properties of a freshwater microalga *Scenedesmus* sp. under different cultivation temperature. Bioresource Technology 2011;102:3098—102.

[165] Converti A, Casazza AA, Ortiz EY, Perego P, Del Borghi M. Effect of temperature and nitrogen concentration on the growth and lipid content of *Nannochloropsis oculata* and *Chlorella vulgaris* for biodiesel production. Chemical Engineering and Processing: Process Intensification 2009;48(6):1146—51.

[166] Khanna S, Goyal A, Moholkar VS. Production of n-butanol from biodiesel derived crude glycerol using *Clostridium pasteurianum* immobilized on Amberlite. Fuel 2013;112:557—61.

[167] Karmee SK, Linardi D, Lee J, Lin CSK. Conversion of lipid from food waste to biodiesel. Waste Management 2015;41:169—73.

[168] Karmee SK, Lin CZK. Lipids from food waste as feedstock for biodiesel production: case Hong Kong. Lipid Technology 2014;26(9):206—9.

[169] Karmee SK, Lin CZK. Valorisation of food waste to biofuel: current trends and technological challenges. Sustainable Chemical Processes 2014;2:1—4.

[170] Pua F-L, Fang Z, Zakaria S, Guo F, Chia C-H. Direct production of biodiesel from high-acid value Jatropha oil with solid acid catalyst derived from lignin. Biotechnology for Biofuels 2011;4(56):1—8.

CHAPTER EIGHTEEN

Potential of Biodiesel as Fuel for Diesel Engine

O.M. Ali[1], R. Mamat[2], M.G. Rasul[3] and G. Najafi[4]

[1]Northern Technical University, Kirkuk, Iraq
[2]Universiti Malaysia Pahang, Pekan, Pahang, Malaysia
[3]Central Queensland University, Rockhampton, QLD, Australia
[4]Tarbiat Modares University, Tehran, Iran

18.1 INTRODUCTION

The global energy consumption has duplicated since 1980s. Fossil fuels are dominant in the global energy mix, as they represent over 80% of the total energy supplies in the world today [1]. A further increase in the utilization of energy sources is expected due to the modern life demands and rapid economic development. This results in the fast depletion of fossil fuel reservoirs as well as unwanted emissions. Oil is the fossil fuel that is most in danger of running out, as it is steadily declining. The sources of this fuel are available only in certain regions of the world, and they are close to their maximum production. A peak in the global oil production is expected in the next few years [2]. According to the World Trade Organization, in 2010, the fuel market was responsible for a 15.8% share of the total trade in merchandising and primary products. Most are due to diesel fuel that is essential for transportation and heavy-duty engines [3]. Besides, the world today is faced with a serious global warming and environmental pollution. The major gas that contributes to the greenhouse phenomena is CO_2 which is mainly emitted from the combustion of fossil fuel. According to the current scenario of increasing CO_2 emissions, the target of controlling the global warming phenomenon is becoming more difficult and costly with each year that passes [4]. The twin crises of fossil fuel depletion and environmental degradation that have arisen since 1970s have underscored the importance of developing alternative sources of liquid-fuel energy. Thus, there is an urgent need to find an alternative energy resource that is renewable, clean, reliable, and yet economically feasible.

Biodiesel, as a replacement for diesel, is easily made from renewable biological sources, such as vegetable oil and animal fats. Chemically, biodiesel is referred to monoalkyl-esters of long-chain fatty acids, and to a variety of ester-based oxygenated fuels. It is well known that transportation almost depends entirely on fossil fuel, particularly petroleum-based fuels such as gasoline, diesel fuel, liquefied petroleum gas (LPG), and natural gas (NG). An alternative fuel to mineral diesel must be technically feasible, economically competitive, environmentally acceptable, and easily available. The current

Clean Energy for Sustainable Development
ISBN 978-0-12-805423-9, http://dx.doi.org/10.1016/B978-0-12-805423-9.00018-1

alternative diesel fuel can be termed biodiesel. It can be used in diesel engines with little or no modification [5,6].

The global warming may be reduced by using alternative fuel—biodiesel is considered as a promising fuel. The best potential future energy source in the transportation sector is biodiesel that mainly emitted carbon monoxide (CO), carbon dioxide (CO_2), oxides of nitrogen (NO_x), sulfur oxides (SO_x), and smoke. The combustion of biodiesel alone provides a high reduction in total unburned hydrocarbons (HCs) and polycyclic aromatic hydrocarbons (PAHs) [7]. Biodiesel further provides significant reductions in particulates and carbon monoxide over mineral diesel fuel. Biodiesel provides slight increases or decreases in NO_x depending on the engine family and testing procedures that follow [8]. Currently, global warming caused by CO_2 is the main climatic problem in the world; therefore, environmental protection is important to ensure a better and safer future. Because biodiesel is made from renewable sources, it presents a convenient way to provide fuel while protecting the environment from unwanted emissions. Biodiesel is an ecological and nonhazardous fuel with low emission values; therefore, it is environmentally useful. Using biodiesel as an alternative fuel is a way to minimize global air pollution, and in particular, it reduces the emission levels that are potential or probable threats to human health [9].

Biodiesel properties are one of the noteworthy issues that restrict the use of biodiesel fuel. It has a direct effect on the main engine elements and parameters, such as fuel handling systems and performance [10,11]. Neat biodiesel contains no petroleum, but it is traditionally blended with mineral diesel to create a biodiesel blend, typically at 20% (by volume), or less. Pure biodiesel is biodegradable, nontoxic, and essentially free of sulfur and aromatics. Biodiesel is the only alternative fuel for compression ignition (CI) that has fully completed the health effects testing requirements of the 1990 Clean Air Act Amendments. Biodiesel, produced to industry and meet the specifications of ASTM D6751, is legally registered with the Environmental Protection Agency as a legal motor fuel for sale and distribution [12].

18.2 DIESEL ENGINE

The first diesel-powered engine appeared in 1898, and since then, diesel engines have become used widely as a power source for power generation, transportation, and different industrial activities. The main difference between the diesel engine and spark-ignition engine that works on gasoline fuel occurs in their fuel ignition process. The diesel engine ignition is caused by spontaneous CI, while the gasoline engine ignition is triggered by a spark from the spark plug. The pressure and temperature rise during compression in the diesel engine provides an appropriate environment for the fuel to be injected, atomized, vaporized, and subsequently burnt. Diesel engines are also known for their increased fuel efficiency and low emissions of CO and unburned HCs.

In the diesel engine, air is compressed with a compression ratio typically between 12 and 24 [13]. This higher compression promotes more efficient fuel combustion due to the longer effective expansion stroke. Furthermore, diesel engines run lean overall, except at full power, with air—fuel ratios as high as 65:1 [13]. Engine speed and power are controlled by changing the amount of fuel in each injection instead of throttling the intake air, which gives diesel engines high thermal efficiency. Due to their lower fuel consumption and CO and HC emissions, diesel engines are used in many trucks and almost all railroad engines and ships, and the power sources use diesel engines for different industrial applications.

In the transportation sector, diesel engine is widely used, due to its simple arrangement design, higher reliability, more power delivered at lower fuel consumption, and higher thermal efficiency [14]. In general, the diesel engine utilization for different applications depends on the engine speed, where engines with speed of 1200 rpm and above are used for transportation. These engines are equipped with pump and small electrical generators. Meanwhile, the medium speed engines from 300 to 1200 rpm are mostly used in large electrical generators, ships, large compressors, and pumps.

18.2.1 Diesel Engine Fuel History

Heat: The concept of using biofuel in diesel engines is not a radically new idea, as an inventor named Rudolph Diesel demonstrated his first developed CI diesel engine using peanut oil as a fuel at the World Exhibition in Paris in 1900 [15,16]. Vegetable oil had been used as a diesel fuel in the 1930s and 1940s, but generally only in emergency conditions, such as World War II [17]. In 1940 the first trial with vegetable oil methyl and ethyl esters were carried out using palm oil ethyl ester as a fuel for buses [18]. However, due to the fact that abundant supply of diesel and vegetable oil fuel were more expensive than diesel, research and development activities in vegetable oil were not seriously pursued [19].

Nowadays, biodiesel fuel returns to gain more and more interest as an attractive fuel when the problems of fossil fuel depletion and rising energy demand together with aggravated environmental pollution had been aggregated in the last decades. Biodiesel represents the viable alternative for CI engines which may be considered being at the forefront of alternative technologies. Therefore, there is a renewed interest at present in biodiesel fuel.

18.3 BIODIESEL FUEL

Biodiesel is the only alternative fuel with properties that make low concentration biodiesel—diesel fuel blends to run well in unmodified conventional CI engines. Understanding these properties provides key toward evaluating and improving diesel engine performance and emissions. The benefits and technical challenge of the biodiesel as a fuel for diesel engine are revealed through the investigation of their properties relevant to engine performance and emissions. Biodiesel has properties comparable to diesel fuel [20,21]. It can be stored anywhere, where the mineral diesel fuel is stored [22,23]. Biodiesel can be made from domestically produced and renewable oilseed crops. The risks of handling, transporting, and storing biodiesel are much lower than those associated with mineral diesel. Biodiesel is safe to handle and transport because it is biodegradable and has a high flash point, unlike the mineral diesel fuel. Biodiesel can be used alone or mixed in any ratio with mineral diesel fuel [24,25].

The biodegradability of biodiesel has been proposed as a solution to waste problems. Biodegradable fuels, such as biodiesel, have an expanding range of potential applications and are environmentally friendly. Therefore, there is growing interest in degradable diesel fuels that degrade more rapidly than the conventional petroleum fuels. Biodiesel is nontoxic and degrades faster than mineral diesel due to its higher oxygen content (typically 11%) [26,27]. Furthermore, the high oxygen content of the biodiesel fuel improves the combustion efficiency due to the increase of the homogeneity of oxygen with the fuel during combustion. Therefore, the combustion efficiency of biodiesel is higher than that of mineral diesel [28,29]. Biodiesel has good lubricant properties compared to mineral diesel oil, in particular very low-sulfhur diesel. This is crucial in reducing wear on the engine parts and the injection system [30,31].

Biodiesel has lower carbon and hydrogen contents compared to diesel fuel, resulting in about a 10% lower mass energy content, and also biodiesel has a higher viscosity and higher density [32,33]. These properties may result in high NO_x emissions, lower engine efficiency, injector coking, and engine compatibility. The high viscosity and low volatility of biodiesel fuel will cause problems in fuel pumping and spray characteristics as well as cause poor combustion in diesel engines. The inefficient mixing of fuel with air contributes to incomplete combustion. Biodiesel has a higher cloud point (CP) and pour point compared to diesel. Therefore, biodiesel fuels are plagued by the growth and agglomeration of paraffin wax crystals when ambient temperatures fall below the fuel's CP, which may cause start-up problems, such as filter clogging in cold climates [34]. While the CP of mineral diesel is reported as $-16°C$, typically biodiesel has a CP higher than $0°C$, and because of that its use is limited to ambient temperatures around freezing [35].

18.3.1 Biodiesel Fuel Trends

Interest in biodiesel is continuing to increase in the whole world. This is motivated primarily by concerns about greenhouse gas (GHG) emissions and global climate change, as well as the desire for renewable/sustainable energy sources, and an interest in developing domestic and more secure fuel supplies. An alternative fuel for mineral diesel must be technically feasible, economically competitive, environmentally acceptable, and easily available. Biodiesel is simple to use in CI diesel engines with few or no modifications. In addition, it can be blended, at any level with mineral diesel to create a biodiesel blend.

In recent years, several countries (and states) have embarked on legislative and/or regulatory pathways that encourage the increased use of biodiesel fuel as an alternative fuel for diesel.

Biodiesel, a promising oxygenated fuel generated from natural and renewable sources, is a fuel comprised of monoalkyl esters of long-chain fatty acids derived from renewable feedstocks. It is increasingly examined as a potential substitute for conventional high-pollutant fuels because it is a biodegradable, nontoxic, and relatively clean-burning fuel. Biodiesel has significantly lower emissions than petroleum-based diesel when burned, whether used in pure form or blended with mineral diesel. It does not contribute to a net rise in the level of CO_2 in the atmosphere and minimizes the intensity of the greenhouse effect [36,37]. In addition, biodiesel is better than diesel fuel in terms of sulfur content, flash point, aromatic content, and biodegradability [38,39].

18.3.2 Current Biodiesel Production Technologies

Biodiesel production is undergoing rapid technological reforms in industries and academia. At present, the main drawback for the commercialization of biodiesel is its higher cost than petroleum-based diesel. Thus in previous years, numerous studies on the use of technologies and different methods to evaluate optimal conditions of biodiesel production technically and economically have been carried out. A number of methods are currently available and have been adopted for reduction of the viscosity of vegetable oils. Four primary ways to make biodiesel are direct use and blending of vegetable oils, microemulsions, thermal cracking (pyrolysis), and transesterification [17]. One of the most common methods used to reduce oil viscosity in the biodiesel industry is called transesterification, which takes place between a vegetable oil or animal fat and an alcohol (methanol, ethanol, or butanol) in the presence of a catalyst (homogeneous or heterogeneous) or without the application of catalysts [40].

18.4 BIODIESEL PRODUCTION PROCEDURE

There are several generally accepted technologies that have been well established for the production of biodiesel fuel. Vegetable oils and animal fats are appropriate to be modified in order to reduce their viscosities so that the product obtained has suitable properties to be used as diesel engine fuels. There are many procedures for this modification to produce a better quality of biodiesel, such as direct use and blending, microemulsions, pyrolysis of vegetable oil, and transesterification [41]. These procedures are briefly reviewed in this section.

18.4.1 Direct Use and Blending of Oils

The use of vegetable oils as alternative fuels has been around since 1900 when the inventor of the diesel engine, Dr. Rudolph Diesel, first tested peanut oil in his compression engine. The direct use of vegetable oils in diesel engines is problematic and has many inherent failings. It has only been researched extensively for the past couple of decades, but has been experimented with for near 100 years. Crude vegetable oils can be mixed directly or diluted with diesel fuel to improve the viscosity so as to solve the problems associated with the use of pure vegetable oils with high viscosities in CI engines [42]. Energy consumption, with the use of pure vegetable oils, was found to be similar to that of diesel fuel. For short term use, ratio of 1:10 to 2:10 oil to diesel fuel has been found to be successful. But, direct use of vegetable oils and/or the use of blends of the oils have generally been considered to be unsatisfactory and impractical for both direct and indirect diesel engines. The high viscosity, acid composition, free fatty acid (FFA) content, as well as gum formation due to oxidation and polymerization during storage and combustion, carbon deposits, and lubricating oil thickening are obvious problems [17]. Heating and blending of vegetable oils may reduce the viscosity and improve volatility of vegetable oils but its molecular structure remains unchanged, hence polyunsaturated character remains.

18.4.2 Microemulsion of Oils

Microemulsification is the formation of microemulsions (co-solvency) which is a potential solution for solving the problem of high vegetable oil viscosity. A microemulsion is defined as a colloidal equilibrium dispersion of optically isotropic fluid microstructures with dimensions generally in the 1—150 nm range formed spontaneously from two normally immiscible liquids and one or more ionic or nonionic amphiphiles [43]. Microemulsion-based fuels are sometimes also termed "hybrid fuels," although blends of conventional diesel fuel with vegetable oils have also been called hybrid fuels [44]. Microemulsions are clear, stable isotropic fluids with three components: an oil phase, an aqueous phase, and a surfactant.

For this purpose, microemulsions with solvents such as methanol, ethanol, and 1-butanol have been studied. All microemulsions with butanol, hexanol, and octanol can meet the maximum viscosity limitation for diesel engines. A microemulsion prepared by blending soybean oil, methanol, 2-octanol, and cetane improver in the ratio of 52.7:13.3:33.3:1.0 has passed the 200 h EMA test. Microemulsion of vegetable oils lowered the viscosity of the oil but resulted in irregular injector needle sticking, heavy carbon deposits, and incomplete combustion during 200 h laboratory screening endurance test [45].

18.4.3 Pyrolysis of Oils

Pyrolysis is the conversion of one organic substance into another by means of heat or by heat with the aid of a catalyst. The pyrolyzed material can be vegetable oil, animal fat, natural fatty acids, or methyl esters of fatty acids [24]. Conversion of vegetable oils and

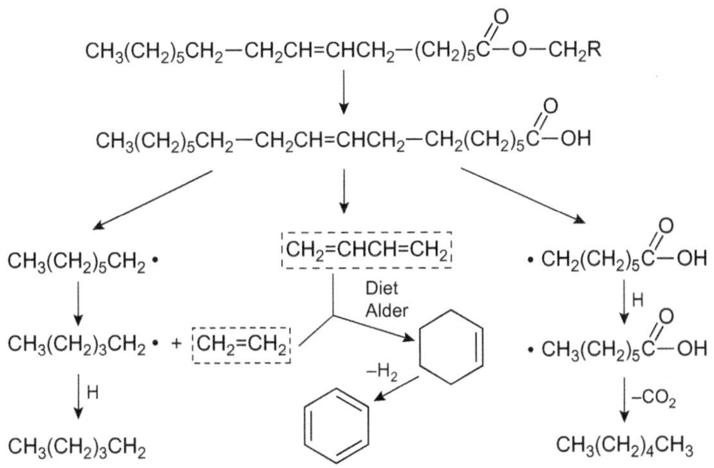
CH3(CH2)5CH2—CH2CH=CHCH2—(CH2)5C—O—CH2R
O
CH3(CH2)5CH2—CH2CH=CHCH2—CH2(CH2)5C—OH
O
CH3(CH2)5CH2 •
CH2=CHCH=CH2
• CH2(CH2)5C—OH
Diet
Alder
H
CH3(CH2)3CH2 • + CH2=CH2
• CH3(CH2)5C—OH
–H2
H
–CO2
CH3(CH2)3CH2
CH3(CH2)4CH3
Figure 18.1 The mechanism of thermal decomposition of triglycerides.

animal fats composed mostly of triglycerides using thermal cracking reactions represents a promising technology for the production of biodiesel. This technology is especially promising in areas where the hydro-processing industry is well established because the technology is very similar to that of conventional petroleum refining [46]. The fuel properties of the liquid product fractions of the thermally decomposed vegetable oil are likely to approach diesel fuels. Many researchers have reported the pyrolysis of triglycerides to obtain products suitable for diesel engines. The research on pyrolysis of triglycerides is divided into catalytic and noncatalytic processes [46]. The mechanism of thermal decomposition of triglycerides is depicted in Fig. 18.1 [47]. Mechanisms for the thermal decomposition of triglycerides are likely to be complex, because of the many structures and multiplicity of possible reactions of mixed triglycerides.

The pyrolysis reactions of soybean, palm tree, and castor oils have been studied in an investigation. The adequate choice of distillation temperature (DT) ranges made it possible to isolate fuels with physical and chemical properties comparable to those specified for petroleum-based fuels [46]. The equipment for thermal cracking and pyrolysis is expensive for modest throughputs. In addition, while the products are chemically similar to petroleum-derived gasoline and diesel fuel, the removal of oxygen during the thermal processing also removes any environmental benefits of using an oxygenated fuel. It produces some low-value materials and, sometimes, more gasoline than diesel fuel [17].

18.4.4 Transesterification of Oils

The most common technology of biodiesel production is transesterification of oils (triglycerides) with alcohol which gives biodiesel (fatty acid alkyl esters, FAAE) as main product and glycerin as a by-product. The basic transesterification is illustrated

$$
\begin{array}{lllll}
CH_2\text{-}OOC\text{-}R_1 & & & CH_2\text{-}OH & R_1\text{-}COO\text{-}R' \\
| & & & | & \\
CH\text{-}OOC\text{-}R_2 & + \quad 3R'OH & \leftrightarrow & CH\text{-}OH & + \quad R_2\text{-}COO\text{-}R' \\
| & & & | & \\
CH_2\text{-}OOC\text{-}R_3 & & & CH_2\text{-}OH & R_3\text{-}COO\text{-}R'
\end{array}
$$

Triglycerides   **Alcohol**   **Glycerin**   **Esters**

Figure 18.2 Transesterification reaction of triglycerides with alcohol.

in Fig. 18.2. The first step is the conversion of triglycerides to diglycerides, which is followed by the conversion of diglycerides to monoglycerides and of monoglycerides to glycerol, yielding one methyl ester molecule from each glyceride at each step [17].

Transesterification, also called alcoholysis, is exchanging of alcohol from an ester by another alcohol in a process similar to hydrolysis, except that an alcohol is used instead of water. The most relevant variables in transesterification process are reaction temperature, reaction time, ratio of alcohol to oil, concentration and type of catalyst, mixing intensity (rpm), and kind of feedstock [48].

18.5 BIODIESEL PRODUCTION TECHNOLOGIES: TRANSESTERIFICATION METHOD

The transesterification reaction proceeds with catalyst or without any catalyst by using primary or secondary monohydric aliphatic alcohols having one to eight carbon atoms [49]. Generally, alcohol and triglycerides (vegetable oil and animal fat) are not miscible to form a single phase of mixture. Hence, the poor surface contact between these two reactants causes transesterification reaction to proceed relatively slow. Introduction of catalysts improves the surface contact and consequently reaction rates and biodiesel yield as it is able to solve the problems of two-phase nature between triglycerides and alcohol. However, without the presence of catalysts, the reaction rate is too slow for it to produce considerable yield of biodiesel. Hence, researchers around the world have been developing numerous alternative technologies that can solve the problems faced due to catalytic reaction by using noncatalytic processes [40].

18.5.1 Catalytic Biodiesel Production

Vegetable oils can be transesterified by heating them with an alcohol and a catalyst. Catalysts used in biodiesel production are divided into two general categories, homogenous and heterogeneous types. If the catalyst remains in the same (liquid) phase as that of the reactants during transesterification, it is homogeneous catalytic transesterification. On the other hand, if the catalyst remains in a different phase (i.e., solid, immiscible liquid, or gaseous) than that of the reactants, the process is called heterogeneous catalytic transesterification [50].

In catalytic methods, the suitable selection of the catalyst is an important parameter to lower the biodiesel production cost. So, Commercial biodiesel is currently produced by transesterification using a homogenous catalyst solution. Another factor affecting the selection of catalyst type is the amount of FFA present in the oil. For oils having lower amount of FFAs, base-catalyzed reaction gives a better conversion in a relatively short time while for higher FFAs containing oils, acid-catalyzed esterification followed by transesterification is suitable. It has been reported that enzymatic reactions are insensitive to FFA and water content in oil. Hence, enzymatic reactions can be used in transesterification of used cooking oil [51]. Various studies have been carried out using different oils as raw material, different alcohols (methanol, ethanol, and butanol), as well as different catalysts, including homogeneous ones, such as sodium hydroxide, potassium hydroxide, and sulfuric acid, and heterogeneous ones, such as lipases, CaO, and MgO [48].

18.5.1.1 Homogeneous Catalytic Transesterification
Homogenous catalysts are categorized into basic and acidic catalysts. The homogenous transesterification process, especially basic type, requires a high purity of raw materials and postreaction separation of catalyst, by-product, and product at the end of the reaction. Both of these requirements drive up the cost of biodiesel. The general form of homogeneous catalytic transesterification process can be seen on a process flow diagram of Fig. 18.3.

18.5.1.1.1 Homogeneous Base Catalytic Transesterification
Currently, biodiesel is commonly produced using homogeneous base catalyst, such as alkaline metal alkoxides and hydroxides, as well as sodium or potassium carbonates.

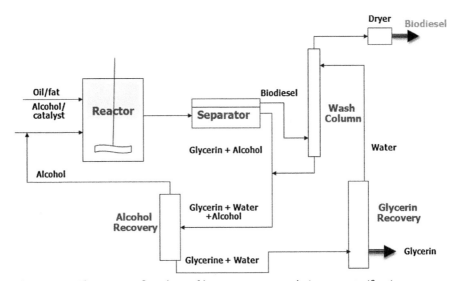

Figure 18.3 The process flowchart of homogeneous catalytic transesterification process.

As a catalyst in the process of basic methanolysis, mostly sodium hydroxide or potassium hydroxide have been used, both in the concentration from 0.4% to 2% w/w of oil. Homogeneous base catalytic catalysts are commonly used in the industries due to several reasons: (1) modest operation condition; (2) high conversion can be achieved in a minimal time, (3) high catalytic activity, and (4) widely available and economical [52]. In general, base catalytic transesterification processes are carried out at low temperatures and pressures (333—338 K and 1.4—4.2 bar) with low catalyst concentrations (0.5—2 wt%) [20].

The limits of this process are due to the sensitivity to purity of reactants, FFA content, as well as to the water concentration of the sample. When the oils contain significant amounts of FFAs and water content, they cannot be converted into biodiesels but to a lot of soap. FFAs of oil react with the basic catalyst to produce soaps that inhibit the separation of biodiesel, glycerin and wash water that results in more wastewater from purification. Because water makes the reaction partially change to saponification, the basic catalyst is consumed in producing soap and reduces catalyst efficiency. The soap causes an increase in viscosity and formation of gels, which reduces ester yield and makes the separation of glycerol difficult [53].

18.5.1.1.2 Homogeneous Acid Catalytic Transesterification

Sulfuric acid, hydrochloric acid, and sulfonic acid are usually preferred as acid catalysts. Acid-catalyzed transesterification starts by mixing the oil directly with the acidified alcohol, so that separation and transesterification occur in single step, with the alcohol acting both as a solvent and as esterification reagent. The use of excess alcohol effects significant reductions in reaction time required for the homogeneous acid-catalyzed reaction. Hence, Bronsted acid—catalyzed transesterification requires high catalyst concentration and a higher molar ratio to reduce the reaction time [53].

One advantage of homogeneous acid catalytic over homogeneous base catalytic transesterification is their low susceptibility to the presence of FFA in the feedstock. However, homogeneous acid catalytic transesterification is especially sensitive to water concentration. It was reported that as little as 0.1 (wt%) water in the reaction mixture was able to affect ester yields in transesterification of vegetable oil with methanol, with the reaction almost completely inhibited at 5 (wt%) water concentration. Another disadvantages of homogeneous acid catalytic transesterification are equipment corrosion, more waste from neutralization, difficult to recycle, higher reaction temperature, long reaction times, and weak catalytic activity [49].

18.5.1.2 Heterogeneous Catalytic Transesterification

In comparison with homogeneous catalysts that act in the same phase as the reaction mixture, heterogeneous catalysts act in a different phase from the reaction mixture. Being in a different phase, heterogeneous catalysts have the advantage of easy separation and

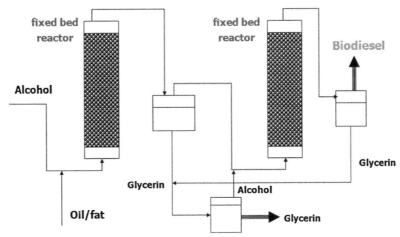

Figure 18.4 The schematic flow diagram of heterogeneous catalytic transesterification process.

reuse. The high consumption of energy and costly separation of the homogeneous catalyst from the reaction mixture, however, have called for development of heterogeneous catalyst. The use of heterogeneous catalyst does not yield soap [54]. The use of heterogeneous catalytic systems in transesterification of triglycerides implies the elimination of several steps of washing/recovery of biodiesel/catalyst, ensuring thereby higher efficiency and profitability of the process as well as lowering its production costs. There is also the possibility of being implemented in a continuous way using a fixed-bed reactor. A typical schematic diagram for heterogeneous catalytic transesterification process is shown in Fig. 18.4.

Compared to homogenous catalytic transesterification process, the heterogeneous catalytic transesterification process can tolerate extreme reaction conditions. The temperature could go from 70°C to as high as 200°C to achieve more than 95% of yield using MgO, CaO, and TiO$_2$ catalysts [54].

18.5.1.2.1 Heterogeneous Solid—Base Catalytic Transesterification

Most of heterogeneous solid catalysts are base or basic oxides coated over large surface area. Solid-base catalysts are more active than solid-acid catalyst. The most common solid-base catalysts are basic zeolites, alkaline earth metal oxides, and hydrotalcites. Solid base can lead to the heterogeneous catalytic process, which promises the cost reasonable for biodiesel production. Using the solid-base catalyst in the form of fixed-bed reactor system causes easier separation of it's from transesterified product. Also, the solid-base catalyst is active in the transesterification at the temperature around the methanol boiling point. Similar to their homogeneous counterparts, solid-base catalysts are more active than solid-acid catalysts [47]. CaO is widely used in solid-base catalyst possesses and has many advantages, such as long catalyst lifetimes, higher activity, lower solubility in methanol, and moderate reaction condition requirements. The reaction rate, however, was slow in producing biodiesel.

18.5.1.2.2 Heterogeneous Solid-Acid Catalytic Transesterification

Despite lower activity, heterogeneous solid-acid catalysts have been used in many industrial processes because they contain a variety of acid sites with different strength of Bronsted or Lewis acidity, compared to the homogenous acid catalysts. Using solid-acid catalysts has following advantages: insensitive to FFA content, simultaneous esterification and transesterification, elimination of purification step of biodiesel, easy separation of the catalyst from the reaction products, and reduction of the corrosion problem, even with the presence of acid species [44].

Heterogeneous solid-acid catalysts, such as Nafion-NR50, sulfated zirconia, and tungstated zirconia, have been chosen to catalyze biodiesel-forming transesterification due to the presence of sufficient acid site strength. Among the solid catalysts, Nafion demonstrated higher selectivity toward the production of methyl ester and glycerol due to its acid strength [49]. However, Nafion has disadvantages of high cost and lower activity compared to liquid acids.

18.5.1.3 Biocatalytic Transesterification

Biocatalysts have been becoming increasingly important in the discussion of biodiesel production recently. It is even hypothesized that these catalysts will eventually have the ability to outperform chemical catalysts. Biocatalysts are naturally occurring lipases that have been identified as having the ability to perform the transesterification reactions that are essential to biodiesel production. These lipases have been isolated from a number of bacterial species: *Pseudomonas fluorescens*, *Pseudomonas cepacia*, *Rhizomucor miehei*, *Rhizopus oryzae*, *Candida rugosa*, *Thermomyces lanuginosus*, and *Candida antarctica*. There are several methods for lipase immobilization, including adsorption, covalent bonding, entrapment, encapsulation, and cross-linking. These immobilization methods have been employed to improve lipase stability for biodiesel production in recent years. Adsorption is still the most widely employed method for lipase immobilization [55]. For transesterification synthesis, at least stoichiometric amount of methanol is required for the complete conversion of triacylglycerols to their corresponding fatty acid methyl esters. However, methanolysis is decreased significantly by adding ½ molar equivalent of methanol at the beginning of the biocatalytic transesterification. This reduction in activity caused by the polar short-chain alcohols was the major obstacle for the enzymatic biodiesel production. In order to solve this problem, researchers use the following three options: methanol stepwise addition, acyl acceptor alterations (methyl acetate and acetate ethyl), and solvent engineering (with t-butanol, 1,4-dioxane, and ionic liquid as solvents) [56].

During the transesterification of triglycerides via biocatalysts, glycerol is produced. The presence of glycerol can then inhibit the reaction by binding to the biocatalyst instead of the triglyceride molecule. To overcome this, an acyl acceptor molecule can be used to bind the glycerol, forming a triglyceride molecule that can no longer bind to the active site of the biocatalyst. When an acyl acceptor is introduced to the reaction, the biocatalyst can exhibit a higher rate of turnover, proving more usefulness. It has also been shown that the triacetylglycerol molecule does not diminish the properties of biodiesel when used as a fuel [24].

Biocatalytic transesterification process has many advantages over the chemical-catalyzed transesterification process, such as generation of zero by-product, no difficulty in product removal, requirement of only moderate process conditions (temperature, 35—45°C), and recycling of the catalysts. Enzymatic reactions can successfully be used for the transesterification of used cooking oil, because enzymatic reactions are insensitive to FFA and water content of the feedstock. These advantages prove that enzyme-catalyzed biodiesel production has high potential to be an ecofriendly process and a promising alternative to the chemical process. However, it still has its fair share of constraints especially when implemented in industrial scale, such as high cost of enzyme, slow reaction rate, and enzyme deactivation [20,57]. A brief comparison between the transesterification process based on homogeneous base catalyst and biocatalyst is shown in Table 18.1.

18.5.2 Noncatalytic Biodiesel Production

Beside catalytic methods, there are two noncatalytic transesterification processes. These are the supercritical alcohol process and BIOX process.

18.5.2.1 Supercritical Alcohol Transesterification

Supercritical alcohol method is a noncatalytic method for biodiesel production in which instead of using catalysts, high-pressure and -temperature are used to carry out the transesterification reaction. The reaction is fast and conversion raises 50—95% for the

Table 18.1 Comparison Between the Homogeneous Base Catalytic and Biocatalytic Process

	Homogeneous Base Catalytic Process	Biocatalytic Process
Reaction temperature	60—70°C	30—40°C
Free fatty acids in feedstock	Saponified products (soap formation)	Methyl esters
Water in raw materials	Interference with the reaction	No influence
Yield of methyl esters	Normal	Higher
Recovery of glycerol	Difficult	Easy
Purification of methyl esters	Repeated washing	None
Catalyst cost	Cheap	Expensive
Rate of reaction	High	Relatively low

Table 18.2 Critical Condition (Temperatures and Pressures) of Various Alcohols

Alcohol	Critical Temperature (°C)	Critical Pressure (MPa)
Methanol	239.2	8.1
Ethanol	243.2	6.4
1–Propanol	264.2	5.1
1–Butanol	287.2	4.9

first 10 min but it requires temperature range of 250–400°C. The transesterfication of triglycerides by supercritical methanol, ethanol, propanol, and butanol has proved to be the most promising process. Table 18.2 shows the critical temperatures and critical pressures of the various alcohols. The vegetable oils were transesterified 1:6–1:40 vegetable oil–alcohol molar ratios in supercritical alcohol conditions [53,58].

The disadvantages of the supercritical methods mostly are high pressure and temperature requirement and high methanol to oil ratios (usually 42) that render the production expensive. Kusudiana and Saka [56,57] studied a noncatalyst process in which vegetable oil was transesterified with supercritical methanol and found that the amount of water in the reaction does not affect the conversion of oil into biodiesel. Conversely, the presence of certain amount of water increases the formation of methyl esters, and esterification of FFAs takes place simultaneously in one stage. Their results showed that the reaction took only 4 min to convert rapeseed oil into biodiesel, even though high temperature (250–400°C) and high pressure (35–60 MPa) were required for making methanol reach the supercritical state.

18.5.2.2 BIOX Cosolvent Transesterification

Due to low solubility of methanol in oil, the rate of conversion of oil into ester is very slow. Another approach to overcome these problems that is now commercialized is the use of cosolvent that is soluble in both methanol and oil. The result is a fast reaction, on the order of 5–10 min, and no catalyst residues exist in either the ester or the glycerol phase. One of such cosolvents is tetrahydrofuran (THF), chosen in part, because it has a boiling point very close to that of methanol and the system requires a rather low operating temperature of 30°C. The Biox (cosolvent) process is a new Canadian process developed originally by Professor David Boocock of the University of Toronto that has attracted considerable attention. In patented Biox production processes, both triglycerides and FFAs are converted in a two-step, single-phase, continuous process at atmospheric pressures and near ambient temperatures, all in less than 90 min. Cosolvent options are available to overcome slow reaction rate caused by the extremely low solubility of the alcohol in the triglyceride phase [59].

THF has been used as a cosolvent and methanol to make mixture one phase [59]. After the completion of the reaction, the biodiesel–glycerol phase separation is clean and both the excess alcohol and the THF cosolvent can be recovered in a single step.

However, because of the possible hazard and toxicity of the cosolvents, they must be completely removed from the glycerol phase as well as the biodiesel phase and the final products should be water-free. The use of a cosolvent, such as THF or methyl tertiary butyl ether, speeds up methanolysis considerably. However, like one-phase butanolysis, one-phase methanolysis initially exhibits a rapid formation of ester, but then slows drastically [60].

The outstanding advantage of the Biox cosolvent process is that it uses inert, recoverable cosolvents in a single pass reaction that takes only seconds at ambient temperature and pressure, and no catalyst residues appear in either the biodiesel phase or the glycerol phase. This process can carry out not only crude vegetable oils but also waste cooking oils and animal fats. The research showed that the recovery of excess alcohol is difficult when using Biox cosolvent process, because the boiling point of the THF cosolvent is very close to that of methanol [55].

18.6 BIODIESEL FUEL STANDARDIZATION

One of the principal means of ensuring satisfactory in-use biodiesel fuel quality is the establishment of a rigorous set of fuel specifications, such as ASTM D6751 (in the United States) and EN 14214 (in the European Union). Numerous other countries have defined their own standards, which in many cases are derived from either ASTM D6751 or EN 14214. Some countries have also worked together to define the guidelines for regional biodiesel standards. For example, a group called the Asia—Pacific Economic Cooperation (APEC) issued a report in 2007 that addressed guidelines for standardizing biodiesel standards within the APEC region [61].

ASTM has established standard specifications for biodiesel fuel blend stocks (B100) for middle distillate fuels, called ASTM D6751 [62], as well as for biodiesel blends of B6 to B20 in mineral diesel, called ASTM D7467 [63]. Blends of B5 and below are permitted under the standard specifications for No. 2 diesel fuel, ASTM D975 [64]. To date, the CEN has only established standard specifications for B100, called EN 14214 [65], but not for mid-level blends, such as B20. The European standard specifications for conventional No. 2 diesel fuel (EN 590) permit blends of B7 and below; and deliberations are underway to allow an increase to B10 [66]. Appendix A provides a side-by-side listing of specifications for biodiesel blend stock (B100; ASTM and CEN) and mid-level biodiesel blends (B6—B20; ASTM only). For each specification, both the limits and the methods are shown [11].

18.7 POTENTIAL OF BIODIESEL

In order to meet the increasing demand of energy and to reduce the emission of CO_2 while ensuring energy security, the world needs to have an effective and sustainable source of energy. Biodiesel is renewable clean bioenergy as it can be produced from vegetable oils, animal fats, and microalgal oil. The property of biodiesel is almost similar to diesel fuel; thus it becomes a promising alternative to diesel fuel. Biodiesel has many benefits, such as it is biodegradable, nontoxic, has a low emission profile (including potential carcinogens), and is a renewable resource [17,67]. In addition, it does not contribute to the increase in CO_2 levels in the atmosphere and thus minimizes the intensity of the greenhouse effect.

Various factors contributing to the cost of biodiesel include raw material, other reactants, nature of purification, its storage, and so on. However, the main factor determining the cost of biodiesel production is the feedstock, which is about 80% of the total operating cost. Therefore, a great economic advantage could be achieved simply by using more economical feedstock, such as waste fats and oils [17,23]. Although the use of renewable energy as an alternative energy source is growing rapidly, it provides only 10% of the world's primary energy consumption in 2010 [68]. It is rather surprising that even in a country such as Malaysia, where biomass can be obtained easily, the use of renewable energy is still very low.

The high oil yield of feedstock is necessary for ensuring large production scale in cheap prices. In terms of the production cost, palm oil stands out as the least expensive oil to be produced per tonne compared with other major vegetable oils as shown in Fig. 18.5 [37]. For instance, the production cost of rapeseed oil in Canada and Europe is more than double the price compared to palm oil. This may be rationalized as oil palm being perennial (it grows every year without annual sowing) and may be considered a low energy—input crop. A major cost of production in the biodiesel market involves

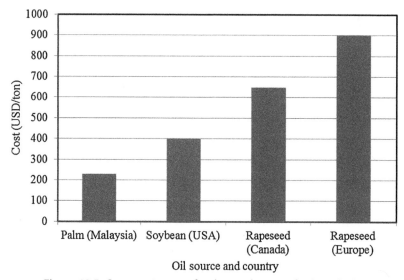

Figure 18.5 Comparative cost for the production of selected oils.

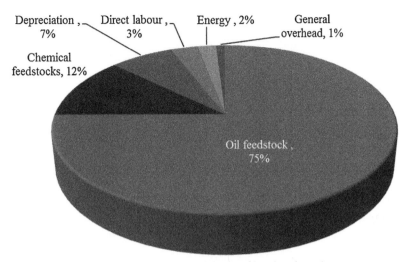

Figure 18.6 General cost breakdown of biodiesel production.

about 75% of raw material cost and the remaining 25% for the other manufacturing and production costs as shown in Fig. 18.6 [69]. Therefore, selecting the cheapest feedstock, such as palm oil—based fuel, is vital to ensure low production cost of biodiesel.

The typical GHG emission saving for the main feedstock of biodiesel is shown in Fig. 18.7 [70]. The life cycle analysis (LCA) conducted on various biodiesel reveals that palm oil—based biodiesel can reduce GHG emission by 62% as compared to soybean oil (40%), rapeseed oil (45%), and sunflower oil (58%) [70]. Palm biodiesel has a higher

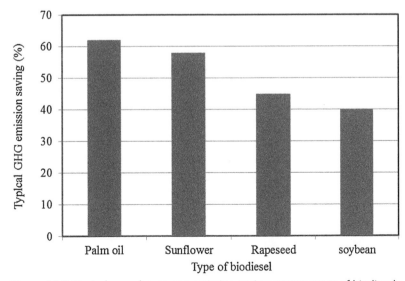

Figure 18.7 Typical greenhouse gas emission saving among types of biodiesels.

cetane index compared to all types of biodiesel derived from vegetable. Its oxidative stability is four times higher than biodiesel from soybean and it has higher lubricity than Diesel.

18.7.1 Biodiesel Fuel Usage Limitations

Using biodiesel as an alternative fuel for CI engines is widely accepted as comparable fuel to diesel. However, many barriers still need to be overcome. Therefore, great efforts are continuously paid to resolve these challenges. Biodiesel blends of 20% and below will work in any diesel engine without the need for modifications. These blends will operate in diesel engines just like mineral diesel. If the blend has been properly treated by the petroleum company, it will work all year round, even in cold climates. B20 also provides similar horsepower, torque, and mileage as diesel [71,72]. Despite the fact that biodiesel can replace diesel satisfactorily, problems related to fuel properties persist at high blending ratio [73,74]. In general, the viscosity of biodiesel is typically higher than that of the mineral diesel often by a factor of two [32,75]. High viscosity can cause larger droplets, poorer vaporization, narrower injection spray angle, and greater in-cylinder penetration of the fuel spray [75,76]. Furthermore, the use of fuel with a high kinematic viscosity can lead to undesired consequences, such as poor fuel atomization during spraying, engine deposits, wear on fuel pump elements and injectors, and additional energy required to pump the fuel [11,77]. Fuel injection systems measure fuels by volume, and thus, the engine output power is influenced by changes in density due to the different injected fuel masses [78,79]. Thus, density is important for various diesel engine performance aspects. The fuel-energy content has a direct impact on the engine power output [80,81]. Biodiesel contains less energy than mineral diesel, the energy content or the net calorific value of biodiesel is about 12% less than that of the mineral diesel fuel on a mass basis, this leads to lower the engine speed and power [10,25].

A key property of biodiesel currently limiting its application to blends of 20% or less is its relatively poor low-temperature properties [35,82]. Biodiesel will gel in cold weather, just like the regular diesel fuel. However, B20 can be treated for winter use; in similar ways that mineral diesel is treated. Using B20 throughout the winter months just takes a little preparation and good fuel management practices. The increase in NO_x emissions from biodiesel combustion, relative to the levels observed from mineral diesel combustion, has been reported by several researchers [7,83]. This increase is of concern in areas that are subject to strict environmental regulations. For universal acceptance of biodiesel, it is desirable to reduce these NO_x emissions at least to the levels observed with mineral diesel combustion.

18.7.2 Biodiesel Fuel Properties

Biodiesel is defined by the ASTM as "a fuel comprised of monoalkyl esters of long-chain fatty acids derived from vegetable oils or animal fats, designated B100" [84]. The chemical composition and properties of biodiesel depend on the length and degree of the unsaturation of the fatty acid alkyl chains. The higher molecular weight and number of double bonds, lack of aromatic compounds, and presence of oxygen in the esters all affect the combustion properties relative to diesel fuel [2]. Each fuel has specific macroscopic properties. These properties are interdependent, meaning that when one property is changed, others are affected. Although fuel properties are interdependent, difficult to isolate, and vary independently, generalities can be made about the influence of a specific fuel property on engine performance and emissions. The properties that most commonly affect the engine performance are viscosity, heating value, cetane number, and flow characteristics.

Biodiesel is an oxygenated fuel that contains 10—15% oxygen by weight [10,20] and lower carbon and hydrogen contents compared to diesel fuel, resulting in about 10% lower mass energy content. However, due to biodiesel's higher fuel density, its volumetric energy content is only about 5—6% lower than the mineral diesel. Typically, biodiesel has somewhat higher molecular weight than mineral diesel. Biodiesel properties can vary substantially from one feedstock to another. This section provides a review of the typical properties of biodiesel from different feedstocks. Furthermore, it is useful to focus on the properties of palm oil biodiesel to evaluate the suitability of this fuel compared with the diesel fuel.

18.7.2.1 Kinematic Viscosity

Viscosity is a measure of resistance to the flow of a liquid due to the internal friction of one part of a fluid moving over another and is based on the molecular structure and temperature [85]. This is a critical property because it affects the behavior of fuel injection, since an accurate amount of fuel is needed for injection. In general, higher viscosity leads to poorer fuel atomization [86]. High viscosity can cause larger droplet sizes, poorer vaporization, narrower injection spray angle, and greater in-cylinder penetration of the fuel spray [76,87]. This can lead to overall poorer combustion, higher emissions, and increased oil dilution. The viscosity of biodiesel is typically higher than that of the mineral diesel often by a factor of two. If the fuel viscosity is high, the injection pump will be unable to supply sufficient fuel to fill the pumping chamber, which will affect in the form of power loss from the engine. On the other hand, if the viscosity is too low, leaking can occur through the seals in the fuel injection system. Furthermore, many of the problems resulting from high viscosity are most noticeable under low ambient temperature and cold-start engine conditions.

18.7.2.2 Density

Fuel density is a key property that affects engine performance. Because fuel injection pumps meter the fuel by volume, not by mass, a greater or lesser mass of fuel is injected depending upon its density. Thus, the air—fuel ratio and energy content within the combustion chamber are influenced by fuel density, which affects the combustion in a diesel engine physically. Furthermore, a lower density fuel requires longer injection duration for the same fuel mass to be injected [88]. In general, densities of biodiesel fuels are slightly higher than those of the mineral diesel (less than 6%) [89].

18.7.2.3 Energy Content

The energy content of fuel is the amount of energy per unit of mass or volume given when combust. While a high-density fuel will have greater energy content per unit volume than a low-density fuel, the low-density fuel has greater energy content per unit mass than a high-density fuel. Fuels of different energy contents will give different power outputs on the same engine unless the fuel injection is individually optimized for each fuel. Due to its high oxygen content, biodiesel has lower mass energy values than the mineral diesel. Neither the US nor the European biodiesel standards include specification for the heating value. Due to its substantial oxygen content, it is generally accepted that biodiesel from all sources has about 10% lower mass energy content (MJ/kg) than mineral diesel [10,76]. The lower energy content for biodiesel demands a higher fuel flow rate than diesel for an engine producing the same amount of power.

18.7.2.4 Cetane Number

The cetane number designated for each fuel is based on the ignition quality. An increase in the cetane number causes a shorter ignition delay, which has the effect of less fuel being injected during the premix burn and more during the diffusion burn portion, thus reducing the cylinder pressure rise, which may result in lower cylinder temperatures [90]. At the same time, a lower cetane fuel advances the ignition timing because of the shorter ignition delay, which increases the combustion pressures and temperatures. Since biodiesel is largely composed of long-chain HC groups (with virtually no branching or aromatic structures), it typically has a higher cetane number than mineral diesel [10]. Biodiesel produced from feedstocks is rich in saturated fatty acids such as palm oil that has higher cetane number than fuels produced from less-saturated feedstocks such as soy bean and rapeseed [11]. Compared to mineral diesel, palm oil biodiesel has higher cetane number by about 16% [11].

18.7.2.5 Cloud and Pour Points

CP is defined as the temperature at which the wax or small crystals in the fuel begin to form. At this temperature, the fuel loses its flow tendency and begins to partially or fully solidify [91]. Meanwhile, the pour point is the lowest temperature at which the fuel flows. All the biodiesel fuels have significantly higher CPs and pour points compared to the diesel fuel [34]. This may cause major problems by plugging the flow line and poses difficulty to use pure biodiesel, particularly in the cold climate. Low-temperature performance is one of the most important considerations for users of biodiesel. Just as with the conventional diesel fuel, precautions must be taken to ensure satisfactory low-temperature operability of biodiesel and its blends. Poor low-temperature performance may be exhibited in several ways, but principally by filter plugging due to wax formation, and engine starving due to reduced fuel flow. There is no single best way to assess low-temperature performance, and the existing fuel standards (both US and European) do not include explicit specifications for cold flow properties for either conventional diesel or biodiesel [11,92]. Feedstocks with highly saturated FA structures, such as palm oil, produce biodiesel fuels with poor cold flow properties; whereas feedstocks with highly unsaturated fatty acid structures, such as rapeseed and safflower oil, produce fuels that have better performance [35]. Compared to mineral diesel, palm oil biodiesel has higher pour point by more than $20°C$ [11].

18.7.2.6 Flash Point

Flash point is inversely related to fuel volatility. The biofuel specifications for flash point are meant to guard against contamination by principally highly volatile impurities and excess methanol remaining after the product stripping processes. Even small amounts of residual methanol in biodiesel will cause a significantly depressed flash point. The flash point value for all biodiesel types is typically well above the flash point of diesel fuel by 25–90%. Compared to mineral diesel, palm oil biodiesel has higher flash point by about 78% [11,26].

18.7.2.7 Lubricity

Biodiesel from all feedstocks is generally regarded as having excellent lubricity, and the lubricity of the ultra-low sulfur diesel (ULSD) can be improved by blending it with biodiesel. Because of its naturally high lubricity, there is no lubricity specification for B100 within either the US or the European biodiesel standards. However, the US standard for B6–B20 blends (ASTM D7467) does include a lubricity specification, as does the conventional diesel fuel standard, ASTM D975. Low blend levels (often just 1–2%) typically provide satisfactory lubricity to the ULSD [22,56]. In part, biodiesel's good lubricity can be attributed to the ester group within the FAME molecules, but a higher degree of lubricity is due to the tracing of impurities in the biodiesel. In particular, FFAs and monoglycerides are highly effective lubricants [93]. It has been noted that the purification of biodiesel by means of distillation reduces its lubricity because these impurities are removed. The effect of unsaturation upon lubricity is unclear, with some researchers reporting positive effects of carbon–carbon double bonds while others report no effect [94,95]. The positive impact of biodiesel impurities upon lubricity is particularly noteworthy, as some of the same impurities (such as monoglycerides) are responsible for poor low-temperature operability problems. Efforts to reduce these impurities (to improve low temperature properties) could have the unintended consequence of worsening the lubricity.

18.7.2.8 Sulfur Content

The amount of total sulfur in motor fuels is determined by ASTM D5453. This standard uses ultraviolet fluorescence to detect trace amounts of sulfur in fuels [96]. For biodiesel, it can be considered a sulfur-free fuel where the allowable sulfur content has an upper limit of 0.05% by weight. For the conventional diesel, the sulfur content has an upper limit of 0.50% by weight [11]. Since the limit on sulfur is 10 times smaller for biodiesel, the amount of pollution derived from sulfur after a fuel has been burned is greatly reduced. This leads to less emission of (SO_x) during combustion. SO_x can have several deleterious effects on people and the environment. For example, sulfur dioxide (SO_2) contributes to respiratory illnesses and aggravates heart and lung diseases, and it also contributes to acid rain [32].

18.8 BIODIESEL FUEL BLENDING

Biodiesel is the only alternative fuel to have fully completed the health effects testing requirements of the Clean Air Act [97,98]. The real advantages for use of biodiesel are in reducing petroleum consumption and GHG effects [36,37]. However, there are many important advantages and disadvantages in using palm biodiesel compared to diesel fuel. The main advantages are the greater cetane number for palm biodiesel compared with diesel fuel and that it is free from sulfur and aromatics. The absence of sulfur prevents sulfur poisoning of aftertreatment systems and reduces corrosion. The high oxygen content of biodiesel fuel leads to reduced carbon emissions [10,20]. On the other hand, the high viscosity and low energy content are the main disadvantages of biodiesel fuel. In addition, it has higher pour point and CPs, limiting operation; lower oxidative stability, shortening storage life; and higher organic carbon emissions. To solve these problems, many researches have been focused on diesel—biodiesel blends [99].

Biodiesel blends are being considered to replace pure mineral diesel in many applications. Biodiesel has combustion characteristics similar to diesel and its blends with diesel has a shorter ignition delay, higher ignition temperature and pressure, as well as peak heat release as compared to the diesel fuel [100,101]. Moreover, the engine power output and brake power efficiency are found to be equivalent to diesel fuel. Biodiesel can be used in its pure form (B100), but more commonly it blends with the conventional diesel fuel. Blends are designated as BXX, where XX indicates the percentage of biodiesel by volume. B0 is conventional diesel. B5 is 5% biodiesel, 95% conventional. B20 is 20% biodiesel, 80% conventional, and so on.

Biodiesel can be blended in any proportion with the diesel fuel, and it can be used in most conventional internal combustion diesel engines with little or no modification [102,103]. It is likely that this value of blend ratio will increase as the biodiesel fuel quality improves and appropriate additives are developed. It has been well proven that the

Table 18.3 Effect of Blending on Fuel Properties

Property	Blending Ratio	Change
Viscosity	10—30	Increased by 1.5—27%
Density	10—30	Increased by 0.35—1.95%
Flash point	10—30	Increased by 2.4—10.6%
Pour point	10—30	Increased by 1—3°C
Cloud point	10—30	Increased by 0.8—4°C
Calorific value	10—30	Decreased by 4—8.3%

presence of higher amount of saturated components increases the CP and pour point of biodiesel. Table 18.3 presents the effect of blending biodiesel from different sources with mineral diesel on various fuel properties that change with the increasing biodiesel ratio in the blend.

Palm biodiesel has higher level of saturated FAME resulting in higher cetane number and higher oxidation stability [104]. Unfortunately, such a biodiesel does not possess good cold flow properties and it is not suitable for the production of winter-grade diesel fuels. Blending sufficient amounts of ordinary diesel can improve the cold flow properties of blended diesel fuel [105].

Numerous experimental studies were conducted to investigate the effect of blending palm oil biodiesel percentage on engine performance and emission. The specific consumption is directly related to the heat value of a fuel [106]. The higher the heat value of fuel, the lower the specific consumption [107,108]. Biodiesel has low heating value (10% lower than diesel) on a weight basis because of the presence of a substantial amount of oxygen in the fuel. As a result, in order to maintain the same brake power output, the brake-specific fuel consumption (BSFC) of palm biodiesel blends would be increased to compensate for the reduced chemical energy in the fuel [109]. Furthermore, the higher BSFC for the biodiesel compared to reference diesel is largely attributed to their higher fuel densities [110]. It is noted that the BSFC is the ratio between the mass of fuel consumption and brake effective power [111]. For a certain volume, as it is calculated on a weight basis, obviously higher densities of fuel will result in higher values for BSFC [112,113]. At all loads; biodiesel and its blends with commercial diesel shows higher specific consumption since these fuels have lower heat value compared to diesel [114,115]. Literature reviewed indicates that commercial diesel fuel shows the lowest specific fuel consumption [102]. It can also be observed that increasing the percentage of biodiesel in the mixture with commercial diesel causes an increase in specific consumption [74,107].

Engine power and brake thermal efficiency for blended biodiesel—diesel fuel is found to be nearly equal to that of diesel [102,108], and the ignition delay is noted to decrease as the palm oil blend percentage becomes higher [110]. Thermal efficiency is the ratio between the power output and the energy introduced through fuel injection, the latter being the product of the injected fuel mass flow rate and the lower heating value. Since it is usual to use the brake power for determining the thermal efficiency in experimental engine studies, the efficiency obtained is really a brake-specific efficiency. This parameter is more appropriate than fuel consumption to compare the performance of different fuels, besides their heating value [116]. In most studies [74,108,115] the brake thermal efficiency is lower for biodiesel than mineral diesel throughout the engine speed range due to the lower calorific values of the blended fuels. However, the brake thermal efficiency for biodiesel is found to be nearly equal to diesel in other experimental works [102,103]. On the other hand, [117] conclude that the lower blends of biodiesel increase the brake thermal efficiency and reduce the fuel consumption compared to diesel. This is a consequence of the tested palm oil biodiesel properties, which have a higher cetane number and lower viscosity value compared to the diesel fuel sample. These contradictory results may be attributed to fuel properties, such as density, viscosity, composition, biodiesel production method, as well as test condition. However, it is difficult to determine the effect of any single fuel characteristics alone on engine performance since many of the characteristics are interrelated.

Biodiesel fuel is biodegradable and nontoxic, and its use provides a reduction of many harmful exhaust emissions. A nearly complete absence of (SO_x) emissions, particulate, and soot, and reduction in unburned HC emissions can be achieved [118]. Life cycle studies have shown that biodiesel contains substantially more energy than what is required for its production and also it significantly reduces net CO_2 emissions. However, unmodified engines using biodiesel blends typically emit higher levels of NO_x [83,119]. The "biodiesel NO_x effect" has been, and continues to be, a subject of a great deal of scientific research where the consensus for the exact reason(s) for this increase has not yet been reached.

The engine emissions trends with increasing biodiesel content are plotted in Fig. 18.8 which show the compilation of data from a 2002 EPA study of a large number of pre-1998 model diesel engines for fuel blends from B0 to B100 [120]. From this figure, it is obvious that the increasing blend fractions of biodiesel generally result in dramatic decreases in particulate matter (PM), CO, and unburned HC, where these reductions are frequently accompanied by significant increases in NO_x. For engines using B100, there is a 10% increase in NO_x. Of even greater concern is the observation that these NO_x increases appear to be more, and not less, dramatic in most modern diesel engines.

Several studies have been conducted to investigate the effect of the blended palm oil biodiesel—diesel fuel on exhaust emissions as compared to diesel. Most of the studies reviewed show that increasing the content of palm oil biodiesel in the blend can reduce the engine emissions except for NO_x, which increases. Increasing palm oil biodiesel in

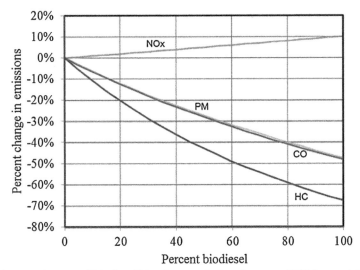

Figure 18.8 Average impact of biodiesel blends on emissions from pre-1998 heavy duty on highway engines.

the blend is efficient to reduce PM and PAHs [106], similarly, CO, HC, and CO_2 emissions are decreased [110,117]. Contrarily, most of the previous literature [107] have shown that CO_2 emissions of palm oil biodiesel and its blends are higher compared to the commercial diesel fuel. Emission of NO_x is increased when the engine is fueled with palm oil biodiesel and its blends with commercial diesel [107,110]. However, Wariwan et al. [117] show that NO_x emission is decreased when the content of palm biodiesel increases in the blend, which is in contrast with those generally found in previous studies.

The majority of the studies reviewed have found sharp reductions in exhaust emissions with palm biodiesel as compared to diesel fuel. The more accepted reasons in the reduction of emissions, particularly CO, CO_2, HCs, SO_2, particulates, and smoke can be attributed to the presence of sufficient oxygen in biodiesel. Oxygen content and cetane number can also be varied based on the biodiesel feedstock and therefore the emission characteristics are different for different biodiesels [29]. As mentioned previously, biodiesel contains oxygen while diesel has no oxygen content. The increased amount of oxygen in the fuel-rich combustion zone is believed to ensure more complete combustion and thereby to ensure that exhaust emissions are reduced [121].

However, most of the researchers have reported slight increase of NO_x emissions for biodiesel. It is quite obvious that with biodiesel, due to the improved combustion, the temperature in the combustion chamber can be expected to be higher, and higher amount of oxygen is also present which leads to the formation of higher quantity of NO_x in palm biodiesel-fueled engines. The most commonly accepted justification for this behavior lies in the higher cetane number of the palm biodiesel that reduces the ignition delay which can increase NO_x emissions [122].

18.9 BIODIESEL FUEL ADDITIVE

Biodiesel blending with diesel fuel is one of the visible methods to solve the problems associated with biodiesel properties. Biodiesel blends are being considered to replace pure mineral diesel in many applications. Biodiesel blends of 20% and below will work in any diesel engine without the need for engine modifications [71,72]. These blends will operate in diesel engines just like mineral diesel and are approved as commercial fuel in many countries. Furthermore, if the blend has been properly treated by a petroleum company, it will work all year round, even in cold climates. B20 also provides similar horsepower, torque, and mileage as diesel. However, there are some technical problems associated with the use of biodiesel fuels at high blending ratio. The use of some of them includes an increase in NO_x exhaust emissions, which have stringent environmental regulations, and relatively poor low-temperature flow properties compared to the mineral diesel. Another problem is the oxidation stability of biodiesel. The esters of unsaturated fatty acids are unstable with respect to light, catalytic systems, and atmospheric oxygen. Since diesel fuels from fossil oil have good oxidation stability, automobile companies have not considered fuel degradation when developing diesel engines and vehicles. It is one of the key issues in using vegetable oil–based fuels, and attention is given to the stability of biodiesel during storage and use. These problems could be circumvented by using additives.

A key property of biodiesel currently limiting its application to blends of 20% or less is its relatively poor low-temperature properties. Petroleum diesel fuels are plagued by the growth and agglomeration of paraffin wax crystals when ambient temperatures fall below the fuel's CP. These solid crystals may cause start-up problems, such as filter clogging, when ambient temperatures drop to around -10 to $-15°C$ [82]. Meanwhile, the CP of mineral diesel is reported to be $-16°C$, with biodiesel typically having a CP of around $0°C$, thereby limiting its use to ambient temperatures above freezing point [85,123].

Although biodiesel has become more attractive as an alternative fuel for diesel engines, its commercial usage is still limited to B5 in Malaysia and B20 in many other countries. It can be concluded that developing fuel additives will become an indispensable tool to make biodiesel fuel economically viable as well as to use cleaner fuels. The additive technical specifications not only cover a wide range of subjects, but also most subjects are inter-dependent. This makes the expertise behind the additive technology indispensable in the global trade of fuels. It is likely that, as energy sources become cleaner and renewable, we might find ourselves facing issues that are quite difficult to overcome, and biodiesel additives may become an indispensable tool worldwide.

Many published research has used different additives as a component of the biodiesel or blends of biodiesel fuels to improve either the combustion properties or the emission of a CI engine. The additives used by different researchers can be classified into chemical additives, solid metallic additives, and commercial additives. Some chemical additives such as methanol can be produced from coal- or petrol-based fuels with low cost production, but it has very limited solubility in the diesel fuel. On the other hand, chemical additive such as ethanol is a biomass-based renewable fuel, which can be produced from vegetable materials, such as corn, sugarcane, sugar beets, sorghum, barley, and cassava, and it has higher miscibility with diesel fuel [124,125]. Ethanol is a low-cost oxygenate with high oxygen content (about 35%) that has been used in biodiesel—ethanol blends. It is reported [126] that the ethanol—diesel—biodiesel fuel blends are stable well below subzero temperature and have equal or superior fuel properties to mineral diesel fuel. Ethanol and methanol, as well as products derived from these alcohols, such as ethers, are under consideration or in use as alternative fuels or as an additive biodiesel fuel. Methanol offers very low particulate emissions, but the problems are their toxicity, low energy density, low cetane number, high aldehyde emissions, and harmful influence on materials used in engine production. Ethanol seems to be the best candidate as a sole fuel as a component of either gasoline or diesel oil [127]. Until now, ethanol has been recognized only as a component of gasoline and not as a component of diesel oils. The properties of ethanol enable it to be applied as a component of diesel oil. The potential of oxygenates as a means of achieving zero net CO_2 renewable fuel has resulted in considerable interest in the production and application of ethanol. In many countries such as the United States of America, Canada, Australia, Brazil, South Africa, Denmark, Sweden, and others ethanol programs are realized. The research on ethanol programs is directed to identify factors that could influence engine performance and exhaust emissions. The understanding of these factors is necessary for the interpretation of the test results.

The fuel type has a direct influence on the engine cycle-to-cycle variations that are associated with power and efficiency losses, in addition to fluctuations in engine speed, torque, and work. Furthermore, the variations in cylinder pressure are correlated with the variations in the brake torque which directly influence engine operation [13]. In general, the fuel additives have a different chemical composition compared to diesel and biodiesel fuel and it may develop a cycle-to-cycle variation when it exceeds certain limits

[118]. These variations might lead to both a reduction of engine output power and higher emissions; so it is necessary to develop effective control strategies for optimum additive ratio through gaining a better understanding of the various factors that affect the overall combustion process.

The selection of additives for the biodiesel fuel depends on economic feasibility, toxicity, fuel blending property, additive solubility, flash point of the blend, viscosity of the blend, solubility of water in the resultant blend, and water partitioning of the additive. At present, concern about environmental regulations has been the major reason to look for alternative fuels. The use of biodiesel has presented a promising alternative in the world. It is not only a renewable energy source, but it can also reduce the dependence on imported oil and support agricultural subsidies in certain regions. The growing interest in this renewable fuel can be illustrated by the number of articles published and patents registered in this area during the past few decades.

18.10 CONCLUSIONS

The chapter presents the potential of biodiesel from different feedstocks as a fuel for diesel engine alternative to mineral diesel fuel. Many methods were used to produce biodiesel fuel with different specifications. The detailed design and procedures of biodiesel production using different techniques has been discussed and compared in detail. The different fuel standard for biodiesel fuel was listed to characterize the fuel specifications and suitability for diesel engine. Cost analysis was performed for the biodiesel production process from various feedstocks. The typical GHG emission saving for the main feedstocks of biodiesel has been discussed in brief detail. The limitations of biodiesel fuel usage was also listed and discussed to indicate the viability of biodiesel as a fuel for existing diesel engine from different feedstocks. Different biodiesel properties have been explained in detail. The effect of each property on engine operation has been discussed compared to fuel standard limits. The different techniques used to introduce biodiesel as a fuel within the standard specification have been discussed in detail. The effect of biodiesel blending with mineral diesel on different fuel properties and engine emissions has been discussed and analyzed. Influence of fuel additives as a visible method to solve the problems associated with biodiesel properties has been discussed. These additives have been introduced with biodiesel to improve fuel properties according to the standard specifications and control engine emissions toward the direct application in diesel engine.

APPENDICES

Appendix A ASTM and EN Biodiesel Fuel Standards Specifications

Properties	ASTM D6751 (B100) Limits	ASTM D6751 (B100) Method	EN14214 (B100) Limits	EN14214 (B100) Method	ASTM D7467 (B6-B20) Limits	ASTM D7467 (B6-B20) Method
Kinematic viscosity at 40°C (mm²/s)	1.9—6.0	D 445	3.5—5	EN 3104/3105	1.9—4.1	D 445
Flash point, closed cup (°C)	93	D 93	101	EN 3679	52	D 93
Water content (vol%, max)	0.05	D2709	0.05	EN12937[g]	0.05	D 2709
Total contamination (mg/kg, max)			24	EN 12662		
Methanol (wt%, max)	0.20[a]	EN 14110	0.20	EN 14110		
Cetane no. (min)	47	D 613	51	EN 5161	40	D 613
Cloud point (°C)	Report[d]	D 2500	Country-specific[d]		Report[d]	D 2500
Sulfated ash (wt%, max)	0.02	D 874	0.02	EN 3987		
Total ash (wt%, max)					0.01	D 482
Gp I metals Na + k (mg/kg, max)	5.0	EN 14538	5.0	EN 14108/14109		
Gp I metals Ca + Mg (mg/kg, max)	5.0	EN 14538	5.0	EN 14538		
Total sulfur (ppm, max)	15[b]	D 5453	10	EN 20846	15	D 5453
Phosphorous (ppm, max)	10	D 4951	4	EN 14107		
Acid no. (mg KOH/g, max)	0.50	D 664	0.50	EN 14104	0.3	D 664
Carbon residue (wt%, max)	0.05	D 4530	0.30[e]	EN 10370	0.35[e]	D 524
Free glycerine (wt%, max)	0.02	D 6584	0.02	EN 14105/14106		
Total glycerine (wt%, max)	0.24	D 6584	0.25	EN 14105		
Monoglyceride (wt%, max)			0.80	EN 14105		
Diglyceride (wt%, max)			0.20	EN 14105		
Triglyceride (wt%, max)			0.20	EN 14105		
Distillation (T90°C, max)	36[c]	D 1160			343	D 86
Copper strip corrosion (3 h at 50°C, max)	No. 3	D 130	No. 1	EN 2160	No. 3	D 130
Oxidation stability (h at 110°C, min)	3.0	EN 14112	6.0	EN 14112	6.0	EN 14112
Linolenic acid methyl ester (wt%, max)			12.0	EN 14103		
Polyunsaturated acid methyl ester (wt%, max)			1.0	Pr EN 15799		
Ester content (wt%, min)			96.5	EN 14103	6—20 vol%	D 6079
Iodine value (g I2/100 g, max)			120	EN 14111		
Density at 15°C (kg/m³)	880	D 1298	860—900	EN 3675	820—858	D 1298/D6890
Lubricity at 60°C, WSD, microns (max)					520	D 6079
Cold soak filterability (s, max)	360[f]	D 7501				

[a]Alternatively, flash point must be >130°C.
[b]For blending with ULSD. For other fuels, higher sulfur levels are allowed.
[c]Atmospheric equivalent T-90 point.
[d]Low-temperature properties are not strictly specified, but should be agreed upon by the fuel supplier or purchaser.
[e]This limit is based on the bottom 10% fraction of the fuel, not the entire fuel.
[f]200 s maximum for use in diesel blends at low temperature (<−12°C).
[g]Method EN 12937 measures total water (in units of μg/g), but not sediment.

REFERENCES

[1] EIA. International energy outlook. Energy Information Administration; 2013.

[2] Demirbas A. Biodiesel: a realistic fuel alternative for diesel engines. London: Springer; 2008.

[3] WTO. Merchandise trade. World Trade Organization; 2010.

[4] IEA. Key world energy statistics. Internationa Energy Agency; 2012.

[5] Ramadhas AS, Jayaraj S, Muraleedharan C. Use of vegetable oils as I.C. engine fuels—a review. Renewable Energy 2004;29:727—42.

[6] Ong HC, Mahlia TMI, Masjuki HH, Norhasyima RS. Comparison of palm oil, *Jatropha curcas* and *Calophyllum inophyllum* for biodiesel: a review. Renewable and Sustainable Energy Reviews 2011;15:3501—15.

[7] Hoekman SK, Robbins C. Review of the effects of biodiesel on NOx emissions. Fuel Processing Technology 2012;96:237—49.

[8] Basha SA, Gopal KR, Jebaraj S. A review on biodiesel production, combustion, emissions and performance. Renewable & Sustainable Energy Reviews 2009;13:1628—34.

[9] Canakci M, Ozsezen AN, Arcaklioglu E, Erdil A. Prediction of performance and exhaust emissions of a diesel engine fueled with biodiesel produced from waste frying palm oil. Expert Systems with Applications 2009;36:9268—80.

[10] Silitonga AS, Masjuki HH, Mahlia TMI, Ong HC, Chong WT, Boosroh MH. Overview properties of biodiesel diesel blends from edible and non-edible feedstock. Renewable and Sustainable Energy Reviews 2013;22:346—60.

[11] Hoekman SK, Broch A, Robbins C, Ceniceros E, Natarajan M. Review of biodiesel composition, properties, and specifications. Renewable and Sustainable Energy Reviews 2012;16:143—69.

[12] NBB. Biodiesel overview. National Biodiesel Board; 2011.

[13] Heywood JB. Internal combustion engine fundamentals. New York: McGraw-Hill; 1988.

[14] McAllister S, Chen J-Y, Fernandez-Pello AC. Fundamentals of combustion processes. New York (USA): Springer; 2011.

[15] Knothe G. Historical perspectives on vegetable oil-based diesel fuels. Inform 2001;12:103—7.

[16] Sidibé SS, Blin J, Vaitilingom G, Azoumah Y. Use of crude filtered vegetable oil as a fuel in diesel engines state of the art: literature review. Renewable and Sustainable Energy Reviews 2010;14:2748—59.

[17] Ma F, Hanna MA. Biodiesel production: a review. Bioresource Technology 1999;70:1—15.

[18] Balat M, Balat H. A critical review of bio-diesel as a vehicular fuel. Energy Conversion and Management 2008;49:2727—41.

[19] Misra RD, Murthy MS. Straight vegetable oils usage in a compression ignition engine—a review. Renewable and Sustainable Energy Reviews 2010;14:3005—13.

[20] Demirbas A. Progress and recent trends in biodiesel fuels. Energy Conversion and Management 2009;50:14—34.

[21] Shahid EM, Jamal Y. Production of biodiesel: a technical review. Renewable and Sustainable Energy Reviews 2011;15:4732—45.

[22] Scharffbillig J, Clark E. A biodiesel blend handling guide. Minnesota Biodiesel Technical Cold Weather 2014.

[23] Demirbas A. Importance of biodiesel as transportation fuel. Energy Policy 2007;35:4661—70.

[24] Yusuf NNAN, Kamarudin SK, Yaakub Z. Overview on the current trends in biodiesel production. Energy Conversion and Management 2011;52:2741—51.

[25] Van Gerpen JH, Peterson CL, Goering CE. Biodiesel: an alternative fuel for compression ignition engines. American Society of Agricultural and Biological Engineers 2007:1—22.

[26] Banga S, Varshney PK. Effect of impurities on performance of biodiesel: a review. Journal of Scientific & Industrial Research 2010;69:575—9.

[27] Sadeghinezhad E, Kazi SN, Badarudin A, Oon CS, Zubir MNM, Mehrali M. A comprehensive review of bio-diesel as alternative fuel for compression ignition engines. Renewable and Sustainable Energy Reviews 2013;28:410—24.

[28] Wang R, Hanna Ma, Zhou W-W, Bhadury PS, Chen Q, Song B-A, et al. Production and selected fuel properties of biodiesel from promising non-edible oils: *Euphorbia lathyris* L., *Sapium sebiferum* L. and *Jatropha curcas* L. Bioresource Technology 2011;102:1194—9.

[29] Yoshimoto Y. Performance and emissions of a diesel engine fueled by biodiesel derived from different vegetable oils and the characteristics of combustion of single droplets. SAE 2009:827—38. http://dx.doi.org/10.4271/2009-01-1812.

[30] Mofijur M, Masjuki HH, Kalam MA, Atabani AE, Shahabuddin M, Palash SM, et al. Effect of biodiesel from various feedstocks on combustion characteristics, engine durability and materials compatibility: a review. Renewable and Sustainable Energy Reviews 2013;28:441—55.

[31] Fazal MA, Haseeb ASMA, Masjuki HH. Biodiesel feasibility study: an evaluation of material compatibility; performance; emission and engine durability. Renewable and Sustainable Energy Reviews 2011;15:1314—24.

[32] Hoekman SK, Gertler AW, Broch A, Robbins C, Natarajan M. Biodistillate transportation fuels 1. Production and properties. SAE 2009;2009:01—27.

[33] Shahid EM, Jamal Y. A review of biodiesel as vehicular fuel. Renewable and Sustainable Energy Reviews 2008;12:2484—94.

[34] Misra RD, Murthy MS. Blending of additives with biodiesels to improve the cold flow properties, combustion and emission performance in a compression ignition engine—A review. Renewable and Sustainable Energy Reviews 2011;15:2413—22.

[35] Echim C, Maes J, Greyt WD. Improvement of cold filter plugging point of biodiesel from alternative feedstocks. Fuel 2012;93:642—8.

[36] Rodrigues MC, Guarieiro LLN, Cardoso MP, Carvalho LS, da Rocha GO, de Andrade JB. Acetaldehyde and formaldehyde concentrations from sites impacted by heavy-duty diesel vehicles and their correlation with the fuel composition: diesel and diesel/biodiesel blends. Fuel 2012;92:258—63.

[37] Lam MK, Tan KT, Lee KT, Mohamed AR. Malaysian palm oil: surviving the food versus fuel dispute for a sustainable future. Renewable and Sustainable Energy Reviews 2009;13:1456—64.

[38] Mekhilef S, Siga S, Saidur R. A review on palm oil biodiesel as a source of renewable fuel. Renewable and Sustainable Energy Reviews 2011;15:1937—49.

[39] Kumar N, Chauhan SR. Performance and emission characteristics of biodiesel from different origins: a review. Renewable and Sustainable Energy Reviews 2013;21:633—58.

[40] Demirbas A. Biodiesel from waste cooking oil via base-catalytic and supercritical methanol transesterification. Energy Conversion and Management 2009;50:923—7.

[41] Leung DYC, Wu X, Leung MKH. A review on biodiesel production using catalyzed transesterification. Applied Energy 2010;87:1083—95.

[42] Koh MY, Ghazi TIM. A review of biodiesel production from Jatropha curcas L. oil. Renewable and Sustainable Energy Reviews 2011;15:2240—51.

[43] Fernandes DM, Serqueira DS, Portela FM, Assunção RMN, Munoz RAA, Terrones MGH. Preparation and characterization of methylic and ethylic biodiesel from cottonseed oil and effect of tert-butylhydroquinone on its oxidative stability. Fuel 2012;97:658—61.

[44] Satyanarayana M, Muraleedharan C. Biodiesel production from vegetable oils: a comparative optimization study. Journal of Biobased Materials and Bioenergy 2009;3:335—41.

[45] Öner C, Altun S. Biodiesel production from inedible animal tallow and an experimental investigation of its use as alternative fuel in a direct injection diesel engine. Applied Energy 2009;86:2114—20.

[46] Maher KD, Bressler DC. Pyrolysis of triglyceride materials for the production of renewable fuels and chemicals. Bioresource Technology 2007;98:2351—68.

[47] Galadima A, Muraza O. Biodiesel production from algae by using heterogeneous catalysts: a critical review. Energy 2014;78:72—83.

[48] Marchetti JM, Miguel VU, Errazu a F. Possible methods for biodiesel production. Renewable and Sustainable Energy Reviews 2007;11:1300—11.

[49] Demirbas A. Biodiesel production from vegetable oils via catalytic and non-catalytic supercritical methanol transesterification methods. Progress in Energy and Combustion Science 2005;31:466—87.

[50] Zabeti M, Wan Daud WMA, Aroua MK. Activity of solid catalysts for biodiesel production: a review. Fuel Processing Technology 2009;90:770—7.

[51] Ragit SS, Mohapatra SK, Kundu K, Gill P. Optimization of neem methyl ester from transesterification process and fuel characterization as a diesel substitute. Biomass and Bioenergy 2011;35:1138—44.

[52] Araujo BQ, Nunes RCDR, de Moura CVR, de Moura EM, Cito AMDGL, dos Santos Junior JR. Synthesis and characterization of beef tallow biodiesel. Energy & Fuels 2010;24:4476—80.

[53] Meher LC, Sagar DV, Naik SN. Technical aspects of biodiesel production by transesterification — a review. Renewable and Sustainable Energy Reviews 2006;10:248—68.

[54] Leu J-H. Biodiesel manufactured from waste cooking oil by alkali transerification reaction and its vehicle application. Journal of Biobased Materials and Bioenergy 2013;7:189—93.

[55] Van Gerpen J, Shanks B, Pruszko R, Clements D, Knoth G. Biodiesel production technology. National Renewable Energy Laboratory; 2004.

[56] Demirbas A. Relationships derived from physical properties of vegetable oil and biodiesel fuels. Fuel 2008;87:1743—8.

[57] Athawale V, Rathi S, Bhabhe M. Novel method for separating fatty esters from partial glycerides in biocatalytic transesterification of oils. Separation and Purification Technology 2000;18:209—15.

[58] Teo SH, Islam A, Yusaf T, Taufiq-Yap YH. Transesterification of *Nannochloropsis oculata* microalga's oil to biodiesel using calcium methoxide catalyst. Energy 2014;78:63—71.

[59] Kusdiana D, Saka S. Effects of water on biodiesel fuel production by supercritical methanol treatment. Bioresource Technology 2004;91:289—95.

[60] Babcock DGB, Konar SK, Mao V. Fast one-phase oil-rich processes for the preparation of vegetable oil methyl esters. Biomass & Bioenergy 1996:11.

[61] Goosen R, Vora K, Vona C. Establishment of the guidelines for the development of biodiesel standards in the APEC region. APEC; 2007.

[62] ASTM. Standard specification for biodiesel fuel blend stock (B100) for middle distillate fuels. 2011.

[63] ASTM. Standard specification for diesel fuel oil, biodiesel blend (B6 to B20). 2013.

[64] ASTM. Standard specification for diesel fuel oils. 2013.

[65] CEN. Automotive fuels — fatty acid methyl esters (FAME) for diesel engines — requirements and test methods carburants. European Standard 2009; DIN EN 142. 2009.

[66] Chollacoop N. Current biodiesel standards. In: NSTDA Annual Conference 2013 (NAC2013), vol. 2013; 2013. p. 0—12.

[67] Chongkhong S, Tongurai C, Chetpattananondh P, Bunyakan C. Biodiesel production by esterification of palm fatty acid distillate. Biomass and Bioenergy 2007;31:563—8.

[68] Lior N. Sustainable energy development: the present (2011) situation and possible paths to the future. Energy 2012;43:174—91.

[69] Mofijur M, Atabani AE, Masjuki HH, Kalam MA, Masum BM. A study on the effects of promising edible and non-edible biodiesel feedstocks on engine performance and emissions production: a comparative evaluation. Renewable and Sustainable Energy Reviews 2013;23:391—404.

[70] Ong HC, Mahlia TMI, Masjuki HH. A review on energy pattern and policy for transportation sector in Malaysia. Renewable and Sustainable Energy Reviews 2012;16:532—42.

[71] NBB. Using-biodiesel/handling-use. National Biodiesel Board; 2014.

[72] Singh B, Guldhe A, Rawat I, Bux F. Towards a sustainable approach for development of biodiesel from plant and microalgae. Renewable and Sustainable Energy Reviews 2014;29:216—45.

[73] Chen L, Chen Y, Hung Y, Chiang T, Tsai C. Fuel properties and combustion characteristics of jatropha oil biodiesel—diesel blends. Journal of the Taiwan Institute of Chemical Engineers 2013;44:214—20.

[74] Deepanraj B, Dhanesh C, Senthil R, Kannan M. Use of palm oil biodiesel blends as a fuel for compression ignition engine. American Journal of Applied Sciences 2011;8:1154—8.

[75] Sarin A. Biodiesel production and properties. Cambridge (UK): Royal Society of Chemistary; 2012.

[76] Pandey RK, Rehman A, Sarviya RM. Impact of alternative fuel properties on fuel spray behavior and atomization. Renewable and Sustainable Energy Reviews 2012;16:1762—78.

[77] Wadumesthrige K, Johnson N, Winston-Galant M, Tang H, Ng KYS, Salley SO. Deterioration of B20 from compression ignition engine operation. SAE 2010:638—49. http://dx.doi.org/10.4271/2010-01-2120.

[78] Liaquat AM, Masjuki HH, Kalam MA, Fattah IMR, Hazrat MA, Varman M, et al. Effect of coconut biodiesel blended fuels on engine performance and emission characteristics. Procedia Engineering 2013;56:583—90.

[79] Ndayishimiye P, Tazerout M. Use of palm oil-based biofuel in the internal combustion engines: performance and emissions characteristics. Energy 2011;36:1790—6.

[80] Özener O, Yüksek L, Ergenç AT, Özkan M. Effects of soybean biodiesel on a DI diesel engine performance, emission and combustion characteristics. Fuel 2014;115:875—83.

[81] Karmakar A, Karmakar S, Mukherjee S. Properties of various plants and animals feedstocks for biodiesel production. Bioresource Technology 2010;101:7201–10.

[82] Smith PC, Ngothai Y, Nguyen QD, O'neill BK. Improving the low-temperature properties of biodiesel: methods and consequences. Renewable Energy 2010;35:1145–51.

[83] Palash SM, Kalam MA, Masjuki HH, Masum BM, Rizwanul Fattah IM, Mofijur M. Impacts of biodiesel combustion on NOx emissions and their reduction approaches. Renewable and Sustainable Energy Reviews 2013;23:473–90.

[84] EPA. Guidance for biodiesel producers and biodiesel blenders/users | guidance for biodiesel producers and biodiesel blenders/users. US Environmental Protection Agency; 2007.

[85] Knothe G, Van Gerpen J, Krahl J. The biodiesel handbook. USA: AOCS Press; 2005.

[86] Hasimoğlu C, Ciniviz M, Özsert İ İY, Parlak A, Sahir Salman M. Performance characteristics of a low heat rejection diesel engine operating with biodiesel. Renewable Energy 2008;33:1709–15.

[87] Ejim CE, Fleck BA, Amirfazli A. Analytical study for atomization of biodiesels and their blends in a typical injector: surface tension and viscosity effects. Fuel 2007;86:1534–44.

[88] Sanjid A, Masjuki HH, Kalam Ma, Rahman SMA, Abedin MJ, Palash SM. Production of palm and jatropha based biodiesel and investigation of palm-jatropha combined blend properties, performance, exhaust emission and noise in an unmodified diesel engine. Journal of Cleaner Production 2013;65:295–303.

[89] Atabani AE, Silitonga AS, Badruddin IA, Mahlia TMI, Masjuki HH, Mekhilef S. A comprehensive review on biodiesel as an alternative energy resource and its characteristics. Renewable and Sustainable Energy Reviews 2012;16:2070–93.

[90] Ng J-H, Ng HK, Gan S. Development of emissions predictor equations for a light-duty diesel engine using biodiesel fuel properties. Fuel 2012;95:544–52.

[91] Lim WH, Ooi TL, Hong HK. Study on low temperature properties of palm oil methyl esters-petrodiesel blends. Journal of Oil Palm Research 2009;21:683–92.

[92] Bhale PV, Deshpande NV, Thombre BS. Improving the low temperature properties of biodiesel fuel. Renewable Energy 2009;34:794–800.

[93] Hu J, Du Z, Li C, Min E. Study on the lubrication properties of biodiesel as fuel lubricity enhancers. Fuel 2005;84:1601–6.

[94] Muñoz M, Moreno F, Monné C, Morea J, Terradillos J. Biodiesel improves lubricity of new low sulphur diesel fuels. Renewable Energy 2011;36:2918–24.

[95] Goodrum JW, Geller DP. Influence of fatty acid methyl esters from hydroxylated vegetable oils on diesel fuel lubricity. Bioresource Technology 2005;96:851–5.

[96] Zappi M, Hernandez R, Sparks D, Horne J, Brough M. A review of the engineering aspects of the biodiesel industry. MSU E-TECH Laboratory Report. 2003. ET-03-003.

[97] Bioenergy. Biodiesel. Michigan State University; 2014.

[98] NBB. Benefits of biodiesel. National Biodiesel Board; 2009.

[99] Moon G, Lee Y, Choi K, Jeong D. Emission characteristics of diesel, gas to liquid, and biodiesel-blended fuels in a diesel engine for passenger cars. Fuel 2010;89:3840–6.

[100] Sarin R, Sharma M, Sinharay S, Malhotra RK. Jatropha–palm biodiesel blends: an optimum mix for Asia. Fuel 2007;86:1365–71.

[101] Alptekin E, Canakci M. Characterization of the key fuel properties of methyl ester–diesel fuel blends. Fuel 2009;88:75–80.

[102] Rao NJ, Sastry MRC, Reddy PN. Comparative analysis of performance and emissions of an engine with palm oil biodiesel blends with diesel. International Journal of Engineering Science and Technology 2011;3:4257–64.

[103] Yadav PKS, Singh O, Singh RP. Performance test of palm fatty acid biodiesel on compression ignition engine. Journal of Petroleum Technology and Alternative Fuels 2010;1:1–9.

[104] Benjumea P, Agudelo J, Agudelo A. Basic properties of palm oil biodiesel–diesel blends. Fuel 2008;87:2069–75.

[105] Sharafutdinov I, Stratiev D, Shishkova I, Dinkov R, Batchvarov A, Petkov P, et al. Cold flow properties and oxidation stability of blends of near zero sulfur diesel from Ural crude oil and FAME from different origin. Fuel 2012;96:556–67.

[106] Lin Y, Lee W, Hou H. PAH emissions and energy efficiency of palm-biodiesel blends fueled on diesel generator. Atmospheric Environment 2006;40:3930–40.

[107] Dutra LM, Teixeira CV, Colaco MJ, Alves LSB, Caldeira AB. Comparative analysis of performance and emissions of an engine operating with palm oil methyl and ethyl esters and their blends with diesel. In: 20th international congress of mechanical engineering, Gramado, RS, Brazil; 2009.

[108] Aziz AA, Said MF, Awang MA. Performance of palm oil-based biodiesel fuels in a single cylinder direct. Palm Oil Developments 1993;42:15−24.

[109] Buyukkaya E. Effects of biodiesel on a DI diesel engine performance, emission and combustion characteristics. Fuel 2010;89:3099−105.

[110] Aziz AA, Said MF, Awang MA, Said M. The effects of Neutralized Palm Oil Methyl Esters (NPOME) on performance and emission of a direct injection diesel engine. 2006.

[111] Qi DH, Chen H, Geng LM, Bian YZ. Experimental studies on the combustion characteristics and performance of a direct injection engine fueled with biodiesel/diesel blends. Energy Conversion and Management 2010;51:2985−92.

[112] Xue J, Grift TE, Hansen AC. Effect of biodiesel on engine performances and emissions. Renewable and Sustainable Energy Reviews 2011;15:1098−116.

[113] Raheman H, Ghadge SV. Performance of compression ignition engine with mahua (*Madhuca indica*) biodiesel. Fuel 2007;86:2568−73.

[114] Yuan C-S, Lin H-Y, Lee W-J, Lin Y-C, Wu T-S, Chen K-F. A new alternative fuel for reduction of polycyclic aromatic hydrocarbon and particulate matter emissions from diesel engines. Journal of the Air & Waste Management Association 2007;57:465−71.

[115] Azar C, Lindgren K, Andersson BA. Global energy scenarios meeting stringent CO_2 constraints—cost-effective fuel choices in the transportation sector. Energy Policy 2003;31:961−76.

[116] Lapuerta M, Armas O, Rodriguezfernandez J. Effect of biodiesel fuels on diesel engine emissions. Progress in Energy and Combustion Science 2008;34:198−223.

[117] Wirawan SS, Tambunan AH, Djamin M, Nabetani H. The effect of palm biodiesel fuel on the performance and emission of the automotive diesel engine. Agricultural Engineering International 2008:1−13.

[118] Ribeiro M, Pinto AC, Quintella CM, Rocha GO, Teixeira LSG, Guarieiro LN, et al. The role of additives for diesel and diesel blended (ethanol or biodiesel) fuels: a review. Energy & Fuels 2007;21:2433−45.

[119] Palash SM, Kalam MA, Masjuki HH, Arbab MI, Masum BM, Sanjid A. Impacts of NOx reducing antioxidant additive on performance and emissions of a multi-cylinder diesel engine fueled with Jatropha biodiesel blends. Energy Conversion and Management 2014;77:577−85.

[120] EPA. A comprehensive analysis of biodiesel impacts on exhaust emissions. US Environmental Protection Agency; 2002.

[121] Murillo S, Míguez JL, Porteiro J, Granada E, Morán JC. Performance and exhaust emissions in the use of biodiesel in outboard diesel engines. Fuel 2007;86:1765−71.

[122] Ban-Weiss GA, Chen JY, Buchholz Ba, Dibble RW. A numerical investigation into the anomalous slight NOx increase when burning biodiesel; a new (old) theory. Fuel Processing Technology 2007;88:659−67.

[123] Dunn RO, Shockley MW, Bagby MO. Improving the low-temperature properties of alternative diesel fuels: vegetable oil-derived methyl esters. Journal of the American Oil Chemists' Society 1996;73:1719−28.

[124] Huang J, Wang Y, Li S, Roskilly AP, Yu H, Li H. Experimental investigation on the performance and emissions of a diesel engine fuelled with ethanol−diesel blends. Applied Thermal Engineering 2009;29:2484−90.

[125] Kass MD, Swartz MM, Lewis SA, Huff SP, Lee DW. Lowering NOx and PM emissions in a light-duty diesel engine with biodiesel-water emulsions. American Society of Agricultural and Biological Engineers 2006. ASAE Annual Meeting.

[126] Labeckas G, Slavinskas S. Influence of fuel additives on performance of direct-injection Diesel engine and exhaust emissions when operating on shale oil. Energy Conversion and Management 2005;46:1731−44.

[127] Chen KS, Lin YC, Te Hsieh L, Lin LF, Wu CC. Saving energy and reducing pollution by use of emulsified palm-biodiesel blends with bio-solution additive. Energy 2010;35:2043−8.

INDEX

'*Note*: Page numbers followed by "f" indicate figures and "t" indicate tables.'